SIMULATION AND THE MONTE CARLO METHOD

WILEY SERIES IN PROBABILITY AND STATISTICS
Established by *Walter A. Shewhart and Samuel S. Wilks*

Editors: *David J. Balding, Noel A. C. Cressie, Garrett M. Fitzmaurice, Geof H. Givens, Harvey Goldstein, Geert Molenberghs, David W. Scott, Adrian F. M. Smith, Ruey S. Tsay, Sanford Weisberg*

Editors Emeriti: *J. Stuart Hunter, Iain M. Johnstone, Joseph B. Kadane, Jozef L. Teugels*

The Wiley Series in Probability and Statistics is well established and authoritative. It covers many topics of current research interest in both pure and applied statistics and probability theory. Written by leading statisticians and institutions, the titles span both state-of-the-art developments in the field and classical methods.

Reflecting the wide range of current research in statistics, the series encompasses applied, methodological and theoretical statistics, ranging from applications and new techniques made possible by advances in computerized practice to rigorous treatment of theoretical approaches. This series provides essential and invaluable reading for all statisticians, whether in academia, industry, government, or research.

A complete list of titles in this series can be found at http://www.wiley.com/go/wsps

SIMULATION AND THE MONTE CARLO METHOD

Third Edition

Reuven Y. Rubinstein
Technion

Dirk P. Kroese
University of Queensland

Published by John Wiley & Sons, Inc., Hoboken, New Jersey.
Published simultaneously in Canada.

For general information on our other products and services or for technical support, please contact our Customer Care Department within the United States at (800) 762-2974, outside the United States at (317) 572-3993 or fax (317) 572-4002.

Wiley also publishes its books in a variety of electronic formats. Some content that appears in print may not be available in electronic formats. For more information about Wiley products, visit our web site at www.wiley.com.

Library of Congress Cataloging-in-Publication Data:

Names: Rubinstein, Reuven Y. | Kroese, Dirk P.
Title: Simulation and the Monte Carlo method.
Description: Third edition / Reuven Rubinstein, Dirk P. Kroese. | Hoboken,
 New Jersey : John Wiley & Sons, Inc., [2017] | Series: Wiley series in
 probability and statistics | Includes bibliographical references and index.
Identifiers: LCCN 2016020742 (print) | LCCN 2016023293 (ebook) | ISBN
 9781118632161 (cloth) | ISBN 9781118632208 (pdf) | ISBN 9781118632383
 (epub)
Subjects: LCSH: Monte Carlo method. | Digital computer simulation. |
 Mathematical statistics. | Sampling (Statistics)
Classification: LCC QA298 .R8 2017 (print) | LCC QA298 (ebook) | DDC
 518/.282--dc23
LC record available at https://lccn.loc.gov/2016020742

10 9 8 7 6 5 4 3 2 1

To the memory of

Reuven Y. Rubinstein (1938–2012)

Reuven Rubinstein passed away during the writing of this third edition. Reuven was one of the pioneers of Monte Carlo simulation and remained at the forefront of research in this area right up to the end of his life. In 2011 he received the highest honor given by INFORMS Simulation Society: The Lifetime Professional Achievement Award, where his achievements were summarized as follows:

> Professor Rubinstein has been a pivotal figure in the theory and practice of simulation as we know it today. His career reflects a high level of creativity and contribution, with a willingness to explore new areas and an amazing ability to suggest surprising new avenues of research and to influence subsequent work.

May his contagious enthusiasm and curiosity live on through his books and the many people whom he inspired.

Dirk P. Kroese

CONTENTS

PREFACE

Since the publication in 2008 of the second edition of *Simulation and the Monte Carlo Method*, significant changes have taken place in the field of Monte Carlo simulation. This third edition gives a fully updated and comprehensive account of the major topics in Monte Carlo simulation.

The book is based on an undergraduate course on Monte Carlo methods given at the Israel Institute of Technology (Technion) and the University of Queensland for the last five years. It is aimed at a broad audience of students in engineering, physical and life sciences, statistics, computer science, mathematics, and simply anyone interested in using Monte Carlo simulation in their study or work. Our aim is to provide an accessible introduction to modern Monte Carlo methods, focusing on the main concepts, while providing a sound foundation for problem solving. For this reason most ideas are introduced and explained via concrete examples, algorithms, and experiments.

Although we assume that the reader has some basic mathematical background, such as an elementary course in probability and statistics, we nevertheless review the basic concepts of probability, stochastic processes, information theory, and convex optimization in Chapter 1.

In a typical stochastic simulation, randomness is introduced into simulation models via independent uniformly distributed random variables. These random variables are then used as building blocks to simulate more general stochastic systems. Chapter 2 deals with the generation of such random numbers, random variables, and stochastic processes.

Many real-world complex systems can be modeled as discrete-event systems. Examples of discrete-event systems include traffic systems, flexible manufacturing systems, computer-communications systems, inventory systems, production lines, coherent lifetime systems, PERT networks, and flow networks. The behavior of such systems is identified via a sequence of discrete "events" that causes the system to change from one "state" to another. We discuss how to model such systems on a computer in Chapter 3.

Chapter 4 treats the statistical analysis of the output data from static and dynamic simulation models. The main difference is that the former do not evolve in time whereas the latter do. For dynamic models, we distinguish between finite-horizon and steady-state simulations. Two popular methods for estimating steady-state performance measures — the batch means and regenerative methods — are discussed as well.

Chapter 5 deals with variance reduction techniques in Monte Carlo simulation, such as antithetic and common random numbers, control random variables, conditional Monte Carlo, stratified sampling, and importance sampling. Using importance sampling, one can often achieve substantial (sometimes dramatic) variance reduction, in particular when estimating rare-event probabilities. While dealing with importance sampling, we present two alternative approaches, called the *variance minimization* and the *cross-entropy* methods. Special attention is paid to importance sampling algorithms in which paths are generated in a sequential manner. Further improvements of such algorithms are obtained by resampling successful paths, giving rise to *sequential importance resampling* algorithms. We illustrate their use via a nonlinear filtering example. In addition, this chapter contains two new importance sampling based methods, called *the transform likelihood ratio method* and *the screening method* for variance reduction. The former presents a simple, convenient and unifying way of constructing efficient importance sampling estimators, whereas the latter ensures lowering of the dimensionality of the importance sampling density. This is accomplished by identifying (screening out) the most important (bottleneck) parameters to be used in the importance sampling distribution. As results, the accuracy of the importance sampling estimator increases substantially.

Chapter 6 gives a concise treatment of the generic *Markov chain Monte Carlo* (MCMC) method for *approximately* generating samples from an arbitrary distribution. We discuss the classic Metropolis–Hastings algorithm and the Gibbs sampler. In the former, one simulates a Markov chain such that its stationary distribution coincides with the target distribution, while in the latter the underlying Markov chain is constructed on the basis of a sequence of conditional distributions. We also deal with applications of MCMC in Bayesian statistics, and explain how MCMC is used to sample from the Boltzmann distribution for the Ising and Potts models, which are extensively used in statistical mechanics. Moreover, we show how MCMC is used in the simulated annealing method to find the global minimum of a multi-extremal function. We also show that both the Metropolis–Hastings and Gibbs samplers can be viewed as special cases of a general MCMC algorithm and then present two more modifications, namely the *slice* and the *reversible jump* samplers.

Chapter 7 is on sensitivity analysis and Monte Carlo optimization of simulated systems. Because of their complexity, the performance evaluation of discrete-event systems is usually studied by simulation, and the simulation is often associated with the estimation of the performance function with respect to some controllable parameters. Sensitivity analysis is concerned with evaluating sensitivities (gradients, Hessians, etc.) of the performance function with respect to system parameters.

This provides guidance to operational decisions and to selecting system parameters that optimize the performance measures. Monte Carlo optimization deals with solving stochastic programs, that is, optimization problems where the objective function and some of the constraints are unknown and need to be obtained via simulation. We deal with sensitivity analysis and optimization of both static and dynamic models. We introduce the celebrated *score function* method for sensitivity analysis, and two alternative methods for Monte Carlo optimization, the so-called *stochastic approximation* and *stochastic counterpart* methods. In particular, in the latter method, we show how using a single simulation experiment one can approximate quite accurately the true unknown optimal solution of the original deterministic program.

Chapter 8 deals with the *cross-entropy* (CE) method, which was introduced by the first author in 1997 as an adaptive algorithm for rare-event estimation using a cross-entropy minimization technique. It was soon realized that the underlying ideas had a much wider range of application than just in rare-event simulation: they could be readily adapted to tackle quite general combinatorial and multi-extremal optimization problems, including many problems associated with learning algorithms and neural computation. We provide a gradual introduction to the CE method, and show its elegance and versatility. In particular, we present a general CE algorithm for the estimation of rare-event probabilities and then slightly modify it for solving combinatorial optimization problems. We discuss applications of the CE method to several combinatorial optimization problems, such as the max-cut problem and the traveling salesman problem, and provide supportive numerical results on its effectiveness. Due to its versatility, tractability, and simplicity, the CE method has potentially a diverse range of applications, for example, in computational biology, DNA sequence alignment, graph theory, and scheduling. Over the last 10 years many hundreds of papers have been written on the theory and applications of CE. For more details see the site www.cemethod.org, our book *The Cross-Entropy Method: A Unified Approach to Combinatorial Optimization, Monte-Carlo Simulation and Machine Learning* (Springer, 2004), and search in the wikipedia under "cross-entropy method". The chapter concludes with a discussion of the minimum cross-entropy (MinxEnt) optimization program.

Chapter 9 introduces the *splitting* method, which uses a sequential sampling plan to decompose a "difficult" problem into a sequence of "easy" problems. The method was originally designed for rare-event simulation, but it has developed into a highly versatile "particle MCMC" algorithm that can be used for rare-event estimation, optimization, and sampling. The chapter presents various splitting algorithms for dynamic and static simulation models, and demonstrates how they can be used to (1) estimate rare-event probabilities, (2) solve hard counting problems, (3) find solutions to challenging optimization problems, and (4) sample from complicated probability distributions. The chapter features a wide variety of case studies and numerical experiments, demonstrating the effectiveness of the method.

Many combinatorial problems can be formulated in terms of searching or counting the total cost of a tree. Chapter 10 presents a new Monte Carlo called *stochastic enumeration* (SE) that is well suited to solve such problems by generating random paths through the tree in a parallel fashion. The SE algorithm can be viewed as a sequential importance sampling method on a "hyper-tree" whose vertices are sets of vertices of the original tree. By combining SE with fast polynomial decision algorithms, we show how it can be used for counting #P-complete problems, such as the number of satisfiability assignments, number of paths in a general network, and the number of perfect matchings in a graph. The usefulness of the method is illustrated via a suite of numerical examples.

The appendix features a variety of supplementary topics, including a brief introduction to exponential families, the discrete-time Kalman filter, and the Cholesky square root method. The computational complexity of randomized algorithms is also discussed. An extensive range of exercises is provided at the end of each chapter.

In addition to two brand-new chapters (Chapters 9 and 10), this third edition offers substantial updates on a range of topics. The material on random number generation has been extensively revised by including state-of-the-art combined multiple recursive generators and the Mersenne Twister. The material on stochastic process generation has been extended by including the simulation of Gaussian processes, Brownian motion, and diffusion processes. The variance reduction chapter now includes a discussion of the novel multi-level Monte Carlo method. Our treatment of sequential importance has been significantly modified by emphasizing the significance of importance resampling. This addition also prepares for the particle MCMC approach in the new splitting chapter. The cross-entropy chapter is further enhanced by adding new insights into likelihood ratio degeneration, leading to the single-level improved CE algorithm. Twenty-five more questions have been added, along with their solutions in the *online solutions manual* that accompanies this book. Finally, to facilitate their implementation, most algorithms have been (re-)written in pseudo-code with flow control.

REUVEN RUBINSTEIN AND DIRK KROESE

Haifa and Brisbane
July 2016

ACKNOWLEDGMENTS

We thank all who contributed to this book. Robert Smith and Zelda Zabinski read and provided useful suggestions on Chapter 6. Alex Shapiro gave a detailed account of the complexity of stochastic programming problems (Section A.8.4). Pierre L'Ecuyer communicated the counterexample in Section 9.10 and Zdravko Botev suggested the multilevel Monte Carlo example in Section 5.6. Jim Spall kindly provided feedback on the second edition, which we have incorporated in the third edition. Josh Chan, Tom Taimre, and Zdravko Botev helped proofread the new material in this book, which is highly appreciated.

We are particularly grateful to the many undergraduate and graduate students at the Technion and the University of Queensland who helped make this book possible and whose valuable ideas and experiments were extremely encouraging and motivating. Qibin Duan, Morgan Grant, Robert Salomone, Rohan Shah, and Erli Wang read through the new material, solved the new exercises, and provided excellent feedback. Special thanks goes out to Slava Vaisman for his help on many computational problems that we encountered, and for the many fruitful discussions we had on the splitting and stochastic enumeration methods.

This book was supported by by the Australian Research Council *Centre of Excellence for Mathematical & Statistical Frontiers*, under grant number CE140100049.

RYR, DPK

CHAPTER 1

PRELIMINARIES

1.1 INTRODUCTION

The purpose of this chapter is to review some basic facts from probability, information theory, and optimization. In particular, Sections 1.2–1.11 summarize the main points from probability theory. Sections 1.12–1.14 describe various fundamental stochastic processes, such as Poisson, Markov, and Gaussian processes. Elements of information theory are given in Section 1.15, and Section 1.16 concludes with an outline of convex optimization theory.

1.2 RANDOM EXPERIMENTS

The basic notion in probability theory is that of a *random experiment*: an experiment whose outcome cannot be determined in advance. The most fundamental example is the experiment where a fair coin is tossed a number of times. For simplicity suppose that the coin is tossed three times. The *sample space*, denoted Ω, is the set of all possible outcomes of the experiment. In this case Ω has eight possible outcomes:

$$\Omega = \{HHH, HHT, HTH, HTT, THH, THT, TTH, TTT\} ,$$

where, for example, HTH means that the first toss is heads, the second tails, and the third heads.

Simulation and the Monte Carlo Method, Third Edition. By R. Y. Rubinstein and D. P. Kroese
Copyright © 2017 John Wiley & Sons, Inc. Published 2017 by John Wiley & Sons, Inc.

Subsets of the sample space are called *events*. For example, the event A that the third toss is heads is

$$A = \{HHH, HTH, THH, TTH\} .$$

We say that event A *occurs* if the outcome of the experiment is one of the elements in A. Since events are sets, we can apply the usual set operations to them. For example, the event $A \cup B$, called the *union* of A and B, is the event that A or B or both occur, and the event $A \cap B$, called the *intersection* of A and B, is the event that A and B both occur. Similar notation holds for unions and intersections of more than two events. The event A^c, called the *complement* of A, is the event that A does not occur. Two events A and B that have no outcomes in common, that is, their intersection is empty, are called *disjoint* events. The main step is to specify the probability of each event.

Definition 1.2.1 (Probability) A *probability* $\mathbb{P}$ is a rule that assigns a number $0 \leqslant \mathbb{P}(A) \leqslant 1$ to each event A, such that $\mathbb{P}(\Omega) = 1$, and such that for any sequence $A_1, A_2, \ldots$ of disjoint events

$$\mathbb{P}\left(\bigcup_i A_i\right) = \sum_i \mathbb{P}(A_i) . \tag{1.1}$$

Equation (1.1) is referred to as the *sum rule* of probability. It states that if an event can happen in a number of different ways, but not simultaneously, the probability of that event is simply the sum of the probabilities of the comprising events.

For the fair coin toss experiment the probability of any event is easily given. Namely, because the coin is fair, each of the eight possible outcomes is equally likely, so that $\mathbb{P}(\{HHH\}) = \cdots = \mathbb{P}(\{TTT\}) = 1/8$. Since any event A is the union of the "elementary" events $\{HHH\}, \ldots, \{TTT\}$, the sum rule implies that

$$\mathbb{P}(A) = \frac{|A|}{|\Omega|} , \tag{1.2}$$

where $|A|$ denotes the number of outcomes in A and $|\Omega| = 8$. More generally, if a random experiment has finitely many and equally likely outcomes, the probability is always of the form (1.2). In that case the calculation of probabilities reduces to counting.

1.3   CONDITIONAL PROBABILITY AND INDEPENDENCE

How do probabilities change when we know that some event $B \subset \Omega$ has occurred? Given that the outcome lies in B, the event A will occur if and only if $A \cap B$ occurs, and the relative chance of A occurring is therefore $\mathbb{P}(A \cap B)/\mathbb{P}(B)$. This leads to the definition of the *conditional probability* of A given B:

$$\mathbb{P}(A \,|\, B) = \frac{\mathbb{P}(A \cap B)}{\mathbb{P}(B)} . \tag{1.3}$$

For example, suppose that we toss a fair coin three times. Let B be the event that the total number of heads is two. The conditional probability of the event A that the first toss is heads, given that B occurs, is $(2/8)/(3/8) = 2/3$.

Rewriting (1.3) and interchanging the role of A and B gives the relation $\mathbb{P}(A \cap B) = \mathbb{P}(A)\,\mathbb{P}(B \mid A)$. This can be generalized easily to the *product rule* of probability, which states that for any sequence of events $A_1, A_2, \ldots, A_n$,

$$\mathbb{P}(A_1 \cdots A_n) = \mathbb{P}(A_1)\,\mathbb{P}(A_2 \mid A_1)\,\mathbb{P}(A_3 \mid A_1 A_2) \cdots \mathbb{P}(A_n \mid A_1 \cdots A_{n-1}) , \qquad (1.4)$$

using the abbreviation $A_1 A_2 \cdots A_k \equiv A_1 \cap A_2 \cap \cdots \cap A_k$.

Suppose that $B_1, B_2, \ldots, B_n$ is a *partition* of Ω. That is, $B_1, B_2, \ldots, B_n$ are disjoint and their union is Ω. Then, by the sum rule, $\mathbb{P}(A) = \sum_{i=1}^{n} \mathbb{P}(A \cap B_i)$ and hence, by the definition of conditional probability, we have the *law of total probability*:

$$\mathbb{P}(A) = \sum_{i=1}^{n} \mathbb{P}(A \mid B_i)\,\mathbb{P}(B_i) . \qquad (1.5)$$

Combining this with the definition of conditional probability gives *Bayes' rule*:

$$\mathbb{P}(B_j \mid A) = \frac{\mathbb{P}(A \mid B_j)\,\mathbb{P}(B_j)}{\sum_{i=1}^{n} \mathbb{P}(A \mid B_i)\,\mathbb{P}(B_i)} . \qquad (1.6)$$

Independence is of crucial importance in probability and statistics. Loosely speaking, it models the lack of information between events. Two events A and B are said to be *independent* if the knowledge that B has occurred does not change the probability that A occurs. That is, A, B independent $\Leftrightarrow \mathbb{P}(A \mid B) = \mathbb{P}(A)$. Since $\mathbb{P}(A \mid B) = \mathbb{P}(A \cap B)/\mathbb{P}(B)$, an alternative definition of independence is

$$A, B \text{ independent} \Leftrightarrow \mathbb{P}(A \cap B) = \mathbb{P}(A)\,\mathbb{P}(B) .$$

This definition covers the case where $B = \emptyset$ (empty set). We can extend this definition to arbitrarily many events.

Definition 1.3.1 (Independence) The events $A_1, A_2, \ldots,$ are said to be *independent* if for any k and any choice of distinct indexes $i_1, \ldots, i_k$,

$$\mathbb{P}(A_{i_1} \cap A_{i_2} \cap \cdots \cap A_{i_k}) = \mathbb{P}(A_{i_1})\,\mathbb{P}(A_{i_2}) \cdots \mathbb{P}(A_{i_k}) .$$

Remark 1.3.1 In most cases independence of events is a model assumption. That is, we assume that there exists a $\mathbb{P}$ such that certain events are independent.

■ **EXAMPLE 1.1**

We toss a biased coin n times. Let p be the probability of heads (for a fair coin $p = 1/2$). Let A_i denote the event that the i-th toss yields heads, $i = 1, \ldots, n$. Then $\mathbb{P}$ should be such that the events $A_1, \ldots, A_n$ are independent, and $\mathbb{P}(A_i) = p$ for all i. These two rules completely specify $\mathbb{P}$. For example, the probability that the first k throws are heads and the last $n - k$ are tails is

$$
\begin{aligned}
\mathbb{P}(A_1 \cdots A_k A_{k+1}^c \cdots A_n^c) &= \mathbb{P}(A_1) \cdots \mathbb{P}(A_k)\,\mathbb{P}(A_{k+1}^c) \cdots \mathbb{P}(A_n^c) \\
&= p^k (1-p)^{n-k} .
\end{aligned}
$$

1.4   RANDOM VARIABLES AND PROBABILITY DISTRIBUTIONS

Specifying a model for a random experiment via a complete description of Ω and $\mathbb{P}$ may not always be convenient or necessary. In practice, we are only interested in certain observations (i.e., numerical measurements) in the experiment. We incorporate these into our modeling process via the introduction of *random variables*, usually denoted by capital letters from the last part of the alphabet (e.g., X, $X_1, X_2, \ldots, Y, Z$).

■ **EXAMPLE 1.2**

We toss a biased coin n times, with p the probability of heads. Suppose that we are interested only in the number of heads, say X. Note that X can take any of the values in $\{0, 1, \ldots, n\}$. The *probability distribution* of X is given by the *binomial formula*

$$\mathbb{P}(X = k) = \binom{n}{k} p^k (1-p)^{n-k}, \quad k = 0, 1, \ldots, n \; . \tag{1.7}$$

Namely, by Example 1.1, each elementary event $\{HTH \cdots T\}$ with exactly k heads and $n - k$ tails has probability $p^k(1 - p)^{n-k}$, and there are $\binom{n}{k}$ such events.

The probability distribution of a general random variable X — identifying such probabilities as $\mathbb{P}(X = x), \mathbb{P}(a \leqslant X \leqslant b)$, and so on — is completely specified by the *cumulative distribution function* (cdf), defined by

$$F(x) = \mathbb{P}(X \leqslant x), \; x \in \mathbb{R} \; .$$

A random variable X is said to have a *discrete* distribution if, for some finite or countable set of values $x_1, x_2, \ldots$, $\mathbb{P}(X = x_i) > 0$, $i = 1, 2, \ldots$ and $\sum_i \mathbb{P}(X = x_i) = 1$. The function $f(x) = \mathbb{P}(X = x)$ is called the *probability mass function* (pmf) of X — but see Remark 1.4.1.

■ **EXAMPLE 1.3**

Toss two fair dice and let M be the largest face value showing. The pmf of M is given by

m	1	2	3	4	5	6	$\sum$
$f(m)$	$\dfrac{1}{36}$	$\dfrac{3}{36}$	$\dfrac{5}{36}$	$\dfrac{7}{36}$	$\dfrac{9}{36}$	$\dfrac{11}{36}$	1

For example, to get $M = 3$, either $(1, 3), (2, 3), (3, 3), (3, 2)$, or $(3, 1)$ has to be thrown, each of which happens with probability $1/36$.

A random variable X is said to have a *continuous* distribution if there exists a positive function f with total integral 1, such that for all a, b,

$$\mathbb{P}(a \leqslant X \leqslant b) = \int_a^b f(u) \, du \; . \tag{1.8}$$

The function f is called the *probability density function* (pdf) of X. Note that in the continuous case the cdf is given by

$$F(x) = \mathbb{P}(X \leqslant x) = \int_{-\infty}^{x} f(u) \, \mathrm{d}u \, ,$$

and f is the derivative of F. We can interpret $f(x)$ as the probability "density" at $X = x$ in the sense that

$$\mathbb{P}(x \leqslant X \leqslant x + h) = \int_{x}^{x+h} f(u) \, \mathrm{d}u \approx h \, f(x) \, .$$

Remark 1.4.1 (Probability Density) Note that we have deliberately used the *same* symbol, f, for both pmf and pdf. This is because the pmf and pdf play very similar roles and can, in more advanced probability theory, both be viewed as particular instances of the general notion of *probability density*. To stress this viewpoint, we will call f in *both* the discrete and continuous case the pdf or (probability) density (function).

1.5 SOME IMPORTANT DISTRIBUTIONS

Tables 1.1 and 1.2 list a number of important continuous and discrete distributions. We will use the notation $X \sim f$, $X \sim F$, or $X \sim$ Dist to signify that X has a pdf f, a cdf F or a distribution Dist. We sometimes write f_X instead of f to stress that the pdf refers to the random variable X. Note that in Table 1.1, Γ is the gamma function: $\Gamma(\alpha) = \int_{0}^{\infty} \mathrm{e}^{-x} x^{\alpha-1} \, \mathrm{d}x, \quad \alpha > 0$.

Table 1.1: Commonly used continuous distributions.

Name	Notation	$f(x)$	$x \in$	Parameters
Uniform	$\mathsf{U}[\alpha, \beta]$	$\dfrac{1}{\beta - \alpha}$	$[\alpha, \beta]$	$\alpha < \beta$
Normal	$\mathsf{N}(\mu, \sigma^2)$	$\dfrac{1}{\sigma \sqrt{2\pi}} \mathrm{e}^{-\frac{1}{2}\left(\frac{x-\mu}{\sigma}\right)^2}$	$\mathbb{R}$	$\sigma > 0, \, \mu \in \mathbb{R}$
Gamma	$\mathsf{Gamma}(\alpha, \lambda)$	$\dfrac{\lambda^\alpha x^{\alpha-1} \mathrm{e}^{-\lambda x}}{\Gamma(\alpha)}$	$\mathbb{R}_+$	$\alpha, \lambda > 0$
Exponential	$\mathsf{Exp}(\lambda)$	$\lambda \, \mathrm{e}^{-\lambda x}$	$\mathbb{R}_+$	$\lambda > 0$
Beta	$\mathsf{Beta}(\alpha, \beta)$	$\dfrac{\Gamma(\alpha + \beta)}{\Gamma(\alpha)\Gamma(\beta)} \, x^{\alpha-1}(1 - x)^{\beta-1}$	$[0, 1]$	$\alpha, \beta > 0$
Weibull	$\mathsf{Weib}(\alpha, \lambda)$	$\alpha\lambda \, (\lambda x)^{\alpha-1} \, \mathrm{e}^{-(\lambda x)^\alpha}$	$\mathbb{R}_+$	$\alpha, \lambda > 0$
Pareto	$\mathsf{Pareto}(\alpha, \lambda)$	$\alpha\lambda \, (1 + \lambda x)^{-(\alpha+1)}$	$\mathbb{R}_+$	$\alpha, \lambda > 0$

Table 1.2: Commonly used discrete distributions.

Name	Notation	$f(x)$	$x \in$	Parameters
Bernoulli	$\mathsf{Ber}(p)$	$p^x(1-p)^{1-x}$	$\{0,1\}$	$0 \leqslant p \leqslant 1$
Binomial	$\mathsf{Bin}(n,p)$	$\binom{n}{x} p^x(1-p)^{n-x}$	$\{0,1,\ldots,n\}$	$0 \leqslant p \leqslant 1,$ $n \in \mathbb{N}$
Discrete uniform	$\mathsf{DU}\{1,\ldots,n\}$	$\dfrac{1}{n}$	$\{1,\ldots,n\}$	$n \in \{1,2,\ldots\}$
Geometric	$\mathsf{G}(p)$	$p(1-p)^{x-1}$	$\{1,2,\ldots\}$	$0 \leqslant p \leqslant 1$
Poisson	$\mathsf{Poi}(\lambda)$	$\mathrm{e}^{-\lambda} \dfrac{\lambda^x}{x!}$	$\mathbb{N}$	$\lambda > 0$

1.6 EXPECTATION

It is often useful to consider different kinds of numerical characteristics of a random variable. One such quantity is the expectation, which measures the mean value of the distribution.

Definition 1.6.1 (Expectation) Let X be a random variable with pdf f. The *expectation* (or expected value or mean) of X, denoted by $\mathbb{E}[X]$ (or sometimes μ), is defined by

$$\mathbb{E}[X] = \begin{cases} \sum_x x\, f(x) & \text{discrete case,} \\ \int_{-\infty}^{\infty} x\, f(x)\, \mathrm{d}x & \text{continuous case.} \end{cases}$$

If X is a random variable, then a function of X, such as X^2 or $\sin(X)$, is again a random variable. Moreover, the expected value of a function of X is simply a weighted average of the possible values that this function can take. That is, for any real function h

$$\mathbb{E}[h(X)] = \begin{cases} \sum_x h(x)\, f(x) & \text{discrete case,} \\ \int_{-\infty}^{\infty} h(x)\, f(x)\, \mathrm{d}x & \text{continuous case.} \end{cases}$$

Another useful quantity is the variance, which measures the spread or dispersion of the distribution.

Definition 1.6.2 (Variance) The *variance* of a random variable X, denoted by $\mathrm{Var}(X)$ (or sometimes σ^2), is defined by

$$\mathrm{Var}(X) = \mathbb{E}[(X - \mathbb{E}[X])^2] = \mathbb{E}[X^2] - (\mathbb{E}[X])^2 \ .$$

The square root of the variance is called the *standard deviation*. Table 1.3 lists the expectations and variances for some well-known distributions.

Table 1.3: Expectations and variances for some well-known distributions.

Dist.	$\mathbb{E}[X]$	$\text{Var}(X)$	Dist.	$\mathbb{E}[X]$	$\text{Var}(X)$
$\text{Bin}(n, p)$	np	$np(1-p)$	$\text{Gamma}(\alpha, \lambda)$	$\dfrac{\alpha}{\lambda}$	$\dfrac{\alpha}{\lambda^2}$
$\text{G}(p)$	$\dfrac{1}{p}$	$\dfrac{1-p}{p^2}$	$\text{N}(\mu, \sigma^2)$	μ	σ^2
$\text{Poi}(\lambda)$	λ	λ	$\text{Beta}(\alpha, \beta)$	$\frac{\alpha}{\alpha+\beta}$	$\frac{\alpha\beta}{(\alpha+\beta)^2(1+\alpha+\beta)}$
$\text{U}(\alpha, \beta)$	$\dfrac{\alpha+\beta}{2}$	$\dfrac{(\beta-\alpha)^2}{12}$	$\text{Weib}(\alpha, \lambda)$	$\frac{\Gamma(1/\alpha)}{\alpha\lambda}$	$\frac{2\Gamma(2/\alpha)}{\alpha} - \left(\frac{\Gamma(1/\alpha)}{\alpha\lambda}\right)^2$
$\text{Exp}(\lambda)$	$\dfrac{1}{\lambda}$	$\dfrac{1}{\lambda^2}$			

The mean and the variance do not give, in general, enough information to completely specify the distribution of a random variable. However, they may provide useful bounds. We discuss two such bounds. Suppose X can only take nonnegative values and has pdf f. For any $x > 0$, we can write

$$
\begin{aligned}
\mathbb{E}[X] &= \int_0^x t f(t)\, \mathrm{d}t + \int_x^\infty t f(t)\, \mathrm{d}t \geq \int_x^\infty t f(t)\, \mathrm{d}t \\
&\geq \int_x^\infty x f(t)\, \mathrm{d}t = x\, \mathbb{P}(X \geq x)\,,
\end{aligned}
$$

from which follows the *Markov inequality:* if $X \geq 0$, then for all $x > 0$,

$$
\mathbb{P}(X \geq x) \leq \frac{\mathbb{E}[X]}{x}\,. \tag{1.9}
$$

If we also know the variance of a random variable, we can give a tighter bound. Namely, for any random variable X with mean μ and variance σ^2, we have

$$
\mathbb{P}(|X - \mu| \geq x) \leq \frac{\sigma^2}{x^2}\,. \tag{1.10}
$$

This is called the *Chebyshev inequality*. The proof is as follows: Let $D^2 = (X - \mu)^2$; then, by the Markov inequality (1.9) and the definition of the variance,

$$
\mathbb{P}(D^2 \geq x^2) \leq \frac{\sigma^2}{x^2}\,.
$$

Also, note that the event $\{D^2 \geq x^2\}$ is equivalent to the event $\{|X - \mu| \geq x\}$, so that (1.10) follows.

1.7 JOINT DISTRIBUTIONS

Often a random experiment is described by more than one random variable. The theory for multiple random variables is similar to that for a single random variable.

Let $X_1, \ldots, X_n$ be random variables describing some random experiment. We can accumulate these into a *random vector* $\mathbf{X} = (X_1, \ldots, X_n)$. More generally, a collection $\{X_t, t \in \mathscr{T}\}$ of random variables is called a *stochastic process*. The set $\mathscr{T}$ is called the *parameter set* or *index set* of the process. It may be discrete (e.g., $\mathbb{N}$ or $\{1, \ldots, 10\}$) or continuous (e.g., $\mathbb{R}_+ = [0, \infty)$ or $[1, 10]$). The set of possible values for the stochastic process is called the *state space*.

The joint distribution of $X_1, \ldots, X_n$ is specified by the *joint cdf*

$$F(x_1, \ldots, x_n) = \mathbb{P}(X_1 \leqslant x_1, \ldots, X_n \leqslant x_n) .$$

The *joint pdf* f is given, in the discrete case, by $f(x_1, \ldots, x_n) = \mathbb{P}(X_1 = x_1, \ldots, X_n = x_n)$, and in the continuous case f is such that

$$\mathbb{P}(\mathbf{X} \in \mathscr{B}) = \int_{\mathscr{B}} f(x_1, \ldots, x_n) \, dx_1 \ldots dx_n$$

for any (measurable) region $\mathscr{B}$ in $\mathbb{R}^n$. The marginal pdfs can be recovered from the joint pdf by integration or summation. For example, in the case of a continuous random vector (X, Y) with joint pdf f, the pdf f_X of X is found as

$$f_X(x) = \int f(x, y) \, dy .$$

Suppose that X and Y are both discrete or both continuous, with joint pdf f, and suppose that $f_X(x) > 0$. Then the *conditional pdf* of Y given $X = x$ is given by

$$f_{Y|X}(y \mid x) = \frac{f(x, y)}{f_X(x)} \quad \text{for all } y .$$

The corresponding *conditional expectation* is (in the continuous case)

$$\mathbb{E}[Y \mid X = x] = \int y \, f_{Y|X}(y \mid x) \, dy .$$

Note that $\mathbb{E}[Y \mid X = x]$ is a function of x, say $h(x)$. The corresponding random variable $h(X)$ is written as $\mathbb{E}[Y \mid X]$. It can be shown (see, for example, [3]) that its expectation is simply the expectation of Y, that is,

$$\mathbb{E}[\mathbb{E}[Y \mid X]] = \mathbb{E}[Y] . \tag{1.11}$$

When the conditional distribution of Y given X is identical to that of Y, X and Y are said to be independent. More precisely:

Definition 1.7.1 (Independent Random Variables) The random variables $X_1, \ldots, X_n$ are called *independent* if for all events $\{X_i \in A_i\}$ with $A_i \subset \mathbb{R}$, $i = 1, \ldots, n$,

$$\mathbb{P}(X_1 \in A_1, \ldots, X_n \in A_n) = \mathbb{P}(X_1 \in A_1) \cdots \mathbb{P}(X_n \in A_n) .$$

A direct consequence of the definition above for independence is that random variables $X_1, \ldots, X_n$ with joint pdf f (discrete or continuous) are independent if and only if

$$f(x_1, \ldots, x_n) = f_{X_1}(x_1) \cdots f_{X_n}(x_n) \tag{1.12}$$

for all $x_1, \ldots, x_n$, where $\{f_{X_i}\}$ are the marginal pdfs.

■ EXAMPLE 1.4 Bernoulli Sequence

Consider the experiment where we flip a biased coin n times, with probability p of heads. We can model this experiment in the following way. For $i = 1, \ldots, n$, let X_i be the result of the i-th toss: $\{X_i = 1\}$ means heads (or success), $\{X_i = 0\}$ means tails (or failure). Also, let

$$\mathbb{P}(X_i = 1) = p = 1 - \mathbb{P}(X_i = 0), \quad i = 1, 2, \ldots, n .$$

Last, assume that $X_1, \ldots, X_n$ are independent. The sequence $\{X_i, i = 1, 2, \ldots\}$ is called a *Bernoulli sequence* or *Bernoulli process* with success probability p. Let $X = X_1 + \cdots + X_n$ be the total number of successes in n trials (tosses of the coin). Denote by $\mathscr{B}$ the set of all binary vectors $\mathbf{x} = (x_1, \ldots, x_n)$ such that $\sum_{i=1}^{n} x_i = k$. Note that $\mathscr{B}$ has $\binom{n}{k}$ elements. We now have

$$
\begin{aligned}
\mathbb{P}(X = k) &= \sum_{\mathbf{x} \in \mathscr{B}} \mathbb{P}(X_1 = x_1, \ldots, X_n = x_n) \\
&= \sum_{\mathbf{x} \in \mathscr{B}} \mathbb{P}(X_1 = x_1) \cdots \mathbb{P}(X_n = x_n) = \sum_{\mathbf{x} \in \mathscr{B}} p^k (1-p)^{n-k} \\
&= \binom{n}{k} p^k (1-p)^{n-k} .
\end{aligned}
$$

In other words, $X \sim \mathsf{Bin}(n, p)$. Compare this with Example 1.2.

Remark 1.7.1 An *infinite* sequence $X_1, X_2, \ldots$ of random variables is called independent if for any finite choice of parameters $i_1, i_2, \ldots, i_n$ (none of them the same) the random variables $X_{i_1}, \ldots, X_{i_n}$ are independent. Many probabilistic models involve random variables $X_1, X_2, \ldots$ that are *independent and identically distributed*, abbreviated as *iid*. We will use this abbreviation throughout this book.

Similar to the one-dimensional case, the expected value of any real-valued function h of $X_1, \ldots, X_n$ is a weighted average of all values that this function can take. Specifically, in the continuous case,

$$\mathbb{E}[h(X_1, \ldots, X_n)] = \int \cdots \int h(x_1, \ldots, x_n) \, f(x_1, \ldots, x_n) \, \mathrm{d}x_1 \ldots \mathrm{d}x_n .$$

As a direct consequence of the definitions of expectation and independence, we have

$$\mathbb{E}[a + b_1 X_1 + b_2 X_2 + \cdots + b_n X_n] = a + b_1 \mu_1 + \cdots + b_n \mu_n \qquad (1.13)$$

for any sequence of random variables $X_1, X_2, \ldots, X_n$ with expectations $\mu_1, \mu_2, \ldots, \mu_n$, where $a, b_1, b_2, \ldots, b_n$ are constants. Similarly, for *independent* random variables, we have

$$\mathbb{E}[X_1 X_2 \cdots X_n] = \mu_1 \mu_2 \cdots \mu_n .$$

The *covariance* of two random variables X and Y with expectations $\mathbb{E}[X] = \mu_X$ and $\mathbb{E}[Y] = \mu_Y$, respectively, is defined as

$$\mathrm{Cov}(X, Y) = \mathbb{E}[(X - \mu_X)(Y - \mu_Y)] .$$

This is a measure for the amount of linear dependency between the variables. A scaled version of the covariance is given by the *correlation coefficient*,

$$\varrho(X, Y) = \frac{\text{Cov}(X, Y)}{\sigma_X \, \sigma_Y},$$

where $\sigma_X^2 = \text{Var}(X)$ and $\sigma_Y^2 = \text{Var}(Y)$. It can be shown that the correlation coefficient always lies between -1 and 1; see Problem 1.13.

For easy reference, Table 1.4 lists some important properties of the variance and covariance. The proofs follow directly from the definitions of covariance and variance and the properties of the expectation.

Table 1.4: Properties of variance and covariance.

1	$\text{Var}(X) = \mathbb{E}[X^2] - (\mathbb{E}[X])^2$
2	$\text{Var}(aX + b) = a^2 \text{Var}(X)$
3	$\text{Cov}(X, Y) = \mathbb{E}[XY] - \mathbb{E}[X]\,\mathbb{E}[Y]$
4	$\text{Cov}(X, Y) = \text{Cov}(Y, X)$
5	$\text{Cov}(aX + bY, Z) = a\,\text{Cov}(X, Z) + b\,\text{Cov}(Y, Z)$
6	$\text{Cov}(X, X) = \text{Var}(X)$
7	$\text{Var}(X + Y) = \text{Var}(X) + \text{Var}(Y) + 2\,\text{Cov}(X, Y)$
8	X and Y indep. $\implies \text{Cov}(X, Y) = 0$

As a consequence of properties 2 and 7, for any sequence of *independent* random variables $X_1, \ldots, X_n$ with variances $\sigma_1^2, \ldots, \sigma_n^2$,

$$\text{Var}(a + b_1 X_1 + b_2 X_2 + \cdots + b_n X_n) = b_1^2 \sigma_1^2 + \cdots + b_n^2 \sigma_n^2 \qquad (1.14)$$

for any choice of constants a and $b_1, \ldots, b_n$.

For random vectors, such as $\mathbf{X} = (X_1, \ldots, X_n)^\top$, it is convenient to write the expectations and covariances in vector notation.

Definition 1.7.2 (Expectation Vector and Covariance Matrix) For any random vector $\mathbf{X}$, we define the *expectation vector* as the vector of expectations

$$\boldsymbol{\mu} = (\mu_1, \ldots, \mu_n)^\top = (\mathbb{E}[X_1], \ldots, \mathbb{E}[X_n])^\top.$$

The *covariance matrix* Σ is defined as the matrix whose (i, j)-th element is

$$\text{Cov}(X_i, X_j) = \mathbb{E}[(X_i - \mu_i)(X_j - \mu_j)].$$

If we define the expectation of a vector (matrix) to be the vector (matrix) of expectations, then we can write

$$\boldsymbol{\mu} = \mathbb{E}[\mathbf{X}]$$

and

$$\Sigma = \mathbb{E}[(\mathbf{X} - \boldsymbol{\mu})(\mathbf{X} - \boldsymbol{\mu})^\top] \, .$$

Note that $\boldsymbol{\mu}$ and Σ take on the same role as μ and σ^2 in the one-dimensional case.

Remark 1.7.2 Note that any covariance matrix Σ is *symmetric*. In fact (see Problem 1.16), it is *positive semidefinite*, that is, for any (column) vector $\mathbf{u}$,

$$\mathbf{u}^\top \Sigma \mathbf{u} \geqslant 0 \, .$$

1.8 FUNCTIONS OF RANDOM VARIABLES

Suppose that $X_1, \ldots, X_n$ are measurements of a random experiment. Often we are only interested in certain *functions* of the measurements rather than the individual measurements. Here are some examples.

■ **EXAMPLE 1.5**

Let X be a continuous random variable with pdf f_X and let $Z = aX + b$, where $a \neq 0$. We wish to determine the pdf f_Z of Z. Suppose that $a > 0$. We have for any z

$$F_Z(z) = \mathbb{P}(Z \leqslant z) = \mathbb{P}(X \leqslant (z - b)/a) = F_X((z - b)/a) \, .$$

Differentiating this with respect to z gives $f_Z(z) = f_X((z - b)/a)/a$. For $a < 0$ we similarly obtain $f_Z(z) = f_X((z - b)/a)/(-a)$. Thus, in general,

$$f_Z(z) = \frac{1}{|a|} f_X\left(\frac{z - b}{a}\right) \, . \tag{1.15}$$

■ **EXAMPLE 1.6**

Generalizing the previous example, suppose that $Z = g(X)$ for some monotonically increasing function g. To find the pdf of Z from that of X we first write

$$F_Z(z) = \mathbb{P}(Z \leqslant z) = \mathbb{P}\left(X \leqslant g^{-1}(z)\right) = F_X\left(g^{-1}(z)\right) \, ,$$

where g^{-1} is the inverse of g. Differentiating with respect to z now gives

$$f_Z(z) = f_X(g^{-1}(z)) \frac{\mathrm{d}}{\mathrm{d}z} g^{-1}(z) = \frac{f_X(g^{-1}(z))}{g'(g^{-1}(z))} \, . \tag{1.16}$$

For monotonically decreasing functions, $\frac{\mathrm{d}}{\mathrm{d}z} g^{-1}(z)$ in the first equation needs to be replaced with its negative value.

■ **EXAMPLE 1.7 Order Statistics**

Let $X_1, \ldots, X_n$ be an iid sequence of random variables with common pdf f and cdf F. In many applications one is interested in the distribution of the *order statistics* $X_{(1)}, X_{(2)}, \ldots, X_{(n)}$, where $X_{(1)}$ is the smallest of the $\{X_i, i = 1, \ldots, n\}$, $X_{(2)}$ is the second smallest, and so on. The cdf of $X_{(n)}$ follows from

$$\mathbb{P}(X_{(n)} \leqslant x) = \mathbb{P}(X_1 \leqslant x, \ldots, X_n \leqslant x) = \prod_{i=1}^{n} \mathbb{P}(X_i \leqslant x) = (F(x))^n .$$

Similarly,

$$\mathbb{P}(X_{(1)} > x) = \mathbb{P}(X_1 > x, \ldots, X_n > x) = \prod_{i=1}^{n} \mathbb{P}(X_i > x) = (1 - F(x))^n .$$

Moreover, because all orderings of $X_1, \ldots, X_n$ are equally likely, it follows that the joint pdf of the ordered sample is, on the wedge $\{(x_1, \ldots, x_n) : x_1 \leqslant x_2 \leqslant \cdots \leqslant x_n\}$, simply $n!$ times the joint density of the unordered sample and zero elsewhere.

1.8.1   Linear Transformations

Let $\mathbf{x} = (x_1, \ldots, x_n)^\top$ be a column vector in $\mathbb{R}^n$ and A an $m \times n$ matrix. The mapping $\mathbf{x} \mapsto \mathbf{z}$, with $\mathbf{z} = A\mathbf{x}$, is called a *linear transformation*. Now consider a *random* vector $\mathbf{X} = (X_1, \ldots, X_n)^\top$, and let

$$\mathbf{Z} = A\mathbf{X} .$$

Then $\mathbf{Z}$ is a random vector in $\mathbb{R}^m$. In principle, if we know the joint distribution of $\mathbf{X}$, then we can derive the joint distribution of $\mathbf{Z}$. Let us first see how the expectation vector and covariance matrix are transformed.

Theorem 1.8.1 *If $\mathbf{X}$ has an expectation vector $\boldsymbol{\mu}_\mathbf{X}$ and covariance matrix $\Sigma_\mathbf{X}$, then the expectation vector and covariance matrix of $\mathbf{Z} = A\mathbf{X}$ are given by*

$$\boldsymbol{\mu}_\mathbf{Z} = A\boldsymbol{\mu}_\mathbf{X} \tag{1.17}$$

and

$$\Sigma_\mathbf{Z} = A\,\Sigma_\mathbf{X}\,A^\top . \tag{1.18}$$

Proof: We have $\boldsymbol{\mu}_\mathbf{Z} = \mathbb{E}[\mathbf{Z}] = \mathbb{E}[A\mathbf{X}] = A\,\mathbb{E}[\mathbf{X}] = A\boldsymbol{\mu}_\mathbf{X}$ and

$$\begin{aligned}
\Sigma_\mathbf{Z} &= \mathbb{E}[(\mathbf{Z} - \boldsymbol{\mu}_\mathbf{Z})(\mathbf{Z} - \boldsymbol{\mu}_\mathbf{Z})^\top] = \mathbb{E}[A(\mathbf{X} - \boldsymbol{\mu}_\mathbf{X})(A(\mathbf{X} - \boldsymbol{\mu}_\mathbf{X}))^\top] \\
&= A\,\mathbb{E}[(\mathbf{X} - \boldsymbol{\mu}_\mathbf{X})(\mathbf{X} - \boldsymbol{\mu}_\mathbf{X})^\top]A^\top \\
&= A\,\Sigma_\mathbf{X}\,A^\top .
\end{aligned}$$

$\square$

Suppose that A is an invertible $n \times n$ matrix. If $\mathbf{X}$ has a joint density $f_\mathbf{X}$, what is the joint density $f_\mathbf{Z}$ of $\mathbf{Z}$? Consider Figure 1.1. For any fixed $\mathbf{x}$, let $\mathbf{z} = A\mathbf{x}$. Hence, $\mathbf{x} = A^{-1}\mathbf{z}$. Consider the n-dimensional cube $C = [z_1, z_1 + h] \times \cdots \times [z_n, z_n + h]$. Let D be the image of C under A^{-1}, that is, the parallelepiped of all points $\mathbf{x}$ such that $A\mathbf{x} \in C$. Then,

$$\mathbb{P}(\mathbf{Z} \in C) \approx h^n\, f_\mathbf{Z}(\mathbf{z}) .$$

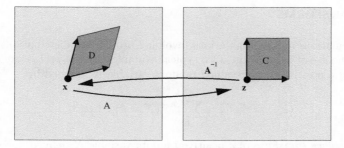

Figure 1.1: Linear transformation.

Now recall from linear algebra (e.g., [5]) that any matrix B linearly transforms an n-dimensional rectangle with volume V into an n-dimensional parallelepiped with volume $V\,|B|$, where $|B| = |\det(B)|$. Thus,

$$\mathbb{P}(\mathbf{Z} \in C) = \mathbb{P}(\mathbf{X} \in D) \approx h^n |A^{-1}|\, f_{\mathbf{X}}(\mathbf{x}) = h^n |A|^{-1} f_{\mathbf{X}}(\mathbf{x}) \ .$$

Letting h go to 0, we obtain

$$f_{\mathbf{Z}}(\mathbf{z}) = \frac{f_{\mathbf{X}}(A^{-1}\mathbf{z})}{|A|}, \quad \mathbf{z} \in \mathbb{R}^n. \tag{1.19}$$

1.8.2 General Transformations

We can apply reasoning similar to that above to deal with general transformations $\mathbf{x} \mapsto \boldsymbol{g}(\mathbf{x})$, written out as

$$\begin{pmatrix} x_1 \\ x_2 \\ \vdots \\ x_n \end{pmatrix} \mapsto \begin{pmatrix} g_1(\mathbf{x}) \\ g_2(\mathbf{x}) \\ \vdots \\ g_n(\mathbf{x}) \end{pmatrix}.$$

For a fixed $\mathbf{x}$, let $\mathbf{z} = \boldsymbol{g}(\mathbf{x})$. Suppose that $\boldsymbol{g}$ is invertible; hence $\mathbf{x} = \boldsymbol{g}^{-1}(\mathbf{z})$. Any infinitesimal n-dimensional rectangle at $\mathbf{x}$ with volume V is transformed into an n-dimensional parallelepiped at $\mathbf{z}$ with volume $V\,|J_{\mathbf{x}}(\boldsymbol{g})|$, where $J_{\mathbf{x}}(\boldsymbol{g})$ is the *matrix of Jacobi* at $\mathbf{x}$ of the transformation $\boldsymbol{g}$, that is,

$$J_{\mathbf{x}}(\boldsymbol{g}) = \begin{pmatrix} \frac{\partial g_1}{\partial x_1} & \cdots & \frac{\partial g_1}{\partial x_n} \\ \vdots & \cdots & \vdots \\ \frac{\partial g_n}{\partial x_1} & \cdots & \frac{\partial g_n}{\partial x_n} \end{pmatrix}.$$

Now consider a random column vector $\mathbf{Z} = \boldsymbol{g}(\mathbf{X})$. Let C be a small cube around $\mathbf{z}$ with volume h^n. Let D be the image of C under $\boldsymbol{g}^{-1}$. Then, as in the linear case,

$$\mathbb{P}(\mathbf{Z} \in C) \approx h^n\, f_{\mathbf{Z}}(\mathbf{z}) \approx h^n |J_{\mathbf{z}}(\boldsymbol{g}^{-1})|\, f_{\mathbf{X}}(\mathbf{x}) \ .$$

Hence we have the transformation rule

$$f_{\mathbf{Z}}(\mathbf{z}) = f_{\mathbf{X}}(\boldsymbol{g}^{-1}(\mathbf{z}))\, |J_{\mathbf{z}}(\boldsymbol{g}^{-1})|, \quad \mathbf{z} \in \mathbb{R}^n. \tag{1.20}$$

(Note: $|J_{\mathbf{z}}(\boldsymbol{g}^{-1})| = 1/|J_{\mathbf{x}}(\boldsymbol{g})|$.)

Remark 1.8.1 In most coordinate transformations, it is $\boldsymbol{g}^{-1}$ that is given — that is, an expression for $\mathbf{x}$ as a function of $\mathbf{z}$ rather than $\boldsymbol{g}$.

1.9 TRANSFORMS

Many calculations and manipulations involving probability distributions are facilitated by the use of transforms. Two typical examples are the *probability generating function* of a positive integer-valued random variable N, defined by

$$G(z) = \mathbb{E}[z^N] = \sum_{k=0}^{\infty} z^k \, \mathbb{P}(N = k) , \quad |z| \leqslant 1 ,$$

and the *Laplace transform* of a positive random variable X defined, for $s \geqslant 0$, by

$$L(s) = \mathbb{E}[e^{-sX}] = \begin{cases} \sum_x e^{-sx} f(x) & \text{discrete case,} \\ \int_0^{\infty} e^{-sx} f(x) \, dx & \text{continuous case.} \end{cases}$$

All transforms share an important *uniqueness property*: two distributions are the same if and only if their respective transforms are the same.

■ **EXAMPLE 1.8**

Let $M \sim \mathsf{Poi}(\mu)$; then its probability generating function is given by

$$G(z) = \sum_{k=0}^{\infty} z^k \, e^{-\mu} \frac{\mu^k}{k!} = e^{-\mu} \sum_{k=0}^{\infty} \frac{(z\mu)^k}{k!} = e^{-\mu} e^{z\mu} = e^{-\mu(1-z)} . \tag{1.21}$$

Now let $N \sim \mathsf{Poi}(\nu)$ independently of M. Then the probability generating function of $M + N$ is given by

$$\mathbb{E}[z^{M+N}] = \mathbb{E}[z^M] \, \mathbb{E}[z^N] = e^{-\mu(1-z)} e^{-\nu(1-z)} = e^{-(\mu+\nu)(1-z)} .$$

Thus, by the uniqueness property, $M + N \sim \mathsf{Poi}(\mu + \nu)$.

■ **EXAMPLE 1.9**

The Laplace transform of $X \sim \mathsf{Gamma}(\alpha, \lambda)$ is given by

$$\begin{aligned} \mathbb{E}[e^{-sX}] &= \int_0^{\infty} \frac{e^{-\lambda x} \lambda^\alpha x^{\alpha-1}}{\Gamma(\alpha)} e^{-sx} \, dx \\ &= \left(\frac{\lambda}{\lambda + s}\right)^\alpha \int_0^{\infty} \frac{e^{-(\lambda+s)x} (\lambda + s)^\alpha x^{\alpha-1}}{\Gamma(\alpha)} \, dx \\ &= \left(\frac{\lambda}{\lambda + s}\right)^\alpha . \end{aligned}$$

As a special case, the Laplace transform of the $\mathsf{Exp}(\lambda)$ distribution is given by $\lambda/(\lambda + s)$. Now let $X_1, \ldots, X_n$ be iid $\mathsf{Exp}(\lambda)$ random variables. The Laplace transform of $S_n = X_1 + \cdots + X_n$ is

$$\mathbb{E}[e^{-sS_n}] = \mathbb{E}[e^{-sX_1} \cdots e^{-sX_n}] = \mathbb{E}[e^{-sX_1}] \cdots \mathbb{E}[e^{-sX_n}] = \left(\frac{\lambda}{\lambda + s}\right)^n ,$$

which shows that $S_n \sim \mathsf{Gamma}(n, \lambda)$.

1.10   JOINTLY NORMAL RANDOM VARIABLES

It is helpful to view normally distributed random variables as simple transformations of *standard normal* — that is, $N(0,1)$-distributed — random variables. In particular, let $X \sim N(0,1)$. Then X has density f_X given by

$$f_X(x) = \frac{1}{\sqrt{2\pi}}\, e^{-\frac{x^2}{2}}.$$

Now consider the transformation $Z = \mu + \sigma X$. Then, by (1.15), Z has density

$$f_Z(z) = \frac{1}{\sqrt{2\pi\sigma^2}}\, e^{-\frac{(z-\mu)^2}{2\sigma^2}}.$$

In other words, $Z \sim N(\mu, \sigma^2)$. We can also state this as follows: if $Z \sim N(\mu, \sigma^2)$, then $(Z-\mu)/\sigma \sim N(0,1)$. This procedure is called *standardization*.

We now generalize this to n dimensions. Let $X_1, \ldots, X_n$ be independent and standard normal random variables. The joint pdf of $\mathbf{X} = (X_1, \ldots, X_n)^\top$ is given by

$$f_{\mathbf{X}}(\mathbf{x}) = (2\pi)^{-n/2} e^{-\frac{1}{2}\mathbf{x}^\top \mathbf{x}}, \quad \mathbf{x} \in \mathbb{R}^n. \tag{1.22}$$

Consider the *affine* transformation (i.e., a linear transformation plus a constant vector)

$$\mathbf{Z} = \boldsymbol{\mu} + B\,\mathbf{X} \tag{1.23}$$

for some $m \times n$ matrix B. Note that, by Theorem 1.8.1, $\mathbf{Z}$ has expectation vector $\boldsymbol{\mu}$ and covariance matrix $\Sigma = BB^\top$. Any random vector of the form (1.23) is said to have a *jointly normal* or *multivariate normal* distribution. We write $\mathbf{Z} \sim N(\boldsymbol{\mu}, \Sigma)$. Suppose that B is an invertible $n \times n$ matrix. Then, by (1.19), the density of $\mathbf{Y} = \mathbf{Z} - \boldsymbol{\mu}$ is given by

$$f_{\mathbf{Y}}(\mathbf{y}) = \frac{1}{|B|\sqrt{(2\pi)^n}}\, e^{-\frac{1}{2}(B^{-1}\mathbf{y})^\top B^{-1}\mathbf{y}} = \frac{1}{|B|\sqrt{(2\pi)^n}}\, e^{-\frac{1}{2}\mathbf{y}^\top (B^{-1})^\top B^{-1}\mathbf{y}}.$$

We have $|B| = \sqrt{|\Sigma|}$ and $(B^{-1})^\top B^{-1} = (B^\top)^{-1} B^{-1} = (BB^\top)^{-1} = \Sigma^{-1}$, so that

$$f_{\mathbf{Y}}(\mathbf{y}) = \frac{1}{\sqrt{(2\pi)^n\, |\Sigma|}}\, e^{-\frac{1}{2}\mathbf{y}^\top \Sigma^{-1}\mathbf{y}}.$$

Because $\mathbf{Z}$ is obtained from $\mathbf{Y}$ by simply adding a constant vector $\boldsymbol{\mu}$, we have $f_{\mathbf{Z}}(\mathbf{z}) = f_{\mathbf{Y}}(\mathbf{z} - \boldsymbol{\mu})$, and therefore

$$f_{\mathbf{Z}}(\mathbf{z}) = \frac{1}{\sqrt{(2\pi)^n\, |\Sigma|}}\, e^{-\frac{1}{2}(\mathbf{z}-\boldsymbol{\mu})^\top \Sigma^{-1}(\mathbf{z}-\boldsymbol{\mu})}, \quad \mathbf{z} \in \mathbb{R}^n. \tag{1.24}$$

Note that this formula is very similar to that of the one-dimensional case.

Conversely, given a covariance matrix $\Sigma = (\sigma_{ij})$, there exists a unique lower triangular matrix

$$B = \begin{pmatrix} b_{11} & 0 & \cdots & 0 \\ b_{21} & b_{22} & \cdots & 0 \\ \vdots & \vdots & & \vdots \\ b_{n1} & b_{n2} & \cdots & b_{nn} \end{pmatrix} \tag{1.25}$$

such that $\Sigma = BB^\top$. This matrix can be obtained efficiently via the *Cholesky square root method*; see Section A.1 of the Appendix.

1.11 LIMIT THEOREMS

We briefly discuss two of the main results in probability: the law of large numbers and the central limit theorem. Both are associated with sums of independent random variables.

Let $X_1, X_2, \ldots$ be iid random variables with expectation μ and variance σ^2. For each n, let $S_n = X_1 + \cdots + X_n$. Since $X_1, X_2, \ldots$ are iid, we have $\mathbb{E}[S_n] = n \mathbb{E}[X_1] = n\mu$ and $\mathrm{Var}(S_n) = n \mathrm{Var}(X_1) = n\sigma^2$.

The law of large numbers states that S_n/n is close to μ for large n. Here is the more precise statement.

Theorem 1.11.1 (Strong Law of Large Numbers) *If $X_1, \ldots, X_n$ are iid with expectation μ, then*

$$\mathbb{P}\left(\lim_{n \to \infty} \frac{S_n}{n} = \mu \right) = 1 \ .$$

The central limit theorem describes the limiting distribution of S_n (or S_n/n), and it applies to both continuous and discrete random variables. Loosely, it states that the random sum S_n has a distribution that is approximately normal, when n is large. The more precise statement is given next.

Theorem 1.11.2 (Central Limit Theorem) *If $X_1, \ldots, X_n$ are iid with expectation μ and variance $\sigma^2 < \infty$, then for all $x \in \mathbb{R}$,*

$$\lim_{n \to \infty} \mathbb{P}\left(\frac{S_n - n\mu}{\sigma \sqrt{n}} \leqslant x \right) = \Phi(x) \ ,$$

where Φ is the cdf of the standard normal distribution.

In other words, S_n has a distribution that is approximately normal, with expectation $n\mu$ and variance $n\sigma^2$. To see the central limit theorem in action, consider Figure 1.2. The left part shows the pdfs of $S_1, \ldots, S_4$ for the case where the $\{X_i\}$ have a $\mathsf{U}[0, 1]$ distribution. The right part shows the same for the $\mathsf{Exp}(1)$ distribution. We clearly see convergence to a bell-shaped curve, characteristic of the normal distribution.

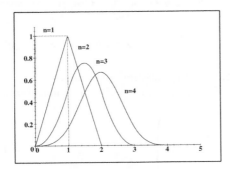 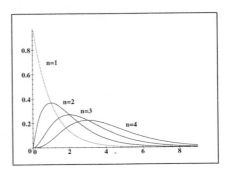

Figure 1.2: Illustration of the central limit theorem for (left) the uniform distribution and (right) the exponential distribution.

A direct consequence of the central limit theorem and the fact that a $\mathsf{Bin}(n,p)$ random variable X can be viewed as the sum of n iid $\mathsf{Ber}(p)$ random variables, $X = X_1 + \cdots + X_n$, is that for large n

$$\mathbb{P}(X \leqslant k) \approx \mathbb{P}(Y \leqslant k) , \qquad (1.26)$$

with $Y \sim \mathsf{N}(np, np(1-p))$. As a rule of thumb, this *normal approximation to the binomial distribution* is accurate if both np and $n(1-p)$ are larger than 5.

There is also a central limit theorem for random vectors. The multidimensional version is as follows: Let $\mathbf{X}_1, \ldots, \mathbf{X}_n$ be iid random vectors with expectation vector $\boldsymbol{\mu}$ and covariance matrix Σ. Then for large n the random vector $\mathbf{X}_1 + \cdots + \mathbf{X}_n$ has approximately a multivariate normal distribution with expectation vector $n\boldsymbol{\mu}$ and covariance matrix $n\Sigma$.

1.12   POISSON PROCESSES

The Poisson process is used to model certain kinds of arrivals or patterns. Imagine, for example, a telescope that can detect individual photons from a faraway galaxy. The photons arrive at random times $T_1, T_2, \ldots$. Let N_t denote the number of arrivals in the time interval $[0, t]$, that is, $N_t = \sup\{k : T_k \leqslant t\}$. Note that the number of arrivals in an interval $I = (a, b]$ is given by $N_b - N_a$. We will also denote it by $N(a, b]$. A sample path of the arrival counting process $\{N_t, t \geqslant 0\}$ is given in Figure 1.3.

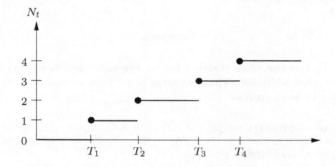

Figure 1.3: A sample path of the arrival counting process $\{N_t, t \geqslant 0\}$.

For this particular arrival process, one would assume that the number of arrivals in an interval (a, b) is independent of the number of arrivals in interval (c, d) when the two intervals do not intersect. Such considerations lead to the following definition:

Definition 1.12.1 (Poisson Process) An arrival counting process $N = \{N_t\}$ is called a *Poisson process* with *rate* $\lambda > 0$ if

(a) The numbers of points in nonoverlapping intervals are independent.

(b) The number of points in interval I has a Poisson distribution with mean $\lambda \times \text{length}(I)$.

Combining (a) and (b) we see that the number of arrivals in any small interval $(t, t + h]$ is independent of the arrival process up to time t and has a $\mathsf{Poi}(\lambda h)$ distribution. In particular, the conditional probability that exactly one arrival occurs during the time interval $(t, t + h]$ is $\mathbb{P}(N(t, t + h] = 1 \mid N_t) = \mathrm{e}^{-\lambda h} \lambda h \approx \lambda h$. Similarly, the probability of no arrivals is approximately $1 - \lambda h$ for small h. In other words, λ is the *rate* at which arrivals occur. Notice also that since $N_t \sim \mathsf{Poi}(\lambda t)$, the expected number of arrivals in $[0, t]$ is λt, that is, $\mathbb{E}[N_t] = \lambda t$. In Definition 1.12.1 N is seen as a random counting measure, where $N(I)$ counts the random number of arrivals in set I.

An important relationship between N_t and T_n is

$$\{N_t \geqslant n\} = \{T_n \leqslant t\} . \tag{1.27}$$

In other words, the number of arrivals in $[0, t]$ is at least n if and only if the n-th arrival occurs at or before time t. As a consequence, we have

$$
\begin{aligned}
\mathbb{P}(T_n \leqslant t) &= \mathbb{P}(N_t \geqslant n) = 1 - \sum_{k=0}^{n-1} \mathbb{P}(N_t = k) \\
&= 1 - \sum_{k=0}^{n-1} \frac{\mathrm{e}^{-\lambda t}(\lambda t)^k}{k!} ,
\end{aligned}
$$

which corresponds exactly to the cdf of the $\mathsf{Gamma}(n, \lambda)$ distribution; see Problem 1.17. Thus

$$T_n \sim \mathsf{Gamma}(n, \lambda) . \tag{1.28}$$

Hence each T_n has the same distribution as the sum of n independent $\mathsf{Exp}(\lambda)$-distributed random variables. This corresponds with the second important characterization of a Poisson process:

> An arrival counting process $\{N_t\}$ is a Poisson process with rate λ if and only if the interarrival times $A_1 = T_1, A_2 = T_2 - T_1, \ldots$ are independent and $\mathsf{Exp}(\lambda)$-distributed random variables.

Poisson and Bernoulli processes are akin, and much can be learned about Poisson processes via the following *Bernoulli approximation*. Let $N = \{N_t\}$ be a Poisson process with parameter λ. We divide the time axis into small time intervals $[0, h), [h, 2h), \ldots$ and count how many arrivals occur in each interval. Note that the number of arrivals in any small time interval of length h is, with high probability, either 1 (with probability $\lambda h \mathrm{e}^{-\lambda h} \approx \lambda h$) or 0 (with probability $\mathrm{e}^{-\lambda h} \approx 1 - \lambda h$). Next, define $X = \{X_n\}$ to be a Bernoulli process with success parameter $p = \lambda h$. Put $Y_0 = 0$ and let $Y_n = X_1 + \cdots + X_n$ be the total number of successes in n trials. $Y = \{Y_n\}$ is called the *Bernoulli approximation* to N. We can view N as a limiting case of Y as we decrease h.

As an example of the usefulness of this interpretation, we now demonstrate that the Poisson property (b) in Definition 1.12.1 follows basically from the *independence* assumption (a). For small h, N_t should have approximately the same distribution as Y_n, where n is the integer part of t/h (we write $n = \lfloor t/h \rfloor$). Hence,

$$
\begin{aligned}
\mathbb{P}(N_t = k) &\approx \mathbb{P}(Y_n = k) \\
&= \binom{n}{k} (\lambda h)^k (1 - (\lambda h))^{n-k} \\
&\approx \binom{n}{k} (\lambda t/n)^k (1 - (\lambda t/n))^{n-k} \\
&\approx \mathrm{e}^{-\lambda t} \frac{(\lambda t)^k}{k!} .
\end{aligned}
\tag{1.29}
$$

Equation (1.29) follows from the Poisson approximation to the binomial distribution; see Problem 1.22.

Another application of the Bernoulli approximation is the following. For the Bernoulli process, given that the total number of successes is k, the positions of the k successes are uniformly distributed over points $1, \ldots, n$. The corresponding property for the Poisson process N is that given $N_t = n$, the arrival times $T_1, \ldots, T_n$ are distributed according to the order statistics $X_{(1)}, \ldots, X_{(n)}$, where $X_1, \ldots, X_n$ are iid $\mathsf{U}[0, t]$.

1.13   MARKOV PROCESSES

Markov processes are stochastic processes whose futures are conditionally independent of their pasts given their present values. More formally, a stochastic process $\{X_t, t \in \mathscr{T}\}$, with $\mathscr{T} \subseteq \mathbb{R}$, is called a *Markov process* if, for every $s > 0$ and t,

$$
(X_{t+s} \mid X_u, u \leqslant t) \quad \sim \quad (X_{t+s} \mid X_t) .
\tag{1.30}
$$

In other words, the conditional distribution of the future variable X_{t+s}, given the entire past of the process $\{X_u, u \leqslant t\}$, is the same as the conditional distribution of X_{t+s} given only the present X_t. That is, in order to predict future states, we only need to know the present one. Property (1.30) is called the *Markov property*.

Depending on the index set $\mathscr{T}$ and state space $\mathscr{E}$ (the set of all values the $\{X_t\}$ can take), Markov processes come in many different forms. A Markov process with a discrete index set is called a *Markov chain*. A Markov process with a discrete state space and a continuous index set (such as $\mathbb{R}$ or $\mathbb{R}_+$) is called a *Markov jump process*.

1.13.1   Markov Chains

Consider a Markov chain $X = \{X_t, t \in \mathbb{N}\}$ with a discrete (i.e., countable) state space $\mathscr{E}$. In this case the Markov property (1.30) is

$$
\mathbb{P}(X_{t+1} = x_{t+1} \mid X_0 = x_0, \ldots, X_t = x_t) = \mathbb{P}(X_{t+1} = x_{t+1} \mid X_t = x_t)
\tag{1.31}
$$

for all $x_0, \ldots, x_{t+1}, \in \mathscr{E}$ and $t \in \mathbb{N}$. We restrict ourselves to Markov chains for which the conditional probabilities

$$
\mathbb{P}(X_{t+1} = j \mid X_t = i), \quad i, j \in \mathscr{E}
\tag{1.32}
$$

are independent of the time t. Such chains are called *time-homogeneous*. The probabilities in (1.32) are called the *(one-step) transition probabilities* of X. The distribution of X_0 is called the *initial distribution* of the Markov chain. The one-step transition probabilities and the initial distribution completely specify the distribution of X. Namely, we have by the product rule (1.4) and the Markov property (1.30),

$$\mathbb{P}(X_0 = x_0, \ldots, X_t = x_t)$$
$$= \mathbb{P}(X_0 = x_0)\,\mathbb{P}(X_1 = x_1 \mid X_0 = x_0) \cdots \mathbb{P}(X_t = x_t \mid X_0 = x_0, \ldots X_{t-1} = x_{t-1})$$
$$= \mathbb{P}(X_0 = x_0)\,\mathbb{P}(X_1 = x_1 \mid X_0 = x_0) \cdots \mathbb{P}(X_t = x_t \mid X_{t-1} = x_{t-1})\,.$$

Since $\mathscr{E}$ is countable, we can arrange the one-step transition probabilities in an array. This array is called the (one-step) *transition matrix* of X. We usually denote it by P. For example, when $\mathscr{E} = \{0, 1, 2, \ldots\}$, the transition matrix P has the form

$$P = \begin{pmatrix} p_{00} & p_{01} & p_{02} & \cdots \\ p_{10} & p_{11} & p_{12} & \cdots \\ p_{20} & p_{21} & p_{22} & \cdots \\ \vdots & \vdots & \vdots & \ddots \end{pmatrix}\,.$$

Note that the elements in every row are positive and sum up to unity.

Another convenient way to describe a Markov chain X is through its *transition graph*. States are indicated by the nodes of the graph, and a strictly positive (> 0) transition probability p_{ij} from state i to j is indicated by an arrow from i to j with weight p_{ij}.

■ EXAMPLE 1.10   Random Walk on the Integers

Let p be a number between 0 and 1. The Markov chain X with state space $\mathbb{Z}$ and transition matrix P defined by

$$P(i, i+1) = p, \quad P(i, i-1) = q = 1 - p, \quad \text{for all } i \in \mathbb{Z}$$

is called a *random walk on the integers*. Let X start at 0; thus, $\mathbb{P}(X_0 = 0) = 1$. The corresponding transition graph is given in Figure 1.4. Starting at 0, the chain takes subsequent steps to the right with probability p and to the left with probability q.

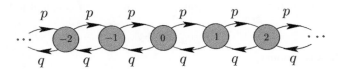

Figure 1.4: Transition graph for a random walk on $\mathbb{Z}$.

We show next how to calculate the probability that, starting from state i at some (discrete) time t, we are in j at (discrete) time $t + s$, that is, the probability $\mathbb{P}(X_{t+s} = j \mid X_t = i)$. For clarity, let us assume that $\mathscr{E} = \{1, 2, \ldots, m\}$ for some fixed m, so that P is an $m \times m$ matrix. For $t = 0, 1, 2, \ldots$, define the row vector

$$\boldsymbol{\pi}^{(t)} = (\mathbb{P}(X_t = 1), \ldots, \mathbb{P}(X_t = m))\,.$$

We call $\boldsymbol{\pi}^{(t)}$ the *distribution vector*, or simply the *distribution*, of X at time t and $\boldsymbol{\pi}^{(0)}$ the *initial distribution* of X. The following result shows that the t-step probabilities can be found simply by matrix multiplication.

Theorem 1.13.1 *The distribution of X at time t is given by*

$$\boldsymbol{\pi}^{(t)} = \boldsymbol{\pi}^{(0)} P^t \tag{1.33}$$

for all $t = 0, 1, \ldots$. (Here P^0 denotes the identity matrix.)

Proof: The proof is by induction. Equality (1.33) holds for $t = 0$ by definition. Suppose that this equality is true for some $t = 0, 1, \ldots$. We have

$$\mathbb{P}(X_{t+1} = k) = \sum_{i=1}^{m} \mathbb{P}(X_{t+1} = k \mid X_t = i)\,\mathbb{P}(X_t = i)\,.$$

But (1.33) is assumed to be true for t, so $\mathbb{P}(X_t = i)$ is the i-th element of $\boldsymbol{\pi}^{(0)} P^t$. Moreover, $\mathbb{P}(X_{t+1} = k \mid X_t = i)$ is the (i, k)-th element of P. Therefore, for every k,

$$\sum_{i=1}^{m} \mathbb{P}(X_{t+1} = k \mid X_t = i)\,\mathbb{P}(X_t = i) = \sum_{i=1}^{m} P(i, k)(\boldsymbol{\pi}^{(0)} P^t)(i)\,,$$

which is just the k-th element of $\boldsymbol{\pi}^{(0)} P^{t+1}$. This completes the induction step, and thus the theorem is proved. $\square$

By taking $\boldsymbol{\pi}^{(0)}$ as the i-th unit vector, $\mathbf{e}_i$, the t-step transition probabilities can be found as $\mathbb{P}(X_t = j \mid X_0 = i) = (\mathbf{e}_i P^t)(j) = P^t(i, j)$, which is the (i, j)-th element of matrix P^t. Thus, to find the t-step transition probabilities, we just have to compute the t-th power of P.

1.13.2   Classification of States

Let X be a Markov chain with discrete state space $\mathscr{E}$ and transition matrix P. We can characterize the relations between states in the following way: If states i and j are such that $P^t(i, j) > 0$ for some $t \geqslant 0$, we say that i *leads to* j and write $i \to j$. We say that i and j *communicate* if $i \to j$ and $j \to i$, and write $i \leftrightarrow j$. Using the relation "$\leftrightarrow$", we can divide $\mathscr{E}$ into *equivalence classes* such that all the states in an equivalence class communicate with each other but not with any state outside that class. If there is only one equivalent class $(= \mathscr{E})$, the Markov chain is said to be *irreducible*. If a set of states $\mathscr{A}$ is such that $\sum_{j \in \mathscr{A}} P(i, j) = 1$ for all $i \in \mathscr{A}$, then $\mathscr{A}$ is called a *closed* set. A state i is called an *absorbing* state if $\{i\}$ is closed. For example, in the transition graph depicted in Figure 1.5, the equivalence classes are $\{1, 2\}$, $\{3\}$, and $\{4, 5\}$. Class $\{1, 2\}$ is the only closed set: the Markov chain cannot escape from it. If state 1 were missing, state 2 would be absorbing. In Example 1.10 the Markov chain is irreducible since all states communicate.

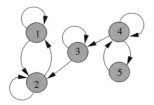

Figure 1.5: A transition graph with three equivalence classes.

Another classification of states is obtained by observing the system from a local point of view. In particular, let T denote the time the chain first visits state j, or first returns to j if it started there, and let N_j denote the total number of visits to j from time 0 on. We write $\mathbb{P}_j(A)$ for $\mathbb{P}(A \mid X_0 = j)$ for any event A. We denote the corresponding expectation operator by $\mathbb{E}_j$. State j is called a *recurrent* state if $\mathbb{P}_j(T < \infty) = 1$; otherwise, j is called *transient*. A recurrent state is called *positive recurrent* if $\mathbb{E}_j[T] < \infty$; otherwise, it is called *null recurrent*. Finally, a state is said to be *periodic, with period δ*, if $\delta \geqslant 2$ is the largest integer for which $\mathbb{P}_j(T = n\delta$ for some $n \geqslant 1) = 1$; otherwise, it is called *aperiodic*. For example, in Figure 1.5 states 1 and 2 are recurrent, and the other states are transient. All these states are aperiodic. The states of the random walk of Example 1.10 are periodic with period 2.

It can be shown that recurrence and transience are class properties. In particular, if $i \leftrightarrow j$, then i recurrent (transient) $\Leftrightarrow j$ recurrent (transient). Thus, in an irreducible Markov chain, one state being recurrent implies that all other states are also recurrent. And if one state is transient, then so are all the others.

1.13.3   Limiting Behavior

The limiting or "steady-state" behavior of Markov chains as $t \to \infty$ is of considerable interest and importance, and this type of behavior is often simpler to describe and analyze than the "transient" behavior of the chain for fixed t. It can be shown (see, for example, [3]) that in an irreducible, aperiodic Markov chain with transition matrix P the t-step probabilities converge to a constant that does not depend on the initial state. More specifically,

$$\lim_{t \to \infty} P^t(i,j) = \pi_j \qquad (1.34)$$

for some number $0 \leqslant \pi_j \leqslant 1$. Moreover, $\pi_j > 0$ if j is positive recurrent and $\pi_j = 0$ otherwise. The intuitive reason behind this result is that the process "forgets" where it was initially if it goes on long enough. This is true for both finite and countably infinite Markov chains. The numbers $\{\pi_j, j \in \mathscr{E}\}$ form the *limiting distribution* of the Markov chain, provided that $\pi_j \geqslant 0$ and $\sum_j \pi_j = 1$. Note that these conditions are not always satisfied: they are clearly not satisfied if the Markov chain is transient, and they may not be satisfied if the Markov chain is recurrent (i.e., when the states are null-recurrent). The following theorem gives a method for obtaining limiting distributions. Here we assume for simplicity that $\mathscr{E} = \{0, 1, 2, \ldots\}$. The limiting distribution is identified with the row vector $\boldsymbol{\pi} = (\pi_0, \pi_1, \ldots)$.

Theorem 1.13.2 *For an irreducible, aperiodic Markov chain with transition matrix P, if the limiting distribution $\boldsymbol{\pi}$ exists, then it is uniquely determined by the solution of*

$$\boldsymbol{\pi} = \boldsymbol{\pi} P \,, \tag{1.35}$$

with $\pi_j \geqslant 0$ and $\sum_j \pi_j = 1$. Conversely, if there exists a positive row vector $\boldsymbol{\pi}$ satisfying (1.35) and summing up to 1, then $\boldsymbol{\pi}$ is the limiting distribution of the Markov chain. Moreover, in that case, $\pi_j > 0$ for all j and all states are positive recurrent.

Proof: (Sketch) For the case where $\mathscr{E}$ is finite, the result is simply a consequence of (1.33). Namely, with $\boldsymbol{\pi}^{(0)}$ being the i-th unit vector, we have

$$P^{t+1}(i,j) = \left(\boldsymbol{\pi}^{(0)} \, P^t \, P \right)(j) = \sum_{k \in \mathscr{E}} P^t(i,k) P(k,j) \,.$$

Letting $t \to \infty$, we obtain (1.35) from (1.34), provided that we can change the order of the limit and the summation. To show uniqueness, suppose that another vector $\mathbf{y}$, with $y_j \geqslant 0$ and $\sum_j y_j = 1$, satisfies $\mathbf{y} = \mathbf{y}P$. Then it is easy to show by induction that $\mathbf{y} = \mathbf{y}P^t$, for every t. Hence, letting $t \to \infty$, we obtain for every j

$$y_j = \sum_i y_i \, \pi_j = \pi_j \,,$$

since the $\{y_j\}$ sum up to unity. We omit the proof of the converse statement. $\square$

◼ EXAMPLE 1.11   Random Walk on the Positive Integers

This is a slightly different random walk than the one in Example 1.10. Let X be a random walk on $\mathscr{E} = \{0, 1, 2, \ldots\}$ with transition matrix

$$P = \begin{pmatrix} q & p & 0 & \cdots & & \\ q & 0 & p & 0 & \cdots & \\ 0 & q & 0 & p & 0 & \cdots \\ \vdots & \ddots & \ddots & \ddots & \ddots & \ddots \end{pmatrix},$$

where $0 < p < 1$ and $q = 1 - p$. X_t could represent, for example, the number of customers who are waiting in a queue at time t.

All states can be reached from each other, so the chain is irreducible and every state is either recurrent or transient. The equation $\boldsymbol{\pi} = \boldsymbol{\pi} P$ becomes

$$\begin{aligned} \pi_0 &= q \, \pi_0 + q \, \pi_1 \,, \\ \pi_1 &= p \, \pi_0 + q \, \pi_2 \,, \\ \pi_2 &= p \, \pi_1 + q \, \pi_3 \,, \\ \pi_3 &= p \, \pi_2 + q \, \pi_4 \,, \end{aligned}$$

and so on. We can solve this set of equation sequentially. If we let $r = p/q$, then we can express the $\pi_1, \pi_2, \ldots$ in terms of π_0 and r as

$$\pi_j = r^j \, \pi_0, \quad j = 0, 1, 2, \ldots \,.$$

If $p < q$, then $r < 1$ and $\sum_{j=0}^{\infty} \pi_j = \pi_0/(1-r)$, and by choosing $\pi_0 = 1 - r$, we can make the sum $\sum \pi_j = 1$. Hence, for $r < 1$, we have found the limiting distribution $\boldsymbol{\pi} = (1-r)(1, r, r^2, r^3, \dots)$ for this Markov chain, and all the states are therefore positive recurrent. However, when $p \geqslant q$, $\sum \pi_j$ is either 0 or infinite, and hence all states are either null-recurrent or transient. (It can be shown that only the case $p = q$ leads to null-recurrent states.)

Let X be a Markov chain with limiting distribution $\boldsymbol{\pi}$. Suppose $\boldsymbol{\pi}^{(0)} = \boldsymbol{\pi}$. Then, combining (1.33) and (1.35), we have $\boldsymbol{\pi}^{(t)} = \boldsymbol{\pi}$. Thus, if the initial distribution of the Markov chain is equal to the limiting distribution, then the distribution of X_t is the same for all t (and is given by this limiting distribution). In fact, it is not difficult to show that for any k the distribution of $X_k, X_{k+1}, X_{k+2} \dots$ is the same as that of $X_0, X_1, \dots$. In other words, when $\boldsymbol{\pi}^{(0)} = \boldsymbol{\pi}$, the Markov chain is a stationary stochastic process. More formally, a stochastic process $\{X_t, t \in \mathbb{N}\}$ is called *stationary* if, for any positive $\tau, t_1, \dots, t_n$, the vector $(X_{t_1}, \dots, X_{t_n})$ has the same distribution as $(X_{t_1+\tau}, \dots, X_{t_n+\tau})$. Similar definitions hold when the index set is $\mathbb{Z}$, $\mathbb{R}_+$, or $\mathbb{R}$. For this reason any distribution $\boldsymbol{\pi}$ for which (1.35) holds is called a *stationary distribution*.

Noting that $\sum_j p_{ij} = 1$, we can rewrite (1.35) as the system of equations

$$\sum_j \pi_i \, p_{ij} = \sum_j \pi_j \, p_{ji} \quad \text{for all } i \in \mathscr{E} . \tag{1.36}$$

These are called the *global balance equations*. We can interpret (1.35) as the statement that the "probability flux" out of i is balanced by the probability flux into i. An important generalization, which follows directly from (1.36), states that the same balancing of probability fluxes holds for an arbitrary set $\mathscr{A}$. That is, for every set $\mathscr{A}$ of states we have

$$\sum_{i \in \mathscr{A}} \sum_{j \notin \mathscr{A}} \pi_i \, p_{ij} = \sum_{i \in \mathscr{A}} \sum_{j \notin \mathscr{A}} \pi_j \, p_{ji} . \tag{1.37}$$

1.13.4 Reversibility

Reversibility is an important notion in the theory of Markov and more general processes. A stationary stochastic process $\{X_t\}$ with index set $\mathbb{Z}$ or $\mathbb{R}$ is said to be *reversible* if, for any positive integer n and for all $t_1, \dots, t_n$, the vector $(X_{t_1}, \dots, X_{t_n})$ has the same distribution as $(X_{-t_1}, \dots, X_{-t_n})$. One way to visualize this is to imagine that we have taken a video of the stochastic process, which we may run in forward and reverse time. If we cannot determine whether the video is running forward or backward, the process is reversible. The main result for reversible Markov chains is that a stationary Markov process is reversible if and only if there exists a collection of positive numbers $\{\pi_i, i \in \mathscr{E}\}$ summing to unity that satisfy the *detailed (or local) balance equations*

$$\pi_i \, p_{ij} = \pi_j \, p_{ji} , \quad i, j \in \mathscr{E}. \tag{1.38}$$

Whenever such a collection $\{\pi_j\}$ exists, it is the stationary distribution of the process.

A good way to think of the detailed balance equations is that they balance the probability flux from state i to state j with that from state j to state i. Contrast this with the equilibrium equations (1.36), which balance the probability flux out of state i with that into state i.

Kolmogorov's criterion is a simple criterion for reversibility based on the transition probabilities. It states that a stationary Markov process is reversible if and only if its transition rates satisfy

$$p(i_1, i_2)\, p(i_2, i_3) \ldots p(i_{n-1}, i_n)\, p(i_n, i_1) = p(i_1, i_n)\, p(i_n, i_{n-1}) \ldots p(i_2, i_1) \qquad (1.39)$$

for all finite loops of states $i_1, \ldots, i_n, i_1$. (For clarity, we have used the notation $p(i, j)$ rather than p_{ij} for the transition probabilities.) The idea is quite intuitive: if the process in forward time is more likely to traverse a certain closed loop in one direction than in the opposite direction, then in backward time it will exhibit the opposite behavior, and hence we have a criterion for detecting the direction of time. If such "looping" behavior does not occur, the process must be reversible.

1.13.5 Markov Jump Processes

A *Markov jump process* $X = \{X_t, t \geqslant 0\}$ can be viewed as a continuous-time generalization of a Markov chain and also of a Poisson process. The Markov property (1.30) now reads

$$\mathbb{P}(X_{t+s} = x_{t+s} \mid X_u = x_u, u \leqslant t) = \mathbb{P}(X_{t+s} = x_{t+s} \mid X_t = x_t) . \qquad (1.40)$$

As in the Markov chain case, one usually assumes that the process is *time-homogeneous*, that is, $\mathbb{P}(X_{t+s} = j \mid X_t = i)$ does not depend on t. Denote this probability by $P_s(i, j)$. An important quantity is the *transition rate* q_{ij} from state i to j, defined for $i \neq j$ as

$$q_{ij} = \lim_{t \downarrow 0} \frac{P_t(i, j)}{t} .$$

The sum of the rates out of state i is denoted by q_i. A typical sample path of X is shown in Figure 1.6. The process jumps at times $T_1, T_2, \ldots$ to states $Y_1, Y_2, \ldots$, staying some length of time in each state.

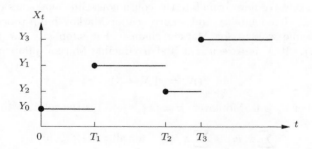

Figure 1.6: A sample path of a Markov jump process $\{X_t, t \geqslant 0\}$.

More precisely, a Markov jump process X behaves (under suitable regularity conditions; see [3]) as follows:

1. Given its past, the probability that X jumps from its current state i to state j is $K_{ij} = q_{ij}/q_i$.

2. The amount of time that X spends in state j has an exponential distribution with mean $1/q_j$, independent of its past history.

The first statement implies that the process $\{Y_n\}$ is in fact a Markov chain, with transition matrix $K = (K_{ij})$.

A convenient way to describe a Markov jump process is through its *transition rate graph*. This is similar to a transition graph for Markov chains. The states are represented by the nodes of the graph, and a transition rate from state i to j is indicated by an arrow from i to j with weight q_{ij}.

■ **EXAMPLE 1.12   Birth-and-Death Process**

A *birth-and-death process* is a Markov jump process with a transition rate graph of the form given in Figure 1.7. Imagine that X_t represents the total number of individuals in a population at time t. Jumps to the right correspond to births, and jumps to the left to deaths. The *birth rates* $\{b_i\}$ and the *death rates* $\{d_i\}$ may differ from state to state. Many applications of Markov chains involve processes of this kind. Note that the process jumps from one state to

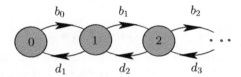

Figure 1.7: The transition rate graph of a birth-and-death process.

the next according to a Markov chain with transition probabilities $K_{0,1} = 1$, $K_{i,i+1} = b_i/(b_i + d_i)$, and $K_{i,i-1} = d_i/(b_i + d_i)$, $i = 1, 2, \ldots$. Moreover, it spends an $\mathsf{Exp}(b_0)$ amount of time in state 0 and $\mathsf{Exp}(b_i + d_i)$ in the other states.

Limiting Behavior   We now formulate the continuous-time analogues of (1.34) and Theorem 1.13.2. Irreducibility and recurrence for Markov jump processes are defined in the same way as for Markov chains. For simplicity, we assume that $\mathscr{E} = \{1, 2, \ldots\}$. If X is a recurrent and irreducible Markov jump process, then regardless of i,

$$\lim_{t \to \infty} \mathbb{P}(X_t = j \mid X_0 = i) = \pi_j \tag{1.41}$$

for some number $\pi_j \geqslant 0$. Moreover, $\boldsymbol{\pi} = (\pi_1, \pi_2, \ldots)$ is the solution to

$$\sum_{j \neq i} \pi_i\, q_{ij} = \sum_{j \neq i} \pi_j\, q_{ji}, \quad \text{for all } i = 1, \ldots, m \tag{1.42}$$

with $\sum_j \pi_j = 1$, if such a solution exists, in which case all states are positive recurrent. If such a solution does not exist, all π_j are 0.

As in the Markov chain case, $\{\pi_j\}$ is called the *limiting distribution* of X and is usually identified with the row vector $\boldsymbol{\pi}$. Any solution $\boldsymbol{\pi}$ of (1.42) with $\sum_j \pi_j = 1$ is called a *stationary distribution*, since taking it as the initial distribution of the Markov jump process renders the process stationary.

Equations (1.42) are again called the *global balance equations* and are readily generalized to (1.37), replacing the transition probabilities with transition rates. More important, if the process is reversible, then, as with Markov chains, the stationary distribution can be found from the *local balance equations*:

$$\pi_i\, q_{ij} = \pi_j\, q_{ji}\,, \quad i, j \in \mathscr{E}\,. \tag{1.43}$$

Reversibility can be easily verified by checking that looping does not occur, that is, via Kolmogorov's criterion (1.39), replacing the probabilities p with rates q.

■ **EXAMPLE 1.13** $M/M/1$ **Queue**

Consider a service facility where customers arrive at certain random times and are served by a single server. Arriving customers who find the server busy wait in the queue. Customers are served in the order in which they arrive. The interarrival times are exponential random variables with rates λ, and the service times of customers are iid exponential random variables with rates μ. Last, the service times are independent of the interarrival times. Let X_t be the number of customers in the system at time t. By the memoryless property of the exponential distribution (see Problem 1.7), it is not difficult to see that $X = \{X_t, t \geqslant 0\}$ is a Markov jump process, and in fact a birth-and-death process with birth rates $b_i = \lambda$, $i = 0, 1, 2, \ldots$ and death rates $d_i = \mu$, $i = 1, 2, \ldots$.

Solving the global balance equations (or, more easily, the local balance equations, since X is reversible), we see that X has a limiting distribution given by

$$\lim_{t \to \infty} \mathbb{P}(X_t = n) = (1 - \varrho)\, \varrho^n, \quad n = 0, 1, 2, \ldots, \tag{1.44}$$

provided that $\varrho = \lambda/\mu < 1$. This means that the expected service time needs to be less than the expected interarrival time for a limiting distribution to exist. In that case, the limiting distribution is also the stationary distribution. In particular, if X_0 is distributed according to (1.44), then X_t has the same distribution for all $t > 0$.

1.14   GAUSSIAN PROCESSES

The normal distribution is also called the *Gaussian* distribution. Gaussian processes are generalizations of multivariate normal random vectors (discussed in Section 1.10). Specifically, a stochastic process $\{X_t, t \in \mathscr{T}\}$ is said to be *Gaussian* if all its finite-dimensional distributions are Gaussian. That is, if for any choice of n and $t_1, \ldots, t_n \in \mathscr{T}$, it holds that

$$(X_{t_1}, \ldots, X_{t_n})^\top \sim \mathsf{N}(\boldsymbol{\mu}, \Sigma) \tag{1.45}$$

for some expectation vector $\boldsymbol{\mu}$ and covariance matrix Σ (both of which depend on the choice of $t_1, \ldots, t_n$). Equivalently, $\{X_t, t \in \mathscr{T}\}$ is Gaussian if any linear combination $\sum_{i=1}^{n} b_i X_{t_i}$ has a normal distribution. Note that a Gaussian process is determined completely by its *expectation function* $\mu_t = \mathbb{E}[X_t]$, $t \in \mathscr{T}$, and *covariance function* $\Sigma_{s,t} = \mathrm{Cov}(X_s, X_t)$, $s, t \in \mathscr{T}$.

■ **EXAMPLE 1.14   Wiener Process (Brownian Motion)**

The quintessential Gaussian process is the *Wiener process* or (standard) *Brownian motion*. It can be viewed as a continuous version of a random walk process. Figure 1.8 gives a typical sample path. The Wiener process plays a central role in probability and forms the basis of many other stochastic processes.

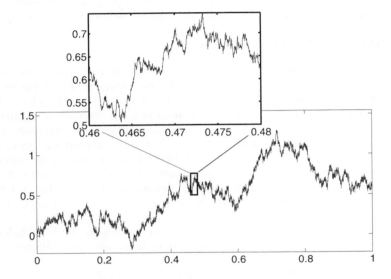

Figure 1.8: A sample path of the Wiener process. The inset shows a magnification of the path over a small time interval.

The Wiener process can be defined as a Gaussian process $\{X_t, t \geqslant 0\}$ with expectation function $\mu_t = 0$ for all t and covariance function $\Sigma_{s,t} = s$ for $0 \leqslant s \leqslant t$. The Wiener process has many fascinating properties (e.g., [11]). For example, it is a Markov process (i.e., it satisfies the Markov property (1.30)) with continuous sample paths that are *nowhere differentiable*. Moreover, the increments $X_t - X_s$ over intervals $[s, t]$ are independent and normally distributed. Specifically, for any $t_1 < t_2 \leqslant t_3 < t_4$,

$$X_{t_4} - X_{t_3} \quad \text{and} \quad X_{t_2} - X_{t_1}$$

are independent random variables, and for all $t \geqslant s \geqslant 0$,

$$X_t - X_s \sim \mathsf{N}(0, t - s) .$$

This leads to a simple simulation procedure for Wiener processes, which is discussed in Section 2.8.

1.15   INFORMATION

In this section we discuss briefly various measures of information in a random experiment. Suppose that we describe the measurements in a random experiment via a random vector $\mathbf{X} = (X_1, \ldots, X_n)$ with pdf f. Then all the information about the experiment (all of our probabilistic knowledge) is obviously contained in the pdf f. However, in most cases we would want to characterize our information about the experiments with just a few key numbers, such as the *expectation* and the *covariance matrix* of $\mathbf{X}$, which provide information about the mean measurements and the variability of the measurements, respectively. Another informational measure comes from coding and communications theory, where the *Shannon entropy* characterizes the average number of bits needed to transmit a message $\mathbf{X}$ over a (binary) communication channel. Yet another approach to information can be found in statistics. Specifically, in the theory of point estimation, the pdf f depends on a parameter vector $\boldsymbol{\theta}$. The question is how well $\boldsymbol{\theta}$ can be estimated via an outcome of $\mathbf{X}$ — in other words, how much information about $\boldsymbol{\theta}$ is contained in the "data" $\mathbf{X}$. Various measures for this type of information are associated with the *maximum likelihood*, the *score*, and the *(Fisher) information matrix*. Finally, the amount of information in a random experiment can often be quantified via a *distance* concept, such as the *Kullback–Leibler* "distance" (divergence), also called the *cross-entropy*.

1.15.1   Shannon Entropy

One of the most celebrated measures of uncertainty in information theory is the *Shannon entropy*, or simply *entropy*. A good reference is [4], where the entropy of a discrete random variable X with density f is defined as

$$
\mathbb{E}\left[\log_2 \frac{1}{f(X)}\right] = -\mathbb{E}\left[\log_2 f(X)\right] = -\sum_{\mathscr{X}} f(x) \log_2 f(x) .
$$

Here X is interpreted as a random character from an alphabet $\mathscr{X}$, such that $X = x$ with probability $f(x)$. We will use the convention $0 \ln 0 = 0$.

It can be shown that the most efficient way to transmit characters sampled from f over a binary channel is to encode them such that the number of bits required to transmit x is equal to $\log_2(1/f(x))$. It follows that $-\sum_{\mathscr{X}} f(x) \log_2 f(x)$ is the expected bit length required to send a random character $X \sim f$; see [4].

A more general approach, which includes continuous random variables, is to define the entropy of a random variable X with density f by

$$
\mathcal{H}(X) = -\mathbb{E}[\ln f(X)] = \begin{cases} -\sum f(x) \ln f(x) & \text{discrete case,} \\ -\int f(x) \ln f(x)\, dx & \text{continuous case.} \end{cases} \tag{1.46}
$$

Definition (1.46) can easily be extended to random vectors $\mathbf{X}$ as (in the continuous case)

$$
\mathcal{H}(\mathbf{X}) = -\mathbb{E}[\ln f(\mathbf{X})] = -\int f(\mathbf{x}) \ln f(\mathbf{x})\, d\mathbf{x} . \tag{1.47}
$$

$\mathcal{H}(\mathbf{X})$ is often called the *joint* entropy of the random variables $X_1, \ldots, X_n$, and it is also written as $\mathcal{H}(X_1, \ldots, X_n)$. In the continuous case, $\mathcal{H}(\mathbf{X})$ is frequently referred to as the *differential entropy* to distinguish it from the discrete case.

■ **EXAMPLE 1.15**

Let X have a Ber(p) distribution for some $0 \leqslant p \leqslant 1$. The density f of X is given by $f(1) = \mathbb{P}(X = 1) = p$ and $f(0) = \mathbb{P}(X = 0) = 1 - p$ so that the entropy of X is

$$\mathcal{H}(X) = -p \ln p - (1 - p) \ln(1 - p) .$$

The graph of the entropy as a function of p is depicted in Figure 1.9. Note that the entropy is maximal for $p = 1/2$, which gives the "uniform" density on $\{0, 1\}$.

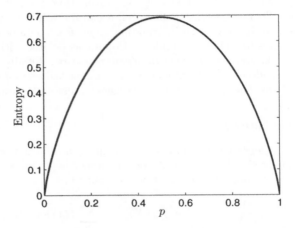

Figure 1.9: The entropy for the Ber(p) distribution as a function of p.

Next, consider a sequence $X_1, \ldots, X_n$ of iid Ber(p) random variables. Let $\mathbf{X} = (X_1, \ldots, X_n)$. The density of $\mathbf{X}$, say g, is simply the product of the densities of the X_i, so that

$$\mathcal{H}(\mathbf{X}) = -\mathbb{E}\left[\ln g(\mathbf{X})\right] = -\mathbb{E}\left[\ln \prod_{i=1}^{n} f(X_i)\right] = \sum_{i=1}^{n} -\mathbb{E}\left[\ln f(X_i)\right] = n\,\mathcal{H}(X) .$$

The properties of $\mathcal{H}(\mathbf{X})$ in the continuous case are somewhat different from those in the discrete one. In particular:

1. The differential entropy can be negative, whereas the discrete entropy is always positive.

2. The discrete entropy is insensitive to invertible transformations, whereas the differential entropy is not. Specifically, if $\mathbf{X}$ is discrete, $\mathbf{Y} = g(\mathbf{X})$, and g is an invertible mapping, then $\mathcal{H}(\mathbf{X}) = \mathcal{H}(\mathbf{Y})$ because $f_{\mathbf{Y}}(\mathbf{y}) = f_{\mathbf{X}}(g^{-1}(\mathbf{y}))$. However, in the continuous case, we have an additional factor due to the Jacobian of the transformation.

It is not difficult to see that of any density f, the one that gives the maximum entropy is the uniform density on $\mathscr{X}$. That is,

$$\mathcal{H}(\mathbf{X}) \text{ is maximal} \iff f(\mathbf{x}) = \frac{1}{|\mathscr{X}|} \text{ (constant)} . \tag{1.48}$$

For two random vectors $\mathbf{X}$ and $\mathbf{Y}$ with joint pdf f, we define the *conditional entropy* of $\mathbf{Y}$ given $\mathbf{X}$ as

$$\mathcal{H}(\mathbf{Y} \mid \mathbf{X}) = -\mathbb{E}\left[\ln \frac{f(\mathbf{X}, \mathbf{Y})}{f_{\mathbf{X}}(\mathbf{X})}\right] = \mathcal{H}(\mathbf{X}, \mathbf{Y}) - \mathcal{H}(\mathbf{X}) , \tag{1.49}$$

where $f_{\mathbf{X}}$ is the pdf of $\mathbf{X}$ and $\frac{f(\mathbf{x}, \mathbf{y})}{f_{\mathbf{X}}(\mathbf{x})}$ is the conditional density of $\mathbf{Y}$ (at $\mathbf{y}$), given $\mathbf{X} = \mathbf{x}$. It follows that

$$\mathcal{H}(\mathbf{X}, \mathbf{Y}) = \mathcal{H}(\mathbf{X}) + \mathcal{H}(\mathbf{Y} \mid \mathbf{X}) = \mathcal{H}(\mathbf{Y}) + \mathcal{H}(\mathbf{X} \mid \mathbf{Y}) . \tag{1.50}$$

It is reasonable to require that any sensible additive measure describing the average amount of uncertainty should satisfy at least (1.50) and (1.48). It follows that the uniform density carries the least amount of information, and the entropy (average amount of uncertainty) of $(\mathbf{X}, \mathbf{Y})$ is equal to the sum of the entropy of $\mathbf{X}$ and the amount of entropy in $\mathbf{Y}$ after the information in $\mathbf{X}$ has been accounted for. It is argued in [10] that any concept of entropy that includes the general properties (1.48) and (1.50) must lead to the definition (1.47).

The *mutual information* of $\mathbf{X}$ and $\mathbf{Y}$ is defined as

$$\mathcal{M}(\mathbf{X}, \mathbf{Y}) = \mathcal{H}(\mathbf{X}) + \mathcal{H}(\mathbf{Y}) - \mathcal{H}(\mathbf{X}, \mathbf{Y}) , \tag{1.51}$$

which, as the name suggests, can be interpreted as the amount of information shared by $\mathbf{X}$ and $\mathbf{Y}$. An alternative expression, which follows from (1.50) and (1.51), is

$$\mathcal{M}(\mathbf{X}, \mathbf{Y}) = \mathcal{H}(\mathbf{X}) - \mathcal{H}(\mathbf{X} \mid \mathbf{Y}) = \mathcal{H}(\mathbf{Y}) - \mathcal{H}(\mathbf{Y} \mid \mathbf{X}) , \tag{1.52}$$

which can be interpreted as the reduction of the uncertainty of one random variable due to the knowledge of the other. It is not difficult to show that the mutual information is always positive. It is also related to the cross-entropy concept, which follows.

1.15.2 Kullback–Leibler Cross-Entropy

Let g and h be two densities on $\mathcal{X}$. The Kullback–Leibler cross-entropy between g and h (compare with (1.47)) is defined (in the continuous case) as

$$\begin{aligned} \mathcal{D}(g, h) &= \mathbb{E}_g\left[\ln \frac{g(\mathbf{X})}{h(\mathbf{X})}\right] \\ &= \int g(\mathbf{x}) \ln g(\mathbf{x}) \, d\mathbf{x} - \int g(\mathbf{x}) \ln h(\mathbf{x}) \, d\mathbf{x} . \end{aligned} \tag{1.53}$$

$\mathcal{D}(g, h)$ is also called the *Kullback–Leibler divergence*, the *cross-entropy*, and the *relative entropy*. If not stated otherwise, we will call $\mathcal{D}(g, h)$ the *cross-entropy* (CE) between g and h. Notice that $\mathcal{D}(g, h)$ is not a distance between g and h in the formal sense, since in general $\mathcal{D}(g, h) \neq \mathcal{D}(h, g)$. Nonetheless, it is often useful to think of $\mathcal{D}(g, h)$ as a distance because

$$\mathcal{D}(g, h) \geqslant 0$$

and $\mathcal{D}(g, h) = 0$ if and only if $g(x) = h(x)$. This follows from Jensen's inequality (if ϕ is a convex function, such as $-\ln$, then $\mathbb{E}[\phi(X)] \geqslant \phi(\mathbb{E}[X])$). Namely

$$\mathcal{D}(g, h) = \mathbb{E}_g \left[-\ln \frac{h(\mathbf{X})}{g(\mathbf{X})} \right] \geqslant -\ln \left\{ \mathbb{E}_g \left[\frac{h(\mathbf{X})}{g(\mathbf{X})} \right] \right\} = -\ln 1 = 0 .$$

It can be readily seen that the mutual information $\mathcal{M}(\mathbf{X}, \mathbf{Y})$ of vectors $\mathbf{X}$ and $\mathbf{Y}$ defined in (1.51) is related to the CE in the following way:

$$\mathcal{M}(\mathbf{X}, \mathbf{Y}) = \mathcal{D}(f, f_{\mathbf{X}} f_{\mathbf{Y}}) = \mathbb{E}_f \left[\ln \frac{f(\mathbf{X}, \mathbf{Y})}{f_{\mathbf{X}}(\mathbf{X}) \, f_{\mathbf{Y}}(\mathbf{Y})} \right] ,$$

where f is the (joint) pdf of $(\mathbf{X}, \mathbf{Y})$ and $f_{\mathbf{X}}$ and $f_{\mathbf{Y}}$ are the (marginal) pdfs of $\mathbf{X}$ and $\mathbf{Y}$, respectively. In other words, the mutual information can be viewed as the CE that measures the distance between the joint pdf f of $\mathbf{X}$ and $\mathbf{Y}$ and the product of their marginal pdfs $f_{\mathbf{X}}$ and $f_{\mathbf{Y}}$, that is, under the assumption that the vectors $\mathbf{X}$ and $\mathbf{Y}$ are *independent*.

1.15.3 Maximum Likelihood Estimator and Score Function

We introduce here the notion of the *score function* (SF) via the classical *maximum likelihood estimator*. Consider a random vector $\mathbf{X} = (X_1, \ldots, X_n)$ that is is distributed according to a fixed pdf $f(\cdot; \boldsymbol{\theta})$ with unknown parameter (vector) $\boldsymbol{\theta} \in \Theta$. Say that we want to estimate $\boldsymbol{\theta}$ on the basis of a given outcome $\mathbf{x}$ (the data) of $\mathbf{X}$. For a given $\mathbf{x}$, the function $\mathcal{L}(\boldsymbol{\theta}; \mathbf{x}) = f(\mathbf{x}; \boldsymbol{\theta})$ is called the *likelihood function*. Note that $\mathcal{L}$ is a function of $\boldsymbol{\theta}$ for a fixed parameter $\mathbf{x}$, whereas for the pdf f it is the other way around. The maximum likelihood *estimate* $\widehat{\boldsymbol{\theta}} = \widehat{\boldsymbol{\theta}}(\mathbf{x})$ of $\boldsymbol{\theta}$ is defined as

$$\widehat{\boldsymbol{\theta}} = \underset{\boldsymbol{\theta} \in \Theta}{\operatorname{argmax}} \, \mathcal{L}(\boldsymbol{\theta}; \mathbf{x}) . \tag{1.54}$$

Because the function $\ln$ is monotone increasing, we also have

$$\widehat{\boldsymbol{\theta}} = \underset{\boldsymbol{\theta} \in \Theta}{\operatorname{argmax}} \ln \mathcal{L}(\boldsymbol{\theta}; \mathbf{x}) . \tag{1.55}$$

The random variable $\widehat{\boldsymbol{\theta}}(\mathbf{X})$ with $\mathbf{X} \sim f(\cdot; \boldsymbol{\theta})$ is the corresponding maximum likelihood *estimator*, which is again written as $\widehat{\boldsymbol{\theta}}$. Note that often the data $X_1, \ldots, X_n$ form a random sample from some pdf $f_1(\cdot; \boldsymbol{\theta})$, in which case $f(\mathbf{x}; \boldsymbol{\theta}) = \prod_{i=1}^{N} f_1(x_i; \boldsymbol{\theta})$ and

$$\widehat{\boldsymbol{\theta}} = \underset{\boldsymbol{\theta} \in \Theta}{\operatorname{argmax}} \sum_{i=1}^{N} \ln f_1(X_i; \boldsymbol{\theta}) . \tag{1.56}$$

If $\mathcal{L}(\boldsymbol{\theta}; \mathbf{x})$ is a continuously differentiable concave function with respect to $\boldsymbol{\theta}$ and the maximum is attained in the interior of Θ, then we can find the maximum likelihood estimator of $\boldsymbol{\theta}$ by solving

$$\nabla_{\boldsymbol{\theta}} \ln \mathcal{L}(\boldsymbol{\theta}; \mathbf{x}) = \mathbf{0} .$$

The function $\mathcal{S}(\cdot; \mathbf{x})$ defined by

$$\mathcal{S}(\boldsymbol{\theta}; \mathbf{x}) = \nabla_{\boldsymbol{\theta}} \ln \mathcal{L}(\boldsymbol{\theta}; \mathbf{x}) = \frac{\nabla_{\boldsymbol{\theta}} f(\mathbf{x}; \boldsymbol{\theta})}{f(\mathbf{x}; \boldsymbol{\theta})} \tag{1.57}$$

is called the *score function*. For the exponential family (A.9) it is easy to see that

$$S(\boldsymbol{\theta}; \mathbf{x}) = \frac{\nabla c(\boldsymbol{\theta})}{c(\boldsymbol{\theta})} + \mathbf{t}(\mathbf{x}) . \tag{1.58}$$

The *random vector* $S(\boldsymbol{\theta}) = S(\boldsymbol{\theta}; \mathbf{X})$ with $\mathbf{X} \sim f(\cdot; \boldsymbol{\theta})$ is called the *(efficient) score*. The expected score is always equal to the zero vector, that is,

$$\mathbb{E}_{\boldsymbol{\theta}}[S(\boldsymbol{\theta})] = \int \nabla_{\boldsymbol{\theta}} f(\mathbf{x}; \boldsymbol{\theta}) \, \mu(\mathrm{d}\mathbf{x}) = \nabla_{\boldsymbol{\theta}} \int f(\mathbf{x}; \theta) \, \mu(\mathrm{d}\mathbf{x}) = \nabla_{\boldsymbol{\theta}} 1 = \mathbf{0} ,$$

where the interchange of differentiation and integration is justified via the bounded convergence theorem.

1.15.4   Fisher Information

The covariance matrix $\mathcal{I}(\boldsymbol{\theta})$ of the score $S(\boldsymbol{\theta})$ is called the *Fisher information matrix*. Since the expected score is always $\mathbf{0}$, we have

$$\mathcal{I}(\boldsymbol{\theta}) = \mathbb{E}_{\boldsymbol{\theta}} \left[S(\boldsymbol{\theta}) S(\boldsymbol{\theta})^\top \right] . \tag{1.59}$$

In the one-dimensional case, we thus have

$$\mathcal{I}(\theta) = \mathbb{E}_{\theta} \left[\left(\frac{\partial \ln f(X; \theta)}{\partial \theta} \right)^2 \right] .$$

Because

$$\frac{\partial^2}{\partial \theta^2} \ln f(x; \theta) = \frac{\frac{\partial^2}{\partial \theta^2} f(x; \theta)}{f(x; \theta)} - \left(\frac{\frac{\partial}{\partial \theta} f(x; \theta)}{f(x; \theta)} \right)^2 ,$$

we see that (under straightforward regularity conditions) the Fisher information is also given by

$$\mathcal{I}(\theta) = -\mathbb{E}_{\theta} \left[\frac{\partial^2 \ln f(X; \theta)}{\partial \theta^2} \right] .$$

In the multidimensional case we have similarly

$$\mathcal{I}(\boldsymbol{\theta}) = -\mathbb{E}_{\boldsymbol{\theta}} [\nabla S(\boldsymbol{\theta})] = -\mathbb{E}_{\boldsymbol{\theta}} \left[\nabla^2 \ln f(\mathbf{X}; \boldsymbol{\theta}) \right] , \tag{1.60}$$

where $\nabla^2 \ln f(\mathbf{X}; \boldsymbol{\theta})$ denotes the *Hessian* of $\ln f(\mathbf{X}; \boldsymbol{\theta})$, that is, the (random) matrix

$$\left(\frac{\partial^2 \ln f(\mathbf{X}; \boldsymbol{\theta})}{\partial \theta_i \partial \theta_j} \right) .$$

The importance of the Fisher information in statistics is corroborated by the famous *Cramér–Rao inequality*, which (in a simplified form) states that the variance of any unbiased estimator Z of $g(\boldsymbol{\theta})$ is bounded from below via

$$\mathrm{Var}(Z) \geqslant (\nabla g(\boldsymbol{\theta}))^\top \mathcal{I}^{-1}(\boldsymbol{\theta}) \, \nabla g(\boldsymbol{\theta}) . \tag{1.61}$$

For more details, see [12].

1.16 CONVEX OPTIMIZATION AND DUALITY

Let $f(x)$, $x \in \mathbb{R}$, be a real-valued function with continuous derivatives — also called a C^1 function. The standard approach to minimizing $f(x)$ is to solve the equation

$$f'(x) = 0 . \tag{1.62}$$

The solutions to (1.62) are called *stationary points*. If, in addition, the function has continuous second derivatives (a so-called C^2 function), the condition

$$f''(x^*) > 0 \tag{1.63}$$

ensures that a stationary point x^* is a *local minimizer*, that is, $f(x^*) < f(x)$ for all x in a small enough neighborhood of x^*.

For a C^1 function on $\mathbb{R}^n$, (1.62) generalizes to

$$\nabla f(\mathbf{x}) \equiv \begin{pmatrix} \frac{\partial f(\mathbf{x})}{\partial x_1} \\ \vdots \\ \frac{\partial f(\mathbf{x})}{\partial x_n} \end{pmatrix} = \mathbf{0} , \tag{1.64}$$

where $\nabla f(\mathbf{x})$ is the *gradient* of f at $\mathbf{x}$. Similarly, a stationary point $\mathbf{x}^*$ is a local minimizer of f if the *Hessian matrix* (or simply *Hessian*) at $\mathbf{x}^*$,

$$\nabla^2 f(\mathbf{x}^*) \equiv \begin{pmatrix} \frac{\partial^2 f(\mathbf{x}^*)}{\partial x_1^2} & \cdots & \frac{\partial^2 f(\mathbf{x}^*)}{\partial x_1 \partial x_n} \\ \vdots & \cdots & \vdots \\ \frac{\partial^2 f(\mathbf{x}^*)}{\partial x_1 \partial x_n} & \cdots & \frac{\partial^2 f(\mathbf{x}^*)}{\partial x_n^2} \end{pmatrix} , \tag{1.65}$$

is *positive definite*, that is, $\mathbf{x}^\top [\nabla^2 f(\mathbf{x}^*)] \mathbf{x} > 0$ for all $\mathbf{x} \neq \mathbf{0}$.

The situation can be further generalized by introducing *constraints*. A general constrained optimization problems can be written as

$$\min_{\mathbf{x} \in \mathbb{R}^n} \quad f(\mathbf{x}) \tag{1.66}$$

$$\text{subject to:} \quad h_i(\mathbf{x}) = 0, \quad i = 1, \dots, m , \tag{1.67}$$

$$g_i(\mathbf{x}) \leqslant 0, \quad i = 1, \dots, k . \tag{1.68}$$

Here f, g_i, and h_i are given functions, $f(\mathbf{x})$ is called the *objective function*, and $h_i(\mathbf{x}) = 0$ and $g_i(\mathbf{x}) \leqslant 0$ represent the *equality* and *inequality* constraints, respectively.

The region of the domain where the objective function is defined and where all the constraints are satisfied is called the *feasible region*. A *global solution* to the optimization problem is a point $\mathbf{x}^* \in \mathbb{R}^n$ such that there exists no other point $\mathbf{x} \in \mathbb{R}^n$ for which $f(\mathbf{x}) < f(\mathbf{x}^*)$. Alternative names are *global minimizer* and *global minimum*, although the latter could be confused with the minimum value of the function. Similarly, for a *local* solution/minimizer, the condition $f(\mathbf{x}) < f(\mathbf{x}^*)$ only needs to hold in some neighborhood of $\mathbf{x}^*$.

Within this formulation fall many of the traditional optimization problems. An optimization problem in which the objective function and the equality and inequality constraints are linear functions, is called a *linear program*. An optimization problem in which the objective function is quadratic, while the constraints are linear functions is called a *quadratic program*. Convexity plays an important role in many practical optimization problems.

Definition 1.16.1 (Convex Set) A set $\mathscr{X} \in \mathbb{R}^n$ is called *convex* if, for all $\mathbf{x}, \mathbf{y} \in \mathscr{X}$ and $\theta \in (0, 1)$, the point $(\theta \mathbf{x} + (1 - \theta)\mathbf{y}) \in \mathscr{X}$.

Definition 1.16.2 (Convex Function) A function $f(\mathbf{x})$ on a convex set $\mathscr{X}$ is called *convex* if, for all $\mathbf{x}, \mathbf{y} \in \mathscr{X}$ and $\theta \in (0, 1)$,

$$f\big(\theta \mathbf{x} + (1 - \theta)\mathbf{y}\big) \leq \theta f(\mathbf{x}) + (1 - \theta)f(\mathbf{y}) . \tag{1.69}$$

If a strict inequality in (1.69) holds, the function is said to be *strictly convex*. If a function f is (strictly) convex, then $-f$ is said to be (strictly) *concave*. Assuming $\mathscr{X}$ is an open set, convexity for $f \in C^1$ is equivalent to

$$f(\mathbf{y}) \geqslant f(\mathbf{x}) + (\mathbf{y} - \mathbf{x})^\top \nabla f(\mathbf{x}) \quad \text{for all } \mathbf{x}, \mathbf{y} \in \mathscr{X}.$$

Moreover, for $f \in C^2$, convexity is equivalent to the Hessian matrix being positive semidefinite for all $\mathbf{x} \in \mathscr{X}$:

$$\mathbf{y}^\top \big[\nabla^2 f(\mathbf{x})\big] \mathbf{y} \geqslant 0, \quad \text{for all } \mathbf{y} \in \mathbb{R}^n .$$

The problem (1.66) is said to be a *convex programming problem* if

1. the objective function f is convex,

2. the inequality constraint functions $\{g_i(\mathbf{x})\}$ are convex, and

3. the equality constraint functions $\{h_i(\mathbf{x})\}$ are *affine*, i.e., of the form $\mathbf{a}_i^\top \mathbf{x} - b_i$.

Note that the last requirement follows from the fact that an equality constraint $h_i(\mathbf{x}) = 0$ can be viewed as a combination of the inequality constraints $h_i(\mathbf{x}) \leqslant 0$ and $-h_i(\mathbf{x}) \leqslant 0$, so that both h_i and $-h_i$ need to be convex. Both the linear and quadratic programs (with positive definite matrix C) are convex.

1.16.1 Lagrangian Method

The main components of the Lagrangian method are the Lagrange multipliers and the Lagrange function. The method was developed by Lagrange in 1797 for the optimization problem (1.66) with equality constraints (1.67). In 1951 Kuhn and Tucker extended Lagrange's method to inequality constraints.

Definition 1.16.3 (Lagrange Function) Given an optimization problem (1.66) containing only equality constraints $h_i(\mathbf{x}) = 0$, $i = 1, \ldots, m$, the *Lagrange function*, or *Lagrangian*, is defined as

$$\mathcal{L}(\mathbf{x}, \boldsymbol{\beta}) = f(\mathbf{x}) + \sum_i \beta_i \, h_i(\mathbf{x}) ,$$

where the coefficients $\{\beta_i\}$ are called the *Lagrange multipliers*.

A necessary condition for a point $\mathbf{x}^*$ to be a local minimizer of $f(\mathbf{x})$ subject to the equality constraints $h_i(\mathbf{x}) = 0$, $i = 1, \ldots, m$, is

$$\nabla_{\mathbf{x}} \mathcal{L}(\mathbf{x}^*, \boldsymbol{\beta}^*) = \mathbf{0} \ ,$$
$$\nabla_{\boldsymbol{\beta}} \mathcal{L}(\mathbf{x}^*, \boldsymbol{\beta}^*) = \mathbf{0} \ ,$$

for some value $\boldsymbol{\beta}^*$. The conditions above are also sufficient if $\mathcal{L}(\mathbf{x}, \boldsymbol{\beta}^*)$ is a convex function of $\mathbf{x}$.

■ **EXAMPLE 1.16 Maximum Entropy Distribution**

Let $p = \{p_i, i = 1, \ldots, n\}$ be a probability distribution. Consider the following program, which maximizes the (Shannon) entropy:

$$\max_{\mathbf{P}} \quad - \sum_{i=1}^{n} p_i \ln p_i$$

$$\text{subject to:} \quad \sum_{i=1}^{n} p_i = 1 \ .$$

The Lagrangian is

$$\mathcal{L}(\mathbf{p}, \beta) = \sum_{i=1}^{n} p_i \ln p_i + \beta \left(\sum_{i=1}^{n} p_i - 1 \right)$$

over the domain $\{(\mathbf{p}, \beta) : p_i \geq 0, i = 1, \ldots, n, \ \beta \in \mathbb{R}\}$. The optimal solution $\mathbf{p}^*$ of the problem is the uniform distribution, that is, $\mathbf{p}^* = (1/n, \ldots, 1/n)$; see Problem 1.35.

Definition 1.16.4 (Generalized Lagrange Function) Given the original optimization problem (1.66), containing both the equality and inequality constraints, the *generalized Lagrange function*, or simply *Lagrangian*, is defined as

$$\mathcal{L}(\mathbf{x}, \boldsymbol{\alpha}, \boldsymbol{\beta}) = f(\mathbf{x}) + \sum_{i=1}^{k} \alpha_i \, g_i(\mathbf{x}) + \sum_{i=1}^{m} \beta_i \, h_i(\mathbf{x}) \ .$$

A necessary condition for a point $\mathbf{x}^*$ to be a local minimizer of $f(\mathbf{x})$ in the optimization problem (1.66) is the existence of an $\boldsymbol{\alpha}^*$ and $\boldsymbol{\beta}^*$ such that

$$\nabla_{\mathbf{x}} \mathcal{L}(\mathbf{x}^*, \boldsymbol{\alpha}^*, \boldsymbol{\beta}^*) = \mathbf{0} \ ,$$
$$\nabla_{\boldsymbol{\beta}} \mathcal{L}(\mathbf{x}^*, \boldsymbol{\alpha}^*, \boldsymbol{\beta}^*) = \mathbf{0} \ ,$$
$$g_i(\mathbf{x}^*) \leqslant 0, \quad i = 1, \ldots, k \ ,$$
$$\alpha_i^* \geqslant 0, \quad i = 1, \ldots, k \ ,$$
$$\alpha_i^* \, g_i(\mathbf{x}^*) = 0, \quad i = 1, \ldots, k \ .$$

These equations are usually referred as the *Karush–Kuhn–Tucker (KKT) conditions*. For *convex* programs we have the following important results:

1. Every local solution $\mathbf{x}^*$ to a convex programming problem is a global solution and the set of global solutions is convex. If, in addition, the objective function is strictly convex, then any global solution is unique.

2. For a strictly convex programming problem with C^1 objective and constraint functions, the KKT conditions are necessary and sufficient for a unique global solution.

1.16.2 Duality

The aim of duality is to provide an alternative formulation of an optimization problem that is often more computationally efficient or has some theoretical significance (see [7], page 219). The original problem (1.66) is referred to as the *primal* problem, whereas the reformulated problem, based on Lagrange multipliers, is referred to as the *dual* problem. Duality theory is most relevant to convex optimization problems. It is well known that if the primal optimization problem is (strictly) convex, then the dual problem is (strictly) concave and has a (unique) solution from which the optimal (unique) primal solution can be deduced.

Definition 1.16.5 (Lagrange Dual Program) The *Lagrange dual program* of the primal program (1.66), is

$$\max_{\alpha, \beta} \quad \mathcal{L}^*(\alpha, \beta)$$

$$\text{subject to:} \quad \alpha \geqslant 0 \,,$$

where $\mathcal{L}^*$ is the *Lagrange dual function*:

$$\mathcal{L}^*(\alpha, \beta) = \inf_{\mathbf{x} \in \mathscr{X}} \mathcal{L}(\mathbf{x}, \alpha, \beta) \,. \tag{1.70}$$

It is not difficult to see that if f^* is the minimal value of the primal problem, then $\mathcal{L}^*(\alpha, \beta) \leqslant f^*$ for any $\alpha \geqslant 0$ and any β. This property is called *weak duality*. The Lagrangian dual program thus determines the best lower bound on f^*. If d^* is the optimal value for the dual problem, then $d^* < f^*$. The difference $f^* - d^*$ is called the *duality gap*.

The duality gap is extremely useful for providing lower bounds for the solutions of primal problems that may be impossible to solve directly. It is important to note that for linearly constrained problems, if the primal is infeasible (does not have a solution satisfying the constraints), then the dual is either infeasible or unbounded. Conversely, if the dual is infeasible, then the primal has no solution. Of crucial importance is the *strong duality* theorem, which states that for convex programs (1.66) with linear constrained functions h_i and g_i the duality gap is zero, and any $\mathbf{x}^*$ and (α^*, β^*) satisfying the KKT conditions are (global) solutions to the primal and dual programs, respectively. In particular, this holds for linear and convex quadratic programs (note that not all quadratic programs are convex).

For a convex primal program with C^1 objective and constraint functions, the Lagrangian dual function (1.70) can be obtained by simply setting the gradient (with respect to $\mathbf{x}$) of the Lagrangian $\mathcal{L}(\mathbf{x}, \alpha, \beta)$ to zero. One can further simplify the dual program by substituting into the Lagrangian the relations between the variables thus obtained.

■ **EXAMPLE 1.17   Linear Programming Problem**

Consider the following linear programming problem:

$$\min_{\mathbf{x}} \quad \mathbf{c}^\top \mathbf{x}$$

$$\text{subject to:} \quad A\mathbf{x} \geqslant \mathbf{b} \, .$$

The Lagrangian is $\mathcal{L}(\mathbf{x}, \boldsymbol{\alpha}) = \mathbf{c}^\top \mathbf{x} - \boldsymbol{\alpha}^\top (A\mathbf{x} - \mathbf{b})$. The Lagrange dual function is the infimum of $\mathcal{L}$ over all $\mathbf{x}$; thus

$$\mathcal{L}^*(\boldsymbol{\alpha}) = \begin{cases} \mathbf{b}^\top \boldsymbol{\alpha} & \text{if } A^\top \boldsymbol{\alpha} = \mathbf{c} \, , \\ -\infty & \text{otherwise,} \end{cases}$$

so that the Lagrange dual program becomes

$$\max_{\boldsymbol{\alpha}} \quad \mathbf{b}^\top \boldsymbol{\alpha}$$

$$\text{subject to:} \quad A^\top \boldsymbol{\alpha} = \mathbf{c} \, ,$$

$$\boldsymbol{\alpha} \geqslant \mathbf{0} \, .$$

An interesting fact to note here is that for the linear programming problem the dual of the dual problem always gives back the primal problem.

■ **EXAMPLE 1.18   Quadratic Programming Problem**

Consider the following quadratic programming problem:

$$\min_{\mathbf{x}} \quad \frac{1}{2} \mathbf{x}^\top C \mathbf{x}$$

$$\text{subject to:} \quad C\mathbf{x} \geqslant \mathbf{b} \, ,$$

where the $n \times n$ matrix C is assumed to be positive definite (for a general quadratic programming problem the matrix C can always be assumed to be symmetric, but it is not necessarily positive definite). The Lagrangian is $\mathcal{L}(\mathbf{x}, \boldsymbol{\alpha}) = \frac{1}{2} \mathbf{x}^\top C \mathbf{x} - \boldsymbol{\alpha}^\top (C\mathbf{x} - \mathbf{b})$. We can minimize this by taking its gradient with respect to $\mathbf{x}$ and setting it to zero. This gives $C\mathbf{x} - C\boldsymbol{\alpha} = C(\mathbf{x} - \boldsymbol{\alpha}) = \mathbf{0}$. The positive definiteness of C implies that $\mathbf{x} = \boldsymbol{\alpha}$. The maximization of the Lagrangian is now reduced to maximizing $\mathcal{L}(\boldsymbol{\alpha}, \boldsymbol{\alpha}) = \frac{1}{2} \boldsymbol{\alpha}^\top C \boldsymbol{\alpha} - \boldsymbol{\alpha}^\top (C\boldsymbol{\alpha} - \mathbf{b}) = -\frac{1}{2} \boldsymbol{\alpha}^\top C \boldsymbol{\alpha} + \boldsymbol{\alpha}^\top \mathbf{b}$ subject to $\boldsymbol{\alpha} \geqslant \mathbf{0}$. Hence we can write the dual problem as

$$\max_{\boldsymbol{\alpha}} \quad -\frac{1}{2} \boldsymbol{\alpha}^\top C \boldsymbol{\alpha} + \boldsymbol{\alpha}^\top \mathbf{b}$$

$$\text{subject to:} \quad \boldsymbol{\alpha} \geqslant \mathbf{0} \, .$$

Notice that the dual problem involves only simple nonnegativity constraints.

Now suppose that we are given the Cholesky factorization $C = BB^\top$. It turns out (see Problem 1.36) that the Lagrange dual of the dual problem above can be written as

$$\min_{\boldsymbol{\mu}} \quad \frac{1}{2} \boldsymbol{\mu}^\top \boldsymbol{\mu} \tag{1.71}$$

$$\text{subject to:} \quad B\boldsymbol{\mu} \geqslant \mathbf{b} \, ,$$

with $\boldsymbol{\mu} = B^\top \boldsymbol{\alpha}$. This is a so-called *least distance* problem, which, provided that we know the Cholesky factorization of C, is easier to solve than the original quadratic programming problem.

A final example of duality is provided by the widely used *minimum cross-entropy method* [9].

■ EXAMPLE 1.19   Minimum Cross-Entropy (MinxEnt) Method

Let $\mathbf{X}$ be a discrete random variable (or vector) taking values $\mathbf{x}_1, \ldots, \mathbf{x}_r$, and let $\mathbf{q} = (q_1, \ldots, q_r)^\top$ and $\mathbf{p} = (p_1, \ldots, p_r)^\top$ be two strictly positive distribution (column) vectors for $\mathbf{X}$. Consider the following primal program of minimizing the cross-entropy of $\mathbf{p}$ and $\mathbf{q}$, that is, $\sum_{i=1}^n p_i \ln(p_i/q_i)$, for a fixed $\mathbf{q}$, subject to linear equality constraints:

$$\min_{\mathbf{p}} \quad \sum_{k=1}^r p_k \ln \frac{p_k}{q_k} \tag{1.72}$$

$$\text{subject to:} \quad \mathbb{E}_{\mathbf{p}}[S_i(\mathbf{X})] = \sum_{k=1}^r S_i(\mathbf{x}_k)\, p_k = \gamma_i, \quad i = 1, \ldots, m \tag{1.73}$$

$$\sum_{k=1}^r p_k = 1 \,, \tag{1.74}$$

where $S_1, \ldots, S_m$ are arbitrary functions.

Here the objective function is convex, since it is a linear combination of functions of the form $p\ln(p/c)$, which are convex on $\mathbb{R}_+$, for any $c > 0$. In addition, the equality constraint functions are affine (of the form $\mathbf{a}^\top \mathbf{p} - \gamma$). Therefore, this problem is convex. To derive the optimal solution $\mathbf{p}^*$ of the primal program above, it is typically easier to solve the associated *dual* program [9]. Below we present the corresponding procedure.

1. The Lagrangian of the primal problem is given by

$$\mathcal{L}(\mathbf{p}, \boldsymbol{\lambda}, \beta) = \sum_{k=1}^r p_k \ln \frac{p_k}{q_k} - \sum_{i=1}^m \lambda_i \left(\sum_{k=1}^r S_i(\mathbf{x}_k)\, p_k - \gamma_i \right) + \beta \left(\sum_{k=1}^r p_k - 1 \right), \tag{1.75}$$

where $\boldsymbol{\lambda} = (\lambda_1, \ldots, \lambda_m)^\top$ is the Lagrange multiplier vector corresponding to (1.73) and β is the Lagrange multiplier corresponding to (1.74). Note that we can use either a plus or a minus sign in the second sum of (1.75). We choose the latter because later we generalize the very same problem to inequality ($\geqslant$) constraints in (1.73), giving rise to a minus sign in the Lagrangian.

2. Solve (for fixed $\boldsymbol{\lambda}$ and β)

$$\min_{\mathbf{p}} \mathcal{L}(\mathbf{p}, \boldsymbol{\lambda}, \beta) \tag{1.76}$$

by solving

$$\nabla_{\mathbf{p}} \mathcal{L}(\mathbf{p}, \boldsymbol{\lambda}, \beta) = \mathbf{0} \,,$$

which gives the set of equations

$$\nabla_{p_k} \mathcal{L}(\mathbf{p}, \boldsymbol{\lambda}, \beta) = \ln \frac{p_k}{q_k} + 1 - \sum_{i=1}^m \lambda_i \, S_i(\mathbf{x}_k) + \beta = 0, \quad k = 1, \ldots, r \,.$$

Denote the optimal solution and the optimal function value obtained from the program (1.76) as $\mathbf{p}(\boldsymbol{\lambda}, \beta)$ and $\mathcal{L}^*(\boldsymbol{\lambda}, \beta)$, respectively. The latter is the Lagrange dual function. So we write

$$p_k(\boldsymbol{\lambda}, \beta) = q_k \exp\left(-\beta - 1 + \sum_{i=1}^{m} \lambda_i S_i(\mathbf{x}_k)\right), \quad k = 1, \ldots, r . \tag{1.77}$$

Since the sum of the $\{p_k\}$ must be 1, we obtain

$$e^\beta = \sum_{k=1}^{r} q_k \exp\left(-1 + \sum_{i=1}^{m} \lambda_i S_i(\mathbf{x}_k)\right) . \tag{1.78}$$

Substituting $\mathbf{p}(\boldsymbol{\lambda}, \beta)$ back into the Lagrangian gives

$$\mathcal{L}^*(\boldsymbol{\lambda}, \beta) = -1 + \sum_{i=1}^{m} \lambda_i \gamma_i - \beta . \tag{1.79}$$

3. Solve the *dual* program

$$\max_{\boldsymbol{\lambda}, \beta} \mathcal{L}^*(\boldsymbol{\lambda}, \beta) . \tag{1.80}$$

Since β and $\boldsymbol{\lambda}$ are related via (1.78), we can solve (1.80) by substituting the corresponding $\beta(\boldsymbol{\lambda})$ into (1.79) and optimizing the resulting function:

$$D(\boldsymbol{\lambda}) = -1 + \sum_{i=1}^{m} \lambda_i \gamma_i - \ln\left\{\sum_{k=1}^{r} q_k \exp\{-1 + \sum_{i=1}^{m} \lambda_i S_i(\mathbf{x}_k)\}\right\} . \tag{1.81}$$

Since $D(\boldsymbol{\lambda})$ is continuously differentiable and concave with respect to $\boldsymbol{\lambda}$, we can derive the optimal solution, $\boldsymbol{\lambda}^*$, by solving

$$\nabla_{\boldsymbol{\lambda}} D(\boldsymbol{\lambda}) = \mathbf{0} , \tag{1.82}$$

which can be written componentwise in the following explicit form:

$$\begin{aligned}
\nabla_{\lambda_j} D(\boldsymbol{\lambda}) &= \gamma_i - \frac{\sum_{k=1}^{r} S_i(\mathbf{x}_k) q_k \exp\left\{-1 + \sum_{j=1}^{m} \lambda_j S_j(\mathbf{x}_k)\right\}}{\sum_{k=1}^{r} q_k \exp\left\{-1 + \sum_{j=1}^{m} \lambda_j S_j(\mathbf{x}_k)\right\}} \\
&= \gamma_i - \frac{\mathbb{E}_{\mathbf{q}}\left[S_i(\mathbf{X}) \exp\left\{-1 + \sum_{j=1}^{m} \lambda_j S_j(\mathbf{X})\right\}\right]}{\mathbb{E}_{\mathbf{q}}\left[\exp\left\{-1 + \sum_{j=1}^{m} \lambda_j S_j(\mathbf{X})\right\}\right]} = 0
\end{aligned} \tag{1.83}$$

for $j = 1, \ldots, m$. The optimal vector $\boldsymbol{\lambda}^* = (\lambda_1^*, \ldots, \lambda_m^*)$ can be found by solving (1.83) numerically. Note that if the primal program has a nonempty interior optimal solution, then the dual program has an optimal solution $\boldsymbol{\lambda}^*$.

4. Finally, substitute $\boldsymbol{\lambda} = \boldsymbol{\lambda}^*$ and $\beta = \beta(\boldsymbol{\lambda}^*)$ back into (1.77) to obtain the solution to the original MinxEnt program.

It is important to note that we do not need to explicitly impose the conditions $p_i \geqslant 0$, $i = 1, \ldots, n$, because the quantities $\{p_i\}$ in (1.77) are automatically strictly positive. This is a crucial property of the CE distance; see also

[1]. It is instructive (see Problem 1.37) to verify how adding the nonnegativity constraints affects the procedure above.

When inequality constraints $\mathbb{E}_{\mathbf{p}}[S_i(\mathbf{X})] \geqslant \gamma_i$ are used in (1.73) instead of equality constraints, the solution procedure remains almost the same. The only difference is that the Lagrange multiplier vector $\boldsymbol{\lambda}$ must now be nonnegative. It follows that the dual program becomes

$$\max_{\boldsymbol{\lambda}} \quad D(\boldsymbol{\lambda})$$

$$\text{subject to:} \quad \boldsymbol{\lambda} \geqslant \mathbf{0} \,,$$

with $D(\boldsymbol{\lambda})$ given in (1.81).

A further generalization is to replace the above discrete optimization problem with a *functional* optimization problem. This topic will be discussed in Chapter 8. In particular, Section 8.9 deals with the MinxEnt method, which involves a functional MinxEnt problem.

PROBLEMS

Probability Theory

1.1   Prove the following results, using the properties of the probability measure in Definition 1.2.1 (here A and B are events):
   a) $\mathbb{P}(A^c) = 1 - \mathbb{P}(A)$.
   b) $\mathbb{P}(A \cup B) = \mathbb{P}(A) + \mathbb{P}(B) - \mathbb{P}(A \cap B)$.

1.2   Prove the product rule (1.4) for the case of three events.

1.3   We draw three balls consecutively from a bowl containing exactly five white and five black balls, without putting them back. What is the probability that all drawn balls will be black?

1.4   Consider the random experiment where we toss a biased coin until heads comes up. Suppose that the probability of heads on any one toss is p. Let X be the number of tosses required. Show that $X \sim \mathsf{G}(p)$.

1.5   In a room with many people, we ask each person his/her birthday (day and month). Let N be the number of people queried until we get a "duplicate" birthday.
   a) Calculate $\mathbb{P}(N > n)$, $n = 0, 1, 2, \ldots$.
   b) For which n do we have $\mathbb{P}(N \leqslant n) \geqslant 1/2$?
   c) Use a computer to calculate $\mathbb{E}[N]$.

1.6   Let X and Y be independent standard normal random variables, and let U and V be random variables that are derived from X and Y via the linear transformation

$$\begin{pmatrix} U \\ V \end{pmatrix} = \begin{pmatrix} \sin \alpha & -\cos \alpha \\ \cos \alpha & \sin \alpha \end{pmatrix} \begin{pmatrix} X \\ Y \end{pmatrix} \,.$$

   a) Derive the joint pdf of U and V.
   b) Show that U and V are independent and standard normally distributed.

1.7   Let $X \sim \mathsf{Exp}(\lambda)$. Show that the *memoryless property* holds: for all $s, t \geqslant 0$,

$$\mathbb{P}(X > t + s \,|\, X > t) = \mathbb{P}(X > s) \,.$$

1.8   Let X_1, X_2, X_3 be independent Bernoulli random variables with success probabilities $1/2$, $1/3$, and $1/4$, respectively. Give their conditional joint pdf, given that $X_1 + X_2 + X_3 = 2$.

1.9   Verify the expectations and variances in Table 1.3.

1.10   Let X and Y have joint density f given by

$$f(x, y) = c\,x\,y, \quad 0 \leqslant y \leqslant x, \quad 0 \leqslant x \leqslant 1 .$$

a) Determine the normalization constant c.
b) Determine $\mathbb{P}(X + 2Y \leqslant 1)$.

1.11   Let $X \sim \mathsf{Exp}(\lambda)$ and $Y \sim \mathsf{Exp}(\mu)$ be independent. Show that

a) $\min(X, Y) \sim \mathsf{Exp}(\lambda + \mu)$,
b) $\mathbb{P}(X < Y \mid \min(X, Y)) = \dfrac{\lambda}{\lambda + \mu}.$

1.12   Verify the properties of variance and covariance in Table 1.4.

1.13   Show that the correlation coefficient always lies between -1 and 1. [Hint: Use the fact that the variance of $aX + Y$ is always nonnegative, for any a.]

1.14   Consider Examples 1.1 and 1.2. Define X as the function that assigns the number $x_1 + \cdots + x_n$ to each outcome $\omega = (x_1, \ldots, x_n)$. The event that there are exactly k heads in n throws can be written as

$$\{\omega \in \Omega : X(\omega) = k\} .$$

If we abbreviate this to $\{X = k\}$, and further abbreviate $\mathbb{P}(\{X = k\})$ to $\mathbb{P}(X = k)$, then we obtain exactly (1.7). Verify that one can always view random variables in this way, that is, as real-valued functions on Ω, and that probabilities such as $\mathbb{P}(X \leqslant x)$ should be interpreted as $\mathbb{P}(\{\omega \in \Omega : X(\omega) \leqslant x\})$.

1.15   Show that

$$\mathrm{Var}\left(\sum_{i=1}^{n} X_i\right) = \sum_{i=1}^{n} \mathrm{Var}(X_i) + 2\sum_{i<j} \mathrm{Cov}(X_i, X_j) .$$

1.16   Let Σ be the covariance matrix of a random column vector $\mathbf{X}$. Write $\mathbf{Y} = \mathbf{X} - \boldsymbol{\mu}$, where $\boldsymbol{\mu}$ is the expectation vector of $\mathbf{X}$. Hence $\Sigma = \mathbb{E}[\mathbf{Y}\mathbf{Y}^\top]$. Show that Σ is positive semidefinite. That is, for any vector $\mathbf{u}$, we have $\mathbf{u}^\top \Sigma \mathbf{u} \geqslant 0$.

1.17   Suppose $Y \sim \mathsf{Gamma}(n, \lambda)$. Show that for all $x \geqslant 0$

$$\mathbb{P}(Y \leqslant x) = 1 - \sum_{k=0}^{n-1} \frac{e^{-\lambda x}(\lambda x)^k}{k!} . \tag{1.84}$$

1.18   Consider the random experiment where we draw uniformly and independently n numbers, $X_1, \ldots, X_n$, from the interval $[0,1]$.

a) Let M be the smallest of the n numbers. Express M in terms of $X_1, \ldots, X_n$.

b) Determine the pdf of M.

1.19 Let $Y = e^X$, where $X \sim N(0, 1)$.
  a) Determine the pdf of Y.
  b) Determine the expected value of Y.

1.20 We select a point (X, Y) from the triangle $(0,0) - (1,0) - (1,1)$ in such a way that X has a uniform distribution on $(0, 1)$ and the conditional distribution of Y given $X = x$ is uniform on $(0, x)$.
  a) Determine the joint pdf of X and Y.
  b) Determine the pdf of Y.
  c) Determine the conditional pdf of X given $Y = y$ for all $y \in (0, 1)$.
  d) Calculate $\mathbb{E}[X \mid Y = y]$ for all $y \in (0, 1)$.
  e) Determine the expectations of X and Y.

Poisson Processes

1.21 Let $\{N_t, t \geqslant 0\}$ be a Poisson process with rate $\lambda = 2$. Find
  a) $\mathbb{P}(N_2 = 1, N_3 = 4, N_5 = 5)$,
  b) $\mathbb{P}(N_4 = 3 \mid N_2 = 1, N_3 = 2)$,
  c) $\mathbb{E}[N_4 \mid N_2 = 2]$,
  d) $\mathbb{P}(N[2, 7] = 4, N[3, 8] = 6)$,
  e) $\mathbb{E}[N[4, 6] \mid N[1, 5] = 3]$.

1.22 Show that for any fixed $k \in \mathbb{N}$, $t > 0$ and $\lambda > 0$,

$$\lim_{n \to \infty} \binom{n}{k} \left(\frac{\lambda t}{n} \right)^k \left(1 - \frac{\lambda t}{n} \right)^{n-k} = \frac{(\lambda t)^k}{k!} e^{-\lambda t} .$$

[Hint: Write out the binomial coefficient and use the fact that $\lim_{n \to \infty} \left(1 - \frac{\lambda t}{n} \right)^n = e^{-\lambda t}$.]

1.23 Consider the Bernoulli approximation in Section 1.12. Let $U_1, U_2, \ldots$ denote the times of success for the Bernoulli process X.
  a) Verify that the "intersuccess" times $U_1, U_2 - U_1, \ldots$ are independent and have a geometric distribution with parameter $p = \lambda h$.
  b) For small h and $n = \lfloor t/h \rfloor$, show that the relationship $\mathbb{P}(A_1 > t) \approx \mathbb{P}(U_1 > n)$ leads in the limit, as $n \to \infty$, to

$$\mathbb{P}(A_1 > t) = e^{-\lambda t}.$$

1.24 If $\{N_t, t \geqslant 0\}$ is a Poisson process with rate λ, show that for $0 \leqslant u \leqslant t$ and $j = 0, 1, 2, \ldots, n$,

$$\mathbb{P}(N_u = j \mid N_t = n) = \binom{n}{j} \left(\frac{u}{t} \right)^j \left(1 - \frac{u}{t} \right)^{n-j},$$

that is, the conditional distribution of N_u given $N_t = n$ is binomial with parameters n and u/t.

Markov Processes

1.25   Determine the (discrete) pdf of each X_n, $n = 0, 1, 2, \ldots$, for the random walk in Example 1.10. Also, calculate $\mathbb{E}[X_n]$ and the variance of X_n for each n.

1.26   Let $\{X_n, n \in \mathbb{N}\}$ be a Markov chain with state space $\{0, 1, 2\}$, transition matrix

$$P = \begin{pmatrix} 0.3 & 0.1 & 0.6 \\ 0.4 & 0.4 & 0.2 \\ 0.1 & 0.7 & 0.2 \end{pmatrix},$$

and initial distribution $\pi = (0.2, 0.5, 0.3)$. Determine
   a) $\mathbb{P}(X_1 = 2)$,
   b) $\mathbb{P}(X_2 = 2)$,
   c) $\mathbb{P}(X_3 = 2 \,|\, X_0 = 0)$,
   d) $\mathbb{P}(X_0 = 1 \,|\, X_1 = 2)$,
   e) $\mathbb{P}(X_1 = 1, X_3 = 1)$.

1.27   Two dogs harbor a total number of m fleas. Spot initially has b fleas and Lassie has the remaining $m - b$. The fleas have agreed on the following immigration policy: at every time $n = 1, 2 \ldots$, a flea is selected at random from the total population and that flea will jump from one dog to the other. Describe the flea population on Spot as a Markov chain and find its stationary distribution.

1.28   Classify the states of the Markov chain with the following transition matrix:

$$P = \begin{pmatrix} 0.0 & 0.3 & 0.6 & 0.0 & 0.1 \\ 0.0 & 0.3 & 0.0 & 0.7 & 0.0 \\ 0.3 & 0.1 & 0.6 & 0.0 & 0.0 \\ 0.0 & 0.1 & 0.0 & 0.9 & 0.0 \\ 0.1 & 0.1 & 0.2 & 0.0 & 0.6 \end{pmatrix}.$$

1.29   Consider the following snakes-and-ladders game. Let N be the number of tosses required to reach the finish using a fair die. Calculate the expectation of N using a computer.

1.30   Ms. Ella Brum walks back and forth between her home and her office every day. She owns three umbrellas, which are distributed over two umbrella stands (one at home and one at work). When it is not raining, Ms. Brum walks without an umbrella. When it is raining, she takes one umbrella from the stand at the place of her departure, provided there is one available. Suppose that the probability that it is raining at the time of any departure is p. Let X_n denote the number of umbrellas available at the place where Ella arrives after walk number n; $n = 1, 2, \ldots$, including the one that she possibly brings with her. Calculate the limiting probability that it rains and no umbrella is available.

1.31 A mouse is let loose in the maze of Figure 1.10. From each compartment the mouse chooses one of the adjacent compartments with equal probability, independent of the past. The mouse spends an exponentially distributed amount of time in each compartment. The mean time spent in each of the compartments 1, 3, and 4 is two seconds; the mean time spent in compartments 2, 5, and 6 is four seconds. Let $\{X_t, t \geqslant 0\}$ be the Markov jump process that describes the position of the mouse for times $t \geqslant 0$. Assume that the mouse starts in compartment 1 at time $t = 0$.

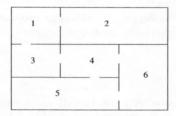

Figure 1.10: A maze.

What are the probabilities that the mouse will be found in each of the compartments $1, 2, \ldots, 6$ at some time t far away in the future?

1.32 In an $M/M/\infty$-queueing system, customers arrive according to a Poisson process with rate a. Every customer who enters is immediately served by one of an infinite number of servers; hence there is no queue. The service times are exponentially distributed, with mean $1/b$. All service and interarrival times are independent. Let X_t be the number of customers in the system at time t. Show that the limiting distribution of X_t, as $t \to \infty$, is Poisson with parameter a/b.

Optimization

1.33 Let $\mathbf{a}$ and let $\mathbf{x}$ be n-dimensional column vectors. Show that $\nabla_{\mathbf{x}} \mathbf{a}^{\top} \mathbf{x} = \mathbf{a}$.

1.34 Let A be a symmetric $n \times n$ matrix and $\mathbf{x}$ be an n-dimensional column vector. Show that $\nabla_{\mathbf{x}} \frac{1}{2} \mathbf{x}^{\top} A \mathbf{x} = A \mathbf{x}$. What is the gradient if A is not symmetric?

1.35 Show that the optimal distribution $\mathbf{p}^*$ in Example 1.16 is given by the uniform distribution.

1.36 Derive the program (1.71).

1.37   Consider the MinxEnt program

$$\min_{\mathbf{p}} \ \sum_{i=1}^{n} p_i \ln \frac{p_i}{q_i}$$

$$\text{subject to:} \quad \mathbf{p} \geqslant \mathbf{0}, \quad A\mathbf{p} = \mathbf{b}, \quad \sum_{i=1}^{n} p_i = 1 \ ,$$

where $\mathbf{p}$ and $\mathbf{q}$ are probability distribution vectors and A is an $m \times n$ matrix.

a) Show that the Lagrangian for this problem is of the form

$$\mathcal{L}(\mathbf{p}, \boldsymbol{\lambda}, \beta, \boldsymbol{\mu}) = \mathbf{p}^\top \boldsymbol{\xi}(\mathbf{p}) - \boldsymbol{\lambda}^\top (A\mathbf{p} - \mathbf{b}) - \boldsymbol{\mu}^\top \mathbf{p} + \beta(\mathbf{1}^\top \mathbf{p} - 1) \ .$$

b) Show that $p_i = q_i \exp(-\beta - 1 + \mu_i + \sum_{j=1}^{m} \lambda_j \, a_{ji})$, for $i = 1, \ldots, n$.

c) Explain why, as a result of the KKT conditions, the optimal $\boldsymbol{\mu}^*$ must be equal to the zero vector.

d) Show that the solution to this MinxEnt program is exactly the same as for the program where the nonnegativity constraints are omitted.

Further Reading

An easy introduction to probability theory with many examples is [13], and a more detailed textbook is [8]. A classical reference is [6]. An accurate and accessible treatment of various stochastic processes is given in [3]. For convex optimization we refer to [2] and [7].

REFERENCES

1. Z. I. Botev, D. P. Kroese, and T. Taimre. Generalized cross-entropy methods for rare-event simulation and optimization. *Simulation: Transactions of the Society for Modeling and Simulation International*, 83(11):785–806, 2007.

2. S. Boyd and L. Vandenberghe. *Convex Optimization*. Cambridge University Press, Cambridge, UK, 2004.

3. E. Çinlar. *Introduction to Stochastic Processes*. Prentice Hall, Englewood Cliffs, NJ, 1975.

4. T. M. Cover and J. A. Thomas. *Elements of Information Theory*. John Wiley & Sons, New York, 1991.

5. C. W. Curtis. *Linear Algebra: An Introductory Approach*. Springer-Verlag, New York, 1984.

6. W. Feller. *An Introduction to Probability Theory and Its Applications*, volume 1. John Wiley & Sons, New York, 2nd edition, 1970.

7. R. Fletcher. *Practical Methods of Optimization*. John Wiley & Sons, New York, 1987.

8. G. R. Grimmett and D. R. Stirzaker. *Probability and Random Processes*. Oxford University Press, Oxford, 3rd edition, 2001.

9. J. N. Kapur and H. K. Kesavan. *Entropy Optimization Principles with Applications*. Academic Press, New York, 1992.

10. A. I. Khinchin. *Information Theory*. Dover Publications, New York, 1957.

11. N. V. Krylov. *Introduction to the Theory of Random Processes*, volume 43 of *Graduate Studies in Mathematics*. American Mathematical Society, Providence, RI, 2002.

12. E. L. Lehmann. *Testing Statistical Hypotheses*. Springer-Verlag, New York, 1997.

13. S. M. Ross. *A First Course in Probability*. Prentice Hall, Englewood Cliffs, NJ, 7th edition, 2005.

CHAPTER 2

RANDOM NUMBER, RANDOM VARIABLE, AND STOCHASTIC PROCESS GENERATION

2.1 INTRODUCTION

This chapter deals with the computer generation of random numbers, random variables, and stochastic processes. In a typical stochastic simulation, randomness is introduced into simulation models via independent uniformly distributed random variables. These random variables are then used as building blocks to simulate more general stochastic systems.

The rest of this chapter is organized as follows. We start, in Section 2.2, with the generation of uniform random variables. Section 2.3 discusses general methods for generating one-dimensional random variables. Section 2.4 presents specific algorithms for generating variables from commonly used continuous and discrete distributions. In Section 2.5 we discuss the generation of random vectors. Sections 2.6 and 2.7 treat the generation of Poisson processes, Markov chains, and Markov jump processes. The generation of Gaussian and diffusion processes is given in Sections 2.8 and 2.9. Finally, Section 2.10 deals with the generation of random permutations.

2.2 RANDOM NUMBER GENERATION

In the early days of simulation, randomness was generated by *manual* techniques, such as coin flipping, dice rolling, card shuffling, and roulette spinning. Later on, *physical devices*, such as noise diodes and Geiger counters, were attached to computers for the same purpose. The prevailing belief held that only mechanical or electronic devices could produce truly random sequences. Although mechanical devices are still widely used in gambling and lotteries, these methods were abandoned by the computer-simulation community for several reasons: (1) mechanical methods were too slow for general use, (2) the generated sequences could not be reproduced, and (3) it was found that the generated numbers exhibit both bias and dependence. Although certain modern physical generation methods are fast and would pass most statistical tests for randomness (e.g., those based on the universal background radiation or the noise of a PC chip), their main drawback remains their lack of repeatability. Most of today's random number generators are not based on physical devices but on simple algorithms that can be easily implemented on a computer. They are fast, require little storage space, and can readily reproduce a given sequence of random numbers. Importantly, a good random number generator captures all the important statistical properties of true random sequences, even though the sequence is generated by a deterministic algorithm. For this reason these generators are sometimes called *pseudorandom*.

Most computer languages already contain a built-in pseudorandom number generator. The user is typically requested only to input the initial seed, X_0, and upon invocation the random number generator produces a sequence of independent, uniform $(0, 1)$ random variables. We therefore assume in this book the availability of such a "black box" that is capable of producing a stream of pseudorandom numbers. In Matlab, for example, this is provided by the **rand** function. The "seed" of the random number generator, which can be set by the **rng** function, determines which random stream is used, and this is very useful for testing purposes.

■ **EXAMPLE 2.1   Generating Uniform Random Variables in Matlab**

This example illustrates the use of the **rand** function in Matlab to generate samples from the $U(0, 1)$ distribution. For clarity we have omitted the "ans = " output in the Matlab session below.

```
>> rand                  % generate a uniform random number
     0.0196
>> rand                  % generate another uniform random number
     0.823
>> rand(1,4)             % generate a uniform random vector
     0.5252   0.2026   0.6721   0.8381
>> rng(1234)             % set the seed to 1234
>> rand                  % generate a uniform random number
     0.1915
>> rng(1234)             % reset the seed to 1234
>> rand
     0.1915              % the previous outcome is repeated
```

The simplest methods for generating pseudorandom sequences use the so-called *linear congruential generators*, introduced in [12]. These generate a deterministic sequence of numbers by means of the recursive formula

$$X_{t+1} = aX_t + c \pmod{m} , \tag{2.1}$$

where the initial value, X_0, is called the *seed* and the a, c, and m (all positive integers) are called the *multiplier*, the *increment*, and the *modulus*, respectively. Note that applying the modulo-m operator in (2.1) means that $aX_i + c$ is divided by m, and the remainder is taken as the value for X_{t+1}. Thus each *state* X_t can only assume a value from the set $\{0, 1, \ldots, m - 1\}$, and the quantities

$$U_t = \frac{X_t}{m} \, , \tag{2.2}$$

called *pseudorandom numbers*, constitute approximations to a true sequence of uniform random variables. Note that the sequence $X_0, X_1, X_2, \ldots$ will repeat itself after at most m steps and will therefore be periodic, with a period not exceeding m. For example, let $a = c = X_0 = 3$ and $m = 5$. Then the sequence obtained from the recursive formula $X_{t+1} = 3X_t + 3 \pmod 5$ is $3, 2, 4, 0, 3$, which has period 4. A careful choice of a, m, and c can lead to a linear congruential generator that passes most standard statistical tests for uniformity and independence. An example is the linear congruential generator with $m = 2^{31} - 1$, $a = 7^4$, and $c = 0$ by Lewis, Goodman, and Miller [13]. However, nowadays linear congruential generators no longer meet the requirements of modern Monte Carlo applications (e.g., [11]) and have been replaced by more general linear recursive algorithms, which are discussed next.

2.2.1 Multiple Recursive Generators

A *multiple-recursive generator* (MRG) of *order* k is determined by a sequence of k-dimensional state vectors $\mathbf{X}_t = (X_{t-k+1}, \ldots, X_t)^\top$, $t = 0, 1, 2, \ldots$, whose components satisfy the linear recurrence

$$X_t = (a_1 X_{t-1} + \cdots + a_k X_{t-k}) \bmod m, \quad t = k, k+1, \ldots \tag{2.3}$$

for some modulus m, multipliers $\{a_i, i = 1, \ldots, k\}$, and a given seed $\mathbf{X}_0 = (X_{-k+1}, \ldots, X_0)$. The maximum period length for this generator is $m^k - 1$. To yield fast algorithms, all but a few of the multipliers should be 0. When m is a large integer, the output stream of random numbers is obtained via $U_t = X_t/m$.

MRGs with very large periods can be implemented efficiently by combining several smaller period MRGs — yielding *combined multiple-recursive generators*.

■ EXAMPLE 2.2

One of the most successful combined MRGs is `MRG32k3a` by L'Ecuyer [9], which employs two MRGs of order 3, with recurrences

$$X_t = (1403580 \, X_{t-2} - 810728 \, X_{t-3}) \bmod m_1 \quad (m_1 = 2^{32} - 209) \, ,$$

$$Y_t = (527612 \, Y_{t-1} - 1370589 \, Y_{t-3}) \bmod m_2 \quad (m_2 = 2^{32} - 22853) \, ,$$

and output

$$U_t = \begin{cases} \dfrac{X_t - Y_t + m_1}{m_1 + 1} & \text{if } X_t \leqslant Y_t \, , \\[2mm] \dfrac{X_t - Y_t}{m_1 + 1} & \text{if } X_t > Y_t \, . \end{cases}$$

The period length is approximately 3×10^{57}. The generator MRG32k3a passes all statistical tests in today's most comprehensive test suit *TestU01* [11] and has been implemented in many software packages, including Matlab, Mathematica, Intel's MKL Library, SAS, VSL, Arena, and Automod. It is also the core generator in L'Ecuyer's SSJ simulation package, and is easily extendable to generate multiple random streams.

2.2.2 Modulo 2 Linear Generators

Good random generators must have very large state spaces. For a linear congruential generator, this means that the modulus m must be a large integer. However, for multiple recursive generators, it is not necessary to take a large modulus, as the period length can be as large as $m^k - 1$. Because binary operations are in general faster than floating point operations (which are in turn faster than integer operations), it makes sense to consider MRGs and other random number generators that are based on linear recurrences modulo 2. A general framework for such random number generators is given in [10], where the state is a k-bit vector $\mathbf{X}_t = (X_{t,1}, \ldots, X_{t,k})^\top$ that is mapped via a linear transformation to a w-bit output vector $\mathbf{Y}_t = (Y_{t,1}, \ldots, Y_{t,w})^\top$, from which the random number $U_t \in (0,1)$ is obtained by *bitwise decimation* as follows:

Algorithm 2.2.1: Generic Linear Recurrence Modulo 2 Generator

   input : Seed distribution μ on state space $\mathcal{S} = \{0,1\}^k$, and sample size N.
   output: Sequence $U_1, \ldots, U_N$ of pseudo-random numbers.
 1 Draw the seed $\mathbf{X}_0$ from the distribution μ.   // initialize
 2 **for** $t = 1$ **to** N **do**
 3 | $\mathbf{X}_t \leftarrow A\mathbf{X}_{t-1}$ // transition
 4 | $\mathbf{Y}_t \leftarrow B\mathbf{X}_t$ // output transformation
 5 | $U_t \leftarrow \sum_{\ell=1}^{w} Y_{t,\ell}\, 2^{-\ell}$ // decimation
 6 **return** $U_1, \ldots, U_N$

Here, A and B are $k \times k$ and $w \times k$ binary matrices, respectively, and all operations are performed modulo 2. In particular, addition corresponds to the bitwise XOR operation (in particular, $1 + 1 = 0$). The integer w can be thought of as the word length of the computer (i.e., $w = 32$ or 64). Usually (but there are exceptions, see [10]) k is taken much larger than w.

■ EXAMPLE 2.3 Mersenne Twister

A popular modulo 2 generator was introduced by Matsumoto and Nishimura [16]. The dimension k of the state vector $\mathbf{X}_t$ in Algorithm 2.2.1 is in this case $k = w\,n$, where w is the word length (default 32) and n a large integer (default 624). The period length for the default choice of parameters can be shown to be $2^{w(n-1)+1} - 1 = 2^{19937} - 1$. Rather than take the state $\mathbf{X}_t$ as a $w\,n \times 1$ vector, it is convenient to consider it as an $n \times w$ matrix with rows $\mathbf{x}_t, \ldots, \mathbf{x}_{t+n-1}$. Starting from the seed rows $\mathbf{x}_0, \ldots, \mathbf{x}_{n-1}$, at each step $t = 0, 1, 2, \ldots$ the $(t+n)$-th row is calculated according to the following rules:

1. Take the first r bits of $\mathbf{x}_t$ and the last $w - r$ bits of $\mathbf{x}_{t+1}$ and concenate them together in a binary vector $\mathbf{x}$.

2. Apply the following binary operation to $\mathbf{x} = (x_1, \ldots, x_w)$ to give a new binary vector $\widetilde{\mathbf{x}}$:

$$\widetilde{\mathbf{x}} = \begin{cases} \mathbf{x} \gg 1 & \text{if } x_w = 0 \,, \\ (\mathbf{x} \gg 1) \oplus \mathbf{a} & \text{if } x_w = 1 \,. \end{cases}$$

3. Let $\mathbf{x}_{t+n} = \mathbf{x}_{t+m} \oplus \widetilde{\mathbf{x}}$.

Here $\oplus$ stands for the XOR operation and $\gg 1$ for the rightshift operation (shift the bits one position to the right, adding a 1 from the left). The binary vector $\mathbf{a}$ and the numbers m and r are specified by the user (see below).

The output at step t of the algorithm is performed as the bitwise decimation of a vector $\mathbf{y}$ that is obtained via the following five steps:

1. $\mathbf{y} = \mathbf{x}_{t+n}$
2. $\mathbf{y} = \mathbf{y} \oplus (\mathbf{y} \gg u)$
3. $\mathbf{y} = \mathbf{y} \oplus ((\mathbf{y} \ll s) \,\&\, \mathbf{b})$
4. $\mathbf{y} = \mathbf{y} \oplus ((\mathbf{y} \ll v) \,\&\, \mathbf{c})$
5. $\mathbf{y} = \mathbf{y} \oplus (\mathbf{y} \gg l)$

Here $\&$ denotes the AND operation (bitwise multiplication), and $(\mathbf{y} \ll s)$ indicates a leftshift by s positions, adding 0s from the right; similarly, $(\mathbf{y} \gg u)$ is a rightshift by u positions, adding 1s from the left. The vectors $\mathbf{b}$ and $\mathbf{c}$ as well as the integers u and s are provided by the user. The recommended parameters for the algorithm are

$$(w, n, m, r) = (32, 624, 397, 31)$$

and

$$(\mathbf{a}, \mathbf{b}, \mathbf{c}, u, s, v, l) = (9908\text{B0DF}_{16}, 9\text{D2C5680}_{16}, \text{EFC6000}_{16}, 11, 7, 15, 18) \,,$$

where the subscript $_{16}$ indicates hexadecimal notation; for example, $7\text{B}_{16} = 0111\ 1101$.

As a concrete example of the workings of the Mersenne twister, suppose that the seed values are as in Table 2.1 (assuming default parameters).

Table 2.1: Initial state of the Mersenne twister.

$\mathbf{x}_0$	0011011011010000101011000101001
$\mathbf{x}_1$	10101010101010101010101010101011
$\mathbf{x}_2$	10001000100000001111100010101011
$\vdots$	$\vdots$
$\mathbf{x}_m$	10010101110100010101011110100011
$\vdots$	$\vdots$
$\mathbf{x}_{n-1}$	00100111001010101110110101100010

Let us generate the first random number. Suppose that $r = 31$, so,

$$\mathbf{x} = \underbrace{0011011011010000101011100010100}_{r \text{ bits of } \mathbf{x}_0} \quad \underbrace{1}_{w - r \text{ bits of } \mathbf{x}_1} \, .$$

The least significant bit (most right bit) of $\mathbf{x}$ is 1, so $\widetilde{\mathbf{x}}$ is obtained by right-shifting $\mathbf{x}$ and XOR-ing the resulting vector with $\mathbf{a}$, so

$\widetilde{\mathbf{x}} = 1001101101101000010101110001010 \oplus 0000000101101010001101001011001$
$= 1001101000000010011000111010110 1.$

We complete the calculation of $\mathbf{x}_n = \mathbf{x}_m \oplus \widetilde{\mathbf{x}}$ and get

$\mathbf{x}_n = 1001010111010001010101111010011 \oplus 10011010000000100110001110101101 =$
$= 0000111111010011001101000000 1110.$

Next, we determine the output vector $\mathbf{y}$ in five steps.

1. $\mathbf{y} = \mathbf{x}_n = 00001111110100110011010000001110.$

2. $\mathbf{y} = \mathbf{y} \oplus (\mathbf{y} \gg 11)$
   $= 00001111110100110011010000001110 \oplus 11111111111000011111101001100110$
   $= 11110000001100101100111001101000.$

3. $\mathbf{y} = \mathbf{y} \oplus ((\mathbf{y} \ll 7) \,\&\, 9D2C5680_{16})$

   $= 11110000001100101100111001101000$
   $\oplus (00011001011001110011010000000000 \,\&\, 00000000011001011001110011001101000)$
   $= 11110000001100101100111001101000 \oplus 00000000001000100000010000000000$
   $= 11110000001000011001010011101000.$

4. $\mathbf{y} = \mathbf{y} \oplus ((\mathbf{y} \ll 15) \,\&\, 00000000000000000110001111110111)$ — similar to 3, which results in
   $$\mathbf{y} = 11110001010100001111110100 \, .$$

5. $\mathbf{y} = \mathbf{y} \oplus (\mathbf{y} \gg 18)$ — similar to 2, which results in
   $$\mathbf{y} = 00001110101011110000011000111100 \, .$$

Having in mind that the decimal representation of final $\mathbf{y}$ is equal to $1.0130 \cdot 10^9$, the algorithm returns $1.0130 \cdot 10^9 / (2^{32} - 1) = 0.2359$ as an output.

For an updated version of the code of the Mersenne twister we refer to
http://www.math.sci.hiroshima-u.ac.jp/~m-mat/MT/emt.html

Note that for initialization, a full $w \times n$ matrix has to be specified. This is often done by running a basic linear generator. The algorithm is very fast when implemented in compiled languages such as C and passes most statistical test but is known to recover too slowly from the states near zero; see [11, Page 23].

2.3 RANDOM VARIABLE GENERATION

In this section we discuss various general methods for generating one-dimensional random variables from a prescribed distribution. We consider the inverse-transform method, the alias method, the composition method, and the acceptance–rejection method.

2.3.1 Inverse-Transform Method

Let X be a random variable with cdf F. Since F is a nondecreasing function, the inverse function F^{-1} may be defined as

$$F^{-1}(y) = \inf\{x : F(x) \geqslant y\} , \quad 0 \leqslant y \leqslant 1 . \tag{2.4}$$

(Readers not acquainted with the notion inf should read min.) It is easy to show that if $U \sim \mathsf{U}(0,1)$, then

$$X = F^{-1}(U) \tag{2.5}$$

has cdf F. That is to say, since F is invertible and $\mathbb{P}(U \leqslant u) = u$, we have

$$\mathbb{P}(X \leqslant x) = \mathbb{P}(F^{-1}(U) \leqslant x) = \mathbb{P}(U \leqslant F(x)) = F(x) . \tag{2.6}$$

Thus, to generate a random variable X with cdf F, draw $U \sim \mathsf{U}(0,1)$ and set $X = F^{-1}(U)$. Figure 2.1 illustrates the inverse-transform method given by the following algorithm:

Algorithm 2.3.1: Inverse-Transform Method

input : Cumulative distribution function F.
output: Random variable X distributed according to F.
1 Generate U from $\mathsf{U}(0,1)$.
2 $X \leftarrow F^{-1}(U)$
3 **return** X

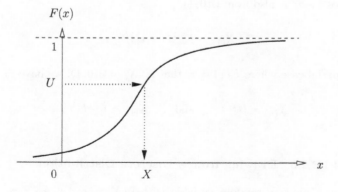

Figure 2.1: Inverse-transform method.

■ **EXAMPLE 2.4**

Generate a random variable from the pdf

$$f(x) = \begin{cases} 2x, & 0 \leqslant x \leqslant 1, \\ 0 & \text{otherwise.} \end{cases} \tag{2.7}$$

The cdf is

$$F(x) = \begin{cases} 0, & x < 0, \\ \int_0^x 2y\, dy = x^2, & 0 \leqslant x \leqslant 1, \\ 1, & x > 1. \end{cases}$$

Applying (2.5), we have

$$X = F^{-1}(U) = \sqrt{U}\,.$$

Therefore, to generate a random variable X from the pdf (2.7), first generate a random variable U from $\mathsf{U}(0,1)$ and then take its square root.

■ **EXAMPLE 2.5   Order Statistics**

Let $X_1,\ldots,X_n$ be iid random variables with cdf F. We wish to generate random variables $X_{(n)}$ and $X_{(1)}$ that are distributed according to the order statistics $\max(X_1,\ldots,X_n)$ and $\min(X_1,\ldots,X_n)$, respectively. From Example 1.7 we see that the cdfs of $X_{(n)}$ and $X_{(1)}$ are $F_n(x) = [F(x)]^n$ and $F_1(x) = 1 - [1 - F(x)]^n$, respectively. Applying (2.5), we get

$$X_{(n)} = F^{-1}(U^{1/n})\,,$$

and, since $1 - U$ is also from $\mathsf{U}(0,1)$,

$$X_{(1)} = F^{-1}(1 - U^{1/n})\,.$$

In the special case where $F(x) = x$, that is, $X_i \sim \mathsf{U}(0,1)$, we have

$$X_{(n)} = U^{1/n} \qquad \text{and} \qquad X_{(1)} = 1 - U^{1/n}\,.$$

■ **EXAMPLE 2.6   Drawing from a Discrete Distribution**

Let X be a discrete random variable with $\mathbb{P}(X = x_i) = p_i$, $i = 1, 2, \ldots,$ with $\sum_i p_i = 1$ and $x_1 < x_2 < \ldots$. The cdf F of X is given by $F(x) = \sum_{i:x_i \leqslant x} p_i$, $i = 1, 2, \ldots$ and is illustrated in Figure 2.2.

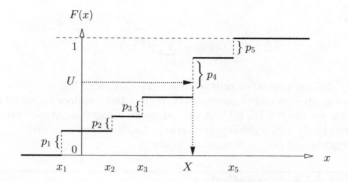

Figure 2.2: Inverse-transform method for a discrete random variable.

Hence the algorithm for generating a random variable from F can be written as follows:

Algorithm 2.3.2: Inverse-Transform Method for a Discrete Distribution

input : Discrete cumulative distribution function F.
output: Discrete random variable X distributed according to F.
1 Generate $U \sim \mathsf{U}(0,1)$.
2 Find the smallest positive integer, k, such that $U \leqslant F(x_k)$. Let $X \leftarrow x_k$.
3 **return** X

Much of the execution time in Algorithm 2.3.2 is spent in making the comparisons of Step 2. This time can be reduced by using efficient search techniques (see [2]).

In general, the inverse-transform method requires that the underlying cdf, F, exist in a form for which the corresponding inverse function F^{-1} can be found analytically or algorithmically. Applicable distributions are, for example, the exponential, uniform, Weibull, logistic, and Cauchy distributions. Unfortunately, for many other probability distributions, it is either impossible or difficult to find the inverse transform, that is, to solve

$$F(x) = \int_{-\infty}^{x} f(t)\,\mathrm{d}t = u$$

with respect to x. Even in the case where F^{-1} exists in an explicit form, the inverse-transform method may not necessarily be the most efficient random variable generation method (see [2]).

2.3.2 Alias Method

An alternative to the inverse-transform method for generating discrete random variables, which does not require time-consuming search techniques as per Step 2 of Algorithm 2.3.2, is the so-called *alias method* [19]. It is based on the fact that an arbitrary discrete n-point pdf f, with

$$f(x_i) = \mathbb{P}(X = x_i), \quad i = 1, \ldots, n\,,$$

can be represented as an equally weighted mixture of n pdfs, $q^{(k)}$, $k = 1, \ldots, n$, each having at most *two* nonzero components. That is, any n-point pdf f can be represented as

$$f(x) = \frac{1}{n} \sum_{k=1}^{n} q^{(k)}(x) \tag{2.8}$$

for suitably defined two-point pdfs $q^{(k)}$, $k = 1, \ldots, n$; see [19].

The alias method is rather general and efficient but requires an initial setup and extra storage for the n pdfs, $q^{(k)}$. A procedure for computing these two-point pdfs can be found in [2]. Once the representation (2.8) has been established, generation from f is simple and can be written as follows:

Algorithm 2.3.3: Alias Method

---

input : Two-point pdfs $q^{(k)}, k = 1, \ldots, n$ representing discrete pdf f.

output: Discrete random variable X distributed according to f.

1 Generate $U \sim \mathsf{U}(0,1)$ and set $K \leftarrow \lceil nU \rceil$.

2 Generate X from the two-point pdf $q^{(K)}$.

3 **return** X

2.3.3 Composition Method

This method assumes that a cdf, F, can be expressed as a *mixture* of cdfs $\{G_i\}$, that is,

$$F(x) = \sum_{i=1}^{m} p_i\, G_i(x) , \tag{2.9}$$

where

$$p_i > 0, \quad \sum_{i=1}^{m} p_i = 1 .$$

Let $X_i \sim G_i$ and let Y be a discrete random variable with $\mathbb{P}(Y = i) = p_i$, independent of X_i, for $1 \leqslant i \leqslant m$. Then a random variable X with cdf F can be represented as

$$X = \sum_{i=1}^{m} X_i\, I_{\{Y=i\}} .$$

It follows that in order to generate X from F, we must first generate the discrete random variable Y and then, given $Y = i$, generate X_i from G_i. We thus have the following method:

Algorithm 2.3.4: Composition Method

---

input : Mixture cdf F.

output: Random variable X distributed according to F.

1 Generate the random variable Y according to $\mathbb{P}(Y = i) = p_i$, $i = 1, \ldots, m$.

2 Given $Y = i$, generate X from the cdf G_i.

3 **return** X

2.3.4   Acceptance–Rejection Method

The inverse-transform and composition methods are direct methods in the sense that they deal directly with the cdf of the random variable to be generated. The acceptance–rejection method, is an indirect method due to Stan Ulam and John von Neumann. It can be applied when the above-mentioned direct methods either fail or turn out to be computationally inefficient.

To introduce the idea, suppose that the *target* pdf f (the pdf from which we want to sample) is bounded on some finite interval $[a, b]$ and is zero outside this interval (see Figure 2.3). Let

$$c = \sup\{f(x) : x \in [a, b]\}\,.$$

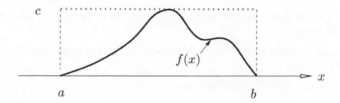

Figure 2.3: The acceptance–rejection method.

In this case, generating a random variable $Z \sim f$ is straightforward, and it can be done using the following acceptance–rejection steps:

1. Generate $X \sim \mathsf{U}(a, b)$.

2. Generate $Y \sim \mathsf{U}(0, c)$ independently of X.

3. If $Y \leqslant f(X)$, return $Z = X$. Otherwise, return to Step 1.

It is important to note that each generated vector (X, Y) is uniformly distributed over the rectangle $[a, b] \times [0, c]$. Therefore the accepted pair (X, Y) is uniformly distributed under the graph f. This implies that the distribution of the accepted values of X has the desired pdf f.

We can generalize this as follows: Let g be an arbitrary density such that $\phi(x) = C\,g(x)$ *majorizes* $f(x)$ for some constant C (Figure 2.4); that is, $\phi(x) \geqslant f(x)$ for all x. Note that of necessity $C \geqslant 1$. We call $g(x)$ the *proposal* pdf and assume that it is easy to generate random variables from it.

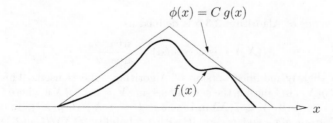

Figure 2.4: The acceptance–rejection method with a majorizing function ϕ.

The acceptance–rejection algorithm can be written as follows:

Algorithm 2.3.5: Acceptance–Rejection Method

---

input : Pdf g and constant C such that $Cg(x) \geqslant f(x)$ for all x.
output: Random variable X distributed according to pdf f.

1 found ← **false**
2 **while not** found **do**
3 | Generate X from $g(x)$.
4 | Generate $Y \sim \mathsf{U}(0, C\, g(X))$.
5 | **if** $Y \leqslant f(X)$ **then** found ← **true**

6 **return** X

The theoretical basis of the acceptance–rejection method is provided by the following theorem:

Theorem 2.3.1 *The random variable generated according to Algorithm 2.3.5 has the desired pdf $f(x)$.*

Proof: Define the following two subsets:

$$\mathscr{A} = \{(x,y) : 0 \leq y \leq Cg(x)\} \quad \text{and} \quad \mathscr{B} = \{(x,y) : 0 \leq y \leq f(x)\}\,, \qquad (2.10)$$

which represent the areas below the curves $Cg(x)$ and $f(x)$, respectively. Note first that Lines 3 and 4 of Algorithm 2.3.5 imply that the random vector (X, Y) is uniformly distributed on $\mathscr{A}$. To see this, let $q(x, y)$ denote the joint pdf of (X, Y), and let $q(y \mid x)$ denote the conditional pdf of Y given $X = x$. Then we have

$$q(x, y) = \begin{cases} g(x)\, q(y \mid x) & \text{if}(x, y) \in \mathscr{A}, \\ 0 & \text{otherwise.} \end{cases} \qquad (2.11)$$

Now Line 4 states that $q(y \mid x)$ equals $1/(Cg(x))$ for $y \in [0, Cg(x)]$ and is zero otherwise. Therefore, $q(x, y) = C^{-1}$ for every $(x, y) \in \mathscr{A}$.

Let (X^*, Y^*) be the first accepted point, that is, the first point that is in $\mathscr{B}$. Since the vector (X, Y) is uniformly distributed on $\mathscr{A}$, the vector (X^*, Y^*) is uniformly distributed on $\mathscr{B}$. Also, since the area of $\mathscr{B}$ equals unity, the joint pdf of (X^*, Y^*) on $\mathscr{B}$ equals unity as well. Thus, the marginal pdf of $Z = X^*$ is

$$\int_0^{f(x)} 1 \, \mathrm{d}y = f(x)\,.$$

$\square$

The *efficiency* of Algorithm 2.3.5 is defined as

$$\mathbb{P}\left((X, Y) \text{ is accepted}\right) = \frac{\text{area } \mathscr{B}}{\text{area } \mathscr{A}} = \frac{1}{C}\,. \qquad (2.12)$$

Often, a slightly modified version of Algorithm 2.3.5 is used. This is because $Y \sim \mathsf{U}(0, C\, g(X))$ in Line 4 is the same as setting $Y = U\, C\, g(X)$, where $U \sim \mathsf{U}(0, 1)$, and we can then write $Y \leqslant f(X)$ in Line 5 as $U \leqslant f(X)/(C\, g(X))$. In other words, generate X from $g(x)$ and accept it with probability $f(X)/(C\, g(X))$; otherwise reject X and try again. Thus the modified version of Algorithm 2.3.5 can be rewritten as follows:

Algorithm 2.3.6: Modified Acceptance–Rejection Method

---

input : Pdf g and constant C such that $Cg(x) \geqslant f(x)$ for all x.
output: Random variable X distributed according to pdf f.
1 found $\leftarrow$ **false**
2 while not found **do**
3 $\quad$ Generate X from $g(x)$.
4 $\quad$ Generate U from $\mathsf{U}(0,1)$ independently of X.
5 $\quad$ **if** $U \leqslant f(X)/(Cg(X))$ **then** found $\leftarrow$ **true**
6 return X

■ **EXAMPLE 2.7** **Example 2.4 (Continued)**

We show how to generate a random variable Z from the pdf

$$f(x) = \begin{cases} 2\,x, & 0 < x < 1, \\ 0 & \text{otherwise,} \end{cases}$$

using the acceptance–rejection method. For simplicity, take $g(x) = 1$, $0 < x < 1$, and $C = 2$. In this case our proposal distribution is simply the uniform distribution on $(0,1)$. Consequently, $f(x)/(C\,g(x)) = x$ and Algorithm 2.3.6 becomes: keep generating X and U independently from $\mathsf{U}(0,1)$ until $U \leqslant X$; then return X. Note that this example is merely illustrative, since it is more efficient to simulate from this pdf using the inverse-transform method.

As a consequence of (2.12), the efficiency of the modified acceptance–rejection method is determined by the acceptance probability $p = \mathbb{P}(U \leqslant f(X)/(Cg(X))) = \mathbb{P}(Y \leqslant f(X)) = 1/C$ for each trial (X, U). Since the trials are independent, the number of trials, N, before a successful pair (Z, U) occurs has the following geometric distribution:

$$\mathbb{P}(N = n) = p\,(1 - p)^{n-1}, \quad n = 1, 2, \ldots, \tag{2.13}$$

with the expected number of trials equal to $1/p = C$.

For this method to be of practical interest, the following criteria must be used in selecting the proposal density $g(x)$:

1. It should be easy to generate a random variable from $g(x)$.

2. The efficiency, $1/C$, of the procedure should be large; that is, C should be close to 1 (which occurs when $g(x)$ is close to $f(x)$).

■ **EXAMPLE 2.8**

Generate a random variable Z from the semicircular density

$$f(x) = \frac{2}{\pi R^2}\sqrt{R^2 - x^2}, \quad -R \leqslant x \leqslant R \ .$$

Take the proposal distribution to be uniform over $[-R, R]$; that is, take $g(x) = 1/(2R)$, $-R \leqslant x \leqslant R$ and choose C as small as possible such that $Cg(x) \geqslant$

$f(x)$; hence $C = 4/\pi$. Then Algorithm 2.3.6 leads to the following generation algorithm:

1. *Generate two independent random variables, U_1 and U_2, from $\mathsf{U}(0,1)$.*

2. *Use U_2 to generate X from $g(x)$ via the inverse-transform method, namely $X = (2U_2 - 1)R$, and calculate*

$$\frac{f(X)}{C\,g(X)} = \sqrt{1 - (2U_2 - 1)^2} \ .$$

3. *If $U_1 \leqslant f(X)/(C\,g(X))$, which is equivalent to $(2U_2 - 1)^2 \leqslant 1 - U_1^2$, return $Z = X = (2U_2 - 1)R$; otherwise, return to Step 1.*

The expected number of trials for this algorithm is $C = 4/\pi$, and the efficiency is $1/C = \pi/4 \approx 0.785$.

2.4   GENERATING FROM COMMONLY USED DISTRIBUTIONS

The next two subsections present algorithms for generating variables from commonly used continuous and discrete distributions. Of the numerous algorithms available (e.g., [2]), we have tried to select those that are reasonably efficient and relatively simple to implement.

2.4.1   Generating Continuous Random Variables

2.4.1.1 Exponential Distribution We start by applying the inverse-transform method to the exponential distribution. If $X \sim \mathsf{Exp}(\lambda)$, then its cdf F is given by

$$F(x) = 1 - e^{-\lambda x}, \qquad x \geqslant 0 \ . \tag{2.14}$$

Hence, solving $u = F(x)$ in terms of x gives

$$F^{-1}(u) = -\frac{1}{\lambda} \ln(1 - u) \ .$$

Because $U \sim \mathsf{U}(0,1)$ implies $1 - U \sim \mathsf{U}(0,1)$, we arrive at the following algorithm:

Algorithm 2.4.1: Generation of an Exponential Random Variable
---
input : $\lambda > 0$
output: Random variable X distributed according to $\mathsf{Exp}(\lambda)$.
1 Generate $U \sim \mathsf{U}(0,1)$.
2 $X \leftarrow -\frac{1}{\lambda} \ln U$
3 **return** X

There are many alternative procedures for generating variables from the exponential distribution. The interested reader is referred to [2].

2.4.1.2 Normal (Gaussian) Distribution If $X \sim \mathsf{N}(\mu, \sigma^2)$, its pdf is given by

$$f(x) = \frac{1}{\sigma\sqrt{2\pi}} \, \exp\left\{-\frac{(x-\mu)^2}{2\sigma^2}\right\} , \qquad -\infty < x < \infty , \qquad (2.15)$$

where μ is the mean (or expectation) and σ^2 the variance of the distribution.

Since inversion of the normal cdf is numerically inefficient, the inverse-transform method is not very suitable for generating normal random variables, so other procedures must be devised. We consider only generation from $\mathsf{N}(0, 1)$ (standard normal variables), since any random $Z \sim \mathsf{N}(\mu, \sigma^2)$ can be represented as $Z = \mu + \sigma X$, where X is from $\mathsf{N}(0, 1)$. One of the earliest methods for generating variables from $\mathsf{N}(0, 1)$ was developed by Box and Muller as follows:

Let X and Y be two independent standard normal random variables; so (X, Y) is a random point in the plane. Let (R, Θ) be the corresponding polar coordinates. The joint pdf $f_{R,\Theta}$ of R and Θ is given by

$$f_{R,\Theta}(r, \theta) = \frac{1}{2\pi} \, \mathrm{e}^{-r^2/2} \, r \quad \text{for } r \geqslant 0 \text{ and } \theta \in [0, 2\pi) .$$

This can be seen by writing x and y in terms of r and θ, to get

$$x = r\cos\theta \quad \text{and} \quad y = r\sin\theta . \qquad (2.16)$$

The Jacobian of this coordinate transformation is

$$\det\begin{pmatrix} \frac{\partial x}{\partial r} & \frac{\partial x}{\partial \theta} \\ \frac{\partial y}{\partial r} & \frac{\partial y}{\partial \theta} \end{pmatrix} = \left| \begin{array}{cc} \cos\theta & -r\sin\theta \\ \sin\theta & r\cos\theta \end{array} \right| = r .$$

The result now follows from the transformation rule (1.20), noting that the joint pdf of X and Y is $f_{X,Y}(x, y) = \frac{1}{2\pi} \mathrm{e}^{-(x^2+y^2)/2}$. It is not difficult to verify that R and Θ are independent, that $\Theta \sim \mathsf{U}[0, 2\pi)$, and that $\mathbb{P}(R > r) = \mathrm{e}^{-r^2/2}$. This means that R has the same distribution as $\sqrt{V}$, with $V \sim \mathsf{Exp}(1/2)$. Namely, $\mathbb{P}(\sqrt{V} > v) = \mathbb{P}(V > v^2) = \mathrm{e}^{-v^2/2}$, $v \geqslant 0$. Thus both Θ and R are easy to generate and are transformed via (2.16) into independent standard normal random variables. This leads to the following algorithm:

Algorithm 2.4.2: Normal Random Variable Generation: Box–Muller Approach

---

output: Independent standard normal random variables X and Y.
1 Generate two independent random variables, U_1 and U_2, from $\mathsf{U}(0, 1)$.
2 $X \leftarrow (-2\ln U_1)^{1/2} \cos(2\pi U_2)$
3 $Y \leftarrow (-2\ln U_1)^{1/2} \sin(2\pi U_2)$
4 **return** X, Y

An alternative generation method for $\mathsf{N}(0, 1)$ is based on the acceptance–rejection method. First, note that in order to generate a random variable Y from $\mathsf{N}(0, 1)$, one can first generate a positive random variable X from the pdf

$$f(x) = \sqrt{\frac{2}{\pi}} \, \mathrm{e}^{-x^2/2}, \qquad x \geqslant 0 , \qquad (2.17)$$

and then assign to X a random sign. The validity of this procedure follows from the symmetry of the standard normal distribution about zero.

To generate a random variable X from (2.17), we bound $f(x)$ by $C\,g(x)$, where $g(x) = \mathrm{e}^{-x}$ is the pdf of the $\mathsf{Exp}(1)$ distribution. The smallest constant C such that $f(x) \leqslant Cg(x)$ is $\sqrt{2\mathrm{e}/\pi}$ (see Figure 2.5). The efficiency of this method is therefore $\sqrt{\pi/2\mathrm{e}} \approx 0.76$.

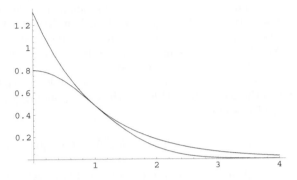

Figure 2.5: Bounding the positive normal density.

The acceptance condition, $U \leqslant f(X)/(C\mathrm{e}^{-X})$, can be written as

$$U \leqslant \exp[-(X-1)^2/2] \,, \tag{2.18}$$

which is equivalent to

$$-\ln U \geqslant \frac{(X-1)^2}{2} \,, \tag{2.19}$$

where X is from $\mathsf{Exp}(1)$. Since $-\ln U$ is also from $\mathsf{Exp}(1)$, the last inequality can be written as

$$V_1 \geqslant \frac{(V_2 - 1)^2}{2} \,, \tag{2.20}$$

where $V_1 = -\ln U$ and $V_2 = X$ are independent and both $\mathsf{Exp}(1)$ distributed.

2.4.1.3 Gamma Distribution

If $X \sim \mathsf{Gamma}(\alpha, \lambda)$ then its pdf is of the form

$$f(x) = \frac{x^{\alpha-1}\lambda^\alpha \mathrm{e}^{-\lambda x}}{\Gamma(\alpha)} \,, \qquad x \geqslant 0 \,. \tag{2.21}$$

The parameters $\alpha > 0$ and $\lambda > 0$ are called the *shape* and *scale* parameters, respectively. Since λ merely changes the scale, it suffices to consider only random variable generation of $\mathsf{Gamma}(\alpha, 1)$. In particular, if $X \sim \mathsf{Gamma}(\alpha, 1)$, then $X/\lambda \sim \mathsf{Gamma}(\alpha, \lambda)$ (see Exercise 2.16). Because the cdf for the gamma distributions does not generally exist in explicit form, the inverse-transform method cannot always be applied to generate random variables from this distribution. Alternative methods are thus called for. We discuss one such method for the case $\alpha \geqslant 1$. Let $f(x) = x^{\alpha-1}\mathrm{e}^{-x}/\Gamma(\alpha)$ and $\psi(x) = d\,(1 + cx)^3$, $x > -1/c$ and zero otherwise, where c and d are positive constants. Note that $\psi(x)$ is a strictly increasing function. Let Y

have density $k(y) = f(\psi(y))\,\psi'(y)\,c_1$, where c_1 is a normalization constant. Then $X = \psi(Y)$ has density f. Namely, by the transformation rule (1.16), we obtain

$$f_X(x) = \frac{k(\psi^{-1}(x))}{\psi'(\psi^{-1}(x))} = f(\psi(\psi^{-1}(x)))\,\frac{\psi'(\psi^{-1}(x))}{\psi'(\psi^{-1}(x))} = f(x)\;.$$

We draw Y via the acceptance–rejection method, using the standard normal distribution as our proposal distribution. We choose c and d such that $k(y) \leqslant C\varphi(y)$, with $C > 1$ close to 1, where φ is the pdf of the $\mathsf{N}(0,1)$ distribution. To find such c and d, we first write $k(y) = c_2\,e^{h(y)}$, where some algebra will show that

$$h(y) = (1 - 3\,\alpha)\ln(1 + c\,y) - d\,(1 + c\,y)^3 + d\;.$$

(Note that $h(0) = 0$.) Next, a Taylor series expansion of $h(y)$ around 0 yields

$$h(y) = c\,(-1 - 3\,d + 3\,\alpha)\,y - \frac{1}{2}\,c^2\,(-1 + 6\,d + 3\,\alpha)\,y^2 + O(y^3)\;.$$

This suggests taking c and d such that the coefficients of y and y^2 in the expansion above are 0 and $-1/2$, respectively, as in the exponent of the standard normal density. So we take $d = \alpha - 1/3$ and $c = \frac{1}{3\sqrt{d}}$. It is not difficult to check that then

$$h(y) \leqslant -\frac{1}{2}\,y^2 \quad \text{for all} \quad y > -\frac{1}{c}\;,$$

and therefore $e^{h(y)} \leqslant e^{-\frac{1}{2}y^2}$, which means that $k(y)$ is dominated by $c_2\sqrt{2\pi}\,\varphi(y)$ for all y. Hence, the acceptance–rejection method for drawing from $Y \sim k$ is as follows: Draw $Z \sim \mathsf{N}(0,1)$ and $U \sim \mathsf{U}(0,1)$ independently. If

$$U < \frac{c_2\,e^{h(Z)}}{c_2\sqrt{2\pi}\,\varphi(Z)}\;,$$

or equivalently, if

$$\ln U < h(Z) + \frac{1}{2}Z^2\;,$$

then return $Y = Z$; otherwise, repeat (we set $h(Z) = -\infty$ if $Z \leqslant -1/c$). The efficiency of this method, $\int_{-1/c}^{\infty} e^{h(y)}dy / \int_{-\infty}^{\infty} e^{-\frac{1}{2}y^2}\,dy$, is greater than 0.95 for all values of $\alpha \geqslant 1$. Finally, we complete the generation of X by taking $X = \psi(Y)$. For the case where $\alpha < 1$, we can use the fact that if $X \sim \mathsf{Gamma}(1 + \alpha, 1)$ and $U \sim \mathsf{U}(0,1)$ are independent, then $XU^{1/\alpha} \sim \mathsf{Gamma}(\alpha, 1)$; see Problem 2.17. Summarizing, we have the following algorithm [15]:

Algorithm 2.4.3: Sampling from the Gamma(α, λ) Distribution

---

1 function Gamrnd(α, λ)

2   **if** $\alpha > 1$ **then**

3   Set $d \leftarrow \alpha - 1/3$ and $c \leftarrow 1/\sqrt{9\,d}$

4   continue $\leftarrow$ **true**

5   **while** continue **do**

6   Generate $Z \sim \mathsf{N}(0, 1)$.

7   **if** $Z > -1/c$ **then**

8          $V \leftarrow (1 + cZ)^3$

9   Generate $U \sim \mathsf{U}(0, 1)$.

10   **if** $\ln U < \frac{1}{2}Z^2 + dV + d\ln(V)$ **then** continue $\leftarrow$ **false**

11    $X \leftarrow dV/\lambda$

12   **else**

13      $X \leftarrow$ Gamrnd$(\alpha + 1, \lambda)$

14   Generate $U \sim \mathsf{U}(0, 1)$.

15      $X \leftarrow XU^{1/\alpha}$

16   **return** X

A gamma distribution with an integer shape parameter, say $\alpha = m$, is also called an *Erlang distribution*, denoted $\mathsf{Erl}(m, \lambda)$. In this case, X can be represented as the sum of iid exponential random variables Y_i. That is, $X = \sum_{i=1}^{m} Y_i$, where the $\{Y_i\}$ are iid exponential variables, each with mean $1/\lambda$; see Example 1.9. Using Algorithm 2.4.1, we can write $Y_i = -\frac{1}{\lambda} \ln U_i$, whence

$$X = -\frac{1}{\lambda} \sum_{i=1}^{m} \ln U_i \; . \qquad (2.22)$$

Equation (2.22) suggests the following generation algorithm:

Algorithm 2.4.4: Generation of an Erlang Random Variable

---

input : Positive integer m and $\lambda > 0$.

output: Random variable $X \sim \mathsf{Erl}(m, \lambda)$.

1 Generate iid random variables $U_1, \ldots, U_m$ from $\mathsf{U}(0, 1)$.

2 $X \leftarrow -\frac{1}{\lambda} \sum_{i=1}^{m} \ln U_i$

3 **return** X

2.4.1.4 *Beta Distribution* If $X \sim \mathsf{Beta}(\alpha, \beta)$, then its pdf is of the form

$$f(x) = \frac{\Gamma(\alpha + \beta)}{\Gamma(\alpha)\Gamma(\beta)} \; x^{\alpha - 1}(1 - x)^{\beta - 1}, \quad 0 \leqslant x \leqslant 1 \; . \qquad (2.23)$$

Both parameters α and β are assumed to be greater than 0. Note that $\mathsf{Beta}(1, 1)$ is simply the $\mathsf{U}(0, 1)$ distribution.

To sample from the beta distribution, let us consider first the case where either α or β equals 1. In that case, we can simply use the inverse-transform method. For example, for $\beta = 1$, the $\mathsf{Beta}(\alpha, 1)$ pdf is

$$f(x) = \alpha x^{\alpha - 1}, \quad 0 \leqslant x \leqslant 1 \; ,$$

and the corresponding cdf becomes

$$F(x) = x^{\alpha}, \qquad 0 \leqslant x \leqslant 1 .$$

Thus a random variable X can be generated from this distribution by drawing $U \sim \mathsf{U}(0,1)$ and returning $X = U^{1/\alpha}$.

A general procedure for generating a $\mathsf{Beta}(\alpha, \beta)$ random variable is based on the fact that if $Y_1 \sim \mathsf{Gamma}(\alpha, 1)$, $Y_2 \sim \mathsf{Gamma}(\beta, 1)$, and Y_1 and Y_2 are independent, then

$$X = \frac{Y_1}{Y_1 + Y_2}$$

is distributed $\mathsf{Beta}(\alpha, \beta)$. The reader is encouraged to prove this assertion (see Problem 2.18). The corresponding algorithm is as follows:

Algorithm 2.4.5: Generation of a Beta Random Variable

input : $\alpha, \beta > 0$
output: Random variable $X \sim \mathsf{Beta}(\alpha, \beta)$.
1 Generate independently $Y_1 \sim \mathsf{Gamma}(\alpha, 1)$ and $Y_2 \sim \mathsf{Gamma}(\beta, 1)$.
2 $X \leftarrow Y_1/(Y_1 + Y_2)$
3 **return** X

For integer $\alpha = m$ and $\beta = n$, another method may be used, based on the theory of order statistics. Let $U_1, \ldots, U_{m+n-1}$ be independent random variables from $\mathsf{U}(0,1)$. Then the m-th order statistic, $U_{(m)}$, has a $\mathsf{Beta}(m, n)$ distribution. This gives the following algorithm.

Algorithm 2.4.6: Generation of a Beta Random Variable with Integer Parameters $\boldsymbol{\alpha = m}$ and $\boldsymbol{\beta = n}$

input : Positive integers m and n.
output: Random variable $X \sim \mathsf{Beta}(m, n)$.
1 Generate $m + n - 1$ iid random variables $U_1, \ldots, U_{m+n-1}$ from $\mathsf{U}(0,1)$.
2 $X \leftarrow U_{(m)}$ // m-th order statistic
3 **return** X

It can be shown that the total number of comparisons needed to find $U_{(m)}$ is $(m/2)(m + 2n - 1)$, so that this procedure loses efficiency for large m and n.

2.4.2 Generating Discrete Random Variables

2.4.2.1 Bernoulli Distribution   If $X \sim \mathsf{Ber}(p)$, its pdf is of the form

$$f(x) = p^x (1-p)^{1-x}, \qquad x = 0, 1 , \tag{2.24}$$

where p is the success probability. Applying the inverse-transform method, we can readily obtain the following generation algorithm:

Algorithm 2.4.7: Generation of a Bernoulli Random Variable

---

input : $p \in (0,1)$
output: Random variable $X \sim \mathsf{Ber}(p)$.
1 Generate $U \sim \mathsf{U}(0,1)$.
2 **if** $U \leqslant p$ **then**
3 $\quad\mid\quad X \leftarrow 1$
4 **else**
5 $\quad\lfloor\quad X \leftarrow 0$
6 **return** X

In Figure 2.6, three typical outcomes (realizations) are given for 100 independent Bernoulli random variables, each with success parameter $p = 0.5$.

Figure 2.6: Results of three experiments with 100 independent Bernoulli trials, each with $p = 0.5$. The dark bars indicate where a success appears.

2.4.2.2 Binomial Distribution   If $X \sim \mathsf{Bin}(n,p)$ then its pdf is of the form

$$f(x) = \binom{n}{x} p^x (1-p)^{n-x}, \quad x = 0, 1, \ldots, n . \tag{2.25}$$

Recall that a binomial random variable X can be viewed as the total number of successes in n independent Bernoulli experiments, each with success probability p; see Example 1.1. Denoting the result of the i-th trial by $X_i = 1$ (success) or $X_i = 0$ (failure), we can write $X = X_1 + \cdots + X_n$, with the $\{X_i\}$ being iid $\mathsf{Ber}(p)$ random variables. The simplest generation algorithm can thus be written as follows:

Algorithm 2.4.8: Generation of a Binomial Random Variable

---

input : Positive integer n and $p \in (0,1)$.
output: Random variable $X \sim \mathsf{Bin}(n,p)$.
1 Generate iid random variables $X_1, \ldots, X_n$ from $\mathsf{Ber}(p)$.
2 $X \leftarrow \sum_{i=1}^{n} X_i$
3 **return** X

Since the execution time of Algorithm 2.4.8 is proportional to n, we may be motivated to use alternative methods for large n. For example, we could consider the normal distribution as an approximation to the binomial. In particular, by the central limit theorem, as n increases, the distribution of X is close to that of $Y \sim \mathsf{N}(np, np(1-p))$; see (1.26). In fact, the cdf of $\mathsf{N}(np - 1/2, np(1-p))$ approximates the cdf of X even better. This is called the *continuity correction*.

Thus, to obtain a binomial random variable, we could generate Y from $\mathsf{N}(np - 1/2, np(1-p))$ and truncate to the nearest nonnegative integer. Equivalently, we could generate $Z \sim \mathsf{N}(0,1)$ and set

$$\max\left\{0, \left\lfloor np + \frac{1}{2} + Z\sqrt{np(1-p)}\right\rfloor\right\} \tag{2.26}$$

as an approximate sample from the $\mathsf{Bin}(n,p)$ distribution. Here $\lfloor \alpha \rfloor$ denotes the integer part of α. One should consider using the normal approximation for $np > 10$ with $p \geqslant \frac{1}{2}$, and for $n(1-p) > 10$ with $p < \frac{1}{2}$.

2.4.2.3 Geometric Distribution
If $X \sim \mathsf{G}(p)$, then its pdf is of the form

$$f(x) = p(1-p)^{x-1}, \qquad x = 1, 2 \dots . \tag{2.27}$$

The random variable X can be interpreted as the number of trials required until the first success occurs in a series of independent Bernoulli trials with success parameter p. Note that $\mathbb{P}(X > m) = (1-p)^m$.

We now present an algorithm based on the relationship between the exponential and geometric distributions. Let $Y \sim \mathsf{Exp}(\lambda)$, with λ such that $1 - p = \mathrm{e}^{-\lambda}$. Then $X = \lfloor Y \rfloor + 1$ has a $\mathsf{G}(p)$ distribution. This is because

$$\mathbb{P}(X > x) = \mathbb{P}(\lfloor Y \rfloor > x - 1) = \mathbb{P}(Y \geqslant x) = \mathrm{e}^{-\lambda x} = (1-p)^x .$$

Hence, to generate a random variable from $\mathsf{G}(p)$, we first generate a random variable from the exponential distribution with $\lambda = -\ln(1-p)$, truncate the obtained value to the nearest integer, and add 1.

Algorithm 2.4.9: Generation of a Geometric Random Variable

input : $p \in (0, 1)$
output: Random variable $X \sim \mathsf{G}(p)$.
1 Generate $Y \sim \mathsf{Exp}(-\ln(1-p))$.
2 $X \leftarrow 1 + \lfloor Y \rfloor$
3 **return** X

2.4.2.4 Poisson Distribution
If $X \sim \mathsf{Poi}(\lambda)$, its pdf is of the form

$$f(n) = \frac{\mathrm{e}^{-\lambda}\lambda^n}{n!} , \qquad n = 0, 1, \dots , \tag{2.28}$$

where λ is the *rate* parameter. There is an intimate relationship between Poisson and exponential random variables, highlighted by the properties of the Poisson process; see Section 1.12. In particular, a Poisson random variable X can be interpreted as the maximal number of iid exponential variables (with parameter λ)

whose sum does not exceed 1. That is,

$$X = \max \left\{ n : \sum_{j=1}^{n} Y_j \leqslant 1 \right\}, \tag{2.29}$$

where the $\{Y_j\}$ are independent and $\mathsf{Exp}(\lambda)$ distributed. Since $Y_j = -\frac{1}{\lambda} \ln U_j$, with $U_j \sim \mathsf{U}(0,1)$, we can rewrite (2.29) as

$$
\begin{aligned}
X &= \max \left\{ n : \sum_{j=1}^{n} -\ln U_j \leqslant \lambda \right\} \\
&= \max \left\{ n : \ln \left(\prod_{j=1}^{n} U_j \right) \geqslant -\lambda \right\} \\
&= \max \left\{ n : \prod_{j=1}^{n} U_j \geqslant e^{-\lambda} \right\}.
\end{aligned}
\tag{2.30}
$$

This leads to the following algorithm:

Algorithm 2.4.10: Generation of a Poisson Random Variable

---

input : $\lambda > 0$
output: Random variable $X \sim \mathsf{Poi}(\lambda)$.

1 Set $n \leftarrow 0$ and $a \leftarrow 1$.
2 **while** $a \geqslant e^{-\lambda}$ **do**
3 | Generate $U \sim \mathsf{U}(0,1)$.
4 | $a \leftarrow aU$
5 | $n \leftarrow n+1$
6 $X \leftarrow n - 1$
7 **return** X

It is readily seen that for large λ, this algorithm becomes slow ($e^{-\lambda}$ is small for large λ, and more random numbers, U_j, are required to satisfy $\prod_{j=1}^{n} U_j < e^{-\lambda}$). Alternative approaches can be found in [2] and [7].

2.5 RANDOM VECTOR GENERATION

Say we need to generate a random vector $\mathbf{X} = (X_1, \ldots, X_n)$ from a given n-dimensional distribution with pdf $f(\mathbf{x})$ and cdf $F(\mathbf{x})$. When the components $X_1, \ldots, X_n$ are *independent*, the situation is easy: we simply apply the inverse-transform method or another generation method of our choice to each component individually.

■ EXAMPLE 2.9

We want to generate uniform random vectors $\mathbf{X} = (X_1, \ldots, X_n)$ from the n-dimensional rectangle $D = \{(x_1, \ldots, x_n) : a_i \leqslant x_i \leqslant b_i, \ i = 1, \ldots, n\}$. It is clear that the components of $\mathbf{X}$ are independent and uniformly distributed: $X_i \sim U[a_i, b_i]$, $i = 1, \ldots, n$. Applying the inverse-transform method to X_i, we can therefore write $X_i = a_i + (b_i - a_i)U_i$, $i = 1, \ldots, n$, where $U_1, \ldots, U_n$ are iid from $\mathsf{U}(0, 1)$.

For *dependent* random variables $X_1, \ldots, X_n$, we can represent the joint pdf $f(\mathbf{x})$, using the product rule (1.4), as

$$f(x_1, \ldots, x_n) = f_1(x_1) f_2(x_2 \,|\, x_1) \cdots f_n(x_n \,|\, x_1, \ldots, x_{n-1}) \,, \qquad (2.31)$$

where $f_1(x_1)$ is the marginal pdf of X_1 and $f_k(x_k \,|\, x_1, \ldots, x_{k-1})$ is the conditional pdf of X_k given $X_1 = x_1, X_2 = x_2, \ldots, X_{k-1} = x_{k-1}$. Thus one way to generate $\mathbf{X}$ is to first generate X_1, then, given $X_1 = x_1$, to generate X_2 from $f_2(x_2 \,|\, x_1)$, and so on, until we generate X_n from $f_n(x_n \,|\, x_1, \ldots, x_{n-1})$.

The applicability of this approach depends, of course, on knowledge of the conditional distributions. In certain models, such as Markov models, this knowledge is easily obtainable.

2.5.1 Vector Acceptance–Rejection Method

The acceptance–rejection Algorithm 2.3.6 is directly applicable to the multidimensional case. We need only to bear in mind that the random variable X (see Line 3 of Algorithm 2.3.6) becomes an n-dimensional random vector $\mathbf{X}$. Consequently, we need a convenient way of generating $\mathbf{X}$ from the multidimensional proposal pdf $g(\mathbf{x})$, for example, by using the vector inverse-transform method. The next example demonstrates the vector version of the acceptance–rejection method.

■ **EXAMPLE 2.10**

We want to generate a random vector $\mathbf{Z}$ that is uniformly distributed over an irregular n-dimensional region G (see Figure 2.7). The algorithm is straightforward:

1. *Generate a random vector, $\mathbf{X}$, uniformly distributed in W, where W is a regular region (multidimensional hypercube, hyperrectangle, hypersphere, hyperellipsoid, etc.).*

2. *If $\mathbf{X} \in G$, accept $\mathbf{Z} = \mathbf{X}$ as the random vector uniformly distributed over G; otherwise, return to Step 1.*

As a special case, let G be the n-dimensional unit ball, that is, $G = \{\mathbf{x} : \sum_i x_i^2 \leqslant 1\}$, and let W be the n-dimensional hypercube $\{-1 \leqslant x_i \leqslant 1\}_{i=1}^n$. To generate a random vector that is uniformly distributed over the interior of the n-dimensional unit ball, we generate a random vector $\mathbf{X}$ that is uniformly distributed over W and then accept or reject it, depending on whether it falls inside or outside the n-dimensional ball. The corresponding algorithm is as follows:

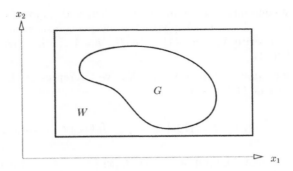

Figure 2.7: The vector acceptance–rejection method.

Algorithm 2.5.1: Generation of a Random Vector Uniformly Distributed Within the n-Dimensional Unit Ball

---

   output: Random vector $\mathbf{X}$ in the unit ball.

1   $R \leftarrow \infty$
2   **while** $R > 1$ **do**
3   Generate $U_1, \ldots, U_n$ as iid random variables from $\mathsf{U}(0,1)$.
4      $X_1 \leftarrow 1 - 2U_1, \ldots, X_n \leftarrow 1 - 2U_n$
5      $R \leftarrow \sum_{i=1}^{n} X_i^2$
6   $\mathbf{X} \leftarrow (X_1, \ldots, X_n)$
7   **return** $\mathbf{X}$

Remark 2.5.1 To generate a random vector that is uniformly distributed over the *surface* of an n-dimensional unit ball — in other words, uniformly over the unit sphere $\{\mathbf{x} : \sum_i x_i^2 = 1\}$ — we need only to scale the vector $\mathbf{X}$ such that it has unit length. That is, we return $\mathbf{Z} = \mathbf{X}/\sqrt{R}$ instead of $\mathbf{X}$.

The efficiency of the vector acceptance–rejection method is equal to the ratio

$$\frac{1}{C} = \frac{\text{volume of the hyperball}}{\text{volume of the hypercube}} = \frac{1}{n \, 2^{n-1}} \frac{\pi^{n/2}}{\Gamma(n/2)} \,,$$

where the volumes of the ball and cube are $\frac{\pi^{n/2}}{(n/2)\Gamma(n/2)}$ and 2^n, respectively. Note that for even n $(n = 2m)$ we have

$$\frac{1}{C} = \frac{\pi^m}{m! \, 2^{2m}} = \frac{1}{m!} \left(\frac{\pi}{2} \right)^m 2^{-m} \to 0 \quad \text{as} \quad m \to \infty.$$

In other words, the acceptance–rejection method grows inefficient in n, and is asymptotically useless.

2.5.2   Generating Variables from a Multinormal Distribution

The key to generating a multivariate normal (or simply multinormal) random vector $\mathbf{Z} \sim \mathsf{N}(\boldsymbol{\mu}, \Sigma)$ is to write it as $\mathbf{Z} = \boldsymbol{\mu} + B\mathbf{X}$, where B a matrix such that $BB^\top = \Sigma$, and $\mathbf{X}$ is a vector of iid $\mathsf{N}(0,1)$ random variables; see Section 1.10. Note that $\boldsymbol{\mu} = (\mu_1, \ldots, \mu_n)$ is the mean vector and Σ is the $(n \times n)$ covariance matrix of $\mathbf{Z}$. For any covariance matrix Σ, such a matrix B can always be found efficiently using the Cholesky square root method; see Section A.1 of the Appendix.

The next algorithm describes the generation of a $\mathsf{N}(\boldsymbol{\mu}, \Sigma)$ distributed random vector $\mathbf{Z}$:

Algorithm 2.5.2: Generation of Multinormal Vectors

input : Mean vector $\boldsymbol{\mu}$ and covariance matrix Σ.
output: Random vector $\mathbf{Z} \sim \mathsf{N}(\boldsymbol{\mu}, \Sigma)$.
1 Generate $X_1, \ldots, X_n$ as iid variables from $\mathsf{N}(0,1)$.
2 Derive the lower Cholesky decomposition $\Sigma = BB^\top$.
3 Set $\mathbf{Z} \leftarrow \boldsymbol{\mu} + B\mathbf{X}$.
4 **return Z**

2.5.3 Generating Uniform Random Vectors over a Simplex

Consider the n-dimensional simplex,

$$\mathscr{Y} = \left\{ \mathbf{y} : y_i \geqslant 0, \quad i = 1, \ldots, n, \quad \sum_{i=1}^{n} y_i \leqslant 1 \right\} . \qquad (2.32)$$

$\mathscr{Y}$ is a simplex on the points $\mathbf{0}, \mathbf{e}_1, \ldots, \mathbf{e}_n$, where $\mathbf{0}$ is the zero vector and $\mathbf{e}_i$ is the i-th unit vector in $\mathbb{R}^n$, $i = 1, \ldots, n$. Let $\mathscr{X}$ be a second n-dimensional simplex:

$$\mathscr{X} = \{ \mathbf{x} : x_i \geqslant 0, \quad i = 1, \ldots, n, \ x_1 \leqslant x_2 \leqslant \cdots \leqslant x_n \leqslant 1 \} .$$

$\mathscr{X}$ is a simplex on the points $\mathbf{0}, \mathbf{e}_n, \mathbf{e}_n + \mathbf{e}_{n-1}, \ldots, \mathbf{1}$, where $\mathbf{1}$ is the sum of all unit vectors (a vector of 1s). Figure 2.8 illustrates the two-dimensional case.

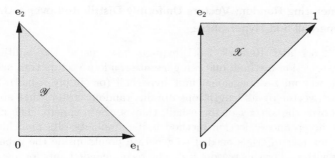

Figure 2.8: Simplexes $\mathscr{Y}$ and $\mathscr{X}$.

Simplex $\mathscr{Y}$ can be obtained from simplex $\mathscr{X}$ by the linear transformation $\mathbf{y} = A\mathbf{x}$ with

$$A = \begin{pmatrix} 1 & 0 & \cdots & 0 \\ -1 & 1 & \cdots & 0 \\ \vdots & \ddots & \ddots & \vdots \\ 0 & \cdots & -1 & 1 \end{pmatrix} .$$

Now, drawing a vector $\mathbf{X} = (X_1, \ldots, X_n)$ according to the uniform distribution on $\mathscr{X}$ is easy: simply take X_i to be the i-th order statistic of iid random variables $U_1, \ldots, U_n$ from $\mathsf{U}(0, 1)$. Since a linear transformation preserves uniformity, applying matrix A to $\mathbf{X}$ yields a vector $\mathbf{Y}$ that is uniformly distributed on $\mathscr{Y}$.

Algorithm 2.5.3: Generating a Vector over a Unit Simplex $\mathscr{Y}$

---

output: Random vector $\mathbf{Y}$ uniformly distributed over the simplex $\mathscr{Y}$.

1 Generate n independent random variables $U_1, \ldots, U_n$ from $\mathsf{U}(0, 1)$.
2 Sort $U_1, \ldots, U_n$ into the order statistics $U_{(1)}, \ldots, U_{(n)}$.
3 $Y_1 \leftarrow U_{(1)}$
4 **for** $i = 2$ **to** n **do** $Y_i = U_{(i)} - U_{(i-1)}$
5 $\mathbf{Y} \leftarrow (Y_1, \ldots, Y_n)$
6 **return** $\mathbf{Y}$

If we define $Y_{n+1} = 1 - \sum_{i=1}^{n} Y_i = 1 - U_{(n)}$, then the resulting $(n+1)$-dimensional vector $(Y_1, \ldots, Y_{n+1})$ will be uniformly distributed over the set

$$\mathscr{F} = \left\{ \mathbf{y} : y_i \geqslant 0, \quad i = 1, \ldots, n+1, \; \sum_{i=1}^{n+1} y_i = 1 \right\} ,$$

that is, over the dominant face of the simplex defined by the points $\mathbf{0}, \mathbf{e}_1, \ldots, \mathbf{e}_{n+1}$.

Finally, in order to generate random vectors uniformly distributed over an n-dimensional simplex defined by arbitrary vertices, say $\mathbf{z}_0, \mathbf{z}_1, \ldots, \mathbf{z}_n$, we simply generate $\mathbf{Y}$ uniformly on $\mathscr{Y}$ and apply the linear transformation

$$\mathbf{Z} = C\mathbf{Y} + \mathbf{z}_0 ,$$

where C is the matrix whose columns are $\mathbf{z}_1 - \mathbf{z}_0, \ldots, \mathbf{z}_n - \mathbf{z}_0$.

2.5.4   Generating Random Vectors Uniformly Distributed over a Unit Hyperball and Hypersphere

Algorithm 2.5.1 and Remark 2.5.1 explain how, using the multidimensional acceptance–rejection method, one can generate random vectors that are uniformly distributed over an n-dimensional unit hyperball (or simply n-ball). By simply dividing each vector by its length, one obtains random vectors that are uniformly distributed over the *surface* of the n-ball, that is, the n-sphere. The main advantage of the acceptance–rejection method is its simplicity. Its main disadvantage is that the number of trials needed to generate points inside the n-ball increases explosively with n. For this reason, it can be recommended only for low dimensions ($n \leqslant 5$). An alternative algorithm is based on the following result:

Theorem 2.5.1 *Let $X_1, \ldots, X_n$ be iid random variables from $\mathsf{N}(0, 1)$, and let $\|\mathbf{X}\| = (\sum_{i=1}^{n} X_i^2)^{\frac{1}{2}}$. Then the vector*

$$\mathbf{Y} = \left(\frac{X_1}{\|\mathbf{X}\|}, \ldots, \frac{X_n}{\|\mathbf{X}\|} \right) \tag{2.33}$$

is distributed uniformly over the n-sphere $\{\mathbf{y} : \|\mathbf{y}\| = 1\}$.

Proof: Note that $\mathbf{Y}$ is simply the projection of $\mathbf{X} = (X_1, \ldots, X_n)$ onto the n-sphere. The fact that $\mathbf{Y}$ is uniformly distributed follows immediately from the fact that the pdf of $\mathbf{X}$ is spherically symmetrical: $f_{\mathbf{X}}(\mathbf{x}) = c\,e^{-\|\mathbf{x}\|^2/2}$. $\square$

To obtain uniform random variables within the n-ball, we simply multiply the vector $\mathbf{Y}$ by $U^{1/n}$, where $U \sim \mathsf{U}(0,1)$. To see this, note that for a random vector $\mathbf{Z} = (Z_1, \ldots, Z_n)$ that is uniformly distributed over the n-ball, the radius $R = \|\mathbf{Z}\|$ satisfies $\mathbb{P}(R \leqslant r) = r^n$. Hence, by the inverse-transform method, we can write $R = U^{1/n}$. This motivates the following alternative:

Algorithm 2.5.4: Generating Uniform Random Vectors inside the n-Ball

---

output: Random vector $\mathbf{Z}$ uniformly distributed within the n-ball.
1 Generate a random vector $\mathbf{X} = (X_1, \ldots, X_n)$ with iid $\mathsf{N}(0,1)$ components.
2 Generate $R = U^{1/n}$, with $U \sim \mathsf{U}(0,1)$.
3 $\mathbf{Z} \leftarrow R\,\mathbf{X}/\|\mathbf{X}\|$
4 **return** $\mathbf{Z}$

2.5.5   Generating Random Vectors Uniformly Distributed inside a Hyperellipsoid

The equation for a hyperellipsoid, centered at the origin, can be written as

$$\mathbf{x}^{\top}\Sigma\mathbf{x} = r^2 \,, \tag{2.34}$$

where Σ is a positive definite and symmetric $(n \times n)$ matrix ($\mathbf{x}$ is interpreted as a column vector). The special case where $\Sigma = I$ (identity matrix) corresponds to a hypersphere of radius r. Since Σ is positive definite and symmetric, there exists a unique lower triangular matrix B such that $\Sigma = BB^{\top}$; see (1.25). We can thus view the set $\mathscr{X} = \{\mathbf{x} : \mathbf{x}^{\top}\Sigma\mathbf{x} \leqslant r^2\}$ as a linear transformation $\mathbf{y} = B^{\top}\mathbf{x}$ of the n-dimensional ball $\mathscr{Y} = \{\mathbf{y} : \mathbf{y}^{\top}\mathbf{y} \leqslant r^2\}$. Since linear transformations preserve uniformity, if the vector $\mathbf{Y}$ is uniformly distributed over the interior of an n-dimensional sphere of radius r, then the vector $\mathbf{X} = (B^{\top})^{-1}\mathbf{Y}$ is uniformly distributed over the interior of a hyperellipsoid (see (2.34)). The corresponding generation algorithm is given below.

Algorithm 2.5.5: Generating Uniform Random Vectors in a Hyperellipsoid

---

input : $\Sigma, r > 0$
output: Random vector $\mathbf{X}$ uniformly distributed within the hyperellipsoid.
1 Generate $\mathbf{Y} = (Y_1, \ldots, Y_n)$ uniformly distributed within the n-ball of radius r.
2 Calculate the (lower Cholesky) matrix B, satisfying $\Sigma = BB^{\top}$.
3 $\mathbf{X} \leftarrow (B^{\top})^{-1}\mathbf{Y}$
4 **return** $\mathbf{X}$

2.6   GENERATING POISSON PROCESSES

This section treats the generation of Poisson processes. Recall from Section 1.12 that there are two different (but equivalent) characterizations of a Poisson process

$\{N_t, t \geqslant 0\}$. In the first (see Definition 1.12.1), the process is interpreted as a counting measure, where N_t counts the number of arrivals in $[0, t]$. The second characterization is that the interarrival times $\{A_i\}$ of $\{N_t, t \geqslant 0\}$ form a *renewal process*, that is, a sequence of iid random variables. In this case, the interarrival times have an $\mathsf{Exp}(\lambda)$ distribution, and we can write $A_i = -\frac{1}{\lambda} \ln U_i$, where the $\{U_i\}$ are iid $\mathsf{U}(0, 1)$ distributed. Using the second characterization, we can generate the arrival times $T_i = A_1 + \cdots + A_i$ during the interval $[0, T]$ as follows:

Algorithm 2.6.1: Generating a Homogeneous Poisson Process

> **input** : Final time T, rate $\lambda > 0$.
> **output:** Number of arrivals N and arrival times $T_1, \ldots, T_N$.
> **1** Set $T_0 \leftarrow 0$ and $N \leftarrow 0$.
> **2 while** $T_N < T$ **do**
> **3** $\quad$ Generate $U \sim \mathsf{U}(0, 1)$.
> **4** $\quad$ $T_{N+1} \leftarrow T_N - \frac{1}{\lambda} \ln U$
> **5** $\quad$ $N \leftarrow N + 1$
> **6 return** $N, T_1, \ldots, T_N$

The first characterization of a Poisson process, that is, as a random counting measure, provides an alternative way of generating such processes, which works also in the multidimensional case. In particular (see the end of Section 1.12), the following procedure can be used to generate a homogeneous Poisson process with rate λ on any set A with "volume" $|A|$:

Algorithm 2.6.2: Generating an n-Dimensional Poisson Process

> **input** : Rate $\lambda > 0$.
> **output:** Number of points N and positions $\mathbf{X}_1, \ldots, \mathbf{X}_N$.
> **1** Generate a Poisson random variable $N \sim \mathsf{Poi}(\lambda\,|A|)$.
> **2** Draw N points $\mathbf{X}_1, \ldots, \mathbf{X}_N$ independently and uniformly in A.
> **3 return** $\mathbf{X}_1, \ldots, \mathbf{X}_N$

A *nonhomogeneous Poisson process* is a counting process $N = \{N_t, t \geqslant 0\}$ for which the number of points in nonoverlapping intervals are independent — similar to the ordinary Poisson process — but the rate at which points arrive is *time dependent*. If $\lambda(t)$ denotes the rate at time t, the number of points in any interval (b, c) has a Poisson distribution with mean $\int_b^c \lambda(t)\,dt$.

Figure 2.9 illustrates a way to construct such processes. We first generate a two-dimensional homogeneous Poisson process on the strip $\{(t, x), t \geqslant 0, 0 \leqslant x \leqslant \lambda\}$, with constant rate $\lambda = \max \lambda(t)$, and then simply project all points below the graph of $\lambda(t)$ onto the t-axis.

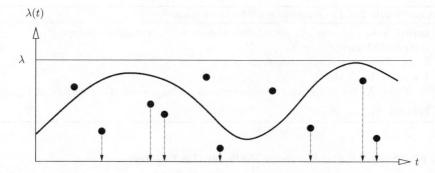

Figure 2.9: Constructing a nonhomogeneous Poisson process.

Note that the points of the two-dimensional Poisson process can be viewed as having a time and space dimension. The arrival epochs form a one-dimensional Poisson process with rate λ, and the positions are uniform on the interval $[0, \lambda]$. This suggests the following alternative procedure for generating nonhomogeneous Poisson processes: each arrival epoch of the one-dimensional homogeneous Poisson process is rejected (thinned) with probability $1 - \frac{\lambda(T_n)}{\lambda}$, where T_n is the arrival time of the n-th event. The surviving epochs define the desired nonhomogeneous Poisson process.

Algorithm 2.6.3: Generating a Nonhomogeneous Poisson Process

    input : Final time T and rate function $\lambda(t)$, with $\max \lambda(t) = \lambda$.
    output: Number of arrivals N and arrival times $T_1, T_2, \ldots, T_N$.
1 $t \leftarrow 0$ and $N \leftarrow 0$
2 **while** $t < T$ **do**
3   Generate $U \sim \mathsf{U}(0, 1)$.
4 $t \leftarrow t - \frac{1}{\lambda} \ln U$
5   Generate $V \sim \mathsf{U}(0, 1)$.
6   **if** $V \leqslant \lambda(t)/\lambda$ **then**
7 $T_{N+1} \leftarrow t$
8 $N \leftarrow N + 1$

9 **return** N and $T_1, \ldots, T_N$

2.7 GENERATING MARKOV CHAINS AND MARKOV JUMP PROCESSES

We now discuss how to simulate a Markov chain $X_0, X_1, X_2, \ldots, X_n$. To generate a Markov chain with initial distribution $\boldsymbol{\pi}^{(0)}$ and transition matrix P, we can use the procedure outlined in Section 2.5 for dependent random variables. That is, first generate X_0 from $\boldsymbol{\pi}^{(0)}$. Then, given $X_0 = x_0$, generate X_1 from the conditional distribution of X_1 given $X_0 = x_0$; in other words, generate X_1 from the x_0-th row of P. Suppose $X_1 = x_1$. Then, generate X_2 from the x_1-st row of P, and so on. The algorithm for a general discrete-state Markov chain with a one-step transition matrix P and an initial distribution vector $\boldsymbol{\pi}^{(0)}$ is as follows:

Algorithm 2.7.1: Generating a Markov Chain

---

input : Sample size N, initial distribution $\boldsymbol{\pi}^{(0)}$, transition matrix P.
output: Markov chain $X_0, \ldots, X_N$.

1 Draw X_0 from the initial distribution $\boldsymbol{\pi}^{(0)}$.
2 **for** $t = 1$ **to** N **do**
3 $\quad$ Draw X_t from the distribution corresponding to the X_{t-1}-th row of P.
4 **return** $X_0, \ldots, X_N$

■ **EXAMPLE 2.11   Random Walk on the Integers**

Consider the random walk on the integers in Example 1.10. Let $X_0 = 0$ (i.e., we start at 0). Suppose the chain is at some discrete time $t = 0, 1, 2 \ldots$ in state i. Then, in Line 3 of Algorithm 2.7.1, we simply need to draw from a two-point distribution with mass p and q at $i + 1$ and $i - 1$, respectively. In other words, we draw $I_t \sim \text{Ber}(p)$ and set $X_{t+1} = X_t + 2I_t - 1$. Figure 2.10 gives a typical sample path for the case where $p = q = 1/2$.

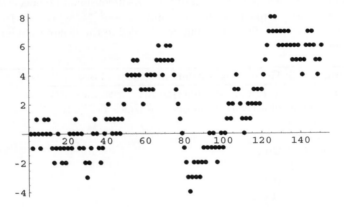

Figure 2.10: Random walk on the integers, with $p = q = 1/2$.

2.7.1   Random Walk on a Graph

As a generalization of Example 2.11, we can associate a random walk with any graph G, whose state space is the vertex set of the graph and whose transition probabilities from i to j are equal to $1/d_i$, where d_i is the degree of i (the number of edges out of i). An important property of such random walks is that they are time-reversible. This can be easily verified from Kolmogorov's criterion (1.39). In other words, there is no systematic "looping". As a consequence, if the graph is connected and if the stationary distribution $\{\pi_i\}$ exists — which is the case when the graph is finite — then the local balance equations hold:

$$\pi_i \, p_{ij} = \pi_j \, p_{ji} \ . \tag{2.35}$$

When $p_{ij} = p_{ji}$ for all i and j, the random walk is said to be *symmetric*. It follows immediately from (2.35) that in this case the equilibrium distribution is uniform over the state space $\mathscr{E}$.

■ EXAMPLE 2.12 Simple Random Walk on an n-Cube

We want to simulate a random walk over the vertices of the n-dimensional hypercube (or simply n-cube); see Figure 2.11 for the three-dimensional case.

Figure 2.11: At each step, one of the three neighbors of the currently visited vertex is chosen at random.

Note that the vertices of the n-cube are of the form $\mathbf{x} = (x_1, \ldots, x_n)$, with x_i either 0 or 1. The set of all 2^n of these vertices is denoted $\{0,1\}^n$. We generate a random walk $\{X_t, t = 0, 1, 2, \ldots\}$ on $\{0,1\}^n$ as follows: Let the initial state X_0 be arbitrary, say $X_0 = (0, \ldots, 0)$. Given $X_t = (x_{t1}, \ldots, x_{tn})$, choose randomly a coordinate J according to the discrete uniform distribution on the set $\{1, \ldots, n\}$. If j is the outcome, then replace x_{jn} with $1 - x_{jn}$. By doing so, we obtain at stage $t + 1$,

$$X_{t+1} = (x_{t1}, \ldots, 1 - x_{tj}, x_{t(j+1)}, \ldots, x_{tn}) \,,$$

and so on.

2.7.2 Generating Markov Jump Processes

The generation of Markov jump processes is quite similar to the generation of Markov chains above. Suppose $X = \{X_t, t \geqslant 0\}$ is a Markov jump process with transition rates $\{q_{ij}\}$. From Section 1.13.5, recall that the Markov jump process jumps from one state to another according to a Markov chain $Y = \{Y_n\}$ (the jump chain), and the time spent in each state i is exponentially distributed with a parameter that may depend on i. The one-step transition matrix, say K, of Y and the parameters $\{q_i\}$ of the exponential holding times can be found directly from the $\{q_{ij}\}$. Namely, $q_i = \sum_j q_{ij}$ (the sum of the transition rates out of i), and $K(i,j) = q_{ij}/q_i$ for $i \neq j$ (thus, the probabilities are simply proportional to the rates). Note that $K(i,i) = 0$. Defining the holding times as $A_1, A_2, \ldots$ and the jump times as $T_1, T_2, \ldots$, we can write the algorithm now as follows:

Algorithm 2.7.2: Generating a Markov Jump Process

---

input : Final time T, initial distribution $\boldsymbol{\pi}^{(0)}$, transition rates $\{q_{ij}\}$.
output: Number of jumps N, jump times $T_1, \ldots, T_N$ and jump states
$\qquad\quad$ $Y_0, \ldots, Y_N$.

1 Draw Y_0 from the initial distribution $\boldsymbol{\pi}^{(0)}$.
2 Initialize $X_0 \leftarrow Y_0$, $T_0 \leftarrow 0$, and $N \leftarrow 0$.
3 **while** $T_N < T$ **do**
4 $\quad$ Draw A_{N+1} from $\mathsf{Exp}(q_{Y_N})$
5 $\quad$ $T_{N+1} \leftarrow T_N + A_{N+1}$
6 $\quad$ Set $X_t \leftarrow Y_N$ for $T_N \leqslant t < T_{N+1}$
7 $\quad$ Draw Y_{N+1} from the distribution corresponding to the Y_N-th row of K.
8 $\quad$ $N \leftarrow N + 1$
9 **return** $N, T_1, \ldots, T_N, Y_0, Y_1, \ldots, Y_N$

2.8 GENERATING GAUSSIAN PROCESSES

Recall from Section 1.14 that, in a Gaussian process $\{X_t, t \in \mathscr{T}\}$, each subvector $(X_{t_1}, \ldots, X_{t_n})^\top$ has a multinormal distribution, $\mathsf{N}(\boldsymbol{\mu}, \Sigma)$, for some expectation vector $\boldsymbol{\mu}$, and covariance matrix Σ, depending on $t_1, \ldots, t_n$. In particular, if the Gaussian process has expectation function μ_t, $t \in \mathscr{T}$ and covariance function $\Sigma_{s,t}, s, t \in \mathscr{T}$, then $\boldsymbol{\mu} = (\mu_{t_1}, \ldots, \mu_{t_n})^\top$ and $\Sigma_{ij} = \Sigma_{t_i,t_j}$, $i, j = 1, \ldots, n$. Consequently, Algorithm 2.5.2 can be used to generate realizations of a Gaussian process at specified times (indexes) $t_1, \ldots, t_n$.

Although it is always possible to find a Cholesky decomposition $\Sigma = BB^\top$ (see Section A.1 of the Appendix), this procedure requires in general $O(n^3)$ floating point operations (for an $n \times n$ matrix). As a result, the generation of high-dimensional Gaussian vectors becomes very time-consuming for large n, unless the process has extra structure.

■ **EXAMPLE 2.13   Wiener Process Generation**

In Example 1.14, the Wiener process was defined as a zero-mean Gaussian process $\{X_t, t \geqslant 0\}$ with covariance function $\Sigma_{s,t} = s$ for $0 \leqslant s \leqslant t$.

It is obviously not possible to generate the complete sample path, as an infinite amount of variables would have to be generated. However, we could use Algorithm 2.5.2 to generate outcomes of the normal random vector $\mathbf{X} = (X_{t_1}, \ldots, X_{t_n})^\top$ for times $t_1 < \cdots < t_n$. The covariance matrix Σ of $\mathbf{X}$ is in this case

$$
\Sigma = \begin{pmatrix}
t_1 & t_1 & t_1 & \cdots & t_1 \\
t_1 & t_2 & t_2 & \cdots & t_2 \\
t_1 & t_2 & t_3 & \cdots & t_3 \\
\vdots & \vdots & \vdots & \ddots & \vdots \\
t_1 & t_2 & t_3 & \cdots & t_n
\end{pmatrix}.
$$

We can easily verify (multiply B with $B^\top$) that the lower Cholesky matrix B is given by

$$B = \begin{pmatrix} \sqrt{t_1} & 0 & 0 & \cdots & 0 \\ \sqrt{t_1} & \sqrt{t_2 - t_1} & 0 & \cdots & 0 \\ \sqrt{t_1} & \sqrt{t_2 - t_1} & \sqrt{t_3 - t_2} & \cdots & 0 \\ \vdots & \vdots & \vdots & \ddots & \vdots \\ \sqrt{t_1} & \sqrt{t_2 - t_1} & \sqrt{t_3 - t_2} & \cdots & \sqrt{t_n - t_{n-1}} \end{pmatrix}.$$

This brings us to the following simulation algorithm:

Algorithm 2.8.1: Generating a Wiener Process

input : Times $0 = t_0 < t_1 < t_2 < \cdots < t_n$.
output: Wiener process random variables at times $t_0, \ldots, t_n$.
1 $X_0 \leftarrow 0$
2 **for** $k = 1$ **to** n **do**
3 $\quad$ Draw $Z \sim \mathsf{N}(0,1)$
4 $\quad$ $X_{t_k} \leftarrow X_{t_{k-1}} + \sqrt{t_k - t_{k-1}}\, Z$
5 **return** $X_{t_0}, \ldots, X_{t_n}$

Figure 1.8 on Page 28 gives a typical realization of a Wiener process on the interval $[0, 1]$, evaluated at grid points $t_n = n\, 10^{-5}, n = 0, 1, \ldots, 10^5$.

By utilizing the properties of the Wiener process (see Example 1.14), we can justify Algorithm 2.8.1 in a more direct way. Specifically, X_{t_1} has a $\mathsf{N}(0, t_1)$ distribution and can therefore be generated as $\sqrt{t_1} Z_1$, where Z_1 has a standard normal distribution. Moreover, for each $k = 2, 3, \ldots, n$, the random variable X_{t_k} is equal to $X_{t_{k-1}} + U_t$, where the increment U_t is independent of $X_{t_{k-1}}$ and has a $\mathsf{N}(t_k - t_{k-1})$ distribution, and can be generated as $\sqrt{t_k - t_{k-1}} Z_k$.

2.9 GENERATING DIFFUSION PROCESSES

The Wiener process, denoted in this section by $\{W_t, t \geqslant 0\}$, plays an important role in probability and forms the basis of so-called *diffusion processes*, denoted here by $\{X_t, t \geqslant 0\}$. These are Markov processes with a continuous time parameter and with continuous sample paths, like the Wiener process.

A diffusion process is often specified as the solution of a *stochastic differential equation* (SDE), which is an expression of the form

$$dX_t = a(X_t, t)\, dt + b(X_t, t)\, dW_t, \tag{2.36}$$

where $\{W_t, t \geqslant 0\}$ is a Wiener process and $a(x, t)$ and $b(x, t)$ are deterministic functions. The coefficient (function) a is called the *drift*, and b^2 is called the *diffusion* coefficient. When a and b are constants, say $a(x, t) = \mu$ and $b(x, t) = \sigma$, the resulting diffusion process is of the form

$$X_t = \mu\, t + \sigma W_t$$

and is called a *Brownian motion* with drift μ and diffusion coefficient σ^2.

A simple technique for approximately simulating diffusion processes is *Euler's method*; see, for example, [4]. The idea is to replace the SDE with the stochastic difference equation

$$Y_{k+1} = Y_k + a(Y_k, kh)\, h + b(Y_k, kh)\sqrt{h}\, Z_{k+1}\,, \quad k = 0, 1, 2, \ldots\,, \tag{2.37}$$

where $Z_1, Z_2, \ldots$ are independent $\mathsf{N}(0, 1)$-distributed random variables. For a small step size h, the process $\{Y_k, k = 0, 1, 2, \ldots\}$ approximates the process $\{X_t, t \geqslant 0\}$ in the sense that $Y_k \approx X_{kh}$, $k = 0, 1, 2, \ldots$.

Algorithm 2.9.1: Euler's Method for SDEs

---

input : Step size h, sample size N, functions a and b.
output: SDE diffusion process approximation at times $0, h, 2h, \ldots, Nh$.
1 Generate Y_0 from the distribution of X_0.
2 **for** $k = 1$ **to** N **do**
3 $\quad$ Draw $Z \sim \mathsf{N}(0, 1)$
4 $\quad Y_{k+1} \leftarrow Y_k + a(Y_k, kh)\, h + b(Y_k, kh)\sqrt{h}\, Z$
5 **return** $Y_0, \ldots, Y_N$

■ **EXAMPLE 2.14   Geometric Brownian Motion**

Geometric Brownian motion is often used in financial engineering to model the price of a risky asset; see, for example, [3]. The corresponding SDE is

$$\mathrm{d}X_t = \mu\, X_t\, \mathrm{d}t + \sigma X_t\, \mathrm{d}W_t\,,$$

with initial value X_0. The Euler approximation is

$$Y_{k+1} = Y_k \left(1 + \mu h + \sigma\sqrt{h}\, Z_{k+1}\right), \quad k = 0, 1, 2, \ldots\,,$$

where $Z_1, Z_2, \ldots$ are independent standard normal random variables.

Figure 2.12 shows a typical path of the geometric Brownian motion starting at $X_0 = 1$, with parameters $\mu = 1$ and $\sigma = 0.2$, generated via the Euler approximation with step size $h = 10^{-4}$. The dashed line is the graph of $t \mapsto \exp[t(\mu - \sigma^2/2)]$ along which the process fluctuates.

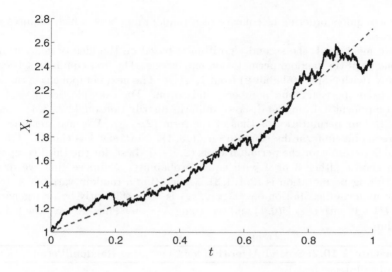

Figure 2.12: Geometric Brownian motion.

More elaborate approximation methods for SDEs can be found, for example, in [4].

2.10 GENERATING RANDOM PERMUTATIONS

Many Monte Carlo algorithms involve generating random permutations, that is, random ordering of the numbers $1, 2, \ldots, n$, for some fixed n. For examples of interesting problems associated with the generation of random permutations, see the traveling salesman problem in Chapter 6, the permanent problem in Chapter 9, and Example 2.15 below.

Suppose that we want to generate each of the $n!$ possible orderings with equal probability. We present two algorithms to achieve this. The first is based on the ordering of a sequence of n uniform random numbers. In the second, we choose the components of the permutation consecutively. The second algorithm is faster than the first.

Algorithm 2.10.1: First Algorithm for Generating Random Permutations

input : Integer $n \geqslant 1$.
output: Random permutation $(X_1, \ldots, X_n)$ of $(1, \ldots, n)$.
1 Generate $U_1, U_2, \ldots, U_n \sim \mathsf{U}(0, 1)$ independently.
2 Arrange these in increasing order.
3 Let $(X_1, \ldots, X_n)$ be the indices of the successive ordered values.
4 **return** $(X_1, \ldots, X_n)$

For example, let $n = 4$ and assume that the generated numbers (U_1, U_2, U_3, U_4) are $(0.7, 0.3, 0.5, 0.4)$. Since $(U_2, U_4, U_3, U_1) = (0.3, 0.4, 0.5, 0.7)$ is the ordered sequence, the resulting permutation is $(2, 4, 3, 1)$. The drawback of this algorithm is that it requires ordering a sequence of n random numbers, which requires $n \ln n$ comparisons.

As we mentioned, the second algorithm is based on the idea of generating the components of the random permutation one by one. The first component is chosen randomly (with equal probability) from $1, \ldots, n$. The next component is randomly chosen from the remaining numbers, and so on. For example, let $n = 4$. We draw component 1 from the discrete uniform distribution on $\{1, 2, 3, 4\}$. Say we obtain 2. Our permutation is thus of the form $(2, \cdot, \cdot, \cdot)$. We next generate from the three-point uniform distribution on $\{1, 3, 4\}$. Now, say 1 is chosen. Thus our intermediate result for the permutation is $(2, 1, \cdot, \cdot)$. Last, for the third component, we can choose either 3 or 4 with equal probability. Suppose that we draw 4. The resulting permutation is $(2, 1, 4, 3)$. Generating a random variable X from a discrete uniform distribution on $\{x_1, \ldots, x_k\}$ is done efficiently by first generating $I = \lfloor k U \rfloor + 1$, with $U \sim \mathsf{U}(0, 1)$ and returning $X = x_I$. Thus we have the following algorithm:

Algorithm 2.10.2: Second Algorithm for Generating Random Permutations

---

   input : Integer $n \geqslant 1$.
   output: Random permutation $(X_1, \ldots, X_n)$ of $(1, \ldots, n)$.
1 Set $\mathscr{P} = \{1, \ldots, n\}$.
2 **for** $i = 1$ **to** n **do**
3 | Generate X_i from the discrete uniform distribution on $\mathscr{P}$.
4 |_ Remove X_i from $\mathscr{P}$.
5 **return** $(X_1, \ldots, X_n)$

Remark 2.10.1 To further raise the efficiency of the second random permutation algorithm, we can implement it as follows: Let $\mathbf{p} = (p_i, \ldots, p_n)$ be a vector that stores the intermediate results of the algorithm at the i-th step. Initially, let $\mathbf{p} = (1, \ldots, n)$. Draw X_1 by uniformly selecting an index $I \in \{1, \ldots, n\}$, and return $X_1 = p_I$. Then, *swap* X_1 and $p_n = n$. In the second step, draw X_2 by uniformly selecting I from $\{1, \ldots, n-1\}$, return $X_2 = p_I$ and swap it with p_{n-1}, and so on. In this way, the algorithm requires the generation of only n uniform random numbers (for drawing from $\{1, 2, \ldots, k\}, k = n, n - 1, \ldots, 2$) and n swap operations.

■ **EXAMPLE 2.15 Generating a Random Tour in a Graph**

Consider a weighted graph G with n nodes, labeled $1, 2, \ldots, n$. The nodes represent cities, and the edges represent the roads between the cities. The problem is to randomly generate a *tour* that visits all the cities exactly once except for the starting city, which is also the terminating city. Without loss of generality, let us assume that the graph is complete, that is, all cities are connected. We can represent each tour via a permutation of the numbers $1, \ldots, n$. For example, for $n = 4$, the permutation $(1, 3, 2, 4)$ represents the tour $1 \to 3 \to 2 \to 4 \to 1$.

More generally, we represent a tour via a permutation $\mathbf{x} = (x_1, \ldots, x_n)$ with $x_1 = 1$, that is, we assume without loss of generality that we start the tour at city number 1. To generate a random tour uniformly on $\mathscr{X}$, we can simply apply Algorithm 2.10.2. Note that the number of all possible tours of elements in the set of all possible tours $\mathscr{X}$ is

$$|\mathscr{X}| = (n-1)! \qquad (2.38)$$

PROBLEMS

2.1   Apply the inverse-transform method to generate a random variable from the discrete uniform distribution with pdf

$$f(x) = \begin{cases} \frac{1}{n+1}, & x = 0, 1, \ldots, n, \\ 0 & \text{otherwise.} \end{cases}$$

2.2   Explain how to generate from the $\mathsf{Beta}(1, \beta)$ distribution using the inverse-transform method.

2.3   Explain how to generate from the $\mathsf{Weib}(\alpha, \lambda)$ distribution using the inverse-transform method.

2.4   Explain how to generate from the $\mathsf{Pareto}(\alpha, \lambda)$ distribution using the inverse-transform method.

2.5   Many families of distributions are of *location-scale* type. That is, the cdf has the form

$$F(x) = F_0\left(\frac{x-\mu}{\sigma}\right),$$

where μ is called the *location* parameter and σ the *scale* parameter, and F_0 is a fixed cdf that does not depend on μ and σ. The $\mathsf{N}(\mu, \sigma^2)$ family of distributions, where F_0 is the standard normal cdf, is a basic example. Write $F(x; \mu, \sigma)$ for $F(x)$. Let $X \sim F_0$ (i.e., $X \sim F(x; 0, 1)$). Prove that $Y = \mu + \sigma X \sim F(x; \mu, \sigma)$. Hence to sample from any cdf in a location-scale family, it suffices to know how to sample from F_0.

2.6   Apply the inverse-transform method to generate random variables from a *Laplace distribution* (i.e., a shifted two-sided exponential distribution) with pdf

$$f(x) = \frac{\lambda}{2}\, e^{-\lambda|x-\theta|}, \quad -\infty < x < \infty \quad (\lambda > 0).$$

2.7   Apply the inverse-transform method to generate a random variable from the *extreme value distribution*, which has cdf

$$F(x) = 1 - e^{-\exp(\frac{x-\mu}{\sigma})}, \quad -\infty < x < \infty \quad (\sigma > 0).$$

2.8   Consider the triangular random variable with pdf

$$f(x) = \begin{cases} 0 & \text{if } x < 2a \text{ or } x \geqslant 2b, \\ \dfrac{x-2a}{(b-a)^2} & \text{if } 2a \leqslant x < a+b, \\ \dfrac{(2b-x)}{(b-a)^2} & \text{if } a+b \leqslant x < 2b. \end{cases}$$

a) Derive the corresponding cdf F.

b) Show that applying the inverse-transform method yields

$$X = \begin{cases} 2a + (b - a)\sqrt{2U} & \text{if } 0 \leqslant U < \frac{1}{2}, \\ 2b + (a - b)\sqrt{2(1 - U)} & \text{if } \frac{1}{2} \leqslant U < 1. \end{cases}$$

2.9   Present an inverse-transform algorithm for generating a random variable from the piecewise-constant pdf

$$f(x) = \begin{cases} C_i, & x_{i-1} \leqslant x \leqslant x_i, \; i = 1, 2, \ldots, n, \\ 0 & \text{otherwise,} \end{cases}$$

where $C_i \geqslant 0$ and $x_0 < x_1 < \cdots < x_{n-1} < x_n$.

2.10   Let

$$f(x) = \begin{cases} C_i\, x, & x_{i-1} \leqslant x < x_i, \; i = 1, \ldots, n, \\ 0 & \text{otherwise,} \end{cases}$$

where $C_i \geqslant 0$ and $x_0 < x_1 < \cdots < x_{n-1} < x_n$.

a) Let $F_i = \sum_{j=1}^{i} \int_{x_{j-1}}^{x_j} C_j\, u \, du$, $i = 1, \ldots, n$. Show that the cdf F satisfies

$$F(x) = F_{i-1} + \frac{C_i}{2}\left(x^2 - x_{i-1}^2\right), \qquad x_{i-1} \leqslant x < x_i, \; i = 1, \ldots, n\,.$$

b) Describe an inverse-transform algorithm for random variable generation from $f(x)$.

2.11   A random variable is said to have a *Cauchy* distribution if its pdf is given by

$$f(x) = \frac{1}{\pi}\frac{1}{1 + x^2}, \qquad x \in \mathbb{R}\,. \tag{2.39}$$

Explain how one can generate Cauchy random variables using the inverse-transform method.

2.12   If X and Y are independent standard normal random variables, then $Z = X/Y$ has a Cauchy distribution. Show this. [Hint: First show that if U and $V > 0$ are continuous random variables with joint pdf $f_{U,V}$, then the pdf of $W = U/V$ is given by $f_W(w) = \int_0^\infty f_{U,V}(w\,v, v)\, v\, dv$.]

2.13   Verify the validity of the composition Algorithm 2.3.4.

2.14   Using the composition method, formulate and implement an algorithm for generating random variables from the following normal (Gaussian) mixture pdf:

$$f(x) = \sum_{i=1}^{3} p_i \frac{1}{b_i}\, \varphi\left(\frac{x - a_i}{b_i}\right),$$

where φ is the pdf of the standard normal distribution and $(p_1, p_2, p_3) = (1/2, 1/3, 1/6)$, $(a_1, a_2, a_3) = (-1, 0, 1)$, and $(b_1, b_2, b_3) = (1/4, 1, 1/2)$.

2.15   Verify that $C = \sqrt{2\mathrm{e}/\pi}$ in Figure 2.5.

2.16   Prove that if $X \sim \mathsf{Gamma}(\alpha, 1)$, then $X/\lambda \sim \mathsf{Gamma}(\alpha, \lambda)$.

2.17   Let $X \sim \mathsf{Gamma}(1 + \alpha, 1)$ and $U \sim \mathsf{U}(0, 1)$ be independent. If $\alpha < 1$, then $XU^{1/a} \sim \mathsf{Gamma}(\alpha, 1)$. Prove this.

2.18   If $Y_1 \sim \mathsf{Gamma}(\alpha, 1)$, $Y_2 \sim \mathsf{Gamma}(\beta, 1)$, and Y_1 and Y_2 are independent, then

$$X = \frac{Y_1}{Y_1 + Y_2}$$

is $\mathsf{Beta}(\alpha, \beta)$ distributed. Prove this.

2.19   Devise an acceptance–rejection algorithm for generating a random variable from the pdf f given in (2.17) using an $\mathsf{Exp}(\lambda)$ proposal distribution. Which λ gives the largest acceptance probability?

2.20   The pdf of the truncated exponential distribution with parameter $\lambda = 1$ is given by

$$f(x) = \frac{e^{-x}}{1 - e^{-a}}, \quad 0 \leqslant x \leqslant a \ .$$

  a) Devise an algorithm for generating random variables from this distribution using the inverse-transform method.
  b) Construct a generation algorithm that uses the acceptance–rejection method with an $\mathsf{Exp}(\lambda)$ proposal distribution.
  c) Find the efficiency of the acceptance–rejection method for the cases $a = 1$, and a approaching zero and infinity.

2.21   Let the random variable X have pdf

$$f(x) = \begin{cases} \frac{1}{4}, & 0 < x < 1, \\ x - \frac{3}{4}, & 1 \leqslant x \leqslant 2. \end{cases}$$

Generate a random variable from $f(x)$, using
  a) the inverse-transform method,
  b) the acceptance–rejection method, using the proposal density

$$g(x) = \frac{1}{2}, \quad 0 \leqslant x \leqslant 2 \ .$$

2.22   Let the random variable X have pdf

$$f(x) = \begin{cases} \frac{1}{2}x, & 0 < x < 1, \\ \frac{1}{2}, & 1 \leqslant x \leqslant \frac{5}{2}. \end{cases}$$

Generate a random variable from $f(x)$, using
  a) the inverse-transform method
  b) the acceptance–rejection method, using the proposal density

$$g(x) = \frac{8}{25}x, \quad 0 \leqslant x \leqslant \frac{5}{2} \ .$$

2.23   Let X have a truncated geometric distribution, with pdf

$$f(x) = c\,p(1 - p)^{x-1}, \quad x = 1, \ldots, n \ ,$$

where c is a normalization constant. Generate a random variable from $f(x)$, using

  a) the inverse-transform method,
  b) the acceptance–rejection method, with $G(p)$ as the proposal distribution. Find the efficiency of the acceptance–rejection method for $n = 2$ and $n = \infty$.

2.24 Generate a random variable $Y = \min_{i=1,\dots,m} \max_{j=1,\dots,r} \{X_{ij}\}$, assuming that the variables X_{ij}, $i = 1,\dots,m$, $j = 1,\dots,r$, are iid with common cdf $F(x)$, using the inverse-transform method. [Hint: Use the results for the distribution of order statistics in Example 2.5.]

2.25 Generate 100 $\mathsf{Ber}(0.2)$ random variables three times and produce bar graphs similar to those in Figure 2.6. Repeat for $\mathsf{Ber}(0.5)$.

2.26 Generate a homogeneous Poisson process with rate 100 on the interval $[0, 1]$. Use this to generate a nonhomogeneous Poisson process on the same interval, with rate function
$$\lambda(t) = 100 \sin^2(10\,t), \quad t \geqslant 0\,.$$

2.27 Generate and plot a realization of the points of a two-dimensional Poisson process with rate $\lambda = 2$ on the square $[0, 5] \times [0, 5]$. How many points fall in the square $[1, 3] \times [1, 3]$? How many do you expect to fall in this square?

2.28 Write a program that generates and displays 100 random vectors that are uniformly distributed within the ellipse
$$5\,x^2 + 21\,x\,y + 25\,y^2 = 9\,.$$

2.29 Implement both random permutation algorithms in Section 2.10. Compare their performance.

2.30 Consider a random walk on the undirected graph in Figure 2.13. For example, if the random walk at some time is in state 5, it will jump to 3, 4, or 6 at the next transition, each with probability $1/3$.

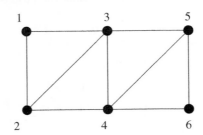

Figure 2.13: A graph.

  a) Find the one-step transition matrix for this Markov chain.
  b) Show that the stationary distribution is given by $\boldsymbol{\pi} = (\frac{1}{9}, \frac{1}{6}, \frac{2}{9}, \frac{2}{9}, \frac{1}{6}, \frac{1}{9})$.
  c) Simulate the random walk on a computer and verify that in the long run, the proportion of visits to the various nodes is in accordance with the stationary distribution.

2.31   Generate various sample paths for the random walk on the integers for $p = 1/2$ and $p = 2/3$.

2.32   Consider the $M/M/1$ queueing system of Example 1.13. Let X_t be the number of customers in the system at time t. Write a computer program to simulate the stochastic process $X = \{X_t\}$ by viewing X as a Markov jump process, and applying Algorithm 2.7.2. Present sample paths of the process for the cases $\lambda = 1$, $\mu = 2$ and $\lambda = 10$, $\mu = 11$.

Further Reading

Classical references on random number generation and random variable generation are [5] and [2]. Other references include [8], [14], and [18] and the tutorial in [17]. A good reference is [1]. The simulation of spatial processes is discussed in [6].

REFERENCES

1. S. Asmussen and P. W. Glynn. *Stochastic Simulation*. Springer-Verlag, New York, 2007.

2. L. Devroye. *Non-Uniform Random Variate Generation*. Springer-Verlag, New York, 1986.

3. P. Glasserman. *Monte Carlo Methods in Financial Engineering*. Springer-Verlag, New York, 2004.

4. P. E. Kloeden and E. Platen. *Numerical Solution of Stochastic Differential Equations*. Springer-Verlag, New York, 1992. corrected third printing.

5. D. E. Knuth. *The Art of Computer Programming*, volume 2: *Seminumerical Algorithms*. Addison-Wesley, Reading, MA, 2nd edition, 1981.

6. D. P. Kroese and Z. I. Botev. Spatial process simulation. In V. Schmidt, editor, *Lectures on Stochastic Geometry, Spatial Statistics and Random Fields*, volume II: Analysis, Modeling and Simulation of Complex Structures. Springer, Berlin, 2014.

7. D. P. Kroese, T. Taimre, and Z. I. Botev. *Handbook of Monte Carlo Methods*. John Wiley & Sons, 2011.

8. A. M. Law and W. D. Kelton. *Simulation Modeling and Analysis*. McGraw-Hill, New York, 3rd edition, 2000.

9. P. L'Ecuyer. Good parameters and implementations for combined multiple recursive random number generators. *Operations Research*, 47(1):159 – 164, 1999.

10. P. L'Ecuyer and F. Panneton. $\mathbb{F}_2$-linear random number generators. In C. Alexopoulos, D. Goldsman, and J. R. Wilson, editors, *Advancing the Frontiers of Simulation: A Festschrift in Honor of George Samuel Fishman*, pages 175–200, New York, 2009. Springer-Verlag.

11. P. L'Ecuyer and R. Simard. TestU01: A C library for empirical testing of random number generators. *ACM Transactions on Mathematical Software*, 33(4), 2007. Article 22.

12. D. H. Lehmer. Mathematical methods in large-scale computing units. *Annals of the Computation Laboratory of Harvard University*, 26:141–146, 1951.

13. P. A. Lewis, A. S. Goodman, and J. M. Miller. A pseudo-random number generator for the system/360. *IBM Systems Journal*, 8(2):136–146, 1969.

14. N. N. Madras. *Lectures on Monte Carlo Methods*. American Mathematical Society, 2002.

15. G. Marsaglia and W. Tsang. A simple method for generating gamma variables. *ACM Transactions on Mathematical Software*, 26(3):363–372, 2000.

16. M. Matsumoto and T. Nishimura. Mersenne twister: A 623-dimensionally equidistributed uniform pseudo-random number generator. *ACM Transactions on Modeling and Computer Simulation*, 8(1):3–30, 1998.

17. B. D. Ripley. Computer generation of random variables: A tutorial. *International Statistical Review*, 51:301–319, 1983.

18. S. M. Ross. *Simulation*. Academic Press, New York, 3rd edition, 2002.

19. A. J. Walker. An efficient method for generating discrete random variables with general distributions. *ACM Transactions on Mathematical Software*, 3:253–256, 1977.

CHAPTER 3

SIMULATION OF DISCRETE-EVENT SYSTEMS

3.1 INTRODUCTION

Computer simulation has long served as an important tool in a wide variety of disciplines: engineering, operations research and management science, statistics, mathematics, physics, economics, biology, medicine, engineering, chemistry, and the social sciences. Through computer simulation, one can study the behavior of real-life systems that are too difficult to examine analytically. Examples can be found in supersonic jet flight, telephone communications systems, wind tunnel testing, large-scale battle management (e.g., to evaluate defensive or offensive weapons systems), or maintenance operations (e.g., to determine the optimal size of repair crews), to mention a few. Recent advances in simulation methodologies, software availability, sensitivity analysis, and stochastic optimization have combined to make simulation one of the most widely accepted and used tools in system analysis and operations research. The sustained growth in size and complexity of emerging real-world systems (e.g., high-speed communication networks and biological systems) will undoubtedly ensure that the popularity of computer simulation continues to grow.

The aim of this chapter is to provide a brief introduction to the art and science of computer simulation, in particular with regard to discrete-event systems. The chapter is organized as follows: Section 3.2 describes basic concepts such as systems, models, simulation, and Monte Carlo methods. Section 3.3 deals with the most fundamental ingredients of discrete-event simulation, namely, the simulation clock and the event list. Section 3.4 explains the ideas behind discrete-event simulation via a number of worked examples.

Simulation and the Monte Carlo Method, Third Edition. By R. Y. Rubinstein and D. P. Kroese

3.2   SIMULATION MODELS

By a *system* we mean a collection of related entities, sometimes called *components* or *elements*, forming a complex whole. For instance, a hospital may be considered a system, with doctors, nurses, and patients as elements. The elements possess certain characteristics or *attributes* that take on logical or numerical values. In our example, an attribute may be the number of beds, the number of X-ray machines, skill level, and so on. Typically, the activities of individual components interact over time. These activities cause changes in the system's state. For example, the state of a hospital's waiting room might be described by the number of patients waiting for a doctor. When a patient arrives at the hospital or leaves it, the system jumps to a new state.

We will be solely concerned with *discrete-event systems*, to wit, those systems in which the state variables change instantaneously through jumps at discrete points in time, as opposed to *continuous systems*, where the state variables change continuously with respect to time. Examples of discrete and continuous systems are, respectively, a bank serving customers and a car moving on the freeway. In the former case, the number of waiting customers is a piecewise constant state variable that changes only when either a new customer arrives at the bank or a customer finishes transacting his business and departs from the bank; in the latter case, the car's velocity is a state variable that can change continuously over time.

The first step in studying a system is to build a model from which to obtain predictions concerning the system's behavior. By a *model* we mean an abstraction of some real system that can be used to obtain predictions and formulate control strategies. Often such models are mathematical (formulas, relations) or graphical in nature. Thus, the actual physical system is translated — through the model — into a mathematical system. In order to be useful, a model must necessarily incorporate elements of two conflicting characteristics: realism and simplicity. On the one hand, the model should provide a reasonably close approximation to the real system and incorporate most of the important aspects of the real system. On the other hand, the model must not be so complex as to preclude its understanding and manipulation.

There are several ways to assess the validity of a model. Usually, we begin testing a model by reexamining the formulation of the problem and uncovering possible flaws. Another check on the validity of a model is to ascertain that all mathematical expressions are dimensionally consistent. A third useful test consists of varying input parameters and checking that the output from the model behaves in a plausible manner. The fourth test is the so-called *retrospective* test. It involves using historical data to reconstruct the past and then determining how well the resulting solution would have performed if it had been used. Comparing the effectiveness of this hypothetical performance with what actually happens then indicates how well the model predicts reality. However, a disadvantage of retrospective testing is that it uses the same data as the model. Unless the past is a representative replica of the future, it is better not to resort to this test at all.

Once a model for the system at hand has been constructed, the next step is to derive a solution from this model. To this end, both *analytical* and *numerical* solutions methods may be invoked. An analytical solution is usually obtained directly from its mathematical representation in the form of formulas. A numerical solution is generally an approximation via a suitable approximation procedure.

Much of this book deals with numerical solution and estimation methods obtained via computer simulation. More precisely, we use *stochastic computer simulation* — often called *Monte Carlo simulation* — which includes some randomness in the underlying model, rather than deterministic computer simulation. The term *Monte Carlo* was used by von Neumann and Ulam during World War II as a code word for secret work at Los Alamos on problems related to the atomic bomb. That work involved simulation of random neutron diffusion in nuclear materials.

Naylor et al. [7] define *simulation* as follows:

> *Simulation is a numerical technique for conducting experiments on a digital computer, which involves certain types of mathematical and logical models that describe the behavior of business or economic systems (or some component thereof) over extended period of real time.*

The following list of typical situations should give the reader some idea of where simulation would be an appropriate tool.

- The system may be so complex that a formulation in terms of a simple mathematical equation may be impossible. Most economic systems fall into this category. For example, it is often virtually impossible to describe the operation of a business firm, an industry, or an economy in terms of a few simple equations. Another class of problems that leads to similar difficulties is that of large-scale, complex queueing systems. Simulation has been an extremely effective tool for dealing with problems of this type.

- Even if a mathematical model can be formulated that captures the behavior of some system of interest, it may not be possible to obtain a solution to the problem embodied in the model by straightforward analytical techniques. Again, economic systems and complex queueing systems exemplify this type of difficulty.

- Simulation may be used as a pedagogical device for teaching both students and practitioners basic skills in systems analysis, statistical analysis, and decision making. Among the disciplines in which simulation has been used successfully for this purpose are business administration, economics, medicine, and law.

- The formal exercise of designing a computer simulation model may be more valuable than the actual simulation itself. The knowledge obtained in designing a simulation study serves to crystallize the analyst's thinking and often suggests changes in the system being simulated. The effects of these changes can then be tested via simulation before implementing them in the real system.

- Simulation can yield valuable insights into the problem of identifying which variables are important and which have negligible effects on the system, and can shed light on how these variables interact; see Chapter 7.

- Simulation can be used to experiment with new scenarios so as to gain insight into system behavior under new circumstances.

- Simulation provides an *in silico* lab, allowing the analyst to discover better control of the system under study.

- Simulation makes it possible to study dynamic systems in either real, compressed, or expanded time horizons.

- Introducing randomness in a system can actually help solve many optimization and counting problems; see Chapters 6 – 10.

As a modeling methodology, simulation is by no means ideal. Some of its shortcomings and various caveats are: Simulation provides *statistical estimates* rather than *exact* characteristics and performance measures of the model. Thus, simulation results are subject to uncertainty and contain experimental errors. Moreover, simulation modeling is typically time-consuming and consequently expensive in terms of analyst time. Finally, simulation results, no matter how precise, accurate, and impressive, provide consistently useful information about the actual system *only* if the model is a valid representation of the system under study.

3.2.1 Classification of Simulation Models

Computer simulation models can be classified in several ways:

1. *Static versus Dynamic Models.* Static models are those that do not evolve over time and therefore do not represent the passage of time. In contrast, dynamic models represent systems that evolve over time (for example, traffic light operation).

2. *Deterministic versus Stochastic Models.* If a simulation model contains *only* deterministic (i.e., nonrandom) components, it is called *deterministic*. In a deterministic model, all mathematical and logical relationships between elements (variables) are fixed in advance and not subject to uncertainty. A typical example is a complicated and analytically unsolvable system of standard differential equations describing, say, a chemical reaction. In contrast, a model with at least one random input variable is called a *stochastic* model. Most queueing and inventory systems are modeled stochastically.

3. *Continuous versus Discrete Simulation Models.* In discrete simulation models the state variable changes instantaneously at discrete points in time, whereas in continuous simulation models the state changes continuously over time. A mathematical model aiming to calculate a numerical solution for a system of differential equations is an example of continuous simulation, while queueing models are examples of discrete simulation.

This chapter deals with discrete simulation and in particular with *discrete-event simulation* (DES) models. The associated systems are driven by the occurrence of discrete events, and their state typically changes over time. We shall further distinguish between so-called *discrete-event static systems* (DESS) and *discrete-event dynamic systems* (DEDS). The fundamental difference between DESS and DEDS is that the former do not evolve over time, whereas the latter do. A queueing network is a typical example of a DEDS. A DESS usually involves evaluating (estimating) complex multidimensional integrals or sums via Monte Carlo simulation.

Remark 3.2.1 (Parallel Computing) Recent advances in computer technology have enabled the use of *parallel* or *distributed* simulation, where discrete-event simulation is carried out on multiple linked (networked) computers, operating simultaneously in a cooperative manner. Such an environment allows simultaneous distribution of different computing tasks among the individual processors, thus reducing the overall simulation time.

3.3   SIMULATION CLOCK AND EVENT LIST FOR DEDS

Recall that DEDS evolve over time. In particular, these systems change their state only at a countable number of time points. State changes are triggered by the execution of simulation events occurring at the corresponding time points. Here, an *event* is a collection of attributes (values, types, flags, etc.), chief among which are the *event occurrence time* — or simply *event time* — the *event type*, and an associated algorithm to execute state changes.

Because of their dynamic nature, DEDS require a time-keeping mechanism to advance the simulation time from one event to another as the simulation evolves over time. The mechanism recording the current simulation time is called the *simulation clock*. To keep track of events, the simulation maintains a list of all pending events. This list is called the *event list*, and its task is to maintain all pending events in *chronological* order. That is, events are ordered by their time of occurrence. In particular, the most imminent event is always located at the head of the event list.

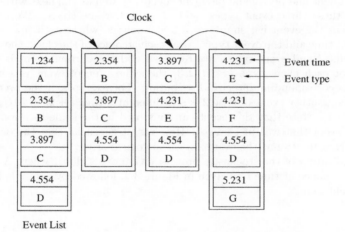

Figure 3.1: The advancement of the simulation clock and event list.

The situation is illustrated in Figure 3.1. The simulation starts by loading the initial events into the event list (chronologically ordered), in this case four events. Next, the most imminent event is unloaded from the event list for execution, and the simulation clock is advanced to its occurrence time, 1.234. After this event is processed and removed, the clock is advanced to the next event, which occurs at time 2.354. In the course of executing a current event, based on its type, the state of the system is updated, and future events are possibly generated and loaded into (or deleted from) the event list. In the example above, the third event — of type C, occurring at time 3.897 — schedules a new event of type E at time 4.231.

The process of unloading events from the event list, advancing the simulation clock, and executing the next most imminent event terminates when some specific stopping condition is met — say, as soon as a prescribed number of customers departs from the system. The following example illustrates this *next-event time advance* approach.

■ EXAMPLE 3.1

Money enters a certain bank account in two ways: via frequent small payments and occasional large payments. Suppose that the times between subsequent frequent payments are independent and uniformly distributed on the continuous interval [7, 10] (in days); and, similarly, the times between subsequent occasional payments are independent and uniformly distributed on [25, 35]. Each frequent payment is exponentially distributed with a mean of 16 units (e.g., one unit is $1000), whereas occasional payments are always of size 100. It is assumed that all payment intervals and sizes are independent. Money is debited from the account at times that form a Poisson process with rate 1 (per day), and the amount debited is normally distributed with mean 5 and standard deviation 1. Suppose that the initial amount of money in the bank account is 150 units.

Note that the state of the system — the account balance — changes only at discrete times. To simulate this DEDS, one need only keep track of when the next frequent and occasional payments occur, as well as the next withdrawal. Denote these three event types by 1, 2, and 3, respectively. We can now implement the event list simply as a 3×2 matrix, where each row contains the event time and the event type. After each advance of the clock, the current event time t and event type i are recorded and the current event is erased. Next, for each event type $i = 1, 2, 3$, the same type of event is scheduled using its corresponding interval distribution. For example, if the event type is 2, then another event of type 2 is scheduled at a time $t + 25 + 10\,U$, where $U \sim \mathsf{U}[0, 1]$. Note that this event can be stored in the same location as the current event that was just erased. However, it is crucial that the event list is then resorted to put the events in chronological order.

A realization of the stochastic process $\{X_t, 0 \leqslant t \leqslant 400\}$, where X_t is the account balance at time t, is given in Figure 3.2, followed by a simple Matlab implementation.

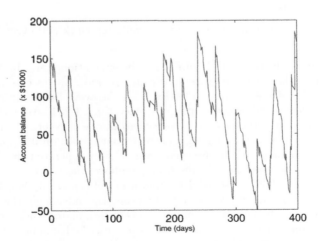

Figure 3.2: A realization of the simulated account balance process.

Matlab Program

```
clear all
T = 400;
x = 150; %initial amount of money.
xx = [150]; tt = [0];
t=0;
ev_list = inf*ones(3,2);            %record time, type
ev_list(1,:) = [7 + 3*rand, 1];     %schedule type 1 event
ev_list(2,:) = [25 + 10*rand,2];    %schedule type 2 event
ev_list(3,:) = [-log(rand),3];      %schedule type 3 event
ev_list = sortrows(ev_list,1);      % sort event list
while t < T
    t = ev_list(1,1);
    ev_type = ev_list(1,2);
    switch ev_type
        case 1
            x = x + 16*-log(rand);
            ev_list(1,:) = [7 + 3*rand + t, 1];
        case 2
            x = x + 100;
            ev_list(1,:) = [25 + 10*rand + t, 2];
        case 3
            x = x - (5 + randn);
            ev_list(1,:) = [-log(rand) + t, 3];
    end
    ev_list = sortrows(ev_list,1); % sort event list
    xx = [xx,x];
    tt = [tt,t];
end
plot(tt,xx)
```

3.4 DISCRETE-EVENT SIMULATION

As mentioned, DES is the standard framework for the simulation of a large class of models in which the system *state* (one or more quantities that describe the condition of the system) needs to be observed only at certain critical epochs (event times). Between these epochs, the system state either stays the same or changes in a predictable fashion. We further explain the ideas behind DES via two more examples.

3.4.1 Tandem Queue

Figure 3.3 depicts a simple queueing system, consisting of two queues in tandem, called a (Jackson) *tandem queue*. Customers arrive at the first queue according to a Poisson process with rate λ. The service time of a customer at the first queue is exponentially distributed with rate μ_1. Customers who leave the first queue enter the second one. The service time in the second queue has an exponential distribution with rate μ_2. All interarrival and service times are independent.

Figure 3.3: A Jackson tandem queue.

Suppose that we are interested in the number of customers, X_t and Y_t, in the first and second queues, respectively, where we regard a customer who is being served as part of the queue. Figure 3.4 depicts a typical realization of the queue length processes $\{X_t,\ t \geqslant 0\}$ and $\{Y_t,\ t \geqslant 0\}$ obtained via DES.

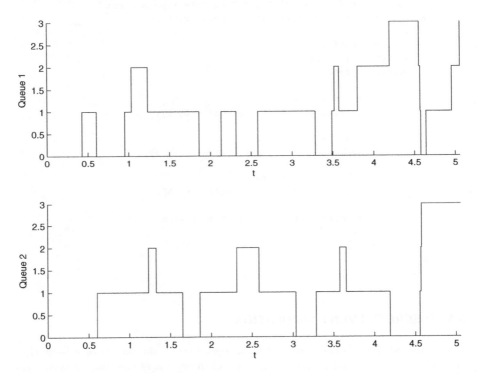

Figure 3.4: A realization of the queue length processes $(X_t, t \geqslant 0)$ and $(Y_t, t \geqslant 0)$.

Before we discuss how to simulate the queue length processes via DES, observe that the system evolves via a sequence of discrete events, as illustrated in Figure 3.5. Specifically, the system state (X_t, Y_t) changes only at times of an arrival at the first queue (indicated by A), a departure from the first queue (indicated by D1), and a departure from the second queue (D2).

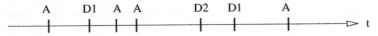

Figure 3.5: Sequence of discrete events (A = arrival, D1 = departure from the first queue, D2 = departure from the second queue).

There are two fundamental approaches to DES, called the *event-oriented* and *process-oriented* approaches. The pseudocode for an event-oriented implementation of the tandem queue is given in Figure 3.6. The program consists of a main subroutine and separate subroutines for each event. In addition, the program maintains an ordered list of scheduled current and future events, the so-called *event list*. Each event in the event list has an event *type* ('A', 'D1', and 'D2') and an event *time* (the time at which the arrival or departure will occur). The role of the main subroutine is primarily to progress through the event list and to call the subroutines that are associated with each event type.

Main

1 **initialize:** $t \leftarrow 0$, $x \leftarrow 0$, $y \leftarrow 0$
2 Schedule 'A' at $t + \mathsf{Exp}(\lambda)$.
3 **while true do**
4   Get the first event in the event list.
5   Let t be the time of this (now current) event.
6   **switch** *current event type* **do**
7   **case** *'A'* : Call **Arrival**
8   **case** *'D1'* : Call **Departure1**
9   **case** *'D2'* : Call **Departure2**
10   Remove the current event from the event list and sort the event list.

Figure 3.6: Main subroutine of an event-oriented simulation program.

The role of the event subroutines is to update the system state and to schedule new events into the event list. For example, an arrival event at time t will trigger another arrival event at time $t + Z$, with $Z \sim \mathsf{Exp}(\lambda)$. We write this, as in the Main routine, in shorthand as $t + \mathsf{Exp}(\lambda)$. Moreover, if the first queue is empty, it will also trigger a departure event from the first queue at time $t + \mathsf{Exp}(\mu_1)$.

Arrival	**Departure1**	**Departure2**
1 Schedule 'A' at $t + \mathsf{Exp}(\lambda)$. 2 **if** $x = 0$ **then** 3 Schedule 'D1' at $t + \mathsf{Exp}(\mu_1)$. 4 $x \leftarrow x + 1$	1 $x \leftarrow x - 1$ 2 **if** $x \neq 0$ **then** 3 Schedule 'D1' at $t + \mathsf{Exp}(\mu_1)$. 4 **if** $y = 0$ **then** 5 Schedule 'D2' at $t + \mathsf{Exp}(\mu_2)$. 6 $y \leftarrow y + 1$	1 $y \leftarrow y - 1$ 2 **if** $y \neq 0$ **then** 3 Schedule 'D2' at $t + \mathsf{Exp}(\mu_2)$.

Figure 3.7: Event subroutines of an event-oriented simulation program.

The process-oriented approach to DES is much more flexible than the event-oriented approach. A process-oriented simulation program closely resembles the actual processes that drive the simulation. Such simulation programs are invariably written in an object-oriented programming language, such as Java or C++. We illustrate the process-oriented approach via our tandem queue example. In contrast to the event-oriented approach, customers, servers, and queues are now actual entities, or *objects* in the program, that can be manipulated. The queues are passive objects that can contain various customers (or be empty), and the customers themselves can contain information such as their arrival and departure times. The servers, however, are active objects (*processes*) that can interact with each other and with the passive objects. For example, the first server takes a client out of the first queue, serves the client, and puts her into the second queue when finished, alerting the second server that a new customer has arrived if necessary. To generate the arrivals, we define a *generator* process that generates a client, puts it in the first queue, alerts the first server if necessary, holds for a random interarrival time (we assume that the interarrival times are iid), and then repeats these actions to generate the next client.

As in the event-oriented approach, there exists an event list that keeps track of the current and pending events. However, this event list now contains *processes*. The process at the top of the event list is the one that is currently active. Processes may ACTIVATE other processes by putting them at the head of the event list. Active processes may HOLD their action for a certain amount of time (such processes are put further up in the event list). Processes may PASSIVATE altogether (temporarily remove themselves from the event list). Figure 3.8 lists the typical structure of a process-oriented simulation program for the tandem queue.

Main

1 **initialize:** create the two queues, the two servers and the generator.
2 ACTIVATE the generator.
3 HOLD(duration of simulation).
4 STOP

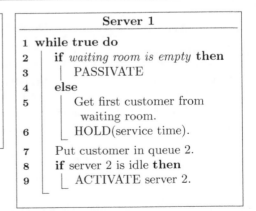

Generator

1 **while true do**
2 Generate new client.
3 Put client in the first queue.
4 **if** server 1 is idle **then**
5 ACTIVATE server 1.
6 HOLD(interarrival time).

Server 1

1 **while true do**
2 **if** *waiting room is empty* **then**
3   PASSIVATE
4 **else**
5   Get first customer from
     waiting room.
6   HOLD(service time).
7 Put customer in queue 2.
8 **if** server 2 is idle **then**
9   ACTIVATE server 2.

Figure 3.8: The structure of a process-oriented simulation program for the tandem queue. The Server 2 process is similar to the Server 1 process, with lines 7–9 replaced with "remove customer from system".

The collection of statistics (for example, the waiting times or queue lengths), can be done by different objects and at various stages in the simulation. For example, customers can record their arrival and departure times and report or record them just before they leave the system. There are many freely available object-oriented simulation environments nowadays, such as SSJ, SimPy, and C++Sim, all inspired by the pioneering simulation language SIMULA.

3.4.2 Repairman Problem

Imagine n machines working simultaneously. The machines are unreliable and fail from time to time. There are $m < n$ identical repairmen who can each work only on one machine at a time. When a machine has been repaired, it is as good as new. Each machine has a fixed lifetime distribution and repair time distribution. We assume that the lifetimes and repair times are independent of each other. Since there are fewer repairmen than machines, it can happen that a machine fails and all repairmen are busy repairing other failed machines. In that case, the failed machine is placed in a queue to be served by the next available repairman. When upon completion of a repair job a repairman finds the failed machine queue empty, he enters the repair pool and remains idle until his service is required again. We assume that machines and repairmen enter their respective queues in a first-in-first-out (FIFO) manner. The system is illustrated in Figure 3.9 for the case of three repairmen and five machines.

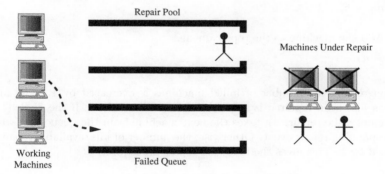

Figure 3.9: The repairman system.

For this particular model the system state could be comprised of the number of available repairmen R_t and the number of failed machines F_t at any time t. In general, the stochastic process $\{(F_t, R_t), t \geqslant 0\}$ is not a Markov process unless the service and lifetimes have exponential distributions.

As with the tandem queue, we first describe an event-oriented and then a process-oriented approach for this model.

3.4.2.1 Event-Oriented Approach There are two types of events: failure events 'F' and repair events 'R'. Each event triggers the execution of the corresponding failure or repair procedure. The task of the main program is to advance the simulation clock and to assign the correct procedure to each event. Denoting by n_f the number of failed machines and by n_r the number of available repairmen, we write the main program in the following form:

MAIN PROGRAM

1 **initialize:** Let $t \leftarrow 0$, $n_r \leftarrow m$ and $n_f \leftarrow 0$. **for** $i = 1$ *to* n. **do**
2 $\quad$ Schedule 'F' of machine i at time $t+$lifetime(i).
3 **while true do**
4 $\quad$ Get the first event in the event list.
5 $\quad$ Let t be the time of this (now current) event.
6 $\quad$ Let i be the machine number associated with this event.
7 $\quad$ **switch** *current event type* **do**
8 $\quad\quad$ **case** *'F'* : Call **Failure**
9 $\quad\quad$ **case** *'R'* : Call **Repair**
10 $\quad$ Remove the current event from the event list.

Upon failure, a repair needs to be scheduled at a time equal to the current time plus the required repair time for the particular machine. However, this is true only if there is a repairman available to carry out the repairs. If this is not the case, the machine is placed in the "failed" queue. The number of failed machines is always increased by 1. The failure procedure is thus as follows:

FAILURE PROCEDURE

1 **if** $n_r > 0$ **then**
2 $\quad$ Schedule 'R' of machine i at time $t +$ repairtime(i).
3 $\quad$ $n_r \leftarrow n_r - 1$
4 **else**
5 $\quad$ Add the machine to the repair queue.
6 $n_f \leftarrow n_f + 1$

Upon repair, the number of failed machines is decreased by 1. The machine that has just been repaired is scheduled for its next failure. If the "failed" queue is not empty, the repairman takes the next machine from the queue and schedules a corresponding repair event. Otherwise, the number of idle/available repairmen is increased by 1. This gives the following repair procedure:

REPAIR PROCEDURE

1 $n_f \leftarrow n_f - 1$
2 Schedule 'F' for machine i at time $t+$lifetime(i).
3 **if** *repair pool not empty* **then**
4 $\quad$ Remove the first machine from the "failed" queue; let j be its number.
5 $\quad$ Schedule 'R' of machine j at time $t+$repairtime(j).
6 **else**
7 $\quad$ $n_r \leftarrow n_r + 1$

3.4.2.2 Process-Oriented Approach To outline a process-oriented approach for any simulation, it is convenient to represent the processes by flowcharts. In this case there are two processes: the repairman process and the machine process. The flowcharts in Figure 3.10 are self-explanatory. Note that the horizontal parallel lines in the flowcharts indicate that the process PASSIVATEs, that is, the process temporarily stops (is removed from the event list), until it is ACTIVATEd by another process. The circled letters A and B indicate how the two interact. A cross in the flowchart indicates that the process is rescheduled in the event list (E.L.). This happens in particular when the process HOLDs for an amount of time. After holding, it resumes from where it left off.

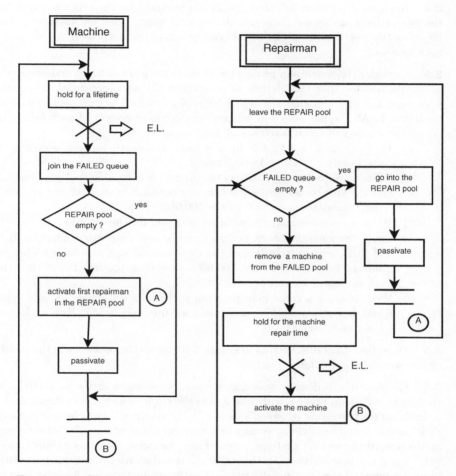

Figure 3.10: Flowcharts for the two processes in the repairman problem.

PROBLEMS

3.1 Consider the $M/M/1$ queueing system in Example 1.13. Let X_t be the number of customers in the system at time t. Write a computer program to simulate the stochastic process $X = \{X_t, t \geq 0\}$ using an event- or process-oriented DES approach. Present sample paths of the process for the cases $\lambda = 1$, $\mu = 2$ and $\lambda = 10$, $\mu = 11$.

3.2   Repeat the above simulation, but now assume $U(0,2)$ interarrival times and $U(0,1/2)$ service times (all independent).

3.3   Run the Matlab program of Example 3.1 (or implement it in the computer language of your choice). Out of 1000 runs, how many lead to a negative account balance during the first 100 days? How does the process behave for large t?

3.4   Implement an event-oriented simulation program for the tandem queue. Let the interarrivals be exponentially distributed with mean 5, and let the service times be uniformly distributed on [3,6]. Plot realizations of the queue length processes of both queues.

3.5   Consider the repairman problem with two identical machines and one repairman. We assume that the lifetime of a machine has an exponential distribution with expectation 5 and that the repair time of a machine is exponential with expectation 1. All the lifetimes and repair times are independent of each other. Let X_t be the number of failed machines at time t.

   a)   Verify that $X = \{X_t, t \geqslant 0\}$ is a birth-and-death process, and give the corresponding birth and death rates.

   b)   Write a program that simulates the process X according to Algorithm 2.7.2 and use this to assess the fraction of time that both machines are out of order. Simulate from $t = 0$ to $t = 100{,}000$.

   c)   Write an event-oriented simulation program for this process.

   d)   Let the exponential life and repair times be uniformly distributed, on [0, 10] and [0, 2], respectively (hence the expectations stay the same as before). Simulate from $t = 0$ to $t = 100{,}000$. How does the fraction of time that both machines are out of order change?

   e)   Now simulate a repairman problem with the above-given life and repair times, but now with five machines and three repairmen. Run again from $t = 0$ to $t = 100{,}000$.

3.6   Draw flow diagrams, such as in Figure 3.10, for all the processes in the tandem queue; see also Figure 3.8.

3.7   Consider the following queueing system. Customers arrive at a circle, according to a Poisson process with rate λ. On the circle, which has circumference 1, a single server travels at constant speed α^{-1}. Upon arrival the customers choose their positions on the circle according to a uniform distribution. The server always moves toward the nearest customer, sometimes clockwise, sometimes counterclockwise. Upon reaching a customer, the server stops and serves the customer according to an exponential service time distribution with parameter μ. When the server is finished, the customer is removed from the circle and the server resumes his journey on the circle. Let $\eta = \lambda \alpha$, and let $X_t \in [0,1]$ be the position of the server at time t. Furthermore, let N_t be the number of customers waiting on the circle at time t. Implement a simulation program for this so-called *continuous poling system with a "greedy" server*, and plot realizations of the processes $\{X_t, t \geqslant 0\}$ and $\{N_t, t \geqslant 0\}$, taking the parameters $\lambda = 1$, $\mu = 2$, for different values of α. Note that although the state space of $\{X_t, t \geqslant 0\}$ is continuous, the system is still a DEDS, since between arrival and service events the system state changes deterministically.

3.8   Consider a *continuous flow line* consisting of three machines in tandem separated by two storage areas, or buffers, through which a continuous (fluid) stream of items flows from one machine to the next; see Figure 3.11.

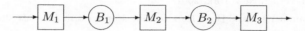

Figure 3.11: A flow line with three machines and two buffers (three-stage flow line).

Each machine $i = 1, 2, 3$ has a specific *machine speed* ν_i, which is the maximum rate at which it can transfer products from its upstream buffer to its downstream buffer. The lifetime of machine i has an exponential distribution with parameter λ_i. The repair of machine i starts immediately after failure and requires an exponential time with parameter μ_i. All life and repair times are assumed to be independent of each other. Failures are operation independent. In particular, the failure rate of a "starved" machine (a machine that is idle because it does not receive input from its upstream buffer) is the same as that of a fully operational machine. The first machine has an unlimited supply.

Suppose all machine speeds are 1, the buffers are of equal size b, and all machines are identical with parameters $\lambda = 1$ and $\mu = 2$.

a)   Implement an event- or process-oriented simulation program for this system.

b)   Assess via simulation the average *throughput* of the system (the long-run amount of fluid that enters/leaves the system per unit of time) as a function of the buffer size b.

Further Reading

One of the first books on Monte Carlo simulation is by Hammersley and Handscomb [3]. Kalos and Whitlock [4] is another classical reference. The event- and process-oriented approaches to discrete-event simulation are elegantly explained in Mitrani [6]. Among the great variety of books on DES, all focusing on different aspects of the modeling and simulation process, we mention [5], [8], [1], and [2]. The choice of computer language in which to implement a simulation program is very subjective. The simple models discussed in this chapter can be implemented in any standard computer language, even Matlab, although the latter does not provide easy event list manipulation. Commercial simulation environments such as ARENA/SIMAN and SIMSCRIPT II.5 make the implementation of larger models much easier. Alternatively, various free SIMULA-like Java packages exist that offer fast implementation of event- and process-oriented simulation programs. Examples are Pierre L'Ecuyer's SSJ http://www.iro.umontreal.ca/~simardr/ssj/, DSOL http://sk-3.tbm.tudelft.nl/simulation/, developed by the Technical University Delft, and the python-based SimPy https://pypi.python.org/pypi/simpy.

REFERENCES

1. J. S. Banks, J. S. Carson II, B. L. Nelson, and D. M. Nicol. *Discrete-Event System Simulation*. Prentice-Hall, Englewood Cliffs, NJ, 4th edition, 2004.

2. G. S. Fishman. *Discrete Event Simulation: Modeling, Programming, and Analysis*. Springer-Verlag, New York, 2001.

3. J. M. Hammersley and D. C. Handscomb. *Monte Carlo Methods*. John Wiley & Sons, New York, 1964.

4. M. H. Kalos and P. A. Whitlock. *Monte Carlo Methods*, volume I: Basics. John Wiley & Sons, New York, 1986.

5. A. M. Law and W. D. Kelton. *Simulation Modeling and Analysis*. McGraw-Hill, New York, 3rd edition, 2000.

6. I. Mitrani. *Simulation Techniques for Discrete Event Systems*. Cambridge University Press, Cambridge, UK, 1982.

7. T. J. Naylor, J. L. Balintfy, D. S. Burdick, and K. Chu. *Computer Simulation Techniques*. John Wiley & Sons, New York, 1966.

8. R. Y. Rubinstein and B. Melamed. *Modern Simulation and Modeling*. John Wiley & Sons, New York, 1998.

CHAPTER 4

STATISTICAL ANALYSIS OF DISCRETE-EVENT SYSTEMS

4.1 INTRODUCTION

An essential part of a simulation study is the statistical analysis of the output data, that is, the data obtained from the simulation model. In this chapter we present several important statistical techniques applied to different types of simulation models. As explained in the previous chapter, simulation models can generally be divided into *static* and *dynamic* models. In both types the behavior of the system is described by the *system state*, which, for all practical purposes, can be thought of as a finite-dimensional random vector $\mathbf{X}$ containing all the information about the system. In static models, the system state does not depend on time. The simulation of such models involves the repeated generation of the system state, and can be implemented using the algorithms in Chapter 2. In dynamic models the system state *does* depend on time, for example, $\mathbf{X}_t$ at time t. The behavior of the system is described by a discrete- or continuous-time stochastic process $\{\mathbf{X}_t\}$.

The rest of this chapter is organized as follows. Section 4.2 gives a brief introduction to point estimation and confidence intervals. Section 4.3 treats the statistical analysis of the output data from static models. Section 4.4 discusses the difference between finite-horizon and steady-state simulation for dynamic models. In Section 4.4.2 we consider steady-state simulation in more detail. Two popular methods for estimating steady-state performance measures — the batch means and regenerative methods — are discussed in Sections 4.4.2.1 and 4.4.2.2, respectively. Finally, in Section 4.5 we present the bootstrap technique.

Simulation and the Monte Carlo Method, Third Edition. By R. Y. Rubinstein and D. P. Kroese

4.2   ESTIMATORS AND CONFIDENCE INTERVALS

Suppose that the objective of a simulation study is to estimate an unknown quantity ℓ based on an estimator $\widehat{\ell}$, which is a function of the data produced by the simulation.

The common situation is when ℓ is the expectation of an output variable Y of the simulation. Suppose repeated runs of the simulation experiment produce independent copies $Y_1, \ldots, Y_N$ of Y. A commonsense estimator of ℓ is then the *sample mean*

$$\widehat{\ell} = \bar{Y} = \frac{1}{N} \sum_{i=1}^{N} Y_i . \tag{4.1}$$

This estimator is *unbiased*, in the sense that $\mathbb{E}[\widehat{\ell}] = \ell$. Moreover, by the law of large numbers $\widehat{\ell}$ converges to ℓ with probability 1 as $N \to \infty$. Notice that an estimator is viewed as a random variable. A particular outcome or observation of an estimator is called an *estimate* (a number), often denoted by the same letter.

In order to specify how *accurate* a particular estimate $\widehat{\ell}$ is, that is, how close it is to the actual unknown parameter ℓ, one needs to provide not only a point estimate $\widehat{\ell}$ but a confidence interval as well. To do so for the sample mean (4.1), observe that by the central limit theorem the estimator $\bar{Y}$ has approximately a $\mathsf{N}(\ell, \sigma^2/N)$ distribution, where σ^2 is the variance of Y — assuming $\sigma^2 < \infty$. Usually, σ^2 is unknown, but it can be estimated with the *sample variance*

$$S^2 = \frac{1}{N-1} \sum_{i=1}^{N} (Y_i - \bar{Y})^2 , \tag{4.2}$$

which (by the law of large numbers) tends to σ^2 as $N \to \infty$. Consequently, for large N, we see that $(\bar{Y} - \ell)\sqrt{N}/S$ is approximately $\mathsf{N}(0,1)$ distributed. Thus, if z_γ denotes the γ-quantile of the $\mathsf{N}(0,1)$ distribution (this is the number such that $\Phi(z_\gamma) = \gamma$, where Φ denotes the standard normal cdf; for example $z_{0.95} = 1.645$, since $\Phi(1.645) = 0.95$), then

$$\mathbb{P}\left(-z_{1-\alpha/2} \leqslant \frac{(\bar{Y} - \ell)\sqrt{N}}{S} \leqslant z_{1-\alpha/2} \right) \approx 1 - \alpha,$$

which after rearranging gives

$$\mathbb{P}\left(\bar{Y} - z_{1-\alpha/2}\frac{S}{\sqrt{N}} \leqslant \ell \leqslant \bar{Y} + z_{1-\alpha/2}\frac{S}{\sqrt{N}} \right) \approx 1 - \alpha .$$

In other words, an approximate $(1 - \alpha)100\%$ *confidence interval* for ℓ is

$$\left(\bar{Y} \pm z_{1-\alpha/2} \frac{S}{\sqrt{N}} \right) , \tag{4.3}$$

where the notation $(a \pm b)$ is shorthand for the interval $(a - b, a + b)$.

Remark 4.2.1 The interpretation of a confidence interval requires some care. It is important to note that (4.3) is a *stochastic* confidence interval that contains ℓ with probability approximately $1-\alpha$. After observing outcomes $y_1, \ldots, y_N$ of the random variables $Y_1, \ldots, Y_N$, we are able to construct a *numerical* confidence interval by replacing the $\{Y_i\}$ with the $\{y_i\}$ in (4.3). However, we can no longer claim that such an interval contains ℓ with probability approximately $1 - \alpha$. This is because ℓ is a *number*, so it either lies in the numerical confidence interval with probability 1 or 0. The interpretation of a 95% numerical confidence interval such as $(1.53, 1.58)$ is thus that it is a particular outcome of a random interval that contains ℓ in 95% of the cases. If we pick at random a ball from an urn with 95 white and 5 black balls *but don't look*, we can be quite confident that the ball in our hand is in fact white. This is how confident we should be that the interval $(1.53, 1.58)$ contains ℓ.

It is common practice in simulation to use and report the *absolute and relative* widths of the confidence interval (4.3), defined as

$$w_a = 2\, z_{1-\alpha/2} \frac{S}{\sqrt{N}} \qquad (4.4)$$

and

$$w_r = \frac{w_a}{\bar{Y}} \,, \qquad (4.5)$$

respectively, provided that $\bar{Y} > 0$. The absolute and relative widths may be used as stopping rules (criteria) to control the length of a simulation run. The relative width is particularly useful when ℓ is very small. For example, if $\ell \approx 10^{-10}$, reporting a result such as $w_a = 0.05$ is almost meaningless, while in contrast, reporting $w_r = 0.05$ is quite meaningful. Another important quantity is the *relative error* (RE) of an estimator $\widehat{\ell}$, defined as

$$\mathrm{RE} = \frac{\sqrt{\mathrm{Var}(\widehat{\ell})}}{\mathbb{E}[\widehat{\ell}]} \,, \qquad (4.6)$$

which, in the case that $\widehat{\ell} = \bar{Y}$, is equal to $\sigma/(\ell\sqrt{N})$. Note that this is equal to w_r divided by $2z_{1-\alpha/2}$ and can be estimated as $S/(\widehat{\ell}\sqrt{N})$.

■ **EXAMPLE 4.1   Estimation of Rare-Event Probabilities**

Consider estimation of the tail probability $\ell = \mathbb{P}(X \geqslant \gamma)$ of some random variable X for a *large* number γ. If ℓ is very small, then the event $\{X \geqslant \gamma\}$ is called a *rare event* and the probability $\mathbb{P}(X \geqslant \gamma)$ is called a *rare-event probability*.

We may attempt to estimate ℓ via (4.1) as

$$\widehat{\ell} = \frac{1}{N} \sum_{i=1}^{N} I_{\{X_i \geqslant \gamma\}} \,, \qquad (4.7)$$

which involves drawing a random sample $X_1, \ldots, X_N$ from the pdf of X and defining the indicators $Y_i = I_{\{X_i \geqslant \gamma\}}$, $i = 1, \ldots, N$. The estimator $\widehat{\ell} = \bar{Y}$ thus defined is called the *crude Monte Carlo* (CMC) estimator. For small ℓ the relative error of the CMC estimator is given by

$$\kappa = \frac{\sqrt{\mathrm{Var}(\widehat{\ell})}}{\mathbb{E}[\widehat{\ell}]} = \sqrt{\frac{1-\ell}{N\ell}} \approx \sqrt{\frac{1}{N\ell}} \,. \qquad (4.8)$$

As a numerical example, suppose that $\ell = 10^{-6}$. In order to estimate ℓ accurately with relative error (say) $\kappa = 0.01$, we need to choose a sample size

$$N \approx \frac{1}{\kappa^2 \ell} = 10^{10} \; .$$

This shows that estimating small probabilities via CMC estimators is computationally meaningless.

4.3   STATIC SIMULATION MODELS

As mentioned in Chapter 3, in a static simulation model the system state does not depend on time. Suppose that we want to determine the expectation

$$\ell = \mathbb{E}[H(\mathbf{X})] = \int H(\mathbf{x}) \, f(\mathbf{x}) \, d\mathbf{x} \; , \tag{4.9}$$

where $\mathbf{X}$ is a random vector with pdf f, and $H(\mathbf{x})$ is a real-valued function called the *performance* function. We assume that ℓ cannot be evaluated analytically and we need to resort to simulation. The situation is exactly as described in Section 4.2, with $Y = H(\mathbf{X})$, and ℓ can be estimated with the sample mean

$$\widehat{\ell} = N^{-1} \sum_{i=1}^{N} H(\mathbf{X}_i) \; , \tag{4.10}$$

where $\mathbf{X}_1, \ldots, \mathbf{X}_N$ is a *random sample* from f; that is, the $\{\mathbf{X}_i\}$ are independent replications of $\mathbf{X} \sim f$.

The following algorithm summarizes how to estimate the expected system performance, $\ell = \mathbb{E}[H(\mathbf{X})]$, and how to calculate the corresponding confidence interval:

Algorithm 4.3.1: Point Estimate and Confidence Interval (Static Model)

---

input : Simulation method for $\mathbf{X} \sim f$, performance function H, sample size N, confidence level $1 - \alpha$.

output: Point estimate and $(1 - \alpha)$-confidence interval for $\ell = \mathbb{E}[H(\mathbf{X})]$.

1 Simulate N replications, $\mathbf{X}_1, \ldots, \mathbf{X}_N$, of $\mathbf{X}$.

2 Let $Y_i \leftarrow H(\mathbf{X}_i)$, $i = 1, \ldots, N$.

3 Calculate the point estimate and a confidence interval of ℓ from (4.1) and (4.3), respectively.

We conclude with two examples where static simulation is used.

■ **EXAMPLE 4.2   Reliability Model**

Consider a system that consists of n components. The operational state of each component $i = 1, \ldots, n$ is represented by $X_i \sim \mathsf{Ber}(p_i)$, where $X_i = 1$ means that the component is working and $X_i = 0$ means that it has failed. Note that the probability that component i is working — its *reliability* — is p_i. The failure behavior of the system is thus represented by the binary random vector $\mathbf{X} = (X_1, \ldots, X_n)$, where it is usually assumed that the $\{X_i\}$ are independent. Suppose that the operational state of the system, say Y, is either functioning or failed, depending on the operational states of the components. In other words, we assume that there exists a function $H : \mathscr{X} \to \{0,1\}$ such that

$$Y = H(\mathbf{X}) \, ,$$

where $\mathscr{X} = \{0,1\}^n$ is the set of all binary vectors of length n.

The function H is called the *structure function* and often can be represented by a graph. In particular, the graph in Figure 4.1 depicts a *bridge network* with five components (links). For this particular model the system works (i.e., $H(\mathbf{X}) = 1$) if the black terminal nodes are connected by working links. The structure function is equal to (see Problem 4.2)

$$H(\mathbf{x}) = 1 - (1 - x_1 \, x_4) \, (1 - x_2 \, x_5) \, (1 - x_1 \, x_3 \, x_5) \, (1 - x_2 \, x_3 \, x_4) \, . \tag{4.11}$$

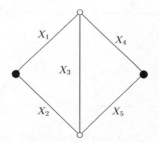

Figure 4.1: A bridge network.

Suppose that we are interested in the reliability ℓ of the general n-component system. We have

$$
\begin{aligned}
\ell = \mathbb{P}(Y = 1) = \mathbb{E}[H(\mathbf{X})] &= \sum_{\mathbf{x} \in \mathscr{X}} H(\mathbf{x}) \, \mathbb{P}(\mathbf{X} = \mathbf{x}) \\
&= \sum_{\mathbf{x} \in \mathscr{X}} H(\mathbf{x}) \prod_{i=1}^{n} \left[p_i^{x_i} \, (1 - p_i)^{1 - x_i} \right] \, .
\end{aligned}
\tag{4.12}
$$

For complex systems with a large number of components and with little structure, it is very time-consuming to compute the system reliability ℓ via (4.12), since this requires the evaluation of $\mathbb{P}(\mathbf{X} = \mathbf{x})$ and $H(\mathbf{x})$ for 2^n vectors $\mathbf{x}$. However, simulation of $\mathbf{X}$ and estimation of ℓ via (4.10) can still be a viable approach, even for large systems, provided that $H(\mathbf{X})$ is readily evaluated. In practice one needs substantially fewer than 2^n samples to estimate ℓ accurately.

■ **EXAMPLE 4.3   Stochastic PERT Network**

The *program evaluation and review technique* (PERT) is a frequently used tool for project management. Typically, a project consists of many activities, some of which can be performed in parallel while others can only be performed after certain preceding activities have been finished. In particular, each activity has a list of *predecessors* that must be completed before it can start. A PERT network is a directed graph where the arcs represent the activities and the vertices represent specific milestones. A milestone is completed when all activities pointing to that milestone are completed. Before an activity can begin, the milestone from which the activity originates must be completed. An example of a precedence list of activities is given in Table 4.3; its PERT graph is given in Figure 4.2.

Table 4.1: Precedence ordering of activities.

Activity	1	2	3	4	5	6	7	8	9	10	11	12
Predecessor(s)	-	-	1	1	2	2	3	3	4,6	5,8	7	9,10

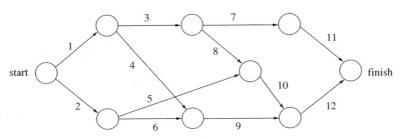

Figure 4.2: A stochastic PERT network.

Suppose that each activity i takes a random time X_i to complete. An important quantity for PERT networks is the maximal project duration, that is, the length of the longest path from start to finish — the so-called *critical path*. Suppose that we are interested in the expected maximal project duration, say ℓ. Letting $\mathbf{X}$ be the vector of activity lengths and $H(\mathbf{X})$ be the length of the critical path, we have

$$\ell = \mathbb{E}[H(\mathbf{X})] = \mathbb{E}\left[\max_{j=1,\dots,p} \sum_{i \in \mathscr{P}_j} X_i\right] , \qquad (4.13)$$

where $\mathscr{P}_j$ is the j-th complete path from start to finish and p is the number of such paths.

4.4   DYNAMIC SIMULATION MODELS

Dynamic simulation models deal with systems that evolve over time. Our goal is (as for static models) to estimate the expected system performance, where the state of the system is now described by a stochastic process $\{\mathbf{X}_t\}$, which may have a continuous or discrete time parameter. For simplicity, we mainly consider the case where $\mathbf{X}_t$ is a scalar random variable; we then write X_t instead of $\mathbf{X}_t$.

We make a distinction between *finite-horizon* and *steady-state* simulation. In finite-horizon simulation, measurements of system performance are defined relative to a specified interval of simulation time $[0, T]$ (where T may be a random variable), while in steady-state simulation, performance measures are defined in terms of certain limiting measures as the time horizon (simulation length) goes to infinity.

The following illustrative example offers further insight into finite-horizon and steady-state simulations. Suppose that the state X_t represents the number of customers in a stable $M/M/1$ queue (see Example 1.13 on Page 27). Let

$$F_{t,m}(x) = \mathbb{P}(X_t \leqslant x \mid X_0 = m) \qquad (4.14)$$

be the cdf of X_t given the initial state $X_0 = m$ (m customers are initially present). $F_{t,m}$ is called the *finite-horizon distribution* of X_t given that $X_0 = m$.

We say that the process $\{X_t\}$ *settles into steady state* (equivalently, that *steady state exists*) if for all m

$$\lim_{t \to \infty} F_{t,m}(x) = F(x) \equiv \mathbb{P}(X \leqslant x) \qquad (4.15)$$

for some random variable X. In other words, *steady state* implies that, as $t \to \infty$, the transient cdf, $F_{t,m}(x)$ (which generally depends on t and m), approaches a steady-state cdf, $F(x)$, which *does not depend* on the initial state, m. The stochastic process, $\{X_t\}$, is said to *converge in distribution* to a random variable $X \sim F$. Such an X can be interpreted as the random state of the system when observed far away in the future. The operational meaning of *steady state* is that after some period of time the transient cdf $F_{t,m}(x)$ comes close to its limiting (steady-state) cdf $F(x)$. It is important to realize that this does *not* mean that at any point in time the realizations of $\{X_t\}$ generated from the simulation run become independent or constant. The situation is illustrated in Figure 4.3, where the dashed curve indicates the expectation of X_t.

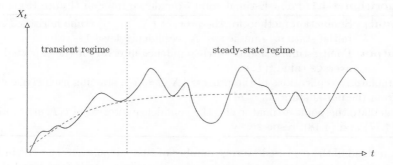

Figure 4.3: The state process for a dynamic simulation model.

The exact distributions (transient and steady-state) are usually available only for simple Markovian models such as the $M/M/1$ queue. For non-Markovian models, usually neither the distributions (transient and steady-state) nor even the associated moments are available via analytical methods. For performance analysis of such models one must resort to simulation.

Note that for some stochastic models, only finite-horizon simulation is feasible, since the steady-state regime either does not exist or the finite-horizon period is so long that the steady-state analysis is computationally prohibitive (e.g., [10]).

4.4.1 Finite-Horizon Simulation

The statistical analysis for finite-horizon simulation models is basically the same as that for static models. To illustrate the procedure, let us suppose that $\{X_t, t \geq 0\}$ is a continuous-time process for which we wish to estimate the expected average value,

$$\ell(T, m) = \mathbb{E}\left[T^{-1} \int_0^T X_t \, dt \right], \tag{4.16}$$

as a function of the time horizon T and the initial state $X_0 = m$. (For a discrete-time process $\{X_t, t = 1, 2, \ldots\}$, the integral $\int_0^T X_t \, dt$ is replaced by the sum $\sum_{t=1}^T X_t$.) For example, if X_t represents the number of customers in a queueing system at time t, then $\ell(T, m)$ is the average number of customers in the system during the time interval $[0, T]$, given $X_0 = m$.

Assume now that N independent replications are performed, each starting at state $X_0 = m$. Then the point estimator and the $(1 - \alpha)$ 100% confidence interval for $\ell(T, m)$ can be written, as in the static case (see (4.10) and (4.3)):

$$\widehat{\ell}(T, m) = N^{-1} \sum_{i=1}^N Y_i \tag{4.17}$$

and

$$\left(\widehat{\ell}(T, m) \pm z_{1-\alpha/2} S N^{-1/2} \right), \tag{4.18}$$

respectively, where $Y_i = T^{-1} \int_0^T X_{ti} \, dt$, X_{ti} is the observation at time t from the i-th replication and S^2 is the sample variance of $\{Y_i\}$. The algorithm for estimating the finite-horizon performance, $\ell(T, m)$, is thus:

Algorithm 4.4.1: Point Estimate and Confidence Interval (Finite Horizon)

---

   input : Simulation method for the process $\{X_t, t \geq 0\}$, time horizon T,
             initial state m, sample size N, confidence level $1 - \alpha$.
  output: Point estimate and $(1 - \alpha)$-confidence interval for the expected
              average value $\ell(T, m)$.
1 Simulate N replications of the process $\{X_t, t \leq T\}$, starting each replication
   from the initial state $X_0 = m$.
2 Calculate the point estimator and the confidence interval of $\ell(T, m)$ from
   (4.17) and (4.18), respectively.

If, instead of the expected average number of customers, we want to estimate the expected *maximum* number of customers in the system during an interval $(0, T]$, the only change required is to replace $Y_i = T^{-1} \int_0^T X_{ti} \, dt$ with $Y_i = \max_{0 \leq t \leq T} X_{ti}$. In the same way, we can estimate other performance measures for this system, such as the probability that the maximum number of customers during $(0, T]$ exceeds some level γ or the expected average period of time that the first k customers spend in the system.

4.4.2 Steady-State Simulation

Steady-state simulation is used for systems that exhibit some form of stationary or long-run behavior. Loosely speaking, we view the system as having started in the infinite past, so that any information about initial conditions and starting times becomes irrelevant. The more precise notion is that the system state is described by a *stationary process*; see also Section 1.13.

■ EXAMPLE 4.4 $M/M/1$ Queue

Consider the birth-and-death process $\{X_t, t \geqslant 0\}$ describing the number of customers in the $M/M/1$ queue; see Example 1.13. When the traffic intensity $\varrho = \lambda/\mu$ is less than 1, this Markov jump process has a limiting distribution,

$$\lim_{t \to \infty} \mathbb{P}(X_t = k) = (1 - \varrho)\varrho^k, \quad k = 0, 1, 2, \ldots,$$

which is also its stationary distribution. When X_0 is distributed according to this limiting distribution, the process $\{X_t, t \geqslant 0\}$ is stationary: it behaves as if it has been going on for an infinite period of time. In particular, the distribution of X_t does not depend on t. A similar result holds for the Markov process $\{Y_n, n = 1, 2, \ldots\}$, describing the number of customers in the system as seen by the n-th arriving customer. It can be shown that under the condition $\varrho < 1$ it has the *same* limiting distribution as $\{X_t, t \geqslant 0\}$. Note that for the $M/M/1$ queue the steady-state expected performance measures are available analytically, while for the $GI/G/1$ queue, to be discussed in Example 4.5, one needs to resort to simulation.

Special care must be taken when making inferences concerning steady-state performances. The reason is that the output data are typically correlated; consequently, the statistical analysis used above, based on independent observations, is no longer applicable.

In order to cancel the effects of the time dependence and the initial distribution, it is common practice to discard the data that are collected during the nonstationary or transient part of the simulation. However, it is not always clear when the process will reach stationarity. If the process is regenerative, then the regenerative method, discussed in Section 4.4.2.2, avoids this transience problem altogether.

From now on, we assume that $\{X_t\}$ is a stationary process. Suppose that we wish to estimate the steady-state expected value $\ell = \mathbb{E}[X_t]$, for example, the expected steady-state queue length, or the expected steady-state sojourn time of a customers in a queue. Then ℓ can be estimated as either

$$\widehat{\ell} = T^{-1} \sum_{t=1}^{T} X_t$$

or

$$\widehat{\ell} = T^{-1} \int_0^T X_t \, dt \,,$$

respectively, depending on whether $\{X_t\}$ is a discrete-time or continuous-time process.

For concreteness, consider the discrete case. The variance of $\widehat{\ell}$ (see Problem 1.15) is given by

$$\text{Var}(\widehat{\ell}) = \frac{1}{T^2} \left(\sum_{t=1}^{T} \text{Var}(X_t) + 2 \sum_{s=1}^{T-1} \sum_{t=s+1}^{T} \text{Cov}(X_s, X_t) \right). \tag{4.19}$$

Since $\{X_t\}$ is stationary, we have $\text{Cov}(X_s, X_t) = \mathbb{E}[X_s X_t] - \ell^2 = R(t - s)$, where R defines the *covariance function* of the stationary process. Note that $R(0) = \text{Var}(X_t)$. As a consequence, we can write (4.19) as

$$T\,\text{Var}(\widehat{\ell}) = R(0) + 2\sum_{t=1}^{T-1}\left(1 - \frac{t}{T}\right)R(t) \,. \tag{4.20}$$

Similarly, if $\{X_t\}$ is a continuous-time process, the sum in (4.20) is replaced with the corresponding integral (from $t = 0$ to T), while all other data remain the same. In many applications $R(t)$ decreases rapidly with t, so that only the first few terms in the sum (4.20) are relevant. These covariances, say $R(0), R(1), \ldots, R(K)$, can be estimated via their (unbiased) sample averages:

$$\widehat{R}(k) = \frac{1}{T - k - 1}\sum_{t=1}^{T-k}(X_t - \widehat{\ell})(X_{t+k} - \widehat{\ell}), \quad k = 0, 1, \ldots, K \,.$$

Thus, for large T the variance of $\widehat{\ell}$ can be estimated as $\tilde{S}^2/T$, where

$$\tilde{S}^2 = \widehat{R}(0) + 2\sum_{t=1}^{K}\widehat{R}(t) \,.$$

To obtain confidence intervals, we use again the central limit theorem; that is, the cdf of $\sqrt{T}(\widehat{\ell} - \ell)$ converges to the cdf of the normal distribution with expectation 0 and variance $\sigma^2 = \lim_{T\to\infty} T\,\text{Var}(\widehat{\ell})$ — the so-called *asymptotic variance* of $\widehat{\ell}$. Using $\tilde{S}^2$ as an estimator for σ^2, we find that an approximate $(1-\alpha)100\%$ confidence interval for ℓ is given by

$$\left(\widehat{\ell} \pm z_{1-\alpha/2}\frac{\tilde{S}}{\sqrt{T}}\right) \,. \tag{4.21}$$

Below we consider two popular methods for estimating steady-state parameters: the *batch means* and *regenerative* methods.

4.4.2.1 Batch Means Method The batch means method is most widely used by simulation practitioners to estimate steady-state parameters from a single simulation run, say of length M. The initial K observations, corresponding to the transient part of the run (called *burn-in*), are deleted, and the remaining $M - K$ observations are divided into N batches, each of length

$$T = \frac{M - K}{N} \,.$$

The deletion serves to eliminate or reduce the initial bias, so that the remaining observations $\{X_t, t > K\}$ are statistically more typical of the steady state.

Suppose we want to estimate the expected steady-state performance $\ell = \mathbb{E}[X_t]$, assuming that the process is stationary for $t > K$. We assume, for simplicity, that $\{X_t\}$ is a discrete-time process. Let X_{ti} denote the t-th observation from the i-th batch. The sample mean of the i-th batch of length T is given by

$$Y_i = \frac{1}{T}\sum_{t=1}^{T}X_{ti}, \quad i = 1, \ldots, N \,.$$

Therefore, the sample mean $\widehat{\ell}$ of ℓ is

$$\widehat{\ell} = \frac{1}{M-K} \sum_{t=K+1}^{M} X_t = \frac{1}{N} \sum_{i=1}^{N} Y_i \,. \tag{4.22}$$

The procedure is illustrated in Figure 4.4.

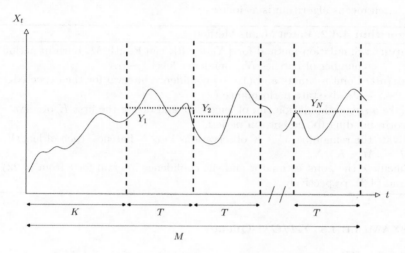

Figure 4.4: Illustration of the batch means procedure.

In order to ensure approximate independence between the batches, their size, T, should be large enough. In order for the central limit theorem to hold approximately, the number of batches, N, should typically be chosen in the range 20–30. In such a case, an approximate confidence interval for ℓ is given by (4.3), where S is the sample standard deviation of the $\{Y_i\}$. In the case where the batch means do exhibit some dependence, we can apply formula (4.21) as an alternative to (4.3).

Next, we discuss briefly how to choose K. In general, this is a very difficult task, since very few analytic results are available. The following queueing example provides some hints on how K should be increased as the traffic intensity in the queue increases.

Let $\{X_t, t \geqslant 0\}$ be the queue length process (not including the customer in service) in an $M/M/1$ queue, and assume that we start the simulation at time zero with an empty queue. It is shown in [1, 2] that in order to be within 1% of the steady-state mean, the length of the initial portion to be deleted, K, should be on the order of $8/(\mu(1-\varrho)^2)$, where $1/\mu$ is the expected service time. Thus, for $\varrho = 0.5,\ 0.8,\ 0.9,$ and 0.95, K equals 32, 200, 800, and 3200 expected service times, respectively.

In general, one can use the following simple rule of thumb.

1. Define the following moving average A_k of length T:

$$A_k = \frac{1}{T} \sum_{t=k+1}^{T+k} X_t \,.$$

2. Calculate A_k for different values of k, say $k = 0, m, 2m, \ldots, rm, \ldots$, where m is fixed, say $m = 10$.

3. Find r such that $A_{rm} \approx A_{(r+1)m} \approx \cdots \approx A_{(r+s)m}$, while $A_{(r-s)m} \not\approx A_{(r-s+1)m}$ $\not\approx \cdots \not\approx A_{rm}$, where $r \geq s$ and $s = 5$, for example.

4. Deliver $K = r m$.

The batch means algorithm is as follows:

Algorithm 4.4.2: Batch Means Method

---

input : Simulation method for $\{X_t, t \geqslant 0\}$, run length M, burn-in period K, number of batches N, confidence level $1 - \alpha$.

output: Point estimate and $(1 - \alpha)$-confidence interval for the expected steady-state performance ℓ.

1 Make a single simulation run of length M and delete the first K observations corresponding to the burn-in period.

2 Divide the remaining $M - K$ observations into N batches, each of length $T = (M - K)/N$.

3 Calculate the point estimator and the confidence interval for ℓ from (4.22) and (4.3), respectively.

■ **EXAMPLE 4.5** *GI/G/*1 **Queue**

The $GI/G/1$ queueing model is a generalization of the $M/M/1$ model discussed in Examples 1.13 and 4.4. The only differences are that (1) the inter-arrival times each have a general cdf F and (2) the service times each have a general cdf G. Let us consider the process $\{Z_n, n = 1, 2, \ldots\}$ describing the number of people in a $GI/G/1$ queue as seen by the n-th arriving customer. Figure 4.5 gives a realization of the batch means procedure for estimating the steady-state queue length. In this example the first $K = 100$ observations are thrown away, leaving $N = 9$ batches, each of size $T = 100$. The batch means are indicated by thick lines.

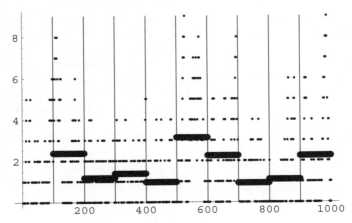

Figure 4.5: The batch means for the process $\{Z_n, n = 1, 2, \ldots\}$.

Remark 4.4.1 (Replication-Deletion Method) In the replication-deletion method, N independent runs are carried out, rather than a single simulation run as in the batch means method. From each replication, one deletes K initial observations corresponding to the finite-horizon simulation and then calculates the point estimator and the confidence interval for ℓ via (4.22) and (4.3), respectively, exactly as in the batch means approach. Note that the confidence interval obtained with the replication-deletion method is unbiased, whereas the one obtained by the batch means method is slightly biased. However, the former requires deletion from *each* replication, as compared to *a single* deletion in the latter. For this reason, the former is not as popular as the latter. For more details on the replication-deletion method, see [10].

4.4.2.2 The Regenerative Method   A stochastic process $\{X_t\}$ is called *regenerative* if there exist random time points $T_0 < T_1 < T_2 < \ldots$ such that at each time point the process restarts probabilistically. More precisely, the process $\{X_t\}$ can be split into iid replicas during intervals, called *cycles*, of lengths $\tau_i = T_i - T_{i-1}$, $i = 1, 2, \ldots$.

■ **EXAMPLE 4.6   Markov Chain**

The standard example of a regenerative process is a Markov chain. Assume that the chain starts from state i. Let $T_0 < T_1 < T_2 < \ldots$ denote the times that it visits state j. Note that at each random time T_n the Markov chain starts afresh, independently of the past. We say that the Markov process *regenerates* itself. For example, consider a two-state Markov chain with transition matrix

$$P = \begin{pmatrix} p_{11} & p_{12} \\ p_{21} & p_{22} \end{pmatrix}. \tag{4.23}$$

Assume that all four transition probabilities p_{ij} are strictly positive and that, starting from state $i = 1$, we obtain the following sample trajectory:

$$(x_0, x_1, x_2, \ldots, x_{10}) = (1, 2, 2, 2, 1, 2, 1, 1, 2, 2, 1).$$

It is readily seen that the transition probabilities corresponding to the sample trajectory above are

$$p_{12}, p_{22}, p_{22}, p_{21}, p_{12}, p_{21}, p_{11}, p_{12}, p_{22}, p_{21}.$$

Taking $j = 1$ as the regenerative state, the trajectory contains four cycles with the following transitions:

$$1 \to 2 \to 2 \to 2 \to 1; \quad 1 \to 2 \to 1; \quad 1 \to 1; \quad 1 \to 2 \to 2 \to 1,$$

and the corresponding cycle lengths are $\tau_1 = 4$, $\tau_2 = 2$, $\tau_3 = 1$, $\tau_4 = 3$.

■ **EXAMPLE 4.7   *GI/G/*1 Queue (Continued)**

Another classic example of a regenerative process is the process $\{X_t, t \geq 0\}$ describing the number of customers in the $GI/G/1$ system, where the regeneration times $T_0 < T_1 < T_2 < \ldots$ correspond to customers arriving at an empty system (see also Example 4.5, where a related discrete-time process is considered). Observe that at each time T_i the process starts afresh, independently of the past; in other words, the process regenerates itself. Figure 4.6 illustrates a typical sample path of the process $\{X_t, t \geqslant 0\}$. Note that here $T_0 = 0$; that is, at time 0 a customer arrives at an empty system.

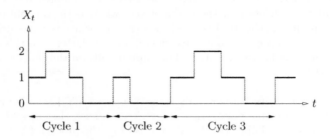

Figure 4.6: A sample path of the process $\{X_t, t \geqslant 0\}$, describing the number of customers in a $GI/G/1$ queue.

■ EXAMPLE 4.8 (s, S) Policy Inventory Model

Consider a continuous-review, single-commodity inventory model supplying external demands and receiving stock from a production facility. When demand occurs, it is either filled or back-ordered (to be satisfied by delayed deliveries). At time t, the *net inventory* (on-hand inventory minus back orders) is N_t, and the *inventory position* (net inventory plus on-order inventory) is X_t. The control policy is an (s, S) policy that operates on the inventory position. Specifically, at any time t when a demand D is received that would reduce the inventory position to less than s (i.e., $X_{t-} - D < s$, where X_{t-} denotes the inventory position just before t), an order of size $S - (X_{t-} - D)$ is placed, which brings the inventory position immediately back to S. Otherwise, no action is taken. The order arrives r time units after it is placed (r is called the *lead* time). Clearly, $X_t = N_t$ if $r = 0$. Both inventory processes are illustrated in Figure 4.7. The dots in the graph of the inventory position (below the s-line) represent what the inventory position would have been if no order was placed.

Let D_i and A_i be the size of the i-th demand and the length of the i-th inter-demand time, respectively. We assume that both $\{D_i\}$ and $\{A_i\}$ are iid sequences, with common cdfs F and G, respectively. In addition, the sequences are assumed to be independent of each other. Under the back-order policy and the assumptions above, both the inventory position process $\{X_t\}$ and the net inventory process $\{N_t\}$ are regenerative. In particular, each process regenerates when it is raised to S. For example, each time an order is placed, the inventory position process regenerates. It is readily seen that the sample path of $\{X_t\}$ in Figure 4.7 contains three regenerative cycles, while the sample path of $\{N_t\}$ contains only two, which occur after the second and third lead times. Note that during these times no order has been placed.

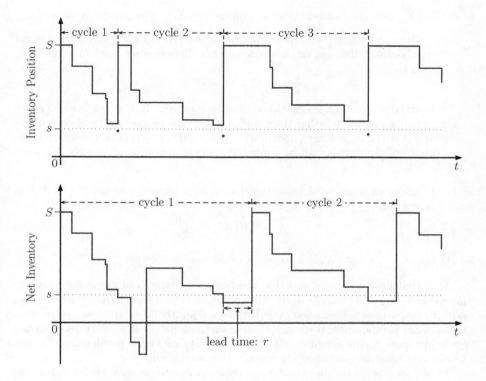

Figure 4.7: Sample paths for the two inventory processes.

The main strengths of the concept of regenerative processes are that the existence of limiting distributions is guaranteed under very mild conditions and the behavior of the limiting distribution depends only on the behavior of the process during a typical cycle.

Let $\{X_t\}$ be a regenerative process with regeneration times $T_0, T_1, T_2, \ldots$. Let $\tau_i = T_i - T_{i-1}$, $i = 1, 2, \ldots$ be the cycle lengths. Depending on whether $\{X_t\}$ is a discrete-time or continuous-time process, define, for some real-valued function H,

$$R_i = \sum_{t=T_{i-1}}^{T_i-1} H(X_t) \tag{4.24}$$

or

$$R_i = \int_{T_{i-1}}^{T_i} H(X_t)\, \mathrm{d}t\,, \tag{4.25}$$

respectively, for $i = 1, 2, \ldots$. We assume, for simplicity, that $T_0 = 0$. We also assume that in the discrete case the cycle lengths are not always a multiple of some integer greater than 1. We can view R_i as the reward (or, alternatively, the cost) accrued during the i-th cycle. Let $\tau = T_1$ be the length of the first regeneration cycle, and let $R = R_1$ be the first reward.

The following properties of regenerative processes will be needed later on (see, e.g., [3]):

(a) If $\{X_t\}$ is regenerative, then the process $\{H(X_t)\}$ is regenerative as well.

(b) If $\mathbb{E}[\tau] < \infty$, then, under mild conditions, the process $\{X_t\}$ has a limiting (or steady-state) distribution, in the sense that there exists a random variable X, such that

$$\lim_{t \to \infty} \mathbb{P}(X_t \leqslant x) = \mathbb{P}(X \leqslant x).$$

In the discrete case, no extra condition is required. In the continuous case, a sufficient condition is that the sample paths of the process are right-continuous and that the cycle length distribution is *non-lattice* — that is, the distribution does not concentrate all its probability mass at points $n\delta$, $n \in \mathbb{N}$, for some $\delta > 0$.

(c) If the conditions in (b) hold, the steady-state expected value, $\ell = \mathbb{E}[H(X)]$, is given by

$$\ell = \mathbb{E}[H(X)] = \frac{\mathbb{E}[R]}{\mathbb{E}[\tau]}. \tag{4.26}$$

(d) (R_i, τ_i), $i = 1, 2, \ldots$, is a sequence of iid random vectors.

Note that property (a) states that the behavior patterns of the system (or any measurable function thereof) during distinct cycles are statistically iid, while property (d) states that rewards and cycle lengths are jointly iid for distinct cycles. Formula (4.26) is fundamental to regenerative simulation. For typical non-Markovian queueing models, the quantity ℓ (the steady-state expected performance) is unknown and must be evaluated via regenerative simulation.

To obtain a point estimate of ℓ, one generates N regenerative cycles, calculates the iid sequence of two-dimensional random vectors (R_i, τ_i), $i = 1, \ldots, N$, and finally estimates ℓ by the *ratio* estimator

$$\widehat{\ell} = \frac{\widehat{R}}{\widehat{\tau}}, \tag{4.27}$$

where $\widehat{R} = N^{-1} \sum_{i=1}^{N} R_i$ and $\widehat{\tau} = N^{-1} \sum_{i=1}^{N} \tau_i$. Note that the estimator $\widehat{\ell}$ is biased; that is, $\mathbb{E}[\widehat{\ell}] \neq \ell$. However, $\widehat{\ell}$ is *strongly consistent*, that is, it converges to ℓ with probability 1 as $N \to \infty$. This follows directly from the fact that, by the law of large numbers, $\widehat{R}$ and $\widehat{\tau}$ converge with probability 1 to $\mathbb{E}[R]$ and $\mathbb{E}[\tau]$, respectively.

The *advantages* of the regenerative simulation method are:

(a) No deletion of transient data is necessary.

(b) It is asymptotically exact.

(b) It is easy to understand and implement.

The *disadvantages* of the regenerative simulation method are:

(a) For many practical cases, the output process, $\{X_t\}$, is either nonregenerative or its regeneration points are difficult to identify. Moreover, in complex systems (e.g., large queueing networks), checking for the occurrence of regeneration points could be computationally expensive.

(b) The estimator $\widehat{\ell}$ is biased.

(c) The regenerative cycles can be very long.

Next, we will establish a confidence interval for ℓ. Let $Z_i = R_i - \ell\tau_i$. It is readily seen that the Z_i are iid random variables, like the random vectors (R_i, τ_i). Letting $\widehat{R}$ and $\widehat{\tau}$ be defined as before, the central limit theorem ensures that

$$\frac{N^{1/2}\left(\widehat{R} - \ell\widehat{\tau}\right)}{\sigma} = \frac{N^{1/2}\left(\widehat{\ell} - \ell\right)}{\sigma/\widehat{\tau}}$$

converges in distribution to the standard normal distribution as $N \to \infty$, where

$$\sigma^2 = \mathrm{Var}(Z) = \mathrm{Var}(R) - 2\ell\,\mathrm{Cov}(R, \tau) + \ell^2\,\mathrm{Var}(\tau) . \tag{4.28}$$

Therefore, a $(1 - \alpha)100\%$ confidence interval for $\ell = \mathbb{E}[R]/\mathbb{E}[\tau]$ is

$$\left(\widehat{\ell} \pm \frac{z_{1-\alpha/2}\,S}{\widehat{\tau}\,N^{1/2}}\right) , \tag{4.29}$$

where

$$S^2 = S_{11} - 2\,\widehat{\ell}\,S_{12} + \widehat{\ell}^{\,2}S_{22} \tag{4.30}$$

is the estimator of σ^2 based on replacing the unknown quantities in (4.28) with their unbiased estimators. That is,

$$S_{11} = \frac{1}{N-1}\sum_{i=1}^{N}(R_i - \widehat{R})^2, \quad S_{22} = \frac{1}{N-1}\sum_{i=1}^{N}(\tau_i - \widehat{\tau})^2$$

and

$$S_{12} = \frac{1}{N-1}\sum_{i=1}^{N}(R_i - \widehat{R})\,(\tau_i - \widehat{\tau}) .$$

Note that (4.29) differs from the standard confidence interval, say (4.3), by having an additional term $\widehat{\tau}$.

The algorithm for estimating the $(1 - \alpha)100\%$ confidence interval for ℓ is as follows:

Algorithm 4.4.3: Regenerative Simulation Method

input : Simulation method for the process $\{X_t\}$, performance function H, number of regenerations N, confidence level $1 - \alpha$.

output: Point estimate and $(1 - \alpha)$-confidence interval for the expected steady-state performance $\ell = \mathbb{E}[H(X)]$.

1 Simulate N regenerative cycles of the process $\{X_t\}$.
2 Compute the sequence $\{(R_i, \tau_i), i = 1, \ldots, N\}$.
3 Calculate the point estimator $\widehat{\ell}$ and the confidence interval of ℓ from (4.27) and (4.29), respectively.

Note that if one uses two independent simulations of length N, one for estimating $\mathbb{E}[R]$ and the other for estimating $\mathbb{E}[\tau]$, then clearly $S^2 = S_{11} + \widehat{\ell}^{\,2}S_{22}$, since $\mathrm{Cov}(R, \tau) = 0$.

Remark 4.4.2 If the reward in each cycle is of the form (4.24) or (4.25), then $\ell = \mathbb{E}[H(X)]$ can be viewed as both the expected steady-state performance and the long-run average performance. This last interpretation is valid even if the reward in each cycle is not of the form (4.24)–(4.25) as long as the $\{(\tau_i, R_i)\}$ are iid. In that case,

$$\ell = \lim_{t \to \infty} \frac{\sum_{i=0}^{N_t - 1} R_i}{t} = \frac{\mathbb{E}[R]}{\mathbb{E}[\tau]} , \tag{4.31}$$

where N_t is the number of regenerations in $[0, t]$.

■ **EXAMPLE 4.9   Markov Chain: Example 4.6 (Continued)**

Consider again the two-state Markov chain with the transition matrix

$$P = \begin{pmatrix} p_{11} & p_{12} \\ p_{21} & p_{22} \end{pmatrix}.$$

Assume, as in Example 4.6, that we start from 1 and obtain the following sample trajectory: $(x_0, x_1, x_2, \ldots, x_{10}) = (1, 2, 2, 2, 1, 2, 1, 1, 2, 2, 1)$, which has four cycles with lengths $\tau_1 = 4$, $\tau_2 = 2$, $\tau_3 = 1$, $\tau_4 = 3$ and corresponding transitions $(p_{12}, p_{22}, p_{22}, p_{21}), (p_{12}, p_{21}), (p_{11}), (p_{12}, p_{22}, p_{21})$. In addition, assume that each transition from i to j incurs a cost (or, alternatively, a reward) c_{ij} and that the related cost matrix is

$$C = (c_{ij}) = \begin{pmatrix} c_{11} & c_{12} \\ c_{21} & c_{22} \end{pmatrix} = \begin{pmatrix} 0 & 1 \\ 2 & 3 \end{pmatrix} .$$

Note that the cost in each cycle is not of the form (4.24) (however, see Problem 4.14) but is given as

$$R_i = \sum_{t=T_{i-1}}^{T_i - 1} c_{X_t, X_{t+1}}, \quad i = 1, 2, \ldots .$$

We illustrate the estimation procedure for the long-run average cost ℓ. First, observe that $R_1 = 1 + 3 + 3 + 2 = 9$, $R_2 = 3$, $R_3 = 0$, and $R_4 = 6$. It follows that $\widehat{R} = 4.5$. Since $\widehat{\tau} = 2.5$, the point estimate of ℓ is $\widehat{\ell} = 1.80$. Moreover, $S_{11} = 15, S_{22} = 5/3, S_{12} = 5$, and $S^2 = 2.4$. This gives a 95% confidence interval for ℓ of $(1.20, 2.40)$.

■ **EXAMPLE 4.10   Example 4.7 (Continued)**

Consider the sample path in Figure 4.6 of the process $\{X_t, t \geqslant 0\}$ describing the number of customers in the $GI/G/1$ system. The corresponding sample path data are given in Table 4.2.

Table 4.2: Sample path data for the $GI/G/1$ queueing process.

$t \in$ interval	X_t	$t \in$ interval	X_t	$t \in$ interval	X_t
$[0.00, 0.80)$	1	$[3.91, 4.84)$	1	$[6.72, 7.92)$	1
$[0.80, 1.93)$	2	$[4.84, 6.72)$	0	$[7.92, 9.07)$	2
$[1.93, 2.56)$	1			$[9.07, 10.15)$	1
$[2.56, 3.91)$	0			$[10.15, 11.61)$	0
Cycle 1		**Cycle 2**		**Cycle 3**	

Notice that the figure and table reveal three complete cycles with the following pairs: $(R_1, \tau_1) = (3.69, 3.91)$, $(R_2, \tau_2) = (0.93, 2.81)$, and $(R_3, \tau_3) = (4.58, 4.89)$. The resultant statistics are (rounded) $\widehat{\ell} = 0.79$, $S_{11} = 3.62$, $S_{22} = 1.08$, $S_{12} = 1.92$, $S^2 = 1.26$, and the 95% confidence interval is (0.79 ± 0.32).

■ **EXAMPLE 4.11   Example 4.8 (Continued)**

Let $\{X_t, t \geqslant 0\}$ be the inventory position process described in Example 4.8. Table 4.3 presents the data corresponding to the sample path in Figure 4.7 for a case where $s = 10$, $S = 40$, and $r = 1$.

Table 4.3: Data for the inventory position process, $\{X_t\}$, with $s = 10$ and $S = 40$. The boxes indicate the regeneration times.

t	X_t	t	X_t	t	X_t
$\boxed{0.00}$	40.00	$\boxed{5.99}$	40.00	$\boxed{9.67}$	40.00
1.79	32.34	6.41	33.91	11.29	32.20
3.60	22.67	6.45	23.93	11.38	24.97
5.56	20.88	6.74	19.53	12.05	18.84
5.62	11.90	8.25	13.32	13.88	13.00
		9.31	10.51	$\boxed{14.71}$	40.00

Based on the data in Table 4.3, we illustrate the derivation of the point estimator and the 95% confidence interval for the steady-state quantity $\ell = \mathbb{P}(X < 30) = \mathbb{E}[I_{\{X<30\}}]$, that is, the probability that the inventory position is less than 30. Table 4.3 shows three complete cycles with the following pairs: $(R_1, \tau_1) = (2.39, 5.99)$, $(R_2, \tau_2) = (3.22, 3.68)$, and $(R_3, \tau_3) = (3.33, 5.04)$, where $R_i = \int_{T_{i-1}}^{T_i} I_{\{X_t<30\}} \, dt$. The resulting statistics are (rounded) $\widehat{\ell} = 0.61$, $S_{11} = 0.26$, $S_{22} = 1.35$, $S_{12} = -0.44$, and $S^2 = 1.30$, which gives a 95% confidence interval (0.61 ± 0.26).

4.5 BOOTSTRAP METHOD

Suppose that we estimate a number ℓ via some estimator $H = H(\mathbf{X})$, where $\mathbf{X} = (X_1, \ldots, X_n)$, and the $\{X_i\}$ form a random sample from some unknown distribution F. It is assumed that H does not depend on the order of the $\{X_i\}$. To assess the quality (e.g., accuracy) of the estimator H, we could draw independent replications $\mathbf{X}_1, \ldots, \mathbf{X}_N$ of $\mathbf{X}$ and find sample estimates for quantities such as the variance of the estimator

$$\mathrm{Var}(H) = \mathbb{E}[H^2] - (\mathbb{E}[H])^2,$$

the *bias* of the estimator

$$\mathrm{Bias} = \mathbb{E}[H] - \ell,$$

and the expected quadratic error, or *mean square error* (MSE)

$$\mathrm{MSE} = \mathbb{E}\left[(H - \ell)^2\right].$$

However, it may be too time-consuming, or simply not feasible, to obtain such replications. An alternative is to *resample* the original data. Specifically, given an outcome $(x_1, \ldots, x_n)$ of $\mathbf{X}$, we draw a random sample $X_1^*, \ldots, X_n^*$ not from F but from an approximation to this distribution. The best estimate that we have about F on the grounds of $\{x_i\}$ is the *empirical distribution*, F_n, which assigns probability mass $1/n$ to each point $x_i, i = 1, \ldots, n$. In the one-dimensional case, the cdf of the empirical distribution is thus given by

$$F_n(x) = \frac{1}{n} \sum_{i=1}^{n} I_{\{x_i \leqslant x\}}.$$

Drawing from this distribution is trivial: for each j, draw $U \sim \mathsf{U}[0,1]$, let $J = \lfloor U\, n \rfloor + 1$, and return $X_j^* = x_J$. Note that if the $\{x_i\}$ are all different, vector $\mathbf{X}^* = (X_1^*, \ldots, X_n^*)$ can take n^n different values.

The rationale behind the resampling idea is that the empirical distribution F_n is close to the actual distribution F and gets closer as n gets larger. Hence, any quantities depending on F, such as $\mathbb{E}_F[h(H)]$, where h is a function, can be approximated by $\mathbb{E}_{F_n}[h(H)]$. The latter is usually still difficult to evaluate, but it can be simply estimated via Monte Carlo simulation as

$$\frac{1}{B} \sum_{i=1}^{B} h(H_i^*),$$

where $H_1^*, \ldots, H_B^*$ are independent copies of $H^* = H(\mathbf{X}^*)$. This seemingly self-referent procedure is called *bootstrapping* — alluding to Baron von Münchhausen, who pulled himself out of a swamp by his own bootstraps. As an example, the bootstrap estimate of the expectation of H is

$$\widehat{\mathbb{E}[H]} = \overline{H}^* = \frac{1}{B} \sum_{i=1}^{B} H_i^*,$$

which is simply the sample mean of $\{H_i^*\}$. Similarly, the bootstrap estimate for $\mathrm{Var}(H)$ is the sample variance

$$\widehat{\mathrm{Var}(H)} = \frac{1}{B-1} \sum_{i=1}^{B} (H_i^* - \overline{H}^*)^2. \tag{4.32}$$

Perhaps of more interest are the bootstrap estimators for the bias and MSE, respectively $\overline{H}^* - H$ and

$$\frac{1}{B} \sum_{i=1}^{B} (H_i^* - H)^2 .$$

Note that for these estimators the unknown quantity ℓ is replaced with the original estimator H. Confidence intervals can be constructed in the same fashion. We discuss two variants: the *normal* method and the *percentile* method. In the normal method, a $(1 - \alpha)100\%$ confidence interval for ℓ is given by

$$(H \pm z_{1-\alpha/2} S^*) ,$$

where S^* is the bootstrap estimate of the standard deviation of H, that is, the square root of (4.32). In the percentile method, the upper and lower bounds of the $(1 - \alpha)100\%$ confidence interval for ℓ are given by the $1 - \alpha/2$ and $\alpha/2$ quantiles of H, which in turn are estimated via the corresponding sample quantiles of the bootstrap sample $\{H_i^*\}$.

PROBLEMS

4.1 We wish to estimate $\ell = \int_{-2}^{2} e^{-x^2/2} \, dx = \int H(x)f(x) \, dx$ via Monte Carlo simulation using two different approaches: (A) defining $H(x) = 4\,e^{-x^2/2}$ and f the pdf of the $\mathsf{U}[-2,2]$ distribution and (B) defining $H(x) = \sqrt{2\pi}\, I_{\{-2 \leqslant x \leqslant 2\}}$ and f the pdf of the $\mathsf{N}(0, 1)$ distribution.

    a) For both cases, estimate ℓ via the estimator $\widehat{\ell}$ in (4.10). Use a sample size of $N = 100$.

    b) For both cases, estimate the relative error of $\widehat{\ell}$, using $N = 100$.

    c) Give a 95% confidence interval for ℓ for both cases, using $N = 100$.

    d) From b), assess how large N should be such that the relative width of the confidence interval is less than 0.001, and carry out the simulation with this N. Compare the result with the true value of ℓ.

4.2 Prove that the structure function of the bridge system in Figure 4.1 is given by (4.11).

4.3 Consider the bridge system in Figure 4.1. Suppose that all link reliabilities are p. Show that the reliability of the system is $p^2(2 + 2\,p - 5\,p^2 + 2\,p^3)$.

4.4 Estimate the reliability of the bridge system in Figure 4.1 via (4.10) if the link reliabilities are $(p_1, \ldots, p_5) = (0.7, 0.6, 0.5, 0.4, 0.3)$. Choose a sample size such that the estimate has a relative error of about 0.01.

4.5 Consider the following sample performance:

$$H(\mathbf{X}) = \min\{X_1 + X_2,\ X_1 + X_4 + X_5,\ X_3 + X_4\}.$$

Assume that the random variables X_i, $i = 1, \ldots, 5$ are iid with common distribution

    (a) $\mathsf{Gamma}(\lambda_i, \beta_i)$, where $\lambda_i = i$ and $\beta_i = i$.

    (b) $\mathsf{Ber}(p_i)$, where $p_i = 1/2i$.

Run a computer simulation with $N = 1000$ replications, and find point estimates and 95% confidence intervals for $\ell = \mathbb{E}[H(\mathbf{X})]$.

4.6   Consider the precedence ordering of activities in Table 4.4. Suppose that durations of the activities (when actually started) are independent of each other, and all have exponential distributions with parameters 1.1, 2.3, 1.5, 2.9, 0.7, and 1.5, for activities 1, ..., 6, respectively.

Table 4.4: Precedence ordering of activities.

Activity	1	2	3	4	5	6
Predecessor(s)	-	-	1	2,3	2,3	5

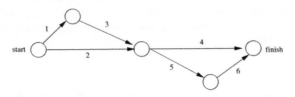

Figure 4.8: The PERT network corresponding to Table 4.4.

a) Verify that the corresponding PERT graph is given by Figure 4.8.
b) Identify the four possible paths from start to finish.
c) Estimate the expected length of the critical path in (4.13) with a relative error of less than 5%.

4.7   Let $\{X_t, t = 0, 1, 2, \ldots\}$ be a random walk on the positive integers; see Example 1.11. Suppose that $p = 0.55$ and $q = 0.45$. Let $X_0 = 0$. Let Y be the maximum position reached after 100 transitions. Estimate the probability that $Y \geqslant 15$ and give a 95% confidence interval for this probability based on 1000 replications of Y.

4.8   Consider the $M/M/1$ queue. Let X_t be the number of customers in the system at time $t \geqslant 0$. Run a computer simulation of the process $\{X_t, t \geqslant 0\}$ with $\lambda = 1$ and $\mu = 2$, starting with an empty system. Let X denote the steady-state number of people in the system. Find point estimates and confidence intervals for $\ell = \mathbb{E}[X]$, using the batch means and regenerative methods as follows:
a) For the batch means method run the system for a simulation time of 10,000, discard the observations in the interval $[0, 100]$, and use $N = 30$ batches.
b) For the regenerative method, run the system for the same amount of simulation time (10,000) and take as regeneration points the times where an arriving customer finds the system empty.
c) For both methods, find the requisite simulation time that ensures a relative width of the confidence interval not exceeding 5%.

4.9   Let Z_n be the number of customers in an $M/M/1$ queueing system, as seen by the n-th arriving customer, $n = 1, 2, \ldots$. Suppose that the service rate is $\mu = 1$ and the arrival rate is $\lambda = 0.6$. Let Z be the steady-state queue length (as seen by an arriving customer far away in the future). Note that $Z_n = X_{T_n-}$, with X_t

as in Problem 4.8, and T_n is the arrival epoch of the n-th customer. Here, "T_n-" denotes the time just before T_n.

a) Verify that $\ell = \mathbb{E}[Z] = 1.5$.

b) Explain how to generate $\{Z_n, n = 1, 2, \ldots\}$ using a random walk on the positive integers, as in Problem 4.7.

c) Find the point estimate of ℓ and a 95% confidence interval for ℓ using the batch means method. Use a sample size of 10^4 customers and $N = 30$ batches, throwing away the first $K = 100$ observations.

d) Do the same as in c) using the regenerative method instead.

e) Assess the minimum length of the simulation run in order to obtain a 95% confidence interval with an absolute width w_a not exceeding 5%.

f) Repeat c), d), and e) with $\varrho = 0.8$ and discuss c), d), and e) as $\varrho \to 1$.

4.10   Table 4.5 displays a realization of a Markov chain, $\{X_t, t = 0, 1, 2, \ldots\}$, with state space $\{0, 1, 2, 3\}$ starting at 0. Let X be distributed according to the limiting distribution of this chain (assuming it has one).

Table 4.5: A realization of the Markov chain.

t	1	2	3	4	5	6	7	8	9	10	11	12	13	14	15
X_t	0	3	0	1	2	1	0	2	0	1	0	1	0	2	0

Find the point estimator, $\widehat{\ell}$, and the 95% confidence interval for $\ell = \mathbb{E}[X]$ using the regenerative method.

4.11   Let W_n be the *waiting time* of the n-th customer in a $GI/G/1$ queue, that is, the total time the customer spends waiting in the queue (thus excluding the service time). The waiting time process $\{W_n, n = 1, 2, \ldots\}$ follows the following well-known *Lindley equation*:

$$W_{n+1} = \max\{W_n + S_n - A_{n+1}, 0\}, \quad n = 1, 2, \ldots, \tag{4.33}$$

where A_{n+1} is the interval between the n-th and $(n+1)$-st arrivals, S_n is the service time of the n-th customer, and $W_1 = 0$ (the first customer does not have to wait and is served immediately).

a) Explain why the Lindley equation holds.

b) Find the point estimate and the 95% confidence interval for the expected waiting time for the 4-th customer in an $M/M/1$ queue with $\varrho = 0.5$, ($\lambda = 1$), starting with an empty system. Use $N = 5000$ replications.

c) Find point estimates and confidence intervals for the expected average waiting time for customers $21, \ldots, 70$ in the same system as in b). Use $N = 5000$ replications. Note that the point estimate and confidence interval required are for the following parameter:

$$\ell = \mathbb{E}\left[\frac{1}{50} \sum_{n=21}^{70} W_n\right].$$

4.12   Run a computer simulation of 1000 regenerative cycles of the (s, S) policy inventory model (see Example 4.8), where demands arrive according to a Poisson process with rate 2 (i.e., $A \sim \mathsf{Exp}(2)$) and the size of each demand follows a Poisson distribution with mean 2 (i.e., $D \sim \mathsf{Poi}(2)$). Take $s = 1$, $S = 6$, lead time $r = 2$, and initial value $X_0 = 4$. Find point estimates and confidence intervals for the quantity $\ell = \mathbb{P}(2 \leqslant X \leqslant 4)$, where X is the steady-state inventory position.

4.13   Simulate the Markov chain $\{X_n\}$ in Example 4.9, using $p_{11} = 1/3$ and $p_{22} = 3/4$ for 1000 regeneration cycles. Obtain a confidence interval for the long-run average cost.

4.14   Consider Example 4.9 again, with $p_{11} = 1/3$ and $p_{22} = 3/4$. Define $Y_i = (X_i, X_{i+1})$ and $H(Y_i) = c_{X_i, X_{i+1}}$, $i = 0, 1, \ldots$. Show that $\{Y_i\}$ is a regenerative process. Find the corresponding limiting/steady-state distribution and calculate $\ell = \mathbb{E}[H(Y)]$, where Y is distributed according to this limiting distribution. Check if ℓ is contained in the confidence interval obtained in Problem 4.13.

4.15   Consider the tandem queue of Section 3.4.1. Let X_t and Y_t denote the number of customers in the first and second queues at time t, including those who are possibly being served. Is $\{(X_t, Y_t), t \geqslant 0\}$ a regenerative process? If so, specify the regeneration times.

4.16   Consider the machine repair problem in Problem 3.5, with three machines and two repair facilities. Each repair facility can take only one failed machine. Suppose that the lifetimes are $\mathsf{Exp}(1/10)$ distributed and the repair times are $\mathsf{U}(0, 8)$ distributed. Let ℓ be the limiting probability that all machines are out of order.

    a) Estimate ℓ via the regenerative estimator $\widehat{\ell}$ in (4.27) using 1000 regeneration cycles. Compute the 95% confidence interval (4.30).

    b) Estimate the bias and MSE of $\widehat{\ell}$ using the bootstrap method with a sample size of $B = 300$. [Hint: The original data are $\mathbf{X} = (X_1, \ldots, X_{100})$, where $X_i = (R_i, \tau_i)$, $i = 1, \ldots, 100$. Resample from these data using the empirical distribution.]

    c) Compute 95% bootstrap confidence intervals for ℓ using the normal and percentile methods with $B = 1000$ bootstrap samples.

Further Reading

The regenerative method in a simulation context was introduced and developed by Crane and Iglehart [4, 5]. A more complete treatment of regenerative processes is given in [3]. Fishman [7] treats the statistical analysis of simulation data in great detail. Gross and Harris [8] is a classical reference on queueing systems. Efron and Tibshirani [6] gives the defining introduction to the bootstrap method. A modern introduction to statistical modeling and computation can be found in [9].

REFERENCES

1. J. Abate and W. Whitt. Transient behavior of regulated Brownian motion, I: Starting at the origin. *Advances in Applied Probability*, 19:560–598, 1987.

2. J. Abate and W. Whitt. Transient behavior of regulated Brownian motion, II: Non-zero initial conditions. *Advances in Applied Probability*, 19:599–631, 1987.

3. S. Asmussen. *Applied Probability and Queues*. John Wiley & Sons, New York, 1987.

4. M. A. Crane and D. L. Iglehart. Simulating stable stochastic systems, I: General multiserver queues. *Journal of the ACM*, 21:103–113, 1974.

5. M. A. Crane and D. L. Iglehart. Simulating stable stochastic systems, II: Markov chains. *Journal of the ACM*, 21:114–123, 1974.

6. B. Efron and R. Tibshirani. *An Introduction to the Bootstrap.* Chapman & Hall, New York, 1994.

7. G. S. Fishman. *Monte Carlo: Concepts, Algorithms and Applications.* Springer-Verlag, New York, 1996.

8. D. Gross and C. M. Harris. *Fundamentals of Queueing Theory.* John Wiley & Sons, New York, 2nd edition, 1985.

9. D. P. Kroese and J. C. C. Chan. *Statistical Modeling and Computation.* Springer, New York, 2014.

10. A. M. Law and W. D. Kelton. *Simulation Modeling and Analysis.* McGraw-Hill, New York, 3rd edition, 2000.

CHAPTER 5

CONTROLLING THE VARIANCE

5.1 INTRODUCTION

This chapter treats basic theoretical and practical aspects of *variance reduction* techniques. Variance reduction can be viewed as a means of utilizing known information about the model in order to obtain more accurate estimators of its performance. Generally, the more we know about the system, the more effective is the variance reduction. One way of gaining this information is through a pilot simulation run of the model. Results from this first-stage simulation can then be used to formulate variance reduction techniques that will subsequently improve the accuracy of the estimators in the second simulation stage. Two of the most effective techniques for variance reduction are *importance sampling* and *conditional Monte Carlo*. Other well-known techniques that can provide moderate variance reduction include the use of common and antithetic variables, control variables, and stratification. The splitting method, discussed in Chapter 9, is another powerful approach to variance reduction.

The chapter is organized as follows. We start, in Sections 5.2–5.5, with common and antithetic variables, control variables, conditional Monte Carlo, and stratified sampling. Section 5.6 introduces the multilevel Monte Carlo method for the estimation of performance measures of diffusion processes. Most of our attention, from Section 5.7 on, is focused on *importance sampling* and *likelihood ratio* techniques. Using importance sampling, one can often achieve substantial (sometimes dramatic) variance reduction, in particular when estimating rare-event probabilities. In Sec-

Simulation and the Monte Carlo Method, Third Edition. By R. Y. Rubinstein and D. P. Kroese
Copyright © 2017 John Wiley & Sons, Inc. Published 2017 by John Wiley & Sons, Inc.

tion 5.7 we present two alternative importance sampling-based techniques, called the *variance minimization* and *cross-entropy* methods. Sections 5.8–5.10 discuss how importance sampling can be carried out sequentially/dynamically. Section 5.11 presents a simple, convenient, and unifying way of constructing efficient importance sampling estimators: the so-called *transform likelihood ratio* (TLR) method. Finally, in Section 5.12 we present the *screening* method for variance reduction, which can also be seen as a dimension-reduction technique. The aim of this method is to identify (screen out) the most important (bottleneck) parameters of the simulated system to be used in an importance sampling estimation procedure.

5.2 COMMON AND ANTITHETIC RANDOM VARIABLES

To motivate the use of common and antithetic random variables in simulation, let us consider a simple example. Let X and Y be random variables with known cdfs, F and G, respectively. Suppose that we need to estimate $\ell = \mathbb{E}[X - Y]$ via simulation. The simplest unbiased estimator for ℓ is $X - Y$. We can simulate X and Y via the IT method:

$$
\begin{aligned}
X &= F^{-1}(U_1) , \quad U_1 \sim \mathsf{U}(0,1) , \\
Y &= G^{-1}(U_2) , \quad U_2 \sim \mathsf{U}(0,1) .
\end{aligned}
\tag{5.1}
$$

It is important to note that X and Y (or U_1 and U_2) *need not be independent*. In fact, since

$$
\mathrm{Var}(X - Y) = \mathrm{Var}(X) + \mathrm{Var}(Y) - 2\,\mathrm{Cov}(X, Y)
\tag{5.2}
$$

and since the marginal cdfs of X and Y have been prescribed, it follows that the variance of $X - Y$ can be minimized by maximizing the covariance in (5.2). We say that *common random variables* are used in (5.1) if $U_2 = U_1$ and *antithetic random variables* are used if $U_2 = 1 - U_1$. Since both F^{-1} and G^{-1} are nondecreasing functions, in using common random variables, we clearly have

$$
\mathrm{Cov}\left(F^{-1}(U), \; G^{-1}(U)\right) \geqslant 0
$$

for $U \sim \mathsf{U}(0,1)$. Consequently, variance reduction is achieved, in the sense that the estimator $F^{-1}(U) - G^{-1}(U)$ has a smaller variance than the *crude Monte Carlo* (CMC) estimator $X - Y$, where X and Y are independent, with cdfs F and G, respectively. In fact, it is well known (e.g., see [44]) that using common random variables maximizes the covariance between X and Y, so that $\mathrm{Var}(X - Y)$ is *minimized*. Similarly, $\mathrm{Var}(X + Y)$ is minimized when antithetic random variables are used.

Now consider minimal variance estimation of $\mathbb{E}[H_1(X) - H_2(Y)]$, where X and Y are unidimensional variables with known marginal cdfs, F and G, respectively, and H_1 and H_2 are real-valued monotone functions. Mathematically, the problem can be formulated as follows:

> Within the set of all two-dimensional joint cdfs of (X, Y), find a joint cdf, F^*, that minimizes $\mathrm{Var}(H_1(X) - H_2(Y))$, subject to X and Y having the prescribed cdfs F and G, respectively.

This problem has been solved by Gal, Rubinstein, and Ziv [14], who proved that if H_1 and H_2 are monotonic in the *same* direction, then the use of common random variables leads to optimal variance reduction, that is,

$$
\min_{F^*} \mathrm{Var}(H_1(X) - H_2(Y)) = \mathrm{Var}\left(H_1[F^{-1}(U)] - H_2[G^{-1}(U)]\right) .
\tag{5.3}
$$

The proof of (5.3) uses the fact that if $H(u)$ is a monotonic function, then $H(F^{-1}(U))$ is monotonic as well, since $F^{-1}(u)$ is. By symmetry, if H_1 and H_2 are monotonic in *opposite* directions, then the use of antithetic random variables (i.e., $U_2 = 1 - U_1$) yields optimal variance reduction.

This result can be further generalized by considering minimal variance estimation of

$$\mathbb{E}[H_1(\mathbf{X}) - H_2(\mathbf{Y})] \,, \tag{5.4}$$

where $\mathbf{X} = (X_1, \ldots, X_n)$ and $\mathbf{Y} = (Y_1, \ldots, Y_n)$ are random vectors with $X_i \sim F_i$ and $Y_i \sim G_i$, $i = 1, \ldots, n$, and the functions H_1 and H_2 are real-valued and monotone in each component of $\mathbf{X}$ and $\mathbf{Y}$. If the pairs $\{(X_i, Y_i)\}$ are independent and H_1 and H_2 are monotonic in the same direction (for each component), then the use of common random variables again leads to minimal variance. That is, we take $X_i = F_i^{-1}(U_i)$ and $Y_i = G_i^{-1}(U_i)$, $i = 1, \ldots, n$, where $U_1, \ldots, U_n$ are independent $\mathsf{U}(0,1)$-distributed random variables, or, symbolically,

$$\mathbf{X} = F^{-1}(\mathbf{U}), \quad \mathbf{Y} = G^{-1}(\mathbf{U}) \,. \tag{5.5}$$

Similarly, if H_1 and H_2 are monotonic in opposite directions, then using antithetic random variables is optimal. Last, if H_1 and H_2 are monotonically increasing with respect to some components and monotonically decreasing with respect to others, then minimal variance is obtained by using the appropriate combination of common and antithetic random variables.

We now describe one of the main applications of antithetic random variables. We want to estimate

$$\ell = \mathbb{E}[H(\mathbf{X})] \,,$$

where $\mathbf{X} \sim F$ is a random vector with independent components and the sample performance function, $H(\mathbf{x})$, is monotonic in each component of $\mathbf{x}$. An example of such a function is given below.

■ EXAMPLE 5.1 Stochastic Shortest Path

Consider the undirected graph in Figure 5.1, depicting a so-called *bridge network*.

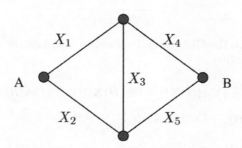

Figure 5.1: Determine the shortest path from A to B in a bridge network.

Our objective is to estimate the expected length ℓ of the shortest path between nodes (vertices) A and B, where the lengths of the links (edges) are random variables $X_1, \ldots, X_5$. We have $\ell = \mathbb{E}[H(\mathbf{X})]$, where

$$H(\mathbf{X}) = \min\{X_1 + X_4,\ X_1 + X_3 + X_5,\ X_2 + X_3 + X_4,\ X_2 + X_5\}\ . \tag{5.6}$$

Note that $H(\mathbf{x})$ is nondecreasing in each component of the vector $\mathbf{x}$.

Similarly, the length of the shortest path $H(\mathbf{X})$ in an arbitrary network with random edge lengths $\{X_i\}$ can be written as

$$H(\mathbf{X}) = \min_{j=1,\ldots,p} \sum_{i \in \mathscr{P}_j} X_i\ , \tag{5.7}$$

where $\mathscr{P}_j$ is the j-th complete path from the source to the sink of the network and p is the number of complete paths in the network. The sample performance is nondecreasing in each of the components.

An unbiased estimator of $\ell = \mathbb{E}[H(\mathbf{X})]$ is the CMC estimator, given by

$$\widehat{\ell} = \frac{1}{N} \sum_{k=1}^{N} H(\mathbf{X}_k)\ , \tag{5.8}$$

where $\mathbf{X}_1, \ldots, \mathbf{X}_N$ is an iid sample from the (multidimensional) cdf F. An alternative unbiased estimator of ℓ, for N even, is

$$\widehat{\ell}^{(a)} = \frac{1}{N} \sum_{k=1}^{N/2} \left\{ H(\mathbf{X}_k) + H(\mathbf{X}_k^{(a)}) \right\}, \tag{5.9}$$

where $\mathbf{X}_k = F^{-1}(\mathbf{U}_k)$ and $\mathbf{X}_k^{(a)} = F^{-1}(\mathbf{1} - \mathbf{U}_k)$, using notation similar to (5.5). The estimator $\widehat{\ell}^{(a)}$ is called the *antithetic estimator* of ℓ. Since $H(\mathbf{X}) + H(\mathbf{X}^{(a)})$ is a particular case of $H_1(\mathbf{X}) - H_2(\mathbf{Y})$ in (5.4) (with $H_2(\mathbf{Y})$ replaced by $-H(\mathbf{X}^{(a)})$), we immediately obtain that $\mathrm{Var}(\widehat{\ell}^{(a)}) \leqslant \mathrm{Var}(\widehat{\ell})$. That is, the antithetic estimator, $\widehat{\ell}^{(a)}$, is more accurate than the CMC estimator, $\widehat{\ell}$.

To compare the efficiencies of $\widehat{\ell}$ and $\widehat{\ell}^{(a)}$, we can consider their *relative time variance*,

$$\varepsilon = \frac{T^{(a)}\,\mathrm{Var}(\widehat{\ell}^{(a)})}{T\,\mathrm{Var}(\widehat{\ell})}\ , \tag{5.10}$$

where $T^{(a)}$ and T are the CPU times required to calculate the estimators $\widehat{\ell}^{(a)}$ and $\widehat{\ell}$, respectively. Note that

$$\mathrm{Var}(\widehat{\ell}^{(a)}) = \frac{N/2}{N^2} \left(\mathrm{Var}(H(\mathbf{X})) + \mathrm{Var}(H(\mathbf{X}^{(a)})) + 2\,\mathrm{Cov}[H(\mathbf{X}), H(\mathbf{X}^{(a)})] \right)$$
$$= \mathrm{Var}(\widehat{\ell}) + \mathrm{Cov}(H(\mathbf{X}), H(\mathbf{X}^{(a)}))/N\ .$$

Also, $T^{(a)} \leqslant T$, since the antithetic estimator, $\widehat{\ell}^{(a)}$, needs only *half* as many random numbers as its CMC counterpart, $\widehat{\ell}$. Neglecting this time advantage, the efficiency measure (5.10) reduces to

$$\varepsilon = \frac{\mathrm{Var}(\widehat{\ell}^{(a)})}{\mathrm{Var}(\widehat{\ell})} = 1 + \frac{\mathrm{Cov}[H(\mathbf{X}), H(\mathbf{X}^{(a)})]}{\mathrm{Var}(H(\mathbf{X}))}\ , \tag{5.11}$$

where the covariance is negative and can be estimated via the corresponding sample covariance.

The use of common/antithetic random variables for the case of dependent components of $\mathbf{X}$ and $\mathbf{Y}$ for strictly monotonic functions, H_1 and H_2, is presented in Rubinstein, Samorodnitsky, and Shaked [39].

■ **EXAMPLE 5.2 Stochastic Shortest Path (Continued)**

We estimate the expected length of the shortest path for the bridge network in Example 5.1 for the case where each link has an exponential weight with parameter 1. Taking a sample size of $N = 10{,}000$ obtains the CMC estimate $\widehat{\ell} = 1.159$ with an estimated variance of $5.6 \cdot 10^{-5}$, whereas the antithetic estimate is $\widehat{\ell} = 1.164$ with an estimated variance of $2.8 \cdot 10^{-5}$. Therefore, the efficiency ε of the estimator $\widehat{\ell}^{(a)}$ relative to the CMC estimator $\widehat{\ell}$ is about 2.0.

■ **EXAMPLE 5.3 Lindley's Equation**

Consider Lindley's equation for the waiting time of the $(n+1)$-st customer in a $GI/G/1$ queue :

$$W_{n+1} = \max\{W_n + U_n, 0\}, \quad W_1 = 0 .$$

See also (4.33). Here $U_n = S_n - A_{n+1}$, where S_n is the service time of the n-th customer, and A_{n+1} is the interarrival time between the n-th and $(n + 1)$-st customer. Since W_n is a monotonic function of each component $A_2, \ldots, A_n$ and $S_1, \ldots, S_{n-1}$, we can obtain variance reduction by using antithetic random variables.

5.3 CONTROL VARIABLES

The *control variables* method is a widely used variance reduction technique. We first consider the one-dimensional case. Let X be an unbiased estimator of μ, to be obtained from a simulation run. A random variable C is called a *control variable* for X if it is correlated with X and its expectation, r, is known. The control variable C is used to construct an unbiased estimator of μ with a variance smaller than that of X. This estimator,

$$X_\alpha = X - \alpha(C - r) , \tag{5.12}$$

where α is a scalar parameter, is called the *linear control variable*. The variance of X_α is given by

$$\text{Var}(X_\alpha) = \text{Var}(X) - 2\alpha \, \text{Cov}(X, C) + \alpha^2 \, \text{Var}(C)$$

(see, e.g., Problem 1.15). Consequently, the value α^* that minimizes $\text{Var}(X_\alpha)$ is

$$\alpha^* = \frac{\text{Cov}(X, C)}{\text{Var}(C)} . \tag{5.13}$$

Typically, α^* is estimated from the corresponding sample covariance and variance. Using α^*, we can write the minimal variance as

$$\text{Var}(X_{\alpha^*}) = (1 - \varrho_{XC}^2)\text{Var}(X) , \tag{5.14}$$

where ϱ_{XC} denotes the correlation coefficient of X and C. Notice that the larger $|\varrho_{XC}|$ is, the greater is the variance reduction.

Formulas (5.12)–(5.14) can be easily extended to the case of multiple control variables. To see this, let $\mathbf{C} = (C_1, \ldots, C_m)^\top$ be a (column) vector of m control variables with known mean vector $\mathbf{r} = \mathbb{E}[\mathbf{C}] = (r_1, \ldots, r_m)^\top$, where $r_i = \mathbb{E}[C_i]$. Then the vector version of (5.12) can be written as

$$X_\alpha = X - \boldsymbol{\alpha}^\top (\mathbf{C} - \mathbf{r}) , \tag{5.15}$$

where $\boldsymbol{\alpha}$ is an m-dimensional vector of parameters. The value $\boldsymbol{\alpha}^*$ that minimizes $\mathrm{Var}(X_\alpha)$ is given by

$$\boldsymbol{\alpha}^* = \Sigma_C^{-1} \boldsymbol{\sigma}_{XC} , \tag{5.16}$$

where Σ_C denotes the $m \times m$ covariance matrix of $\mathbf{C}$ and $\boldsymbol{\sigma}_{XC}$ denotes the $m \times 1$ vector whose i-th component is the covariance of X and C_i, $i = 1, \ldots, m$. The corresponding minimal variance evaluates to

$$\mathrm{Var}(X_{\alpha^*}) = (1 - R_{XC}^2)\mathrm{Var}(X) , \tag{5.17}$$

where

$$R_{XC}^2 = (\boldsymbol{\sigma}_{XC})^\top \Sigma_C^{-1} \boldsymbol{\sigma}_{XC} / \mathrm{Var}(X)$$

is the square of the so-called *multiple correlation coefficient* of X and $\mathbf{C}$. Again the larger $|R_{XC}|$ is, the greater is the variance reduction. The case where $\mathbf{X}$ is a vector with dependent components and the vector $\boldsymbol{\alpha}$ is replaced by a corresponding matrix is treated in Rubinstein and Marcus [36].

The following examples illustrate various applications of the control variables method.

■ **EXAMPLE 5.4   Stochastic Shortest Path (Continued)**

Consider again the stochastic shortest path estimation problem for the bridge network in Example 5.1. Among the control variables we can use are the lengths of the paths $\mathscr{P}_j$, $j = 1, \ldots, 4$, that is, any (or all) of

$$\begin{aligned}
C_1 &= X_1 + X_4 \\
C_2 &= X_1 + X_3 + X_5 \\
C_3 &= X_2 + X_3 + X_4 \\
C_4 &= X_2 + X_5 .
\end{aligned}$$

The expectations of the $\{C_i\}$ are easy to calculate, and each C_i is positively correlated with the length of the shortest path $H(\mathbf{X}) = \min\{C_1, \ldots, C_4\}$.

■ **EXAMPLE 5.5   Lindley's Equation (Continued)**

Consider Lindley's equation for the waiting time process $\{W_n, n = 1, 2, \ldots\}$ in the $GI/G/1$ queue; see Example 5.3. As a control variable for W_n we can take C_n, defined by the recurrence relation

$$C_{n+1} = C_n + U_n, \qquad C_1 = 0 ,$$

where $U_n = S_n - A_{n+1}$, as in the waiting time process. Obviously, C_n and W_n are highly correlated. Moreover, the expectation $r_n = \mathbb{E}[C_n]$ is known. It is $r_n = (n-1)(\mathbb{E}[S] - \mathbb{E}[A])$, where $\mathbb{E}[S]$ and $\mathbb{E}[A]$ are the expected service and interarrival times, respectively. The corresponding linear control process is

$$Y_n = W_n - \alpha(C_n - r_n) .$$

■ **EXAMPLE 5.6   Queueing Networks**

Now we return to the estimation of the expected steady-state performance $\ell = \mathbb{E}[X]$ in a queueing network. For example, suppose that X is the steady-state number of customers in the system. As a linear control random process, one may take

$$Y_t = X_t - \alpha(C_t - r_t) ,$$

where X_t is the number of customers in the original system, and C_t is the number of customers in an auxiliary *Markovian* network for which the steady-state distribution is known. The latter network must be synchronized in time with the original network.

In order to produce high correlations between the two processes, $\{X_t\}$ and $\{C_t\}$, it is desirable that both networks have similar topologies and similar loads. In addition, they must use a common stream of random numbers for generating the input variables. Expressions for the expected steady-state performance $r = \mathbb{E}[C]$, such as the expected number in the system in a Markovian network, may be found in [19].

5.4   CONDITIONAL MONTE CARLO

Let

$$\ell = \mathbb{E}[H(\mathbf{X})] = \int H(\mathbf{x}) f(\mathbf{x}) \, d\mathbf{x} \tag{5.18}$$

be some expected performance measure of a computer simulation model, where $\mathbf{X}$ is the input random variable (vector) with a pdf $f(\mathbf{x})$ and $H(\mathbf{X})$ is the sample performance measure (output random variable). Suppose that there is a random variable (or vector), $\mathbf{Y} \sim g(\mathbf{y})$, such that the conditional expectation $\mathbb{E}[H(\mathbf{X}) \mid \mathbf{Y} = \mathbf{y}]$ can be computed analytically. Since, by (1.11),

$$\ell = \mathbb{E}[H(\mathbf{X})] = \mathbb{E}[\,\mathbb{E}[H(\mathbf{X}) \mid \mathbf{Y}]\,] , \tag{5.19}$$

it follows that $\mathbb{E}[H(\mathbf{X}) \mid \mathbf{Y}]$ is an unbiased estimator of ℓ. Furthermore, it is readily seen that

$$\text{Var}(\mathbb{E}[H(\mathbf{X}) \mid \mathbf{Y}]) \leqslant \text{Var}(H(\mathbf{X})) , \tag{5.20}$$

so using the random variable $\mathbb{E}[H(\mathbf{X}) \mid \mathbf{Y}]$, instead of $H(\mathbf{X})$, leads to variance reduction. Thus conditioning *always* leads to variance reduction. To prove (5.20), we use the property (see Problem 5.7) that for any pair of random variables (U, V),

$$\text{Var}(U) = \mathbb{E}[\,\text{Var}(U \mid V)\,] + \text{Var}(\,\mathbb{E}[U \mid V]\,) . \tag{5.21}$$

Since both terms on the right-hand side are nonnegative, (5.20) immediately follows. The conditional Monte Carlo idea is sometimes referred to as *Rao-Blackwellization*. The conditional Monte Carlo algorithm is given next.

Algorithm 5.4.1: Conditional Monte Carlo

---

input : Method to generate $\mathbf{Y} \sim g$, performance function H, sample size N.
output: Estimator $\widehat{\ell}_c$ of $\ell = \mathbb{E}[H(\mathbf{X})]$.
1 Generate an iid sample $\mathbf{Y}_1, \dots \mathbf{Y}_N$ from g.
2 Calculate $\mathbb{E}[H(\mathbf{X}) \,|\, \mathbf{Y}_k]$, $k = 1, \dots, N$ analytically.
3 Set $\widehat{\ell}_c \leftarrow \frac{1}{N} \sum_{k=1}^{N} \mathbb{E}[H(\mathbf{X}) \,|\, \mathbf{Y}_k]$.
4 **return** $\widehat{\ell}_c$

Algorithm 5.4.1 requires that a random variable $\mathbf{Y}$ be found, such that $\mathbb{E}[H(\mathbf{X}) \,|\, \mathbf{Y} = \mathbf{y}]$ is known analytically for all $\mathbf{y}$. Moreover, for Algorithm 5.4.1 to be of practical use, the following conditions must be met:

(a) $\mathbf{Y}$ should be easy to generate.

(b) $\mathbb{E}[H(\mathbf{X}) \,|\, \mathbf{Y} = \mathbf{y}]$ should be readily computable for all values of $\mathbf{y}$.

(c) $\mathbb{E}[\text{Var}(H(\mathbf{X}) \,|\, \mathbf{Y})]$ should be large relative to $\text{Var}(\mathbb{E}[H(\mathbf{X}) \,|\, \mathbf{Y}])$.

■ **EXAMPLE 5.7 Random Sums**

Consider the estimation of

$$\ell = \mathbb{P}(S_R \leqslant x) = \mathbb{E}[I_{\{S_R \leqslant x\}}] \,,$$

where

$$S_R = \sum_{i=1}^{R} X_i \,,$$

R is a random variable with a given distribution and the $\{X_i\}$ are iid with $X_i \sim F$ and independent of R. Let F^r be the cdf of the random variable S_r for fixed $R = r$. Noting that

$$F^r(x) = \mathbb{P}\left(\sum_{i=1}^{r} X_i \leqslant x\right) = \mathbb{E}\left[F\left(x - \sum_{i=2}^{r} X_i\right)\right],$$

we obtain

$$\ell = \mathbb{E}\left[\mathbb{E}\left[I_{\{S_R \leqslant x\}} \,\bigg|\, \sum_{i=2}^{R} X_i\right]\right] = \mathbb{E}\left[F\left(x - \sum_{i=2}^{R} X_i\right)\right].$$

Thus, we can take the following estimator of ℓ based on conditioning:

$$\widehat{\ell}_c = \frac{1}{N} \sum_{k=1}^{N} F\left(x - \sum_{i=2}^{R_k} X_{ki}\right). \tag{5.22}$$

5.4.1 Variance Reduction for Reliability Models

Next, we present two variance reduction techniques for reliability models based on conditioning. As in Example 4.2 on page 110, we are given an unreliable system of n components, each of which can be either functioning or failed, with a structure function H that determines the state of the system (working or failed) as a function of the states of the components. The component states $X_1, \ldots, X_n$ are assumed to be independent, with reliabilities $\{p_i\}$ and unreliabilities $\{q_i\}$, where $q_i = 1 - p_i$. The probability of system failure — the unreliability of the system — is thus $\bar{r} = \mathbb{P}(H(\mathbf{X}) = 0)$. In typical applications the unreliability is very small and is difficult to estimate via CMC.

5.4.1.1 Permutation Monte Carlo

Permutation Monte Carlo is a conditional Monte Carlo technique for network reliability estimation (see Elperin et al. [12]). Here the components are unreliable links in a network, such as in Example 4.2. The system state $H(\mathbf{X})$ is the indicator of the event that certain preselected nodes are connected by functioning links. Suppose that we need to estimate the system's unreliability $\bar{r} = \mathbb{P}(H(\mathbf{X}) = 0)$.

To apply the conditional Monte Carlo idea, we view the static network as a snapshot of a *dynamic* network at time $t = 1$. In this dynamic system, the links are repaired independently of each other with an exponential repair time rate of $\mu_e = -\ln(q_e), e = 1, \ldots, n$. At time $t = 0$ all links are failed. The state of the links at time t is given by the vector $\mathbf{X}_t$. Note that $\{\mathbf{X}_t, t \geqslant 0\}$ is a Markov jump process with state space $\{0, 1\}^n$. Since the probability of each link e being operational at time $t = 1$ is p_e, the reliability of the dynamic network at time $t = 1$ is exactly the same as the reliability of the original network.

Let Π denote the *order* in which the links become operational, and let $S_0, S_0 + S_1, \ldots, S_0 + \cdots + S_{n-1}$ be the times at which those links are constructed. Π is a random variable that takes values in the space of permutations of the set of links $\mathcal{E} = \{1, \ldots, n\}$ — hence, the name *permutation Monte Carlo*. For any permutation $\pi = (e_1, e_2, \ldots, e_n)$, define $\mathcal{E}_0 = \mathcal{E}$ and $\mathcal{E}_i = \mathcal{E}_{i-1} \setminus \{e_i\}$, $1 \leqslant i \leqslant n - 1$. Thus $\mathcal{E}_i$ corresponds to the set of links that are still failed after i links have been repaired. Let $b = b(\pi)$ be the number of repairs required (in the order defined by π) to bring the network up. This is called the *critical number* or *construction anchor* for π.

From the theory of Markov jump processes (see Section 1.13.5), it follows that

$$\mathbb{P}(\Pi = \pi) = \prod_{i=1}^{n} \frac{\mu_{e_i}}{\lambda_{i-1}}, \tag{5.23}$$

where $\lambda_i = \sum_{e \in \mathcal{E}_i} \mu_e$. More important, conditional on Π the sojourn times $S_0, \ldots, S_{n-1}$ are independent and each S_i is exponentially distributed with parameter λ_i, $i = 0, \ldots, n - 1$. By conditioning on Π, we have

$$\bar{r} = \sum_{\pi} \mathbb{P}[H(\mathbf{X}_1) = 0 \,|\, \Pi = \pi] \, \mathbb{P}[\Pi = \pi] = \mathbb{E}[g(\Pi)], \tag{5.24}$$

with

$$g(\pi) = \mathbb{P}[H(\mathbf{X}_1) = 0 \,|\, \Pi = \pi]. \tag{5.25}$$

From the definitions of S_i and b, we see that $g(\pi)$ is equal to the probability that the sum of b independent exponential random variables with rates $\lambda_i, i = 0, 1, \ldots, b-1$

exceeds 1. This can be computed exactly, for example, by using convolutions. Specifically, we have

$$g(\pi) = 1 - F_0 \star \cdots \star F_{b-1}(1) \,,$$

where F_i is the cdf of the $\mathsf{Exp}(\lambda_i)$ distribution, and $\star$ means convolution; that is,

$$F \star G(t) = \int_0^t F(t - x) \, \mathrm{d}G(x) \,.$$

Alternatively, it can be shown (e.g., see [30]) that

$$g(\pi) = (1, 0, \ldots, 0) \, \mathrm{e}^A \, (1, \ldots, 1)^\top \,, \tag{5.26}$$

where A is the matrix with diagonal elements $-\lambda_0, \ldots, -\lambda_{b-1}$ and upper-diagonal elements $\lambda_0, \ldots, \lambda_{b-2}$ and 0 elsewhere. Here e^A is defined as the *matrix exponential* $\sum_{k=0}^\infty A^k / k!$.

Let $\Pi_1, \ldots, \Pi_N$ be iid random permutations, each distributed according to Π; then

$$\widehat{\overline{r}} = \frac{1}{N} \sum_{k=1}^N g(\Pi_k) \tag{5.27}$$

is an unbiased estimator for $\overline{r}$. This leads to the following algorithm for estimating the unreliability $\overline{r}$:

Algorithm 5.4.2: Permutation Monte Carlo

---

input : Structure function H, component unreliabilities $\{q_i\}$, sample size N.
output: Estimator $\widehat{\overline{r}}$ of the system unreliability $\overline{r}$.

1 for $k = 1$ **to** N **do**
2   Draw a random permutation Π according to (5.23). A simple way, similar to Algorithm 2.10.1, is to draw $Y_e \sim \mathsf{Exp}(\mu_e)$, $e = 1, \ldots, n$ independently and return Π as the indices of the (increasing) ordered values.
3   Determine the critical number b and the rates $\lambda_i, i = 1, \ldots, b - 1$.
4   Evaluate the conditional probability $g(\Pi)$ exactly, for example, via (5.26).
5 Deliver (5.27) as the estimator for $\overline{r}$.
6 return $\widehat{\overline{r}}$

5.4.1.2 Conditioning Using Minimal Cuts The second method used to estimate unreliability efficiently, developed by Ross [33], employs the concept of a minimal cut. A state vector $\mathbf{x}$ is called a *cut vector* if $H(\mathbf{x}) = 0$. If in addition $H(\mathbf{y}) = 1$ for all $\mathbf{y} > \mathbf{x}$, then $\mathbf{x}$ is called the *minimal cut vector*. Note that $\mathbf{y} > \mathbf{x}$ means that $y_i \geq x_i$, $i = 1, \ldots, n$, with $y_i > x_i$ for some i. If $\mathbf{x}$ is a minimal cut vector, the set $C = \{i : x_i = 0\}$ is called a *minimal cut set*. That is, a minimal cut set is a minimal set of components whose *failure* ensures the failure of the system. If $C_1, \ldots, C_m$ denote all the minimal cut sets, the system is functioning if and only if at least one component in each of the cut sets is functioning. It follows that $H(\mathbf{x})$ can be written as

$$H(\mathbf{x}) = \prod_{j=1}^m \max_{i \in C_j} x_i = \prod_{j=1}^m \left(1 - \prod_{i \in C_j} (1 - x_i) \right) \,. \tag{5.28}$$

To proceed, we need the following proposition, which is adapted from [33].

Proposition 5.4.1 Let $Y_1, \ldots, Y_m$ be Bernoulli random variables (possibly dependent) with success parameters $a_1, \ldots, a_m$. Define $S = \sum_{j=1}^m Y_j$ and let $a = \mathbb{E}[S] = \sum_{j=1}^m a_j$. Let J be a discrete uniform random variable on $\{1, \ldots, m\}$ independent of $Y_1, \ldots, Y_m$. Last, let R be any random variable that is independent of J. Then

$$\mathbb{P}(J = j \,|\, Y_J = 1) = \frac{a_j}{a}, \quad j = 1, \ldots, m , \tag{5.29}$$

and

$$\mathbb{E}[SR] = \mathbb{E}[S]\, \mathbb{E}[R \,|\, Y_J = 1] . \tag{5.30}$$

Proof: To derive formula (5.29) write, using Bayes' formula

$$\mathbb{P}(J = j \,|\, Y_J = 1) = \frac{\mathbb{P}(Y_J = 1 \,|\, J = j)\, \mathbb{P}(J = j)}{\sum_{i=1}^m \mathbb{P}(Y_J = 1 \,|\, J = i)\, \mathbb{P}(J = i)} .$$

Taking into account that $\mathbb{P}(Y_J = 1 \,|\, J = j) = \mathbb{P}(Y_j = 1 \,|\, J = j) = \mathbb{P}(Y_j = 1) = a_j$, the result follows. To prove (5.30), we write

$$\mathbb{E}[SR] = \sum_{j=1}^m \mathbb{E}[R Y_j] = \sum_{j=1}^m \mathbb{E}[R \,|\, Y_j = 1]\, \mathbb{P}(Y_j = 1)$$

$$= a \sum_{j=1}^m \mathbb{E}[R \,|\, Y_j = 1]\, \frac{a_j}{a} .$$

Since $a = \mathbb{E}[S]$ and, by (5.29), $\{a_j/a\}$ is the conditional distribution of J given $Y_J = 1$, (5.30) follows. $\qquad\square$

We will apply Proposition 5.4.1 to the estimation of the unreliability $\bar{r} = \mathbb{P}(H(\mathbf{X}) = 0)$. Let $Y_j = \prod_{i \in C_j}(1 - X_i)$, $j = 1, \ldots, m$, where, as before, $\{C_j\}$ denotes the collection of minimal cut sets. Thus, Y_j is the indicator of the event that all components in C_j are *failed*. Note that $Y_j \sim \mathsf{Ber}(a_j)$, with

$$a_j = \prod_{i \in C_j} q_i . \tag{5.31}$$

Let $S = \sum_{j=1}^m Y_j$ and $a = \mathbb{E}[S] = \sum_{j=1}^m a_j$. By (5.28) we have $\bar{r} = \mathbb{P}(S > 0)$, and by (5.30) it follows that

$$\bar{r} = \mathbb{E}[S]\, \mathbb{E}\left[\frac{I_{\{S>0\}}}{S} \,\bigg|\, Y_J = 1 \right] = \mathbb{E}\left[\frac{a}{S} \,\bigg|\, Y_J = 1 \right],$$

where conditional on $Y_J = 1$ the random variable J takes the value j with probability a_j/a for $j = 1, \ldots, m$. This leads to the following algorithm for estimating the unreliability $\bar{r}$:

Algorithm 5.4.3: Conditioning via Minimal Cuts

---

input : Minimal cut sets $C_1, \ldots, C_m$, component reliabilities $\{p_i\}$, sample size N.

output: Estimator $\widehat{\bar{r}}$ of the system unreliability $\bar{r}$.

1 for $k = 1$ **to** N **do**

2 | Simulate random variable J according to $\mathbb{P}(J = j) = a_j/a$, $j = 1, \ldots, m$.

3 | Set X_i equal to 0 for all $i \in C_J$ and generate the values of all other X_i, $i \notin C_J$ from their corresponding $\text{Ber}(p_i)$ distributions.

4 | Let S_k be the number of minimal cut sets that have all their components failed (note that $S_k \geqslant 1$).

5 Set $\widehat{\bar{r}} \leftarrow N^{-1} \sum_{k=1}^{N} a/S_k$ as an estimator of $\bar{r} = \mathbb{P}(S > 0)$.

6 return $\widehat{\bar{r}}$

It is readily seen that when a, the mean number of failed minimal cuts, is very small, the resulting estimator $\frac{a}{S}$ will have a very small variance. In addition, we could apply importance sampling to the conditional estimator $\frac{a}{S}$ to further reduce the variance.

5.5 STRATIFIED SAMPLING

Stratified sampling is closely related to both the composition method of Section 2.3.3 and the conditional Monte Carlo method discussed in the previous section. As always, we wish to estimate

$$\ell = \mathbb{E}[H(\mathbf{X})] = \int H(\mathbf{x})f(\mathbf{x})\,\mathrm{d}\mathbf{x}\,.$$

Suppose that $\mathbf{X}$ can be generated via the composition method. Thus, we assume that there exists a random variable Y taking values in $\{1, \ldots, m\}$, say, with known probabilities $\{p_i, i = 1, \ldots, m\}$, and we assume that it is easy to sample from the conditional distribution of $\mathbf{X}$ given Y. The events $\{Y = i\}, i = 1, \ldots, m$ form disjoint subregions, or *strata* (singular: stratum), of the sample space Ω, hence the name *stratification*. We use the conditioning formula (1.11) and write

$$\ell = \mathbb{E}[\mathbb{E}[H(\mathbf{X}) \,|\, Y]] = \sum_{i=1}^{m} p_i\, \mathbb{E}[H(\mathbf{X}) \,|\, Y = i]\,. \tag{5.32}$$

This representation suggests that we can estimate ℓ via the following *stratified sampling estimator*

$$\widehat{\ell}^{\,s} = \sum_{i=1}^{m} p_i \frac{1}{N_i} \sum_{j=1}^{N_i} H(\mathbf{X}_{ij})\,, \tag{5.33}$$

where $\mathbf{X}_{ij}$ is the j-th observation from the conditional distribution of $\mathbf{X}$ given $Y = i$. Here N_i is the sample size assigned to the i-th stratum. The variance of the stratified sampling estimator is given by

$$\mathrm{Var}\left(\widehat{\ell}^{\,s}\right) = \sum_{i=1}^{m} \frac{p_i^2}{N_i} \mathrm{Var}(H(\mathbf{X}) \,|\, Y = i) = \sum_{i=1}^{m} \frac{p_i^2 \sigma_i^2}{N_i}\,, \tag{5.34}$$

where $\sigma_i^2 = \mathrm{Var}(H(\mathbf{X}) \,|\, Y = i)$.

How the strata should be chosen depends very much on the problem at hand. However, for a given particular choice of the strata, the sample sizes $\{N_i\}$ can be obtained in an optimal manner, as given in the next theorem.

Theorem 5.5.1 (Stratified Sampling) *Assuming that a maximum number of N samples can be collected, that is, $\sum_{i=1}^{m} N_i = N$, the optimal value of N_i is given by*

$$N_i^* = N \, \frac{p_i \, \sigma_i}{\sum_{j=1}^{m} p_j \, \sigma_j} \,, \tag{5.35}$$

which gives a minimal variance of

$$\mathrm{Var}(\widehat{\ell}^{*s}) = \frac{1}{N} \left[\sum_{i=1}^{m} p_i \, \sigma_i \right]^2 . \tag{5.36}$$

Proof: The proof is straightforward and uses Lagrange multipliers; it is left as an exercise to the reader (see Problem 5.10). $\qquad\square$

Theorem 5.5.1 asserts that the minimal variance of $\widehat{\ell}^s$ is attained for sample sizes N_i that are proportional to $p_i \, \sigma_i$. A difficulty is that although the probabilities p_i are assumed to be known, the standard deviations $\{\sigma_i\}$ are usually unknown. In practice, one would estimate the $\{\sigma_i\}$ from "pilot" runs and then proceed to estimate the optimal sample sizes, N_i^*, from (5.35).

A simple stratification procedure that can achieve variance reduction without requiring prior knowledge of σ_i^2 and $H(\mathbf{X})$, is presented next.

Proposition 5.5.1 *Let the sample sizes N_i be proportional to p_i, that is, $N_i = p_i \, N$, $i = 1, \ldots m$. Then*

$$\mathrm{Var}(\widehat{\ell}^s) \leqslant \mathrm{Var}(\widehat{\ell}) \,.$$

Proof: Substituting $N_i = p_i \, N$ in (5.34) yields $\mathrm{Var}(\widehat{\ell}^s) = \frac{1}{N} \sum_{i=1}^{m} p_i \, \sigma_i^2$. The result now follows from

$$N \, \mathrm{Var}(\widehat{\ell}) = \mathrm{Var}(H(\mathbf{X})) \geqslant \mathbb{E}[\mathrm{Var}(H(\mathbf{X}) \,|\, Y)] = \sum_{i=1}^{m} p_i \, \sigma_i^2 = N \, \mathrm{Var}(\widehat{\ell}^s),$$

where we have used (5.21) in the inequality. $\qquad\square$

Proposition 5.5.1 states that the estimator $\widehat{\ell}^s$ is more accurate than the CMC estimator $\widehat{\ell}$. It effects stratification by favoring those events $\{Y = i\}$ whose probabilities p_i are largest. Intuitively, this cannot, in most cases, be an optimal assignment, since information on σ_i^2 and $H(\mathbf{X})$ is ignored.

In the special case of equal weights ($p_i = 1/m$ and $N_i = N/m$), the estimator (5.33) reduces to

$$\widehat{\ell}^s = \frac{1}{N} \sum_{i=1}^{m} \sum_{j=1}^{N/m} H(\mathbf{X}_{ij}) \,, \tag{5.37}$$

and the method is known as the *systematic sampling method* (e.g., see Cochran [9]).

5.6 MULTILEVEL MONTE CARLO

When estimating the expected value $\ell = \mathbb{E}[Y]$ of a functional $Y = H(\mathbf{X})$ of a diffusion process $\mathbf{X} = \{X_t\}$ (see Section 2.9) by simulation, there are typically two sources of error: estimation error and bias. If an iid samples $Y_1, \ldots, Y_N$ can be simulated *exactly* from the distribution of Y, then their sample mean is an estimator of $\mathbb{E}[Y]$, and the estimation error, as usual, is expressed in terms of the sample variance of the $\{Y_i\}$. However, it is often not possible to obtain exact copies of Y because Y could depend on the whole (continuous) path of the diffusion process. For example, $\mathbf{X}$ could be a Wiener process on the interval $[0,1]$ and Y its maximum value. Simulating the whole process is not feasible, but it is easy to simulate the process at any grid points $0, 1/n, 2/n, \ldots, n/n$, via Algorithm 2.8.1 and approximate $Y = \max_{0 \leqslant t \leqslant 1} X_t$ with $\widetilde{Y} = \max_{i=0,\ldots,n} X_{i/n}$. However, since $\mathbb{E}[\widetilde{Y}] < \mathbb{E}[Y]$, this introduces a bias. This bias goes to zero as the grid gets finer; that is, as $n \to \infty$. A similar bias is introduced when simulating a stochastic differential equation via Euler's method; see Algorithm 2.9.1.

Thus, to obtain a small overall error, both the number of grid points, $n+1$, and the sample size, N, have to be large enough. However, taking both n and N large, say $N = 10^6$ and $n = 10^4$, may lead to very slow simulations. The multilevel Monte Carlo method of Giles [16] significantly reduces the computational effort required to obtain both small bias and estimation error, by simulating the process at *multiple* grids. As such, the method resembles the multigrid methods in numerical analysis [43].

To explain the multilevel methodology, we consider first the two-level case, which uses a fine and a coarse grid. Let $\mathbf{X}^{(2)}$ be an approximation of $\mathbf{X}$ simulated on the fine grid, and let $Y^{(2)} = H(\mathbf{X}^{(2)})$ be the corresponding performance. Using the *same* path we can generate an approximation $\mathbf{X}^{(1)}$ of $\mathbf{X}$ on the coarse grid. For example, Figure 5.2 shows two simulated paths, $\mathbf{X}^{(2)}$ (solid, black) and $\mathbf{X}^{(1)}$ (dashed, red), evaluated at a fine grid and a coarse grid. The coarse grid is a subset of the fine grid, and $\mathbf{X}^{(1)}$ is simply taken as the subvector of $\mathbf{X}^{(2)}$ evaluated at the coarse grid points. Let $Y^{(1)}$ be the corresponding performance.

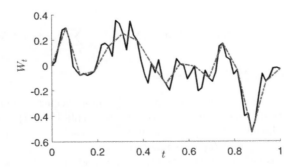

Figure 5.2: The solid (black) line represents an approximation of the Wiener process evaluated at a grid with step size $1/64$. The dashed (red) path shows the coarsened version on a grid with step size $1/16$. The performances $Y^{(1)}$ and $Y^{(2)}$ could, for example, be the maximum heights of the paths.

Obviously, $Y^{(2)}$ and $Y^{(1)}$ are highly positively correlated. Hence, the latter could be used as a *control variable* for the former. This suggests the control variable estimator

$$\frac{1}{N_2} \sum_{i=1}^{N_2} \left\{ Y_i^{(2)} - \alpha \left(Y_i^{(1)} - \mathbb{E}\left[Y^{(1)} \right] \right) \right\},$$

where $\{(Y_i^{(1)}, Y_i^{(2)})\}$ are N_2 independent copies of $(Y^{(1)}, Y^{(2)})$. Note that any value for α will lead to variance reduction if the expectation $\mathbb{E}[Y^{(1)}]$ is known. Unfortunately, the expectation is not known. However, it can be replaced with an estimator

$$\frac{1}{N_1} \sum_{i=1}^{N_1} \widetilde{Y}_i^{(1)}$$

evaluated at the coarser level, and independent of $\{(Y_i^{(1)}, Y_i^{(2)})\}$. If, further, α is set to 1, then the estimator for $\ell = \mathbb{E}[Y]$ reduces to

$$\frac{1}{N_1} \sum_{i=1}^{N_1} \widetilde{Y}_i^{(1)} + \frac{1}{N_2} \sum_{i=1}^{N_2} \left(Y_i^{(2)} - Y_i^{(1)} \right) \stackrel{\text{def}}{=} Z_1 + Z_2 \,,$$

where Z_1 and Z_2 are independent. The first term, Z_1, estimates the expected performance at the coarse level, $\mathbb{E}[Y^{(1)}]$, and Z_2 estimates the expected bias between the fine and coarse level performances; that is, $\mathbb{E}[Y^{(2)}] - \mathbb{E}[Y^{(1)}]$. Since, the difference between $\mathbb{E}[Y^{(2)}]$ and $\mathbb{E}[Y]$ may still be significant, it may be advantageous to use $K > 2$ grids, with step sizes $1/n_1 > \cdots > 1/n_K$, say. This leads to the K-level estimator

$$\widehat{\ell} = Z_1 + Z_2 + \cdots + Z_K,$$

where $Z_1, \ldots, Z_K$ are independent of each other, with

$$Z_1 = \frac{1}{N_1} \sum_{i=1}^{N_1} Y_i^{(1)}$$

and

$$Z_k = \frac{1}{N_k} \sum_{i=1}^{N_k} \left(Y_i^{(k)} - Y_i^{(k-1)} \right), \quad k = 2, \ldots, K \,,$$

where each $Y_i^{(k)}$ is distributed as $Y^{(k)} = H(\mathbf{X}^{(k)})$, and $\mathbf{X}^{(k)}$ is a path obtained using the k-th level grid. Note that both Z_k and Z_{k-1} contain variables $\{Y_i^{(k-1)}\}$. It is important to realize that these are obtained from *different* and independent simulations. For the 2-level case we emphasized this difference by using $Y_i^{(1)}$ and $\widetilde{Y}_i^{(1)}$, respectively.

For simplicity, we assume that the process $\mathbf{X}$ is to be simulated on the interval $[0, 1]$ and that the k-level grid points are $0, 1/n_k, 2/n_k, \ldots, 1$, with $n_k = M^k$, where M is a small integer, such as $M = 4$. It is useful to take the sample size $\{N_k\}$ proportional to the step sizes $\{1/n_k\}$; but see also Remark 5.6.1. This leads to the following multilevel Monte Carlo algorithm:

Algorithm 5.6.1: Multilevel Monte Carlo

---

    input : Mesh factor M, number of levels K, simulation effort N.
    output: Estimator $\widehat{\ell}$ of $\ell = \mathbb{E}[H(\mathbf{X})]$.

1 for $k = 1$ **to** K **do**
2 $n_k \leftarrow M^k$
3 $N_k \leftarrow \lceil N/n_k \rceil$

4 for $i = 1$ **to** N_1 **do**
5   Generate $\mathbf{X}_i^{(1)}$ `// the i-th path at the coarsest level`
6 $Y_i^{(1)} \leftarrow H(\mathbf{X}_i^{(1)})$

7 $Z_1 \leftarrow \sum_{i=1}^{N_1} Y_i^{(1)}/N_1$
8 for $k = K$ **to** 2 **do**
9   **for** $i = 1$ **to** N_k **do**
10   Generate $\mathbf{X}_i^{(k)}$ `// the i-th path at level k`
11   Coarsen $\mathbf{X}_i^{(k)}$ to $\mathbf{X}_i^{(k-1)}$
12 $Y_i^{(k)} \leftarrow H(\mathbf{X}_i^{(k)})$
13 $Y_i^{(k-1)} \leftarrow H(\mathbf{X}_i^{(k-1)})$

14 $Z_k \leftarrow \sum_{i=1}^{N_k} \left(Y_i^{(k)} - Y_i^{(k-1)} \right)/N_k$

15 $\widehat{\ell} \leftarrow \sum_{k=1}^{K} Z_k$
16 return $\widehat{\ell}$

■ EXAMPLE 5.8

We estimate the expected maximum of a Wiener process in the interval $[0,1]$ via Algorithm 5.6.1, using the following parameters: $M = 4$, $K = 10$, and $N = 10^6$. Figure 5.2 depicts typical pairs of paths that are generated at level $k = 3$ of the algorithm. The solid (black) path is generated on a grid with step size $1/4^3 = 1/64$ and the dashed (red) path is the coarsened version, on the subgrid with step size $1/4^2 = 1/16$.

    Typical outcomes of $Z_1, \ldots, Z_{10}$ are $0.557, 0.1094, 0.0629, 0.0326, 0.0178,$ $0.0086, 0.00429, 0.0020, 0.0023,$ and 0, which suggests that $K = 10$ is high enough to eliminate most of the bias. Adding the $\{Z_k\}$ gives the estimate $\widehat{\ell} = 0.7967$, which is close the exact value $\ell = \sqrt{2/\pi} \approx 0.7978$ (e.g., [24]). The simulation time was 1.5 seconds. Note that the majority of paths (250,000 out of a total of 333,337) are simulated on the coarsest grid, with $M + 1 = 5$ points, whereas the finest grid has only 1 simulation. In contrast, simulating 333,337 paths at the highest resolution (1,048,577 points) would take a very long time.

Remark 5.6.1 (Choice of Parameters) The optimal choice for the number of levels K, grid factor M, and simulation effort N is largely problem dependent. Commonly used values for M are $M = 4$ and $M = 2$. For a given M, the number of levels K should be chosen large enough such that $\mathbb{E}[Y^{(K)}]$ is close enough to $\mathbb{E}[Y]$. This can be assessed by investigating how fast $\{Z_k, k = 2, 3, \ldots, K\}$ converges to zero. The parameter N regulates the overall simulation effort. By increasing N, the accuracy of the estimator is improved. In [16] an asymptotic complexity analysis is given for the multilevel Monte Carlo estimator $\widehat{\ell}$ as $K \to \infty$. The analysis suggests that the optimal sample size N_k should be chosen proportional to $\sqrt{V_k/n_k}$, where V_k is the variance of Z_k. The $\{V_k\}$ could be estimated via a pilot run.

5.7   IMPORTANCE SAMPLING

The most fundamental variance reduction technique is *importance sampling*. As we will see below, importance sampling often leads to a dramatic variance reduction (sometimes on the order of millions, in particular when estimating rare event probabilities), while with all of the above-mentioned variance reduction techniques only a moderate reduction, typically up to 10-fold, can be achieved. Importance sampling involves choosing a sampling distribution that favors important samples. Let, as before,

$$\ell = \mathbb{E}_f[H(\mathbf{X})] = \int H(\mathbf{x})\, f(\mathbf{x})\, d\mathbf{x}\,, \tag{5.38}$$

where H is the sample performance and f is the probability density of $\mathbf{X}$. For reasons that will become clear shortly, we add a subscript f to the expectation to indicate that it is taken with respect to the density f.

Let g be another probability density such that $H f$ is *dominated* by g. That is, $g(\mathbf{x}) = 0 \Rightarrow H(\mathbf{x}) f(\mathbf{x}) = 0$. Using the density g, we can represent ℓ as

$$\ell = \int H(\mathbf{x})\, \frac{f(\mathbf{x})}{g(\mathbf{x})}\, g(\mathbf{x})\, d\mathbf{x} = \mathbb{E}_g\left[H(\mathbf{X})\, \frac{f(\mathbf{X})}{g(\mathbf{X})}\right]\,, \tag{5.39}$$

where the subscript g means that the expectation is taken with respect to g. Such a density is called the *importance sampling* density, *proposal* density, or *instrumental* density (as we use g as an instrument to obtain information about ℓ). Consequently, if $\mathbf{X}_1, \ldots, \mathbf{X}_N$ is a *random sample* from g, that is, $\mathbf{X}_1, \ldots, \mathbf{X}_N$ are iid random vectors with density g, then

$$\widehat{\ell} = \frac{1}{N} \sum_{k=1}^{N} H(\mathbf{X}_k)\, \frac{f(\mathbf{X}_k)}{g(\mathbf{X}_k)} \tag{5.40}$$

is an unbiased estimator of ℓ. This estimator is called the *importance sampling estimator*. The ratio of densities,

$$W(\mathbf{x}) = \frac{f(\mathbf{x})}{g(\mathbf{x})}\,, \tag{5.41}$$

is called the *likelihood ratio*. For this reason the importance sampling estimator is also called the *likelihood ratio estimator*. In the particular case where there is no change of measure, that is, $g = f$, we have $W = 1$, and the likelihood ratio estimator in (5.40) reduces to the usual CMC estimator.

5.7.1   Weighted Samples

The likelihood ratios need only be known *up to a constant*, that is, $W(\mathbf{X}) = c\, w(\mathbf{X})$ for some known function $w(\cdot)$. Since $\mathbb{E}_g[W(\mathbf{X})] = 1$, we can write $\ell = \mathbb{E}_g[H(\mathbf{X})\, W(\mathbf{X})]$ as

$$\ell = \frac{\mathbb{E}_g[H(\mathbf{X})\, W(\mathbf{X})]}{\mathbb{E}_g[W(\mathbf{X})]}\,.$$

This suggests, as an alternative to the standard likelihood ratio estimator (5.41), the following *weighted sample estimator*:

$$\widehat{\ell}_w = \frac{\sum_{k=1}^{N} H(\mathbf{X}_k)\, w_k}{\sum_{k=1}^{N} w_k}\,. \tag{5.42}$$

Here, the $\{w_k\}$, with $w_k = w(\mathbf{X}_k)$, are interpreted as *weights* of the random sample $\{\mathbf{X}_k\}$, and the sequence $\{(\mathbf{X}_k, w_k)\}$ is called a *weighted (random) sample* from $g(\mathbf{x})$. Similar to the regenerative ratio estimator in Chapter 4, the weighted sample estimator (5.42) introduces some bias, which tends to 0 as N increases. Loosely speaking, we may view the weighted sample $\{(\mathbf{X}_k, w_k)\}$ as a representation of $f(\mathbf{x})$ in the sense that $\ell = \mathbb{E}_f[H(\mathbf{X})] \approx \widehat{\ell}_w$ for any function $H(\cdot)$.

5.7.2   Variance Minimization Method

Since the choice of the importance sampling density g is crucially linked to the variance of the estimator $\widehat{\ell}$ in (5.40), we consider next the problem of minimizing the variance of $\widehat{\ell}$ with respect to g, that is,

$$\min_{g} \operatorname{Var}_g \left(H(\mathbf{X}) \frac{f(\mathbf{X})}{g(\mathbf{X})} \right). \tag{5.43}$$

It is not difficult to prove (e.g., see Rubinstein and Melamed [37] and Problem 5.14) that the solution of the problem (5.43) is

$$g^*(\mathbf{x}) = \frac{|H(\mathbf{x})|\, f(\mathbf{x})}{\int |H(\mathbf{x})|\, f(\mathbf{x})\, \mathrm{d}\mathbf{x}}. \tag{5.44}$$

In particular, if $H(\mathbf{x}) \geqslant 0$ — which we will assume from now on — then

$$g^*(\mathbf{x}) = \frac{H(\mathbf{x})\, f(\mathbf{x})}{\ell} \tag{5.45}$$

and

$$\operatorname{Var}_{g^*}(\widehat{\ell}) = \operatorname{Var}_{g^*}(H(\mathbf{X})W(\mathbf{X})) = \operatorname{Var}_{g^*}(\ell) = 0.$$

The density g^* as per (5.44) and (5.45) is called the *optimal importance sampling density*.

■ **EXAMPLE 5.9**

Let $X \sim \mathsf{Exp}(u^{-1})$ and $H(X) = I_{\{X \geqslant \gamma\}}$ for some $\gamma > 0$. Let f denote the pdf of X. Consider the estimation of

$$\ell = \mathbb{E}_f[H(X)] = \int_{\gamma}^{\infty} u^{-1}\, \mathrm{e}^{-x\, u^{-1}}\, \mathrm{d}x = \mathrm{e}^{-\gamma\, u^{-1}}.$$

We have

$$g^*(x) = H(x)\, f(x)\, \ell^{-1} = I_{\{x \geqslant \gamma\}}\, u^{-1}\, \mathrm{e}^{-x\, u^{-1}}\, \mathrm{e}^{\gamma u^{-1}} = I_{\{x \geqslant \gamma\}}\, u^{-1}\, \mathrm{e}^{-(x-\gamma)\, u^{-1}}.$$

Thus, the optimal importance sampling distribution of X is the *shifted* exponential distribution. Note that Hf is dominated by g^* but f itself is not dominated by g^*. Since g^* is optimal, the likelihood ratio estimator $\widehat{\ell}$ is constant. Namely, with $N = 1$,

$$\widehat{\ell} = H(X)\, W(X) = \frac{H(X)f(X)}{H(X)f(X)/\ell} = \ell.$$

It is important to realize that, although (5.40) is an unbiased estimator for *any* pdf g dominating Hf, not all such pdfs are appropriate. One of the main rules for choosing a good importance sampling pdf is that the estimator (5.40) should have finite variance. This is equivalent to the requirement that

$$\mathbb{E}_g\left[H^2(\mathbf{X})\frac{f^2(\mathbf{X})}{g^2(\mathbf{X})}\right] = \mathbb{E}_f\left[H^2(\mathbf{X})\frac{f(\mathbf{X})}{g(\mathbf{X})}\right] < \infty . \tag{5.46}$$

This suggests that g should not have a "lighter tail" than f and that, preferably, the likelihood ratio, f/g, should be bounded.

In general, implementation of the optimal importance sampling density g^* as per (5.44) and (5.45) is problematic. The main difficulty lies in the fact that, to derive $g^*(\mathbf{x})$, we need to know ℓ. But ℓ is precisely the quantity we want to estimate from the simulation!

In most simulation studies the situation is even worse, since the analytical expression for the sample performance H is unknown in advance. To overcome this difficulty, we could perform a pilot run with the underlying model, obtain a sample $H(\mathbf{X}_1), \ldots, H(\mathbf{X}_N)$, and then use it to estimate g^*. It is important to note that sampling from such an artificially constructed density can be a very complicated and time-consuming task, especially when g is a high-dimensional density.

Remark 5.7.1 (Degeneracy of the Likelihood Ratio Estimator) The likelihood ratio estimator $\widehat{\ell}$ in (5.40) suffers from a form of degeneracy in the sense that the distribution of $W(\mathbf{X})$ under the importance sampling density g may become increasingly skewed as the dimensionality n of $\mathbf{X}$ increases. That is, $W(\mathbf{X})$ may take values close to 0 with high probability, but may also take very large values with a small though significant probability. As a consequence, the variance of $W(\mathbf{X})$ under g may become very large for large n. As an example of this degeneracy, assume, for simplicity, that the components in $\mathbf{X}$ are iid, under both f and g. Hence, both $f(\mathbf{x})$ and $g(\mathbf{x})$ are the products of their marginal pdfs. Suppose that the marginal pdfs of each component X_i are f_1 and g_1, respectively. We can then write $W(\mathbf{X})$ as

$$W(\mathbf{X}) = \exp\sum_{i=1}^{n} \ln \frac{f_1(X_i)}{g_1(X_i)} . \tag{5.47}$$

Using the law of large numbers, the random variable $\sum_{i=1}^{n} \ln\left(f_1(X_i)/g_1(X_i)\right)$ is approximately equal to $n\,\mathbb{E}_{g_1}[\ln\left(f_1(X)/g_1(X)\right)]$ for large n. Hence,

$$W(\mathbf{X}) \approx \exp\left\{-n\,\mathbb{E}_{g_1}\left[\ln\left(\frac{g_1(X)}{f_1(X)}\right)\right]\right\} . \tag{5.48}$$

Since $\mathbb{E}_{g_1}[\ln(g_1(\mathbf{X})/f_1(\mathbf{X}))]$ is nonnegative (see page 31), the likelihood ratio $W(\mathbf{X})$ tends to 0 as $n \to \infty$. However, by definition, the expectation of $W(\mathbf{X})$ under g is always 1. This indicates that the distribution of $W(\mathbf{X})$ becomes increasingly skewed when n gets large. Several methods have been introduced to prevent this degeneracy. Examples are the heuristics of Doucet et al. [11], Liu [27], and Robert and Casella [32] and the so-called screening method. The last will be presented in Sections 5.12 and 8.2.2 and can be considered as a dimension-reduction technique.

When the pdf f belongs to some parametric family of distributions, it is often convenient to choose the importance sampling distribution from the *same* family.

In particular, suppose that $f(\cdot) = f(\cdot; \mathbf{u})$ belongs to the family

$$\mathscr{F} = \{f(\cdot; \mathbf{v}), \ \mathbf{v} \in \mathscr{V}\} \ .$$

Then the problem of finding an optimal importance sampling density in this class reduces to the following *parametric* minimization problem:

$$\min_{\mathbf{v} \in \mathscr{V}} \mathrm{Var}_{\mathbf{v}} \left(H(\mathbf{X}) \, W(\mathbf{X}; \mathbf{u}, \mathbf{v})\right), \tag{5.49}$$

where $W(\mathbf{X}; \mathbf{u}, \mathbf{v}) = f(\mathbf{X}; \mathbf{u})/f(\mathbf{X}; \mathbf{v})$. We will call the vector $\mathbf{v}$ the *reference parameter vector* or *tilting vector*. Since under $f(\cdot; \mathbf{v})$ the expectation $\ell = \mathbb{E}_{\mathbf{v}}[H(\mathbf{X}) \, W(\mathbf{X}; \mathbf{u}, \mathbf{v})]$ is constant, the optimal solution of (5.49) coincides with that of

$$\min_{\mathbf{v} \in \mathscr{V}} V(\mathbf{v}) \ , \tag{5.50}$$

where

$$V(\mathbf{v}) = \mathbb{E}_{\mathbf{v}}[H^2(\mathbf{X}) \, W^2(\mathbf{X}; \mathbf{u}, \mathbf{v})] = \mathbb{E}_{\mathbf{u}}[H^2(\mathbf{X}) \, W(\mathbf{X}; \mathbf{u}, \mathbf{v})] \ . \tag{5.51}$$

We will call either of the equivalent problems (5.49) and (5.50) the *variance minimization* (VM) problem, and we will call the parameter vector $_*\mathbf{v}$ that minimizes programs (5.49)–(5.50) the *optimal VM reference parameter vector*. We refer to $\mathbf{u}$ as the *nominal* parameter.

The sample average version of (5.50)–(5.51) is

$$\min_{\mathbf{v} \in \mathscr{V}} \widehat{V}(\mathbf{v}) \ , \tag{5.52}$$

where

$$\widehat{V}(\mathbf{v}) = \frac{1}{N} \sum_{k=1}^{N} [H^2(\mathbf{X}_k) \, W(\mathbf{X}_k; \mathbf{u}, \mathbf{v})] \ , \tag{5.53}$$

and the sample $\mathbf{X}_1, \ldots, \mathbf{X}_N$ is from $f(\mathbf{x}; \mathbf{u})$. Note that as soon as the sample $\mathbf{X}_1, \ldots, \mathbf{X}_N$ is available, the function $\widehat{V}(\mathbf{v})$ becomes a deterministic one.

Since in typical applications both functions $V(\mathbf{v})$ and $\widehat{V}(\mathbf{v})$ are convex and differentiable with respect to $\mathbf{v}$, and since one can typically interchange the expectation and differentiation operators (see Rubinstein and Shapiro [38]), the solutions of programs (5.50)–(5.51) and (5.52)–(5.53) can be obtained by solving (with respect to $\mathbf{v}$) the following system of equations:

$$\mathbb{E}_{\mathbf{u}}[H^2(\mathbf{X}) \, \nabla W(\mathbf{X}; \mathbf{u}, \mathbf{v})] = \mathbf{0} \tag{5.54}$$

and

$$\frac{1}{N} \sum_{k=1}^{N} H(\mathbf{X}_k) \, \nabla W(\mathbf{X}_k; \mathbf{u}, \mathbf{v}) = \mathbf{0} \ , \tag{5.55}$$

respectively, where

$$\nabla W(\mathbf{X}; \mathbf{u}, \mathbf{v}) = \nabla \frac{f(\mathbf{X}; \mathbf{u})}{f(\mathbf{X}; \mathbf{v})} = [\nabla \ln f(\mathbf{X}; \mathbf{v})] \, W(\mathbf{X}; \mathbf{u}, \mathbf{v}) \ ,$$

the gradient is with respect to $\mathbf{v}$ and the function $\nabla \ln f(\mathbf{x}; \mathbf{v})$ is the score function; see (1.57). Note that the system of nonlinear equations (5.55) is typically solved using numerical methods.

■ **EXAMPLE 5.10**

Consider estimating $\ell = \mathbb{E}[X]$, where $X \sim \mathsf{Exp}(u^{-1})$. Choosing $f(x;v) = v^{-1}\exp(-xv^{-1}), x \geqslant 0$ as the importance sampling pdf, the program (5.50) reduces to

$$\min_v V(v) = \min_v \frac{v}{u^2} \int_0^\infty x^2 \, \mathrm{e}^{-(2u^{-1}-v^{-1})x} \, \mathrm{d}x = \min_{v \geqslant u/2} \frac{2uv^4}{(2v-u)^3} \, .$$

The optimal reference parameter $_*v$ is given by

$$_*v = 2\,u \, .$$

We see that $_*v$ is exactly two times larger than u. Solving the sample average version (5.55) (numerically), one should find that, for large N, its optimal solution $_*\widehat{v}$ will be close to the true parameter $_*v$.

■ **EXAMPLE 5.11   Example 5.9 (Continued)**

Consider again estimating $\ell = \mathbb{P}_u(X \geqslant \gamma) = \exp(-\gamma u^{-1})$. In this case, using the family $\{f(x;v), v > 0\}$ defined by $f(x;v) = v^{-1}\exp(xv^{-1}), x \geqslant 0$, we can reduce the program (5.50) to

$$\min_v V(v) = \min_v \frac{v}{u^2} \int_\gamma^\infty \mathrm{e}^{-(2u^{-1}-v^{-1})x} \, \mathrm{d}x = \min_{v \geqslant u/2} \frac{v^2}{u} \frac{\mathrm{e}^{-\gamma(2u^{-1}-v^{-1})}}{(2v-u)} \, .$$

The optimal reference parameter $_*v$ is given by

$$_*v = \frac{1}{2}\left\{\gamma + u + \sqrt{\gamma^2 + u^2}\right\} = \gamma + \frac{u}{2} + \mathcal{O}((u/\gamma)^2) \, ,$$

where $\mathcal{O}(x^2)$ is a function of x such that

$$\lim_{x \to 0} \frac{\mathcal{O}(x^2)}{x^2} = \text{constant} \, .$$

We see that for $\gamma \gg u$, $_*v$ is approximately equal to γ.

It is important to note that in this case the sample version (5.55) (or (5.52) – (5.53)) is meaningful only for small γ, in particular for those γ for which ℓ is *not a rare-event probability*, say where $\ell > 10^{-4}$. For very small ℓ, a tremendously large sample N is needed (because of the indicator function $I_{\{X \geqslant \gamma\}}$), and thus the importance sampling estimator $\widehat{\ell}$ is useless. We will discuss the estimation of rare-event probabilities in more detail in Chapter 8.

Observe that the VM problem (5.50) can also be written as

$$\min_{\mathbf{v} \in \mathscr{V}} V(\mathbf{v}) = \min_{\mathbf{v} \in \mathscr{V}} \mathbb{E}_{\mathbf{w}}\left[H^2(\mathbf{X})\,W(\mathbf{X};\mathbf{u},\mathbf{v})\,W(\mathbf{X};\mathbf{u},\mathbf{w})\right], \tag{5.56}$$

where $\mathbf{w}$ is an arbitrary reference parameter. Note that (5.56) is obtained from (5.51) by multiplying and dividing the integrand by $f(\mathbf{x};\mathbf{w})$. We now replace the expected value in (5.56) by its sample (stochastic) counterpart and then take the optimal solution of the associated Monte Carlo program as an estimator of $_*\mathbf{v}$. Specifically, the stochastic counterpart of (5.56) is

$$\min_{\mathbf{v}\in\mathscr{V}} \widehat{V}(\mathbf{v}) = \min_{\mathbf{v}\in\mathscr{V}} \frac{1}{N} \sum_{k=1}^{N} H^2(\mathbf{X}_k)\, W(\mathbf{X}_k\,;\mathbf{u},\mathbf{v})\, W(\mathbf{X}_k\,;\mathbf{u},\mathbf{w})\,, \qquad (5.57)$$

where $\mathbf{X}_1,\ldots,\mathbf{X}_N$ is an iid sample from $f(\cdot\,;\mathbf{w})$ and $\mathbf{w}$ is an appropriately chosen *trial* parameter. Solving the stochastic program (5.57) thus yields an estimate, say $\widehat{_*\mathbf{v}}$, of $_*\mathbf{v}$. In some cases it may be useful to *iterate* this procedure, that is, use $\widehat{_*\mathbf{v}}$ as a trial vector in (5.57), to obtain a better estimate.

Once the reference parameter $\mathbf{v} = \widehat{_*\mathbf{v}}$ is determined, ℓ is estimated via the likelihood ratio estimator

$$\widehat{\ell} = \frac{1}{N} \sum_{k=1}^{N} H(\mathbf{X}_k)\, W(\mathbf{X}_k;\mathbf{u},\mathbf{v})\,, \qquad (5.58)$$

where $\mathbf{X}_1,\ldots,\mathbf{X}_N$ is a random sample from $f(\cdot\,;\mathbf{v})$. Typically, the sample size N in (5.58) is larger than that used for estimating the reference parameter. We call (5.58) the *standard likelihood ratio* (SLR) estimator.

5.7.3 Cross-Entropy Method

An alternative approach for choosing an "optimal" reference parameter vector in (5.58) is based on the Kullback–Leibler cross-entropy, or simply *cross-entropy* (CE), mentioned in (1.53). For clarity, we repeat that the CE distance between two pdfs g and h is given (in the continuous case) by

$$\begin{aligned}
\mathcal{D}(g,h) &= \mathbb{E}_g\left[\ln\frac{g(\mathbf{X})}{h(\mathbf{X})}\right] = \int g(\mathbf{x})\ln\frac{g(\mathbf{x})}{h(\mathbf{x})}\,\mathrm{d}\mathbf{x} \\
&= \int g(\mathbf{x})\ln g(\mathbf{x})\,\mathrm{d}\mathbf{x} - \int g(\mathbf{x})\ln h(\mathbf{x})\,\mathrm{d}\mathbf{x}\,.
\end{aligned} \qquad (5.59)$$

Recall that $\mathcal{D}(g,h) \geqslant 0$, with equality if and only if $g = h$.

The general idea is to choose the importance sampling density, say h, such that the CE distance between the optimal importance sampling density g^* in (5.44) and h is minimal. We call this the *CE optimal pdf*. Thus, this pdf solves the following *functional* optimization program:

$$\min_{h} \mathcal{D}\left(g^*,h\right).$$

If we optimize over all densities h, then it is immediate from $\mathcal{D}(g^*,h) \geqslant 0$ that the CE optimal pdf coincides with the VM optimal pdf g^*.

As with the VM approach in (5.49) and (5.50), we will restrict ourselves to the parametric family of densities $\{f(\cdot\,;\mathbf{v}),\mathbf{v}\in\mathscr{V}\}$ that contains the "nominal" density $f(\cdot\,;\mathbf{u})$. The CE method now aims to solve the *parametric* optimization problem

$$\min_{\mathbf{v}} \mathcal{D}\left(g^*,f(\cdot\,;\mathbf{v})\right).$$

Since the first term on the right-hand side of (5.59) does not depend on $\mathbf{v}$, minimizing the Kullback–Leibler distance between g^* and $f(\cdot\,;\mathbf{v})$ is equivalent to *maximizing* with respect to $\mathbf{v}$,

$$\int H(\mathbf{x}) f(\mathbf{x}; \mathbf{u}) \ln f(\mathbf{x}; \mathbf{v}) \, d\mathbf{x} = \mathbb{E}_{\mathbf{u}} \left[H(\mathbf{X}) \ln f(\mathbf{X}; \mathbf{v}) \right],$$

where we have assumed that $H(\mathbf{x})$ is nonnegative. Arguing as in (5.50), we find that the CE optimal reference parameter vector $\mathbf{v}^*$ can be obtained from the solution of the following simple program:

$$\max_{\mathbf{v}} D(\mathbf{v}) = \max_{\mathbf{v}} \mathbb{E}_{\mathbf{u}} \left[H(\mathbf{X}) \ln f(\mathbf{X}; \mathbf{v}) \right]. \qquad (5.60)$$

Since typically $D(\mathbf{v})$ is convex and differentiable with respect to $\mathbf{v}$ (see Rubinstein and Shapiro [38]), the solution to (5.60) may be obtained by solving

$$\mathbb{E}_{\mathbf{u}} \left[H(\mathbf{X}) \nabla \ln f(\mathbf{X}; \mathbf{v}) \right] = \mathbf{0} , \qquad (5.61)$$

provided that the expectation and differentiation operators can be interchanged. The sample counterpart of (5.61) is

$$\frac{1}{N} \sum_{k=1}^{N} H(\mathbf{X}_k) \nabla \ln f(\mathbf{X}_k; \mathbf{v}) = \mathbf{0} . \qquad (5.62)$$

By analogy to the VM program (5.50), we call (5.60) the *CE program*, and we call the parameter vector $\mathbf{v}^*$ that minimizes the program (5.63) the *optimal CE reference parameter vector*.

Arguing as in (5.56), it is readily seen that (5.60) is equivalent to the following program:

$$\max_{\mathbf{v}} D(\mathbf{v}) = \max_{\mathbf{v}} \mathbb{E}_{\mathbf{w}} \left[H(\mathbf{X}) W(\mathbf{X}; \mathbf{u}, \mathbf{w}) \ln f(\mathbf{X}; \mathbf{v}) \right], \qquad (5.63)$$

where $W(\mathbf{X}; \mathbf{u}, \mathbf{w})$ is again the likelihood ratio and $\mathbf{w}$ is an *arbitrary* tilting parameter. Similar to (5.57), we can estimate $\mathbf{v}^*$ as the solution of the stochastic program

$$\max_{\mathbf{v}} \widehat{D}(\mathbf{v}) = \max_{\mathbf{v}} \frac{1}{N} \sum_{k=1}^{N} H(\mathbf{X}_k) W(\mathbf{X}_k; \mathbf{u}, \mathbf{w}) \ln f(\mathbf{X}_k; \mathbf{v}) , \qquad (5.64)$$

where $\mathbf{X}_1, \ldots, \mathbf{X}_N$ is a random sample from $f(\cdot; \mathbf{w})$. As in the VM case, we mention the possibility of *iterating* this procedure, that is, using the solution of (5.64) as a trial parameter for the next iteration.

Since in typical applications the function $\widehat{D}$ in (5.64) is convex and differentiable with respect to $\mathbf{v}$ (see [38]), the solution of (5.64) may be obtained by solving (with respect to $\mathbf{v}$) the following system of equations:

$$\frac{1}{N} \sum_{k=1}^{N} H(\mathbf{X}_k) W(\mathbf{X}_k; \mathbf{u}, \mathbf{w}) \nabla \ln f(\mathbf{X}_k; \mathbf{v}) = \mathbf{0} , \qquad (5.65)$$

where the gradient is with respect to $\mathbf{v}$.

Our extensive numerical studies show that for moderate dimensions n, say $n \leq 50$, the optimal solutions of the CE programs (5.63) and (5.64) (or (5.65)) and their VM counterparts (5.56) and (5.57) are typically nearly the same. However, for high-dimensional problems $(n > 50)$, we found numerically that the importance sampling estimator $\hat{\ell}$ in (5.58) based on VM updating of $\mathbf{v}$ outperforms its CE counterpart in both variance and bias. The latter is caused by the degeneracy of W, to which, we found, CE is more sensitive.

The advantage of the CE program is that it can often be solved *analytically*. In particular, this happens when the distribution of $\mathbf{X}$ belongs to an *exponential family* of distributions; see Section A.3 of the Appendix. Specifically (see (A.16)), for a one-dimensional exponential family parameterized by the mean, the CE optimal parameter is *always*

$$v^* = \frac{\mathbb{E}_u[H(X)\,X]}{\mathbb{E}_u[H(X)]} = \frac{\mathbb{E}_w[W(X;u,w)\,H(X)\,X]}{\mathbb{E}_w[H(X)\,W(X;u,w)]}\,, \tag{5.66}$$

and the corresponding sample-based updating formula is

$$\hat{v} = \frac{\sum_{k=1}^{N} H(X_k)\,W(X_k;u,w)\,X_k}{\sum_{k=1}^{N} H(X_k)\,W(X_k;u,w)}\,, \tag{5.67}$$

respectively, where $X_1, \ldots, X_N$ is a random sample from the density $f(\cdot;w)$ and w is an arbitrary parameter. The multidimensional version of (5.67) is

$$\hat{v}_i = \frac{\sum_{k=1}^{N} H(\mathbf{X}_k)\,W(\mathbf{X}_k;\mathbf{u},\mathbf{w})\,X_{ki}}{\sum_{k=1}^{N} H(\mathbf{X}_k)\,W(\mathbf{X}_k;\mathbf{u},\mathbf{w})} \tag{5.68}$$

for $i = 1, \ldots, n$, where X_{ki} is the i-th component of vector $\mathbf{X}_k$ and $\mathbf{u}$ and $\mathbf{w}$ are parameter vectors.

Observe that for $\mathbf{u} = \mathbf{w}$ (no likelihood ratio term W), (5.68) reduces to

$$\hat{v}_i = \frac{\sum_{k=1}^{N} H(\mathbf{X}_k)\,X_{ki}}{\sum_{k=1}^{N} H(\mathbf{X}_k)}\,, \tag{5.69}$$

where $\mathbf{X}_k \sim f(\mathbf{x};\mathbf{u})$.

Observe also that because of the degeneracy of W, one would always prefer the estimator (5.69) to (5.68), especially for high-dimensional problems. But as we will see below, this is not always feasible, particularly when estimating rare-event probabilities in Chapter 8.

■ **EXAMPLE 5.12   Example 5.10 (Continued)**

Consider again the estimation of $\ell = \mathbb{E}[X]$, where $X \sim \mathsf{Exp}(u^{-1})$ and $f(x;v) = v^{-1}\exp(xv^{-1}), x \geqslant 0$. Solving (5.61), we find that the optimal reference parameter v^* is equal to

$$v^* = \frac{\mathbb{E}_u[X^2]}{\mathbb{E}_u[X]} = 2u\,.$$

Thus, v^* is exactly the same as $_*v$. For the sample average of (5.61), we should find that for large N its optimal solution $\hat{v}^*$ is close to the optimal parameter $v^* = 2u$.

■ **EXAMPLE 5.13   Example 5.11 (Continued)**

Consider again the estimation of $\ell = \mathbb{P}_u(X \geqslant \gamma) = \exp(-\gamma u^{-1})$. In this case, we readily find from (5.66) that the optimal reference parameter is $v^* = \gamma + u$. Note that similar to the VM case, for $\gamma \gg u$, the optimal reference parameter is approximately γ.

Note that in the preceding example, similar to the VM problem, the CE sample version (5.65) is meaningful only when γ is chosen such that ℓ is *not a rare-event probability*, say when $\ell > 10^{-4}$. In Chapter 8 we present a general procedure for estimating rare-event probabilities of the form $\ell = \mathbb{P}_\mathbf{u}(S(\mathbf{X}) \geqslant \gamma)$ for an arbitrary function $S(\mathbf{x})$ and level γ.

■ **EXAMPLE 5.14   Finite Support Discrete Distributions**

Let X be a discrete random variable with finite support, that is, X can only take a finite number of values, say $a_1, \ldots, a_m$. Let $u_i = \mathbb{P}(X = a_i), i = 1, \ldots, m$ and define $\mathbf{u} = (u_1, \ldots, u_m)$. The distribution of X is thus trivially parameterized by the vector $\mathbf{u}$. We can write the density of X as

$$f(x; \mathbf{u}) = \sum_{i=1}^m u_i \, I_{\{x=a_i\}} \, .$$

From the discussion at the beginning of this section, we know that the optimal CE and VM parameters *coincide*, since we optimize over *all* densities on $\{a_1, \ldots, a_m\}$. From (5.44) the VM (and CE) optimal density is given by

$$\begin{aligned}
f(x; \mathbf{v}^*) &= \frac{H(x) \, f(x; \mathbf{u})}{\sum_x H(x) \, f(x; \mathbf{u})} \\
&= \frac{\sum_{i=1}^m H(a_i) \, u_i \, I_{\{x=a_i\}}}{\mathbb{E}_\mathbf{u}[H(X)]} \\
&= \sum_{i=1}^m \frac{H(a_i) \, u_i}{\mathbb{E}_\mathbf{u}[H(X)]} \, I_{\{x=a_i\}} \\
&= \sum_{i=1}^m \left(\frac{\mathbb{E}_u[H(X) \, I_{\{X=a_i\}}]}{\mathbb{E}_\mathbf{u}[H(X)]} \right) I_{\{x=a_i\}} \, ,
\end{aligned}$$

so that

$$v_i^* = \frac{\mathbb{E}_\mathbf{u}[H(X) \, I_{\{X=a_i\}}]}{\mathbb{E}_\mathbf{u}[H(X)]} = \frac{\mathbb{E}_\mathbf{w}[H(X) \, W(X; \mathbf{u}, \mathbf{w}) I_{\{X=a_i\}}]}{\mathbb{E}_\mathbf{w}[H(X) \, W(X; \mathbf{u}, \mathbf{w})]} \tag{5.70}$$

for any reference parameter $\mathbf{w}$, provided that $\mathbb{E}_\mathbf{w}[H(X) \, W(X; \mathbf{u}, \mathbf{w})] > 0$.

The vector $\mathbf{v}^*$ can be estimated from the stochastic counterpart of (5.70), that is, as

$$\widehat{v}_i = \frac{\sum_{k=1}^N H(X_k) \, W(X_k; \mathbf{u}, \mathbf{w}) I_{\{X_k=a_i\}}}{\sum_{k=1}^N H(X_k) \, W(X_k; \mathbf{u}, \mathbf{w})} \, , \tag{5.71}$$

where $X_1, \ldots, X_N$ is an iid sample from the density $f(\cdot; \mathbf{w})$.

A similar result holds for a random vector $\mathbf{X} = (X_1, \ldots, X_n)$, where $X_1, \ldots, X_n$ are independent discrete random variables with finite support, characterized by the parameter vectors $\mathbf{u}_1, \ldots, \mathbf{u}_n$. Because of the independence assumption, the CE problem (5.63) separates into n subproblems of the form above, and all the components of the optimal CE reference parameter $\mathbf{v}^* = (\mathbf{v}_1^*, \ldots, \mathbf{v}_n^*)$, which is now a vector of vectors, follow from (5.71). Note that in this case the optimal VM and CE reference parameters are usually not equal, since we are not optimizing the CE over all densities. See, however, Proposition 4.2 in Rubinstein and Kroese [35] for an important case where they *do* coincide and yield a zero-variance likelihood ratio estimator.

The updating rule (5.71), which involves discrete finite support distributions, and in particular the Bernoulli distribution, will be extensively used for combinatorial optimization problems later on in the book.

■ **EXAMPLE 5.15   Example 5.1 (Continued)**

Consider the bridge network in Figure 5.1, and let

$$S(\mathbf{X}) = \min\{X_1 + X_4,\ X_1 + X_3 + X_5,\ X_2 + X_3 + X_4,\ X_2 + X_5\}.$$

We now want to estimate the probability that the shortest path from node A to node B has a length of at least γ; that is, with $H(\mathbf{x}) = I_{\{S(\mathbf{x}) \geqslant \gamma\}}$, we want to estimate

$$\ell = \mathbb{E}[H(\mathbf{X})] = \mathbb{P}_{\mathbf{u}}(S(\mathbf{X}) \geqslant \gamma) = \mathbb{E}_{\mathbf{u}}[I_{\{S(\mathbf{x}) \geqslant \gamma\}}].$$

We assume that the components $\{X_i\}$ are independent, that $X_i \sim \mathsf{Exp}(u_i^{-1})$, $i = 1, \ldots, 5$, and that γ is chosen such that $\ell \geq 10^{-2}$. Thus here the CE updating formula (5.68) and its particular case (5.69) (with $\mathbf{w} = \mathbf{u}$) apply. We will show that this yields substantial variance reduction. The likelihood ratio in this case is

$$W(\mathbf{x}; \mathbf{u}, \mathbf{v}) = \frac{f(\mathbf{x}; \mathbf{u})}{f(\mathbf{x}; \mathbf{v})} = \frac{\prod_{i=1}^5 \frac{1}{u_i} e^{-x_i/u_i}}{\prod_{i=1}^5 \frac{1}{v_i} e^{-x_i/v_i}}$$

$$= \exp\left(-\sum_{i=1}^5 x_i \left(\frac{1}{u_i} - \frac{1}{v_i}\right)\right) \prod_{i=1}^5 \frac{v_i}{u_i}. \tag{5.72}$$

As a concrete example, let the *nominal* parameter vector $\mathbf{u}$ be equal to $(1, 1, 0.3,\ 0.2, 0.1)$ and let $\gamma = 1.5$. We will see that this probability ℓ is approximately 0.06.

Note that the typical length of a path from A to B is smaller than $\gamma = 1.5$; hence, using importance sampling instead of CMC should be beneficial. The idea is to estimate the optimal parameter vector $\mathbf{v}^*$ *without* using likelihood ratios, that is, using (5.69), since likelihood ratios, as in (5.68) (with quite arbitrary $\mathbf{w}$, say by guessing an initial trial vector $\mathbf{w}$), would typically make the estimator of $\mathbf{v}^*$ unstable, especially for high-dimensional problems.

Denote by $\widehat{\mathbf{v}}_1$ the CE estimator of $\mathbf{v}^*$ obtained from (5.69). We can iterate (repeat) this procedure, say for T iterations, using (5.68), and starting with $\mathbf{w} = \widehat{\mathbf{v}}_1, \widehat{\mathbf{v}}_2, \ldots$. Once the final reference vector $\widehat{\mathbf{v}}_T$ is obtained, we then estimate ℓ via a *larger* sample from $f(\mathbf{x}; \widehat{\mathbf{v}}_T)$, say of size N_1, using the SLR estimator (5.58). Note, however, that for high-dimensional problems, iterating in this way could lead to an unstable final estimator $\widehat{\mathbf{v}}_T$. In short, a single iteration with (5.69) might often be the best alternative.

Table 5.1 presents the performance of the estimator (5.58), starting from $\mathbf{w} = \mathbf{u} = (1, 1, 0.3, 0.2, 0.1)$ and then iterating (5.68) three times. Note again that in the first iteration we generate a sample $\mathbf{X}_1, \dots, \mathbf{X}_N$ from $f(\mathbf{x}; \mathbf{u})$ and then apply (5.69) to obtain an estimate $\widehat{\mathbf{v}} = (\widehat{v}_1, \dots, \widehat{v}_5)$ of the CE optimal reference parameter vector $\mathbf{v}^*$. The sample sizes for updating $\widehat{\mathbf{v}}$ and calculating the estimator $\widehat{\ell}$ were $N = 10^3$ and $N_1 = 10^5$, respectively. In the table RE denotes the estimated relative error.

Table 5.1: Iterating the five-dimensional vector $\widehat{\mathbf{v}}$.

Iteration	$\widehat{\mathbf{v}}$					$\widehat{\ell}$	RE
0	1	1	0.3	0.2	0.1	0.0643	0.0121
1	2.4450	2.3274	0.2462	0.2113	0.1030	0.0631	0.0082
2	2.3850	2.3894	0.3136	0.2349	0.1034	0.0644	0.0079
3	2.3559	2.3902	0.3472	0.2322	0.1047	0.0646	0.0080

Note that $\widehat{\mathbf{v}}$ already converged after the first iteration, so using likelihood ratios in iterations 2 and 3 did not add anything to the quality of $\widehat{\mathbf{v}}$. It also follows from the results of Table 5.1 that CE outperforms CMC (compare the relative errors 0.008 and 0.0121 for CE and CMC, respectively). To obtain a similar relative error of 0.008 with CMC would require a sample size of approximately $2.5 \cdot 10^5$ instead of 10^5; we thus obtained a reduction by a factor of 2.5 when using the CE estimation procedure. As we will see in Chapter 8 for smaller probabilities, a variance reduction of several orders of magnitude can be achieved.

5.8 SEQUENTIAL IMPORTANCE SAMPLING

Sequential importance sampling (SIS), also called *dynamic importance sampling*, is simply importance sampling carried out in a sequential manner. To explain the SIS procedure, consider the expected performance ℓ in (5.38) and its likelihood ratio estimator $\widehat{\ell}$ in (5.40), with $f(\mathbf{x})$ the "target" and $g(\mathbf{x})$ the importance sampling, or proposal, pdf. Suppose that (1) $\mathbf{X}$ is decomposable, that is, it can be written as a vector $\mathbf{X} = (X_1, \dots, X_n)$, where each of the X_i may be multidimensional, and (2) it is easy to sample from $g(\mathbf{x})$ sequentially. Specifically, suppose that $g(\mathbf{x})$ is of the form

$$g(\mathbf{x}) = g_1(x_1)\, g_2(x_2 \mid x_1) \cdots g_n(x_n \mid x_1, \dots, x_{n-1}), \qquad (5.73)$$

where it is easy to generate X_1 from density $g_1(x_1)$, and conditional on $X_1 = x_1$, the second component from density $g_2(x_2 \mid x_1)$, and so on, until we obtain a single random vector $\mathbf{X}$ from $g(\mathbf{x})$. Repeating this independently N times, each time sampling from $g(\mathbf{x})$, we obtain a random sample $\mathbf{X}_1, \dots, \mathbf{X}_N$ from $g(\mathbf{x})$ and an estimator of ℓ according to (5.40). To further simplify the notation, we abbreviate $(x_1, \ldots, x_t)$ to $\mathbf{x}_{1:t}$ for all t. In particular, $\mathbf{x}_{1:n} = \mathbf{x}$. Typically, t can be viewed as a (discrete) time parameter and $\mathbf{x}_{1:t}$ as a path or trajectory. By the product rule of probability (1.4), the target pdf $f(\mathbf{x})$ can also be written sequentially, that is,

$$f(\mathbf{x}) = f(x_1) f(x_2 \,|\, x_1) \cdots f(x_n \,|\, \mathbf{x}_{1:n-1}). \tag{5.74}$$

From (5.73) and (5.74) it follows that we can write the likelihood ratio in product form as

$$W(\mathbf{x}) = \frac{f(x_1) f(x_2 \,|\, x_1) \cdots f(x_n \,|\, \mathbf{x}_{1:n-1})}{g_1(x_1) g_2(x_2 \,|\, x_1) \cdots g_n(x_n \,|\, \mathbf{x}_{1:n-1})} \tag{5.75}$$

or, if $W_t(\mathbf{x}_{1:t})$ denotes the likelihood ratio up to time t, recursively as

$$W_t(\mathbf{x}_{1:t}) = u_t \, W_{t-1}(\mathbf{x}_{1:t-1}), \quad t = 1, \ldots, n \ , \tag{5.76}$$

with initial weight $W_0(\mathbf{x}_{1:0}) = 1$ and *incremental weights* $u_1 = f(x_1)/g_1(x_1)$ and

$$u_t = \frac{f(x_t \,|\, \mathbf{x}_{1:t-1})}{g_t(x_t \,|\, \mathbf{x}_{1:t-1})} = \frac{f(\mathbf{x}_{1:t})}{f(\mathbf{x}_{1:t-1}) \, g_t(x_t \,|\, \mathbf{x}_{1:t-1})} \ , \quad t = 2, \ldots, n \ . \tag{5.77}$$

In order to update the likelihood ratio recursively, as in (5.77), we need to known the marginal pdfs $f(\mathbf{x}_{1:t})$. This may not be easy when f does not have a Markov structure, as it requires integrating $f(\mathbf{x})$ over all $x_{t+1}, \ldots, x_n$. Instead, we can introduce a sequence of *auxiliary* pdfs $f_1, f_2, \ldots, f_n$ that are easily evaluated and such that each $f_t(\mathbf{x}_{1:t})$ is a good approximation to $f(\mathbf{x}_{1:t})$. The terminating pdf f_n must be equal to the original f. Since

$$f(\mathbf{x}) = \frac{f_1(x_1)}{1} \frac{f_2(\mathbf{x}_{1:2})}{f_1(x_1)} \cdots \frac{f_n(\mathbf{x}_{1:n})}{f_{n-1}(\mathbf{x}_{1:n-1})} \ , \tag{5.78}$$

we have as a generalization of (5.77) the incremental updating weight

$$u_t = \frac{f_t(\mathbf{x}_{1:t})}{f_{t-1}(\mathbf{x}_{1:t-1}) \, g_t(x_t \,|\, \mathbf{x}_{1:t-1})} \tag{5.79}$$

for $t = 1, \ldots, n$, where we put $f_0(\mathbf{x}_{1:0}) = 1$. Summarizing, the SIS method can be written as follows:

Algorithm 5.8.1: SIS Method

---

input : Sample size N, pdfs $\{f_t\}$ and $\{g_t\}$, performance function H.
output: Estimator of $\ell = \mathbb{E}\left[H(\mathbf{X})\right] = \mathbb{E}[H(\mathbf{X}_{1:n})]$.

1 **for** $k = 1$ **to** N **do**
2 $\quad X_1 \sim g_1(x_1)$
3 $\quad W_1 \leftarrow \frac{f_1(X_1)}{g_1(X_1)}$
4 $\quad$ **for** $t = 2$ **to** n **do**
5 $\quad\quad X_t \sim g_t(x_t \,|\, \mathbf{X}_{1:t-1})$ $\qquad$ // simulate next component
6 $\quad\quad W_t \leftarrow W_{t-1} \frac{f_t(\mathbf{X}_{1:t})}{f_{t-1}(\mathbf{X}_{1:t-1}) \, g_t(X_t \,|\, \mathbf{X}_{1:t-1})}$
7 $\quad W^{(k)} \leftarrow W_n$
8 $\quad \mathbf{X}^{(k)} \leftarrow \mathbf{X}_{1:n}$

9 **return** $N^{-1} \sum_{k=1}^{N} H(\mathbf{X}^{(k)}) \, W_n^{(k)}$

Remark 5.8.1 Note that the incremental weights u_t only need to be defined *up to a constant*, say c_t, for each t. In this case the likelihood ratio $W(\mathbf{x})$ is known up to a constant as well, say $W(\mathbf{x}) = C\,w(\mathbf{x})$, where $1/C = \mathbb{E}_g[w(\mathbf{X})]$ can be estimated via the corresponding sample mean. In other words, when the normalization constant is unknown, we can still estimate ℓ using the weighted sample estimator (5.42) rather than the likelihood ratio estimator (5.40).

■ **EXAMPLE 5.16   Random Walk on the Integers**

Consider the random walk on the integers of Example 1.10 (on Page 20), with probabilities p and q for jumping up or down, respectively. Suppose that $p < q$, so that the walk has a drift toward $-\infty$. Our goal is to estimate the rare-event probability ℓ of reaching state K before state 0, starting from state $0 < k \ll K$, where K is a large number. As an intermediate step, consider first the probability of reaching K in exactly n steps, that is, $\mathbb{P}(X_n = K) = \mathbb{E}[I_{A_n}]$, where $A_n = \{X_n = K\}$. We have

$$f(\mathbf{x}_{1:n}) = f(x_1\,|\,k)\,f(x_2\,|\,x_1)\,f(x_3\,|\,x_2)\ldots f(x_n\,|\,x_{n-1})\,,$$

where the conditional probabilities are either p (for upward jumps) or q (for downward jumps). If we simulate the random walk with *different* upward and downward probabilities, $\tilde{p}$ and $\tilde{q}$, then the importance sampling pdf $g(\mathbf{x}_{1:n})$ has the same form as $f(\mathbf{x}_{1:n})$ above. Thus the importance weight after Step t is updated via the incremental weight

$$u_t = \frac{f(x_t\,|\,x_{t-1})}{g(x_t\,|\,x_{t-1})} = \begin{cases} p/\tilde{p} & \text{if } x_t = x_{t-1} + 1\,, \\ q/\tilde{q} & \text{if } x_t = x_{t-1} - 1\,. \end{cases}$$

The probability $\mathbb{P}(A_n)$ can now be estimated via importance sampling as

$$\frac{1}{N}\sum_{i=1}^{N} W_{i,n}\,I_{\{X_{i,n}=K\}}\,, \tag{5.80}$$

where the paths $\mathbf{X}_{i,1:n}, i = 1, \ldots, N$, are generated via g, rather than f and $W_{i,n}$ is the likelihood ratio of the i-th such path. Returning to the estimation of ℓ, let τ be the first time that either 0 or K is reached. Writing $I_{\{X_t=K\}} = H(\mathbf{X}_{1:t})$, we have

$$\ell = \mathbb{E}_f[I_{\{X_\tau=K\}}] = \mathbb{E}_f[H(\mathbf{X}_{1:\tau})] = \sum_{n=1}^{\infty}\mathbb{E}[H(\mathbf{X}_{1:n})\,I_{\{\tau=n\}}]$$

$$= \sum_{n=1}^{\infty}\sum_{\mathbf{x}} H(\mathbf{x}_{1:n})\,I_{\{\tau=n\}}\,f(\mathbf{x}_{1:n})$$

$$= \sum_{n=1}^{\infty}\sum_{\mathbf{x}} \underbrace{\frac{f(\mathbf{x}_{1:n})}{g(\mathbf{x}_{1:n})}I_{\{x_n=K\}}\,I_{\{\tau=n\}}}_{\tilde{H}(\mathbf{x}_{1:n})}g(\mathbf{x}_{1:n})$$

$$= \mathbb{E}_g[\tilde{H}(\mathbf{X}_{1:\tau})] = \mathbb{E}_g[W_\tau\,I_{\{X_\tau=K\}}]\,,$$

with W_τ the likelihood ratio of $\mathbf{X}_{1:\tau}$, which can be updated at each time t by multiplying with either $p/\tilde{p}$ or $q/\tilde{q}$ for upward and downward steps, respectively. Note that $I_{\{\tau=n\}}$ is indeed a function of $\mathbf{x}_n = (x_1, \ldots, x_n)$. This leads to the same estimator as (5.80) with the deterministic n replaced by the stochastic τ. It can be shown (e.g., see [5]) that choosing $\tilde{p} = q$ and $\tilde{q} = p$, that is, *interchanging* the probabilities, gives an efficient estimator for ℓ.

■ **EXAMPLE 5.17   Counting Self-Avoiding Walks**

The self-avoiding random walk, or simply *self-avoiding walk*, is a basic mathematical model for polymer chains. For simplicity, we will deal only with the two-dimensional case. Each self-avoiding walk is represented by a path $\mathbf{x} = (x_1, x_2, \ldots, x_{n-1}, x_n)$, where x_i represents the two-dimensional position of the i-th molecule of the polymer chain. The distance between adjacent molecules is fixed at 1, and the main requirement is that the chain does not self-intersect. We assume that the walk starts at the origin. An example of a self-avoiding walk walk of length 130 is given in Figure 5.3.

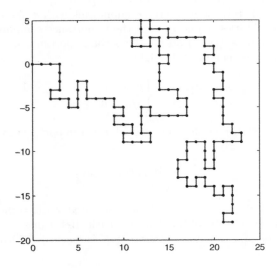

Figure 5.3: A self-avoiding random walk of length $n = 130$.

One of the main questions regarding the self-avoiding walk model is: how many self-avoiding walks are there of length n? Let $\mathscr{X}^*$ be the set of self-avoiding walks of length n. The exact number of self-avoiding walks up to $n = 72$ can be found in `http://www.ms.unimelb.edu.au/~iwan/saw/series/sqsaw.ser`. The first 20 entries are as follows:

n	$|\mathscr{X}^*|$	n	$|\mathscr{X}^*|$	n	$|\mathscr{X}^*|$	n	$|\mathscr{X}^*|$
0	1	5	284	10	44100	15	6416596
1	4	6	780	11	120292	16	17245332
2	12	7	2172	12	324932	17	46466676
3	36	8	5916	13	881500	18	124658732
4	100	9	16268	14	2374444	19	335116620

We wish to estimate $|\mathscr{X}^*|$ via Monte Carlo. The crude Monte Carlo approach is to use acceptance–rejection in the follow way:

1. Generate a random sample $\mathbf{X}^{(1)}, \ldots, \mathbf{X}^{(N)}$ uniformly distributed over the set $\mathscr{X}$ of all random walks of length n. This set has $|\mathscr{X}| = 4^n$ elements. Generating the samples from $\mathscr{X}$ is easy.

2. Estimate the desired number $|\mathscr{X}^*|$ as

$$\widehat{|\mathscr{X}^*|} = |\mathscr{X}| \frac{1}{N} \sum_{k=1}^{N} I_{\{\mathbf{X}^{(k)} \in \mathscr{X}^*\}} , \tag{5.81}$$

where $I_{\{\mathbf{X}^{(k)} \in \mathscr{X}^*\}}$ denotes the indicator of the event $\{\mathbf{X}^{(k)} \in \mathscr{X}^*\}$. Note that according to (5.81) we accept the generated point $\mathbf{X}^{(k)}$ if $\mathbf{X}^{(k)} \in \mathscr{X}^*$ and reject it otherwise.

Unfortunately, for large n the event $\{\mathbf{X}^{(k)} \in \mathscr{X}^*\}$ is very rare. Acceptance–rejection is meaningless if there are no acceptable samples. Instead, we could opt to use importance sampling. In particular, let g be an importance sampling pdf defined on some set $\mathscr{X}$ and let $\mathscr{X}^* \subset \mathscr{X}$; then $|\mathscr{X}^*|$ can be written as

$$|\mathscr{X}^*| = \sum_{\mathbf{x} \in \mathscr{X}} I_{\{\mathbf{x} \in \mathscr{X}^*\}} \frac{g(\mathbf{x})}{g(\mathbf{x})} = \mathbb{E}_g \left[\frac{I_{\{\mathbf{x} \in \mathscr{X}^*\}}}{g(\mathbf{X})} \right] . \tag{5.82}$$

To estimate $|\mathscr{X}^*|$ via Monte Carlo, we draw a random sample $\mathbf{X}_1, \ldots, \mathbf{X}_N$ from g and take the estimator

$$\widehat{|\mathscr{X}^*|} = \frac{1}{N} \sum_{k=1}^{N} I_{\{\mathbf{X}^{(k)} \in \mathscr{X}^*\}} \frac{1}{g(\mathbf{X}^{(k)})} . \tag{5.83}$$

The best choice for g is $g^*(\mathbf{x}) = 1/|\mathscr{X}^*|, \mathbf{x} \in \mathscr{X}^*$; in words, $g^*(\mathbf{x})$ is the uniform pdf over the discrete set $\mathscr{X}^*$. Under g^* the estimator has zero variance, so that only *one sample is required*. Clearly, such g^* is infeasible. Fortunately, the SAW counting problem presents a natural *sequential* importance sampling density g. This pdf is defined by the following *one-step-look-ahead* procedure:

Algorithm 5.8.2: One-Step–Look-Ahead

---

input : Length of path n.
output: Self-avoiding walk of length n, or $\emptyset$ (no such path found).

1 Let $X_0 \leftarrow (0,0)$ and $t \leftarrow 1$.
2 **for** $t = 1$ **to** n **do**
3 | Let d_t be the number of neighbors of X_{t-1} that have not yet been visited.
4 | **if** $d_t > 0$ **then**
5 | | Choose X_t with probability $1/d_t$ from its neighbors.
6 | **else**
7 | | **return** $\emptyset$ // no SAW of length n found
8 **return** $X_1, \ldots, X_n$

Note that the procedure generates either a self-avoiding walk $\mathbf{x}$ of length n or $\emptyset$. Let $g(\mathbf{x})$ be the corresponding discrete pdf. Then, for any self-avoiding walk $\mathbf{x}$ of length n, we have by the product rule (1.4) that

$$g(\mathbf{x}) = \frac{1}{d_1}\frac{1}{d_2}\cdots\frac{1}{d_n} = \frac{1}{w(\mathbf{x})},$$

where

$$w(\mathbf{x}) = d_1 \cdots d_n . \tag{5.84}$$

The self-avoiding walk counting algorithm below now follows directly from (5.83).

Algorithm 5.8.3: Counting Self-Avoiding Walks

---

input : Length of path n.
output: Estimator $\widehat{|\mathscr{X}^*|}$ of the number of self-avoiding walks of length n.

1 Generate independently N paths $\mathbf{X}^{(1)}, \ldots, \mathbf{X}^{(N)}$ via the one-step-look-ahead procedure.
2 For each self-avoiding walk $\mathbf{X}^{(k)}$, compute $w(\mathbf{X}^{(k)})$ as in (5.84). If $\emptyset$ is returned, set $w(\mathbf{X}^{(k)}) \leftarrow 0$.
3 **return** $\frac{1}{N} \sum_{k=1}^{N} w(\mathbf{X}^{(k)})$

The efficiency of the simple one-step-look-ahead method deteriorates rapidly as n becomes large. It becomes impractical to simulate walks of length more than 200. This is due to the fact that, if at any one step t the point x_{t-1} does not have unoccupied neighbors ($d_t = 0$), then the "weight" $w(\mathbf{x})$ is zero and contributes nothing to the final estimate of $|\mathscr{X}^*|$. This problem can occur early in the simulation, rendering any subsequent sequential buildup useless. Better-performing algorithms do not restart from scratch but reuse successful partial walks to build new walks. These methods usually split the self-avoiding partial walks into a number of copies and continue them as if they were independently built up from scratch. We refer to [27] for a discussion of these more advanced algorithms. We will revisit this example in Chapter 9, where the splitting method is used to estimate the number of SAWs.

5.9   SEQUENTIAL IMPORTANCE RESAMPLING

A common problem with sequential importance sampling (Algorithm 5.8.1) is that the distribution of the importance weight W_t becomes very skewed as t increases, resulting in a high probability of a very small weight and a small probability of a very large weight; see also Remark 5.7.1. As a consequence, most of the N samples will not contribute significantly to the final estimator $\widehat{\ell}$ in (5.41).

One way to rectify this issue is to *resample* high-weight samples. To explain the resampling procedure, we first give the parallel version of Algorithm 5.8.1. Instead of simulating all n components of $\mathbf{X} = (X_1, \ldots, X_n)$ and repeating this process N times, we can simulate N copies of the first component X_1, then simulate N copies of the second components $\mathbf{X}_{1:2} = (X_1, X_2)$, and so on. To further enhance the generality of Algorithm 5.8.1, we assume that each auxiliary pdf f_t is known *up to a normalization constant c_t*; see also Remark 5.8.1. In particular, we assume that the product $c_t f_t(\mathbf{x}_{1:t})$ can be explicitly evaluated, whereas c_t and $f_t(\mathbf{x}_{1:t})$ may not. If f_n is known, we can set $c_n = 1$ and use the ordinary likelihood ratio estimator (5.41) to estimate ℓ. If c_n is unknown, we must use the weighted estimator (5.42). That is, in Line 8 below return

$$\left(\sum_{k=1}^{N} W_n^{(k)} \right)^{-1} \sum_{k=1}^{N} H(\mathbf{X}_{1:n}^{(k)}) \, W_n^{(k)} \ .$$

Algorithm 5.9.1: Parallel SIS

input : Sample size N, unnormalized pdfs $\{c_t f_t\}$ and pdfs $\{g_t\}$, performance function H.

output: Estimator of $\ell = \mathbb{E}\left[H\left(\mathbf{X}\right)\right] = \mathbb{E}[H(\mathbf{X}_{1:n})]$.

1 **for** $k = 1$ **to** N **do**

2 $\quad$ $X_1^{(k)} \sim g_1(x_1)$ $\qquad\qquad\qquad\qquad\qquad$ `// simulate first component`

3 $\quad$ $W_1^{(k)} \leftarrow \frac{c_1 f_1(X_1^{(k)})}{g_1(X_1^{(k)})}$

4 **for** $t = 2$ **to** n **do**

5 $\quad$ **for** $k = 1$ **to** N **do**

6 $\quad\quad$ $X_t^{(k)} \sim g_t(x_t \mid \mathbf{X}_{1:t-1}^{(k)})$ $\qquad\qquad$ `// simulate next component`

7 $\quad\quad$ $W_t^{(k)} \leftarrow W_{t-1}^{(k)} \dfrac{c_t f_t(\mathbf{X}_{1:t}^{(k)})}{c_{t-1} f_{t-1}(\mathbf{X}_{1:t-1}^{(k)}) \, g_t(X_t^{(k)} \mid \mathbf{X}_{1:t-1}^{(k)})}$

8 **return** $N^{-1} \displaystyle\sum_{k=1}^{N} H(\mathbf{X}_{1:n}^{(k)}) \, W_n^{(k)}$

Note that at any stage $t = 1, \ldots, n$ the "weighted particles" $\{(\mathbf{X}_{1:t}^{(k)}, W_t^{(k)})\}_{k=1}^N$ can provide the unbiased estimator $\sum_{k=1}^{N} H_t(\mathbf{X}_{1:t}^{(k)}) W_t^{(k)}$ of $\mathbb{E}_{f_t}[H_t(\mathbf{X}_{1:t})]$ for any function H_t of $\mathbf{X}_{1:t}$. Let $\{\mathbf{Y}_t^{(k)}\}_{k=1}^N$ be a sample of size N chosen with replacement from $\{\mathbf{X}_{1:t}^{(k)}\}_{k=1}^N$ with probabilities proportional to $\{W_t^{(k)}\}_{k=1}^N$, and let $\overline{W}_t = N^{-1} \sum_{k=1}^{N} W_t^{(k)}$. Then,

$$
\mathbb{E}\left[\sum_{k=1}^{N} H_t(\mathbf{Y}_t^{(k)})\,\overline{W}_t\right] = \mathbb{E}\left[\overline{W}_t \sum_{k=1}^{N} \mathbb{E}\left[H_t(\mathbf{Y}_t^{(k)}) \mid \mathbf{X}_{1:t}^{(1)}, \ldots, \mathbf{X}_{1:t}^{(N)}, W_t^{(1)}, \ldots, W_t^{(N)}\right]\right]
$$

$$
= \mathbb{E}\left[\overline{W}_t \sum_{k=1}^{N} \sum_{j=1}^{N} \frac{H_t(\mathbf{X}_{1:t}^{(j)})\,W_t^{(j)}}{N\overline{W}_t}\right] = \mathbb{E}\left[\sum_{j=1}^{N} H_t(\mathbf{X}_{1:t}^{(j)})\,W_t^{(j)}\right]. \tag{5.85}
$$

This suggests that we replace the variables $\{(\mathbf{X}_{1:t}^{(k)}, W_t^{(k)})\}_{k=1}^{N}$ by $\{(\mathbf{Y}_t^{(k)}, \overline{W}_t)\}_{k=1}^{N}$ and continue the sequential importance sampling algorithm. This type of resampling is called *bootstrap resampling*. When the importance weights are all identical, this corresponds to *simple random sampling with replacement*.

Adding such a resampling step to Algorithm 5.9.1 for every t results in *sequential importance resampling* (SIR) Algorithm 5.9.2. It can be shown that the weighted estimator returned by the algorithm is asymptotically unbiased and asymptotically normal [7].

Note that the addition of a resampling step can result in a worse estimator. For example, if H is a positive function, then the optimal importance sampling density is $g \propto Hf$ and the resulting importance sampling estimator has zero variance. If a resampling step is added, then the resulting estimator can have nonzero variance.

Algorithm 5.9.2: SIR Algorithm with Bootstrap Resampling

---

    input : Sample size N, unnormalized pdfs $\{c_t f_t\}$ and pdfs $\{g_t\}$, performance
              function H.

    output: Estimator of $\ell = \mathbb{E}[H(\mathbf{X})] = \mathbb{E}[H(\mathbf{X}_{1:n})]$.

1 **for** $k = 1$ **to** N **do**

2     $X_1^{(k)} \sim g_1(x_1)$ // simulate first component

3     $W_1^{(k)} \leftarrow \dfrac{c_1 f_1(X_1^{(k)})}{g_1(X_1^{(k)})}$

4 **for** $t = 2$ **to** n **do**

5     $\mathbf{Y}_{t-1}^{(1)}, \ldots, \mathbf{Y}_{t-1}^{(N)} \leftarrow$ iid samples from $\mathbf{X}_{1:t-1}^{(1)}, \ldots, \mathbf{X}_{1:t-1}^{(N)}$, with probabilities
         proportional to $W_{t-1}^{(1)}, \ldots, W_{t-1}^{(N)}$ // resample

6     $\overline{W}_{t-1} \leftarrow N^{-1} \sum_{k=1}^{N} W_{t-1}^{(k)}$ // compute average weight

7   **for** $k = 1$ **to** N **do**

8        $X_t^{(k)} \sim g_t(x_t \mid \mathbf{Y}_{t-1}^{(k)})$ // simulate next component

9        $\mathbf{X}_{1:t}^{(k)} \leftarrow (\mathbf{Y}_{t-1}^{(k)}, X_t^{(k)})$

10       $W_t^{(k)} \leftarrow \overline{W}_{t-1} \dfrac{c_t f_t(\mathbf{X}_{1:t}^{(k)})}{c_{t-1} f_{t-1}(\mathbf{Y}_{t-1}^{(k)})\, g_t(X_t^{(k)} \mid \mathbf{Y}_{t-1}^{(k)})}$ // update weight

11 **return** $\left(\sum_{k=1}^{N} W_n^{(k)}\right)^{-1} \displaystyle\sum_{k=1}^{N} H(\mathbf{X}_{1:n}^{(k)}) W_n^{(k)}$

There are various other ways in which the resampling step can be carried out. For example, in the *enrichment* method of [42] resampling is performed by making r_t copies of $\mathbf{X}_{1:t}^{(k)}$ for some integer $r_t > 0$. A natural generalization of enrichment is to split every particle $\mathbf{X}_{1:t}^{(k)}$ into a random number $R_t^{(k)}$ of copies, e.g., with some fixed expectation r_t that does not have to be an integer. The natural choice is to take $R_t^{(k)} = \lfloor r_t \rfloor + B_t^{(k)}$, where the $\{B_t^{(k)}\}$ are independent Bernoulli random variables with parameter $r_t - \lfloor r_t \rfloor$. The variability in the total number of copies made can be reduced by letting the $\{B_t^{(k)}\}_{k=1}^N$ be such that the total number of resampled particles is fixed. It is also possible to combine bootstrap resampling and enrichment, as in the *Pruned-Enriched Rosenbluth* method of [18]. Chapter 9 further explores the merits of such "particle Monte Carlo" methods in which a sequential sampling scheme is combined with a resampling/splitting step.

5.10 NONLINEAR FILTERING FOR HIDDEN MARKOV MODELS

This section describes an application of SIS and SIR to nonlinear filtering. Many problems in engineering, applied sciences, statistics, and econometrics can be formulated as *hidden Markov models* (HMMs). In its simplest form, an HMM is a stochastic process $\{(X_t, Y_t)\}$, where X_t (which may be multidimensional) represents the *true* state of some system and Y_t represents the *observed* state of the system at a discrete time t. It is usually assumed that $\{X_t\}$ is a Markov chain, say with initial distribution $f(x_0)$ and one-step transition probabilities $f(x_t \,|\, x_{t-1})$. It is important to note that the actual state of the Markov chain remains *hidden*, hence the name HMM. All information about the system is conveyed by the process $\{Y_t\}$. We assume that, given $X_0, \ldots, X_t$, the observation Y_t depends only on X_t via some conditional pdf $f(y_t \,|\, x_t)$. Note that we have used here a Bayesian style of notation in which all (conditional) probability densities are represented by the *same* symbol f. We will use this notation throughout the rest of this section. We denote by $\mathbf{X}_{1:t} = (X_1, \ldots, X_t)$ and $\mathbf{Y}_{1:t} = (Y_1, \ldots, Y_t)$ the unobservable and observable sequences up to time t, respectively — and similarly for their lowercase equivalents.

The HMM is represented graphically in Figure 5.4. This is an example of a *Bayesian network*. The idea is that edges indicate the dependence structure between two variables. For example, given the states $X_1, \ldots, X_t$, the random variable Y_t is conditionally independent of $X_1, \ldots, X_{t-1}$, because there is no direct edge from Y_t to any of these variables. We thus have $f(y_t \,|\, \mathbf{x}_{1:t}) = f(y_t \,|\, x_t)$, and more generally

$$f(\mathbf{y}_{1:t} \,|\, \mathbf{x}_{1:t}) = f(y_1 \,|\, x_1) \cdots f(y_t \,|\, x_t) = f(\mathbf{y}_{1:t-1} \,|\, \mathbf{x}_{1:t-1}) \, f(y_t \,|\, x_t) \,. \qquad (5.86)$$

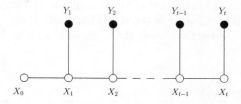

Figure 5.4: A graphical representation of the HMM.

Summarizing, we have

$$\begin{aligned} X_t &\sim f(x_t \,|\, x_{t-1}) &&\text{(state equation)}, \\ Y_t &\sim f(y_t \,|\, x_t) &&\text{(observation equation)}. \end{aligned} \qquad (5.87)$$

■ **EXAMPLE 5.18**

An example of (5.87) is the following popular model:

$$X_t = \varphi_1(X_{t-1}) + \varepsilon_{1t} ,$$
$$Y_t = \varphi_2(X_t) + \varepsilon_{2t} , \tag{5.88}$$

where $\varphi_1(\cdot)$ and $\varphi_2(\cdot)$ are given vector functions and ε_{1t} and ε_{2t} are independent d-dimensional Gaussian random vectors with zero mean and covariance matrices C_1 and C_2, respectively.

Our goal, based on an outcome $\mathbf{y}_{1:t}$ of $\mathbf{Y}_{1:t}$, is to determine, or estimate *on-line*, the following quantities:

1. The joint conditional pdf $f(\mathbf{x}_{1:t} \mid \mathbf{y}_{1:t})$ and, as a special case, the marginal conditional pdf $f(x_t \mid \mathbf{y}_{1:t})$, which is called the *filtering* pdf.

2. The expected performance

$$\ell = \mathbb{E}_{f(\mathbf{x}_{1:t} \mid \mathbf{y}_{1:t})}[H(\mathbf{X}_{1:t})] = \int H(\mathbf{x}_{1:t}) \, f(\mathbf{x}_{1:t} \mid \mathbf{y}_{1:t}) \, d\mathbf{x}_{1:t} . \tag{5.89}$$

It is well known [11] that the conditional pdf $f(\mathbf{x}_{1:t} \mid \mathbf{y}_{1:t})$ or the filtering pdf $f(x_t \mid \mathbf{y}_{1:t})$ can be found explicitly only for the following two particular cases:

(a) When $\varphi_1(x)$ and $\varphi_2(x)$ in (5.88) are linear, the filtering pdf is obtained from the celebrated *Kalman filter*. The Kalman filter is explained in Section A.6 of the Appendix.

(b) When the $\{x_t\}$ can take only a finite number, say K, of possible values, for example, as in binary signals, we can calculate $f(x_t \mid \mathbf{y}_{1:t})$ efficiently with complexity $\mathcal{O}(K^2 t)$. Applications can be found in digital communication and speech recognition; see, for example, Section A.7 of the Appendix.

Because the target pdf $f(\mathbf{x}_{1:t} \mid \mathbf{y}_{1:t})$ for the general state space model (5.87) is difficult to obtain exactly, one needs to resort to Monte Carlo methods. To put the nonlinear filtering problem in the sequential Monte Carlo framework of Section 5.8, we first write $f(\mathbf{x}_{1:t} \mid \mathbf{y}_{1:n})$ in sequential form, similar to (5.78). A natural candidate for the "auxiliary" pdf at time t is the conditional pdf $f(\mathbf{x}_{1:t} \mid \mathbf{y}_{1:t})$. That is, only the observations up to time t are used. By Bayes' rule we have for each $t = 1, \ldots, n$,

$$\frac{f(\mathbf{x}_{1:t} \mid \mathbf{y}_{1:t})}{f(\mathbf{x}_{1:t-1} \mid \mathbf{y}_{1:t-1})}$$

$$= \frac{f(\mathbf{y}_{1:t} \mid \mathbf{x}_{1:t}) f(\mathbf{x}_{1:t})}{f(\mathbf{y}_{1:t})} \frac{f(\mathbf{y}_{1:t-1})}{f(\mathbf{y}_{1:t-1} \mid \mathbf{x}_{1:t-1}) f(\mathbf{x}_{1:t-1})}$$

$$= \frac{f(\mathbf{y}_{1:t-1} \mid \mathbf{x}_{1:t-1}) \, f(y_t \mid x_t) \, f(\mathbf{x}_{1:t-1}) \, f(x_t \mid x_{t-1})}{f(\mathbf{y}_{1:t-1}) \, f(y_t \mid \mathbf{y}_{1:t-1})} \frac{f(\mathbf{y}_{1:t-1})}{f(\mathbf{y}_{1:t-1} \mid \mathbf{x}_{1:t-1}) f(\mathbf{x}_{1:t-1})}$$

$$= \frac{f(y_t \mid x_t) \, f(x_t \mid x_{t-1})}{f(y_t \mid \mathbf{y}_{1:t-1})} , \tag{5.90}$$

where we have also used (5.86) and the fact that $f(x_t \mid \mathbf{x}_{1:t-1}) = f(x_t \mid x_{t-1})$, $t = 1, 2, \ldots$, by the Markov property.

This result is of little use for an exact calculation of $f(\mathbf{x}_{1:n} \mid \mathbf{y}_{1:n})$, since it requires computation of $f(y_t \mid \mathbf{y}_{1:t-1})$, which involves the evaluation of complicated integrals. However, if both functions (pdfs) $f(x_t \mid x_{t-1})$ and $f(y_t \mid x_t)$ can be evaluated exactly (which is a reasonable assumption), then SIS can be used to approximately simulate from $f(\mathbf{x}_{1:t} \mid \mathbf{y}_{1:t})$ as follows: Let $g_t(\mathbf{x}_{1:t} \mid \mathbf{y}_{1:t})$ be the importance sampling pdf. We assume that, similar to (5.73), we can write $g_t(\mathbf{x}_{1:t} \mid \mathbf{y}_{1:t})$ recursively as

$$g_t(\mathbf{x}_{1:t} \mid \mathbf{y}_{1:t}) = g_0(x_0 \mid y_0) \prod_{s=1}^{t} g_s(x_s \mid \mathbf{x}_{s-1}, \mathbf{y}_s) . \tag{5.91}$$

Then, by analogy to (5.76), and using (5.90) (dropping the normalization constant $f(y_t \mid \mathbf{y}_{1:t-1})$), we can write the importance weight W_t of a path $\mathbf{x}_{1:t}$ generated from $g_t(\mathbf{x}_{1:t} \mid \mathbf{y}_{1:t})$ recursively as

$$W_t = W_{t-1} \frac{f(y_t \mid x_t) f(x_t \mid x_{t-1})}{g_t(x_t \mid \mathbf{x}_{1:t-1}, \mathbf{y}_{1:t})} = W_{t-1} u_t . \tag{5.92}$$

A natural choice for the importance sampling pdf is

$$g_t(x_t \mid \mathbf{x}_{1:t-1}, \mathbf{y}_{1:t}) = f(x_t \mid x_{t-1}) , \tag{5.93}$$

in which case the incremental weight simplifies to

$$u_t = f(y_t \mid x_t). \tag{5.94}$$

With this choice of sampling distribution, we are simply guessing the values of the hidden process $\{X_t\}$ without paying attention to the observed values.

Once the importance sampling density is chosen, sampling from the target pdf $f(\mathbf{x}_{1:t} \mid \mathbf{y}_{1:t})$ proceeds as described in Section 5.8. For more details, the interested reader is referred to [11], [27], and [32].

■ EXAMPLE 5.19   Bearings-Only Tracking

Suppose that we want to track an object (e.g., a submarine) via a radar device that only reports the *angle* to the object (see Figure 5.5). In addition, the angle measurements are noisy. We assume that the initial position and velocity are known and that the object moves at a constant speed.

Let $X_t = (p_{1t}, v_{1t}, p_{2t}, v_{2t})^\top$ be the vector of positions and (discrete) velocities of the target object at time $t = 0, 1, 2, \ldots$, and let Y_t be the measured angle. The problem is to track the unknown state of the object X_t based on the measurements $\{Y_t\}$ and the initial conditions.

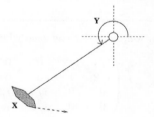

Figure 5.5: Track the object via noisy measurements of the angle.

The process $(X_t, Y_t), t = 0, 1, 2, \ldots$ is described by the following system:

$$X_t = A X_{t-1} + \varepsilon_{1t}$$
$$Y_t = \arctan(p_{1t}, p_{2t}) + \varepsilon_{2t} .$$

Here $\arctan(u, v)$ denotes the four-quadrant arc-tangent, that is, $\arctan(v/u) + c$, where c is either 0, $\pm\pi$, or $\pm\pi/2$, depending on the quadrant in which (u, v) lies. The random noise vectors $\{\varepsilon_{1t}\}$ are assumed to be $\mathsf{N}(\mathbf{0}, C_1)$ distributed, and the measurement noise ε_{2t} is $\mathsf{N}(0, \sigma_2^2)$ distributed. All noise variables are independent of each other. The matrix A is given by

$$A = \begin{pmatrix} 1 & 1 & 0 & 0 \\ 0 & 1 & 0 & 0 \\ 0 & 0 & 1 & 1 \\ 0 & 0 & 0 & 1 \end{pmatrix} .$$

The problem is to find the conditional pdf $f(x_t \mid \mathbf{y}_{1:t})$ and, in particular, the expected system state $\mathbb{E}[X_t \mid \mathbf{y}_{1:t}]$.

We indicate how this problem can be solved via SIS. Using (5.93) for the sampling distribution means simply that X_t is drawn from a $\mathsf{N}(Ax_{t-1}, C_1)$ distribution. As a consequence of (5.94), the incremental weight, $u_t = f(y_t \mid x_t)$, is equal to the value at y_t of the normal pdf with mean $\arctan(p_{1t}, p_{2t})$ and variance σ_2^2. The corresponding SIS procedure is summarized below. Note that the parallel SIS procedure is given, in which the $\{W_t^{(k)}\}$ and $\{X_t^{(k)}\}$ are computed at the same time by running N parallel processes.

Algorithm 5.10.1: SIS Procedure for Bearings-Only Tracking

---

input : Sample size N, matrices A and C_1, parameter σ_2, and distribution of the starting state.

output: Estimator $\widehat{x}_t$ of the expected position $\mathbb{E}[X_t \mid \mathbf{y}_{1:t}]$.

1 Initialize $X_0^{(k)}$ and set $W_0^{(k)} \leftarrow 0$, $k = 1, \ldots, N$

2 **for** $t = 1$ **to** n **do**

3   **for** $k = 1$ **to** N **do**

4          $X_t^{(k)} \sim \mathsf{N}(AX_{t-1}^{(k)}, C_1)$

5          $u_t \leftarrow \dfrac{1}{\sigma_2 \sqrt{2\pi}} \exp\left\{ -\dfrac{1}{2} \left(\dfrac{y_t - \arctan(p_{1t}, p_{2t})}{\sigma_2} \right)^2 \right\}$

6          $W_t^{(k)} \leftarrow W_{t-1}^{(k)} u_t$

7 $\widehat{x}_t \leftarrow (\sum_{k=1}^{N} W_t^{(k)} X_t^{(k)}) / \sum_{k=1}^{N} W_t^{(k)}$

8 **return** $\widehat{x}_t$

As a numerical illustration, consider the case where $\sigma_2 = 0.005$ and

$$C_1 = \sigma_1^2 \begin{pmatrix} 1/4 & 1/2 & 0 & 0 \\ 1/2 & 1 & 0 & 0 \\ 0 & 0 & 1/4 & 1/2 \\ 0 & 0 & 1/2 & 1 \end{pmatrix} ,$$

with $\sigma_1 = 0.001$. Let $X_0 \sim \mathsf{N}(\boldsymbol{\mu}_0, \Sigma_0)$, with $\boldsymbol{\mu}_0 = (-0.05, 0.001, 0.2, -0.055)^{\top}$, and

$$\Sigma_0 = 0.1^2 \begin{pmatrix} 0.5^2 & 0 & 0 & 0 \\ 0 & 0.005^2 & 0 & 0 \\ 0 & 0 & 0.3^2 & 0 \\ 0 & 0 & 0 & 0.01^2 \end{pmatrix}.$$

The left panel in Figure 5.6 shows how the estimated process $\{\widehat{x}_t\}$, obtained via SIS, tracks the actual process $\{X_t\}$ over $n = 25$ time steps, using a sample size of $N = 10,000$. In the right panel the result of SIR with bootstrap resampling (Algorithm 5.9.2) is shown, for the same sample size as in the SIS case. We see that the actual position is tracked over time more accurately.

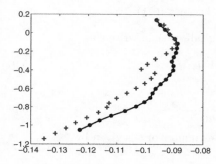 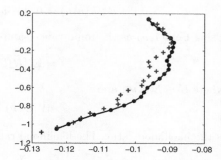

Figure 5.6: Comparison of the performance of the SIS (left) and SIR (right) algorithms for the bearings-only tracking problem, using a sample size of $N = 10^4$ over 25 time steps.

For both SIS and SIR, as time increases, the tracking rapidly becomes more unstable. This is a consequence of the degeneracy of the likelihood ratio. Indeed, after a few iterations, only a handful of samples contain the majority of the importance weight. This yields high variability between many runs and provides less reliable estimates. The resampling step mitigates some of this degeneracy. Several other heuristic resampling techniques have been proposed (e.g., see [11]).

5.11 TRANSFORM LIKELIHOOD RATIO METHOD

The *transform likelihood ratio* (TLR) method is a simple, convenient, and *unifying* way of constructing efficient importance sampling estimators. To motivate the TLR method, we consider the estimation of

$$\ell = \mathbb{E}[H(\mathbf{X})], \tag{5.95}$$

where $\mathbf{X} \sim f(\mathbf{x})$. Consider first the case where $\mathbf{X}$ is one-dimensional (we write X instead of $\mathbf{X}$). Let F be the cdf of X. According to the IT method, we can write

$$X = F^{-1}(U), \tag{5.96}$$

where $U \sim \mathsf{U}(0, 1)$ and F^{-1} is the inverse of the cdf F. Substituting $X = F^{-1}(U)$ into $\ell = \mathbb{E}[H(X)]$, we obtain

$$\ell = \mathbb{E}[H(F^{-1}(U))] = \mathbb{E}[\widetilde{H}(U)].$$

Note that in contrast to $\ell = \mathbb{E}[H(X)]$, where the expectation is taken with respect to $f(x)$, in $\ell = [\widetilde{H}(U)]$, the expectation is taken with respect to the uniform $\mathsf{U}(0,1)$ distribution. The extension to the multidimensional case is simple.

Let $h(u;\nu)$ be another density on $(0,1)$, parameterized by some reference parameter ν, with $h(u;\nu) > 0$ for all $0 \leqslant u \leqslant 1$ (note that u is a variable and not a parameter). An example is the $\mathsf{Beta}(\nu,1)$ distribution, with density

$$h(u;\nu) = \nu\, u^{\nu-1}, \quad u \in (0,1) ,$$

with $\nu > 0$, or the $\mathsf{Beta}(1,\nu)$ distribution, with density

$$h(u;\nu) = \nu\,(1-u)^{\nu-1}, \quad u \in (0,1) .$$

Using $\mathsf{Beta}(1,\nu)$ as the importance sampling pdf, we can write ℓ as

$$\ell = \mathbb{E}_\nu[\widetilde{H}(U)\,\widetilde{W}(U;\nu)] , \tag{5.97}$$

where $U \sim h(u;\nu)$, and

$$\widetilde{W}(U;\nu) = \frac{1}{h(U;\nu)} \tag{5.98}$$

is the likelihood ratio. The likelihood ratio estimator of ℓ is given by

$$\widehat{\ell} = N^{-1}\sum_{k=1}^{N}\widetilde{H}(U_k)\,\widetilde{W}(U_k;\nu) , \tag{5.99}$$

where $U_1,\ldots,U_N$ is a random sample from $h(u;\nu)$, We call (5.99) the *inverse-transform likelihood ratio* (ITLR) estimator; see Kroese and Rubinstein [23].

Suppose, for example, that $X \sim \mathsf{Weib}(\alpha,\lambda)$, which is to say that X has the density

$$f(x;\alpha,\lambda) = \alpha\lambda(\lambda x)^{\alpha-1}\mathrm{e}^{-(\lambda x)^\alpha}. \tag{5.100}$$

Note that a Weibull random variable can be generated using the transformation

$$X = \lambda^{-1}\,Z^{1/\alpha}, \tag{5.101}$$

where the random variable Z has an $\mathsf{Exp}(1)$ distribution. Applying the IT method, we obtain

$$X = F^{-1}(U) = \lambda^{-1}(-\ln(1-U))^{1/\alpha}, \tag{5.102}$$

and $\widetilde{H}(U_i)\,\widetilde{W}(U_i;\nu)$ in (5.99) reduces to $H(\lambda^{-1}\,(-\ln(1-U_i))^{1/\alpha})/h(U_i;\nu)$.

The TLR method is a natural extension of the ITLR method. It comprises two steps. The first is a simple *change of variable* step, and the second involves an application of the SLR technique to the transformed pdf.

To apply the first step, we simply write $\mathbf{X}$ as a function of another random vector, say as

$$\mathbf{X} = G(\mathbf{Z}) . \tag{5.103}$$

If we define

$$\widetilde{H}(\mathbf{Z}) = H(G(\mathbf{Z})) ,$$

then estimating (5.95) is equivalent to estimating

$$\ell = \mathbb{E}[\widetilde{H}(\mathbf{Z})] . \tag{5.104}$$

Note that the expectations in (5.95) and (5.104) are taken with respect to the original density of $\mathbf{X}$ and the transformed density of $\mathbf{Z}$. As an example, consider again a one-dimensional case and let $X \sim \mathsf{Weib}(\alpha, \lambda)$. Recalling (5.101), we have $\widetilde{H}(Z) = H(\lambda^{-1} Z^{1/\alpha})$, and thus $\ell = \mathbb{E}[H(\lambda^{-1} Z^{1/\alpha})]$.

To apply the second step, we assume that $\mathbf{Z}$ has a density $h(\mathbf{z}; \boldsymbol{\theta})$ in some class of densities $\{h(\mathbf{z}; \boldsymbol{\eta})\}$. Then we can seek to estimate ℓ efficiently via importance sampling, for example, using the standard likelihood ratio method. In particular, by analogy to (5.58), we obtain the following estimator:

$$\widehat{\ell} = \frac{1}{N} \sum_{k=1}^{N} \widetilde{H}(\mathbf{Z}_k) \, \widetilde{W}(\mathbf{Z}_k; \boldsymbol{\theta}, \boldsymbol{\eta}) \, , \qquad (5.105)$$

where

$$\widetilde{W}(\mathbf{Z}_k; \boldsymbol{\theta}, \boldsymbol{\eta}) = \frac{h(\mathbf{Z}_k; \boldsymbol{\theta})}{h(\mathbf{Z}_k; \boldsymbol{\eta})}$$

and $\mathbf{Z}_k \sim h(\mathbf{z}; \boldsymbol{\eta})$. We will call the SLR estimator (5.105) based on the transformation (5.103), the *TLR estimator*. As an example, consider again the $\mathsf{Weib}(\alpha, \lambda)$ case. Using (5.101), we could take $h(z; \eta) = \eta \, e^{-\eta z}$ as the sampling pdf, with $\eta = \theta = 1$ as the nominal parameter. Hence, in this case, $\widehat{\ell}$ in (5.105) reduces to

$$\widehat{\ell} = \frac{1}{N} \sum_{k=1}^{N} \widetilde{H} \left(\lambda^{-1} Z_k^{1/\alpha} \right) \widetilde{W}(Z_k; \theta, \eta) \, , \qquad (5.106)$$

with

$$\widetilde{W}(Z_k; \theta, \eta) = \frac{h(Z_k; \theta)}{h(Z_k; \eta)} = \frac{\theta \, e^{-\theta Z_k}}{\eta \, e^{-\eta Z_k}}$$

and $Z_k \sim \mathsf{Exp}(\eta)$.

To find the optimal parameter vector $\boldsymbol{\eta}^*$ of the TLR estimator (5.105), we can solve, by analogy to (5.63), the CE program

$$\max_{\boldsymbol{\eta}} D(\boldsymbol{\eta}) = \max_{\boldsymbol{\eta}} \mathbb{E}_{\boldsymbol{\tau}} \left[\widetilde{H}(\mathbf{Z}) \, \widetilde{W}(\mathbf{Z}; \boldsymbol{\theta}, \boldsymbol{\tau}) \ln h(\mathbf{Z}; \boldsymbol{\eta}) \right] \, , \qquad (5.107)$$

and similarly for the stochastic counterpart of (5.107).

Since $\mathbf{Z}$ can be distributed quite arbitrarily, its distribution is typically chosen from an exponential family of distributions (see Section A.3 of the Appendix), for which the optimal solution $\boldsymbol{\eta}^*$ of (5.107) can be obtained analytically in a convenient and simple form. Below we present the TLR algorithm for estimating $\ell = \mathbb{E}_f[H(\mathbf{X})]$, assuming that $\mathbf{X}$ is a random vector with independent, continuously distributed components. The key is to find a transformation function G such that $\mathbf{X} = G(\mathbf{Z})$, with $\mathbf{Z} \sim h(\mathbf{z}; \boldsymbol{\theta})$. For example, we can take $\mathbf{Z}$ with all components being iid and distributed according to an exponential family (e.g., $\mathsf{Exp}(1)$).

Algorithm 5.11.1: Transform likelihood ratio (TLR) method

---

input : Function G such that $\mathbf{X} = G(\mathbf{Z})$, with $\mathbf{Z} \sim h(\mathbf{z}; \boldsymbol{\theta})$. Sample sizes N and N_2. Initial parameter $\boldsymbol{\tau}$.

output: Estimator $\widehat{\ell}$ of $\ell = \mathbb{E}[H(\mathbf{X})] = \mathbb{E}[\widetilde{H}(\mathbf{Z})]$.

1 Generate a random sample $\mathbf{Z}_1, \ldots, \mathbf{Z}_N$ from $h(\cdot; \boldsymbol{\tau})$.

2 Solve the stochastic counterpart of the program (5.107) (for a one-parameter exponential family parameterized by the mean, apply directly the analytic solution (5.68). Iterate if necessary. Denote the solution by $\widehat{\boldsymbol{\eta}}$.

3 Generate a (larger) random sample $\mathbf{Z}_1, \ldots, \mathbf{Z}_{N_1}$ from $h(\cdot; \widehat{\boldsymbol{\eta}})$ and estimate $\ell = \mathbb{E}[H(G(\mathbf{Z}))]$ via the TLR estimator (5.105), taking $\boldsymbol{\eta} = \widehat{\boldsymbol{\eta}}$.

4 **return** $\widehat{\ell}$

The TLR Algorithm 5.11.1 ensures that as soon as the transformation $\mathbf{X} = G(\mathbf{Z})$ is chosen, one can estimate ℓ using the TLR estimator (5.105) instead of the SLR estimator (5.58). Although the accuracy of both estimators (5.105) and (5.58) is the same (Rubinstein and Kroese [35]), the advantage of the former is its universality and its ability to avoid the computational burden while directly delivering the analytical solution of the stochastic counterpart of the program (5.107).

5.12 PREVENTING THE DEGENERACY OF IMPORTANCE SAMPLING

In this section, we show how to prevent the degeneracy of importance sampling estimators via the *screening method*. The degeneracy of likelihood ratios in high-dimensional Monte Carlo simulation problems is one of the central topics in Monte Carlo simulation; see also Remark 5.7.1.

To motivate the screening method, consider again Example 5.15 and observe that only the first two importance sampling parameters of the five-dimensional vector $\widehat{\mathbf{v}} = (\widehat{v}_1, \widehat{v}_2, \widehat{v}_3, \widehat{v}_4, \widehat{v}_5)$ are substantially different from those in the nominal parameter vector $\mathbf{u} = (u_1, u_2, u_3, u_4, u_5)$. The reason is that the partial derivatives of ℓ with respect to u_1 and u_2 are significantly larger than those with respect to u_3, u_4, and u_5. We call such elements u_1 and u_2 *bottleneck elements*. Based on this observation, we could use instead of the importance sampling vector $\widehat{\mathbf{v}} = (\widehat{v}_1, \widehat{v}_2, \widehat{v}_3, \widehat{v}_4, \widehat{v}_5)$ the vector $\widehat{\mathbf{v}} = (\widehat{v}_1, \widehat{v}_2, u_3, u_4, u_5)$, reducing the number of importance sampling parameters from five to two. This not only has computational advantages — we would then solve a two-dimensional variance or CE minimization program instead of a five-dimensional one — but also leads to further variance reduction, since the likelihood ratio term W with two product terms is less "noisy" than the one with five product terms. Bottleneck elements can be identified as those elements i that have the largest *relative perturbation* $\delta_i = (\widehat{v}_i - u_i)/u_i$.

Algorithm 5.12.1 presents the two-stage screening algorithm for estimating

$$\ell = \mathbb{E}_{\mathbf{u}}[H(\mathbf{X})] = \int H(\mathbf{x}) f(\mathbf{x}; \mathbf{u}) \, d\mathbf{x} ,$$

based on CE, and denoted as CE-SCR. Its VM counterpart, the VM-SCR algorithm, is similar. For simplicity, we assume that the components of $\mathbf{X}$ are independent and that each component is distributed according to a one-dimensional exponential family that is parameterized by the mean — the dependent case could be treated similarly. Moreover, $H(\mathbf{x})$ is assumed to be a monotonically increasing function in each component of $\mathbf{x}$. A consequence of the assumptions above is that the parameter vector $\mathbf{v}$ has dimension n.

At the first stage of the algorithm (Lines 1–9), we identify and estimate the bottleneck parameters, leaving the nonbottleneck parameters as they are, and at the second stage (Lines 10–12), we compute and return the importance sampling estimator

$$\widehat{\ell}_B = \frac{1}{N} \sum_{k=1}^N H(\mathbf{X}_k)\, W_B(\mathbf{X}_{kB}; \widehat{\mathbf{v}}_B)\,, \qquad (5.108)$$

where $\mathbf{X}_{kB}$ is the subvector of bottleneck elements of $\mathbf{X}_k$ and W_B is the corresponding likelihood ratio. Note that, by analogy to Proposition A.4.2 in the Appendix, if each component of the random vector $\mathbf{X}$ is from a one-parameter exponential family parameterized by the mean and if $H(\mathbf{x})$ is a monotonically increasing function in each component of $\mathbf{x}$, then each element of the CE optimal parameter $\mathbf{v}^*$ is at least as large as the corresponding one of $\mathbf{u}$. We leave the proof as an exercise for the reader.

Algorithm 5.12.1: CE-SCR Two-Stage Screening Algorithm

> **input** : Sample size N, performance function H, tolerance δ, number of repetitions d.
> **output:** Estimator $\widehat{\ell}_B$ of the expected performance $\mathbb{E}[H(\mathbf{X})]$.

1   Initialize $B \leftarrow \{1, \ldots, n\}$.
2   **for** $t = 1$ **to** d **do**
3   Generate a sample $\mathbf{X}_1, \ldots, \mathbf{X}_N$ from $f(\mathbf{x}; \mathbf{u})$.
4   **for** $i = 1$ **to** n **do**
5         $\widehat{v}_i \leftarrow \dfrac{\sum_{k=1}^N H(\mathbf{X}_k)\, X_{ki}}{\sum_{k=1}^N H(\mathbf{X}_k)}$
6         $\delta_i \leftarrow \dfrac{\widehat{v}_i - u_i}{u_i}$ `// calculate the relative perturbation`
7   **if** $\delta_i < \delta$ **then**
8            $\widehat{v}_i = u_i$
9            $B \leftarrow B \setminus \{i\}$ `// remove i from the bottleneck set`

10 Generate a sample $\mathbf{X}_1, \ldots, \mathbf{X}_N$ from $f(\mathbf{x}; \widehat{\mathbf{v}})$
11 $\widehat{\ell}_B \leftarrow \frac{1}{N} \sum_{k=1}^N H(\mathbf{X}_k)\, W_B(\mathbf{X}_{kB}; \widehat{\mathbf{v}}_B)$.
12 **return** $\widehat{\ell}_B$

It is important to note the following:

1. As mentioned, under the present assumptions (independent components, each from a one-parameter exponential family parameterized by the mean, and $H(\mathbf{x})$ monotonically increasing in each component), the components of the $\mathbf{v}^*$ are at least as large as the corresponding elements of $\mathbf{u}$. Algorithm 5.12.1 takes this into account and always identifies all elements i corresponding to $\delta_i < 0$ as nonbottleneck ones.

2. Recall that Lines 3–9 are purposely performed d times. This allows us to better determine the nonbottleneck parameters, since it is likely that they will fluctuate around their nominal value u_i and therefore δ_i will become negative or very small in one of the replications.

In general, large-dimensional, complex simulation models contain both bottleneck and nonbottleneck parameters. The number of bottleneck parameters is typically smaller than the number of nonbottleneck parameters. Imagine a situation where the size (dimension) of the vector **u** is large, say 100, and the number of bottleneck elements is only about 10–15. Then, clearly, an importance sampling estimator based on bottleneck elements alone will not only be much more accurate than its standard importance sampling counterpart involving all 100 likelihood ratios (containing both bottleneck and nonbottleneck ones) but, in contrast to the latter, will not be degenerated.

The bottleneck phenomenon often occurs when one needs to estimate the probability of a nontypical event in the system, like a rare-event probability. This will be treated in Chapter 8. For example, if one observes a failure in a reliability system with highly reliable elements, then it is very likely that several elements (typically the less reliable ones) forming a minimal cut in the model all fail simultaneously. Another example is the estimation of a buffer overflow probability in a queueing network, that is, the probability that the total number of customers in all queues exceeds some large number. Again, if a buffer overflow occurs, it is quite likely that this has been caused by a buildup in the bottleneck queue, which is the most congested one in the network.

5.12.0.1 Numerical Results   We next present numerical studies with Algorithm 5.12.1 for a generalization of the bridge system in Example 5.1, depicted in Figure 5.7. We will implement screening for both CE and VM methods. Recall that for the CE method the parameter vector $\widehat{\mathbf{v}}$ (and $\widehat{\mathbf{v}}_B$) can often be updated analytically, in particular when the sampling distribution comes from an exponential family. In contrast, for VM the updating typically involves a numerical procedure.

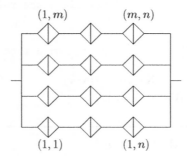

Figure 5.7: An $m \times n$ bridge system.

The system consists of $m \times n$ bridges arranged in a grid, and all bridges are of the form in Figure 5.1. Denote the lengths of the edges within the (i, j)-th bridge by $X_{ij1}, \ldots, X_{ij5}$. Then the length of the shortest path through bridge (i, j) is

$$Y_{ij} = \min\{X_{ij1} + X_{ij4},\ X_{ij2} + X_{ij5}, \\ X_{ij1} + X_{ij3} + X_{ij5},\ X_{ij2} + X_{ij3} + X_{ij4}\} . \tag{5.109}$$

Suppose that we want to estimate the expected maximal length ℓ of the shortest paths in all rows, that is, $\ell = \mathbb{E}[H(\mathbf{X})]$, with

$$H(\mathbf{X}) = \max\{Y_{11} + \cdots + Y_{1n}, \ldots, Y_{m1} + \cdots + Y_{mn}\} . \qquad (5.110)$$

In our numerical results, we assume that the components X_{ijk} of the random vector $\mathbf{X}$ are independent and that each component has a Weib(α, u) distribution, which is to say X_{ijk} has the density

$$f(x; \alpha, u) = \alpha u(ux)^{\alpha-1} e^{-(ux)^{\alpha}},$$

with $u = u_{ijk}$. Recall that such a Weibull random variable can be generated using the transformation $X = u^{-1} Z^{1/\alpha}$, where Z is a random variable distributed Exp(1). We also assume that only u is controllable, while α is fixed and equals 0.2. We purposely selected some elements of $\mathbf{u}$ to be bottleneck ones and set $\delta = 0.1$. It is important to note that the $\{X_{ijk}\}$ are here *not* parameterized by the mean. However, by taking $1/u_{ijk}$ as the parameter, we are in the framework described above. In particular, the relative perturbations are carried out with respect to $\alpha/\widehat{v}_{ijk}$ and α/u_{ijk}.

Table 5.2 presents the performance of Algorithm 5.12.1 for the 1×1 (single-bridge) model (5.110). Here $u_{111} = 1$ and $u_{112} = 1$ are chosen to be the bottleneck parameters, whereas the remaining (nonbottleneck) ones are set equal to 2. The notations in Table 5.2 are as follows:

1. *Mean, max*, and *min* $\widehat{\ell}$ denote the sample mean, maximum, and minimum values of 10 independently generated estimates of $\widehat{\ell}$.

2. *RE* denotes the sample relative error for $\widehat{\ell}$, averaged over the 10 runs.

3. *CPU* denotes the average CPU time in seconds based on 10 runs.

Table 5.2: Performance of Algorithm 5.12.1 for the single-bridge model with samples $N = N_1 = 500$.

	CMC	CE	VM	CE-SCR	VM-SCR
Mean $\widehat{\ell}$	4.052	3.970	3.734	3.894	3.829
Max $\widehat{\ell}$	8.102	4.327	4.201	4.345	4.132
Min $\widehat{\ell}$	1.505	3.380	3.395	3.520	3.278
RE	0.519	0.070	0.078	0.076	0.068
CPU	0.00	0.04	0.21	0.05	0.13

From the results of Table 5.2, it follows that for this relatively small model both CE and VM perform similarly to their screening counterparts. We will further see (in Chapter 8) that as the complexity of the model increases, VM-SCR outperforms its three alternatives, in particular CE-SCR. Note that for this model, both CE and VM detected correctly, at the first stage, the two bottleneck parameters. In particular, Table 5.3 presents a typical dynamics of detecting the two bottleneck parameters at the first stage of Algorithm 5.12.1 for a single-bridge model having a total of 5 parameters. In Table 5.3, t denotes the replication number at the first stage, while the 0s and 1s indicate whether the corresponding parameters are identified as nonbottleneck or bottleneck parameters, respectively. As can be seen, after two replications four bottleneck parameters are left, after six replications three are identified as bottleneck parameters, and after seven replications the process stabilizes, detecting correctly the two true bottleneck parameters.

Table 5.3: Typical dynamics for detecting the bottleneck parameters at the first stage of Algorithm 5.12.1 for the bridge model.

t	u_1	u_2	u_3	u_4	u_5	t	u_1	u_2	u_3	u_4	u_5
0	1	1	1	1	1	5	1	1	0	1	0
1	1	1	0	1	1	6	1	1	0	1	0
2	1	1	0	1	1	7	1	1	0	0	0
3	1	1	0	1	0	8	1	1	0	0	0
4	1	1	0	1	0	9	1	1	0	0	0

Table 5.4 presents a typical evolution of $\{\widehat{\mathbf{v}}_t\}$ in the single-bridge model for the VM and VM-SCR methods at the second stage of Algorithm 5.12.1.

Table 5.4: Typical evolution of $\{\widehat{\mathbf{v}}_t\}$ for the VM and VM-SCR methods.

	VM						VM-SCR				
t	$\widehat{v}_1$	$\widehat{v}_2$	$\widehat{v}_3$	$\widehat{v}_4$	$\widehat{v}_5$	t	$\widehat{v}_1$	$\widehat{v}_2$	$\widehat{v}_3$	$\widehat{v}_4$	$\widehat{v}_5$
	1.000	1.000	2.000	2.000	2.000		1.000	1.000	2	2	2
1	0.537	0.545	2.174	2.107	1.615	1	0.555	0.599	2	2	2
2	0.346	0.349	2.071	1.961	1.914	2	0.375	0.402	2	2	2
3	0.306	0.314	1.990	1.999	1.882	3	0.315	0.322	2	2	2

As is clearly seen, the bottleneck parameters decrease about three times after the third iteration, while the nonbottleneck ones fluctuate about their nominal value $u = 2$.

Table 5.5 presents the performance of Algorithm 5.12.1 for the 3×10 bridge model with six bottlenecks corresponding to the elements u_{111}, u_{112}, u_{211}, u_{212}, u_{311}, u_{312}. We set $u_{111} = u_{112} = u_{211} = u_{212} = u_{311} = u_{312} = 1$, while the remaining (nonbottlenecks) values are set equal to 2. Note again that in this case both CE and VM found the true six bottlenecks.

Table 5.5: Performance of Algorithm 5.12.1 for the 3×10 model with six bottleneck elements and sample size $N = N_1 = 1000$.

	CMC	CE	VM	CE-SCR	VM-SCR
Mean $\widehat{\ell}$	16.16	16.11	14.84	16.12	15.67
Max $\widehat{\ell}$	22.65	26.85	16.59	18.72	17.20
Min $\widehat{\ell}$	11.13	7.007	12.59	14.63	14.80
RE	0.20	0.34	0.075	0.074	0.049
CPU	0.00	0.49	68.36	0.73	27.54

From the results in Table 5.5, it follows that without screening even the naive Monte Carlo outperforms the standard CE. However, using screening results in substantial improvement of CE. Finally, VM-SCR outperforms all four remaining alternatives.

PROBLEMS

5.1   Consider the integral $\ell = \int_a^b H(x)\,dx = (b-a)\,\mathbb{E}[H(X)]$, with $X \sim \mathsf{U}(a,b)$. Let $X_1,\ldots,X_N$ be a random sample from $\mathsf{U}(a,b)$. Consider the estimators $\widehat{\ell} = \frac{1}{N}\sum_{i=1}^N H(X_i)$ and $\widehat{\ell}_1 = \frac{1}{2N}\sum_{i=1}^N \{H(X_i) + H(b + a - X_i)\}$. Prove that if $H(x)$ is monotonic in x, then

$$\mathrm{Var}(\widehat{\ell}_1) \leqslant \frac{1}{2}\mathrm{Var}(\widehat{\ell})\ .$$

In other words, using antithetic random variables is more accurate than using CMC.

5.2   Estimate the expected length of the shortest path for the bridge network in Example 5.1. Use both the CMC estimator (5.8) and the antithetic estimator (5.9). For both cases, take a sample size of $N = 100,000$. Suppose that the lengths of the links $X_1,\ldots,X_5$ are exponentially distributed, with means $1,1,0.5,2,1.5$. Compare the results.

5.3   Common random variables (CRVs) are often used when estimating derivatives or gradients of functions. As a toy example, consider the estimation of the derivative $\ell'(u)$ of the function $\ell(u) = \mathbb{E}[Y]$, where $Y \sim \mathsf{Exp}(2u)$. Hence, $\ell'(u) = -1/(2u^2)$. A simple way to estimate $\ell'(u)$ is to first approximate it with the *forward difference*

$$\frac{\ell(u+h) - \ell(u)}{h}$$

for small h, and then to estimate both $\ell(u+h)$ and $\ell(u)$ via Monte Carlo simulation. For example, generate $X_1,\ldots,X_N$ from $\mathsf{Exp}(u + h)$ and $Y_1,\ldots,Y_n$ from $\mathsf{Exp}(u)$ independently, and take $(\bar{X} - \bar{Y})/h$ as an estimator of the forward difference (and hence as a biased estimator of $\ell'(u)$). However, it is better to draw each pair (X_i, Y_i) with CRNs, for example, by letting $X_i = -\ln(U_i)/(2(u + h))$ and $Y_i = -\ln(U_i)/(2u)$, where $U_i \sim \mathsf{U}(0,1), i = 1,\ldots,N$.

   a)   Implement the two forward difference estimators for the case $u = 1$ and $h = 10^{-2}$, taking $N = 10^6$. Estimate and compare the relative errors of both estimators.

   b)   Compute the relative errors exactly (not using simulation).

5.4   Use the batch means method to estimate the expected stationary waiting time in a $GI/G/1$ queue via Lindley's equation for the case where the interarrival times are $\mathsf{Exp}(1/2)$ distributed and the service times are $\mathsf{U}[0.5, 2]$ distributed. Take a simulation run of $M = 10,000$ customers, discarding the first $K = 100$ observations. Examine to what extent variance reduction can be achieved by using antithetic random variables.

5.5   Run the stochastic shortest path problem in Example 5.4 and estimate the performance $\ell = \mathbb{E}[H(\mathbf{X})]$ from 1000 independent replications, using the given (C_1, C_2, C_3, C_4) as the vector of control variables, assuming that $X_i \sim \mathsf{Exp}(1)$, $i = 1,\ldots,5$. Compare the results with those obtained with the CMC method.

5.6   Estimate the expected waiting time of the fourth customer in a $GI/G/1$ queue for the case where the interarrival times are $\mathsf{Exp}(1/2)$ distributed and the service times are $\mathsf{U}[0.5, 2]$ distributed. Use Lindley's equation and control variables, as described in Example 5.5. Generate $N = 1000$ replications of W_4 and provide a 95% confidence interval for $\mathbb{E}[W_4]$.

5.7   Prove that for any pair of random variables (U, V),

$$\mathrm{Var}(U) = \mathbb{E}[\mathrm{Var}(U \mid V)] + \mathrm{Var}(\mathbb{E}[U \mid V]) \ .$$

[Hint: Use the facts that $\mathbb{E}[U^2] = \mathbb{E}[\mathbb{E}[U^2 \mid V]]$ and $\mathrm{Var}(X) = \mathbb{E}[X^2] - (\mathbb{E}[X])^2$.]

5.8   Let $R \sim \mathsf{G}(p)$ and define $S_R = \sum_{i=1}^{R} X_i$, where $X_1, X_2, \ldots$ is a sequence of iid $\mathsf{Exp}(\lambda)$ random variables that are independent of R.

   a)   Show, that $S_R \sim \mathsf{Exp}(\lambda p)$. [Hint: The easiest way is to use transform methods and conditioning.]

   b)   For $\lambda = 1$ and $p = 1/10$, estimate $\mathbb{P}(S_R > 10)$ using CMC with a sample size of $N = 1000$.

   c)   Repeat b), now using the conditional Monte Carlo estimator (5.22). Compare the results with those of a) and b).

5.9   Consider the random sum S_R in Problem 5.8, with parameters $p = 0.25$ and $\lambda = 1$. Estimate $\mathbb{P}(S_R > 10)$ via stratification using strata corresponding to the partition of events $\{R = 1\}$, $\{R = 2\}$, $\ldots, \{R = 7\}$, and $\{R > 7\}$. Allocate a total of $N = 10,000$ samples via both $N_i = p_i N$ and the optimal N_i^* in (5.35), and compare the results. For the second method, use a simulation run of size 1000 to estimate the standard deviations $\{\sigma_i\}$.

5.10   Show that the solution to the minimization program

$$\min_{N_1,\ldots,N_m} \sum_{i=1}^{m} \frac{p_i^2 \, \sigma_i^2}{N_i} \quad \text{such that} \quad N_1 + \cdots + N_m = N \ ,$$

is given by (5.35). This justifies the stratified sampling Theorem 5.5.1.

5.11   Use Algorithm 5.4.2 and (5.26) to estimate the reliability of the bridge reliability network in Example 4.2 on page 110 via permutation Monte Carlo. Consider two cases, where the link reliabilities are given by $\mathbf{p} = (0.3, 0.1, 0.8, 0.1, 0.2)$ and $\mathbf{p} = (0.95, 0.95, 0.95, 0.95, 0.95)$, respectively. Take a sample size of $N = 2000$.

5.12   Repeat Problem 5.11, using Algorithm 5.4.3. Compare the results.

5.13   This exercise discusses the counterpart of Algorithm 5.4.3 involving minimal paths rather than minimal cuts. A state vector $\mathbf{x}$ in the reliability model of Section 5.4.1 is called a *path vector* if $H(\mathbf{x}) = 1$. If in addition $H(\mathbf{y}) = 0$ for all $\mathbf{y} < \mathbf{x}$, then $\mathbf{x}$ is called the *minimal path vector*. The corresponding set $A = \{i : x_i = 1\}$ is called the *minimal path set*; that is, a minimal path set is a minimal set of components whose *functioning* ensures the functioning of the system. If $A_1, \ldots, A_m$ denote all the minimal paths sets, then the system is functioning if and only if all the components of at least one minimal path set are functioning.

   a)   Show that

$$H(\mathbf{x}) = \max_{k} \prod_{i \in A_k} x_i = 1 - \prod_{k=1}^{m} \left(1 - \prod_{i \in A_k} x_i\right). \tag{5.111}$$

b) Define

$$Y_k = \prod_{i \in A_k} X_i, \ \ k = 1, \dots, m,$$

that is, Y_k is the indicator of the event that all components in A_i are functioning. Apply Proposition 5.4.1 to the sum $S = \sum_{k=1}^{m} Y_k$ and devise an algorithm similar to Algorithm 5.4.3 to estimate the reliability $r = \mathbb{P}(S > 0)$ of the system.

c) Test this algorithm on the bridge reliability network in Example 4.2.

5.14 Prove (see (5.44)) that the solution of

$$\min_g \operatorname{Var}_g \left(H(\mathbf{X}) \frac{f(\mathbf{X})}{g(\mathbf{X})} \right)$$

is

$$g^*(\mathbf{x}) = \frac{|H(\mathbf{x})| \, f(\mathbf{x})}{\int |H(\mathbf{x})| \, f(\mathbf{x}) \, d\mathbf{x}} \ .$$

5.15 Let $Z \sim \mathsf{N}(0, 1)$. Estimate $\mathbb{P}(Z > 4)$ via importance sampling, using the following shifted exponential sampling pdf:

$$g(x) = e^{-(x-4)}, \quad x \geqslant 4 \ .$$

Choose N large enough to obtain accuracy to at least three significant digits and compare with the exact value.

5.16 Pearson's χ^2 discrepancy measure between densities g and h is defined as

$$d(g, \, h) = \frac{1}{2} \int \frac{[g(\mathbf{x}) - h(\mathbf{x})]^2}{h(\mathbf{x})} \, d\mathbf{x} \ .$$

Verify that the VM program (5.43) is equivalent to minimizing the Pearson χ^2 discrepancy measure between the zero-variance pdf g^* in (5.45) and the importance sampling density g. In this sense, the CE and VM methods are similar, since the CE method minimizes the Kullback–Leibler distance between g^* and g.

5.17 Repeat Problem 5.2 using importance sampling, where the lengths of the links are exponentially distributed with means $v_1, \dots, v_5$. Write down the deterministic CE updating formulas, and estimate these via a simulation run of size 1000 using $\mathbf{w} = \mathbf{u}$.

5.18 Consider the natural exponential family ((A.9) in the Appendix). Show that (5.61), with $\mathbf{u} = \boldsymbol{\theta}_0$ and $\mathbf{v} = \boldsymbol{\theta}$, reduces to solving

$$\mathbb{E}_{\boldsymbol{\theta}_0} \left[H(\mathbf{X}) \left(\frac{\nabla c(\boldsymbol{\theta})}{c(\boldsymbol{\theta})} + \mathbf{t}(\mathbf{X}) \right) \right] = \mathbf{0} \ . \tag{5.112}$$

5.19 As an application of (5.112), suppose that we want to estimate the expectation of $H(X)$, with $X \sim \mathsf{Exp}(\lambda_0)$. Show that the corresponding CE optimal parameter is

$$\lambda^* = \frac{\mathbb{E}_{\lambda_0}[H(X)]}{\mathbb{E}_{\lambda_0}[H(X)X]} \ .$$

Compare with (A.15) in the Appendix. Explain how to estimate λ^* via simulation.

5.20   Let $X \sim \mathsf{Weib}(\alpha, \lambda_0)$. We wish to estimate $\ell = \mathbb{E}_{\lambda_0}[H(X)]$ via the SLR method, generating samples from $\mathsf{Weib}(\alpha, \lambda)$ — thus changing the scale parameter λ but keeping the scale parameter α fixed. Use (5.112) and Table A.1 in the Appendix to show that the CE optimal choice for λ is

$$\lambda^* = \left(\frac{\mathbb{E}_{\lambda_0}[H(X)]}{\mathbb{E}_{\lambda_0}[H(X) X^\alpha]} \right)^{1/\alpha} .$$

Explain how we can estimate λ^* via simulation.

5.21   Let $X_1, \ldots, X_n$ be independent $\mathsf{Exp}(1)$ distributed random variables. Let $\mathbf{X} = (X_1, \ldots, X_n)$ and $S(\mathbf{X}) = X_1 + \cdots + X_n$. We wish to estimate $\mathbb{P}(S(\mathbf{X}) \geqslant \gamma)$ via importance sampling, using $X_i \sim \mathsf{Exp}(\theta)$, for all i. Show that the CE optimal parameter θ^* is given by

$$\theta^* = \frac{\mathbb{E}[I_{\{S(\mathbf{X}) \geqslant \gamma\}}]}{\mathbb{E}[I_{\{S(\mathbf{X}) \geqslant \gamma\}} \, \overline{X}]} ,$$

with $\overline{X} = (X_1 + \cdots + X_n)/n$ and $\mathbb{E}$ indicating the expectation under the original distribution (where each $X_i \sim \mathsf{Exp}(1)$).

5.22   Consider Problem 5.20. Define $G(z) = z^{1/\alpha}/\lambda_0$ and $\widetilde{H}(z) = H(G(z))$.
   a)   Show that if $Z \sim \mathsf{Exp}(1)$, then $G(Z) \sim \mathsf{Weib}(\alpha, \lambda_0)$.
   b)   Explain how to estimate ℓ via the TLR method.
   c)   Show that the CE optimal parameter for Z is given by

$$\theta^* = \frac{\mathbb{E}_\eta[\widetilde{H}(Z) \, W(Z; 1, \eta)]}{\mathbb{E}_\eta[\widetilde{H}(Z) \, Z \, W(Z; 1, \eta)]},$$

   where $W(Z; 1, \eta)$ is the ratio of the $\mathsf{Exp}(1)$ and $\mathsf{Exp}(\eta)$ pdfs.

5.23   Assume that the expected performance can be written as $\ell = \sum_{i=1}^m a_i \, \ell_i$, where $\ell_i = \int H_i(\mathbf{x}) \, d\mathbf{x}$, and the a_i, $i = 1, \ldots, m$ are known coefficients. Let $Q(\mathbf{x}) = \sum_{i=1}^m a_i \, H_i(\mathbf{x})$. For any pdf g dominating $Q(\mathbf{x})$, the random variable

$$L = \sum_{i=1}^m a_i \frac{H_i(\mathbf{X})}{g(\mathbf{X})} = \frac{Q(\mathbf{X})}{g(\mathbf{X})} ,$$

where $\mathbf{X} \sim g$, is an unbiased estimator of ℓ — note that there is only one sample. Prove that L attains the smallest variance when $g = g^*$, with

$$g^*(\mathbf{x}) = |Q(\mathbf{x})| / \int |Q(\mathbf{x})| \, d\mathbf{x} ,$$

and that

$$\mathrm{Var}_{g^*}(L) = \left(\int |Q(\mathbf{x})| \, d\mathbf{x} \right)^2 - \ell^2 .$$

5.24   The Hit-or-Miss Method.   Suppose that the sample performance function, H, is bounded on the interval $[0, b]$, say, $0 \leqslant H(x) \leqslant c$ for $x \in [0, b]$. Let

$\ell = \int H(x)\,\mathrm{d}x = b\,\mathbb{E}[H(X)]$, with $X \sim \mathsf{U}[0,b]$. Define an estimator of ℓ by

$$\widehat{\ell^h} = \frac{bc}{N} \sum_{i=1}^{N} I_{\{Y_i < H(X_i)\}}\,,$$

where $\{(X_i, Y_i) : j = 1, \ldots, N\}$ is a sequence of points uniformly distributed over the rectangle $[0,b] \times [0,c]$ (see Figure 5.8). The estimator $\widehat{\ell^h}$ is called the *hit-or-miss estimator*, since a point (X, Y) is accepted or rejected depending on whether that point falls inside or outside the shaded area in Figure 5.8, respectively. Show that the hit-or-miss estimator has a larger variance than the CMC estimator,

$$\widehat{\ell} = \frac{b}{N} \sum_{i=1}^{N} H(X_i)\,,$$

with $X_1, \ldots, X_N$ a random sample from $\mathsf{U}[0,b]$.

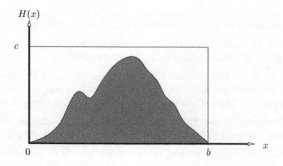

Figure 5.8: The hit-or-miss method.

Further Reading

The fundamental paper on variance reduction techniques is Kahn and Marshal [20]. There are a plenty of good Monte Carlo textbooks with chapters on variance reduction techniques. Among them are [13], [17], [21], [22], [25], [29], [32], [33], and [40]. For a comprehensive study of variance reduction techniques, see Fishman [13] and Rubinstein [34]. Asmussen and Glynn [2] provide a modern treatment of variance reduction and rare-event simulation. See also Chapter 9 of [24], which explains in detail the wide variety of variance reduction techniques that can be used to solve a single estimation problem. Influential books on sequential importance resampling are [27] and [11]. Multilevel Monte Carlo is a burgeoning area of research. The recent paper by Rhee and Glynn [31] explains how randomization of the levels of an infinite-level multilevel Monte Carlo algorithm can achieve theoretically a zero-bias estimator. Botev [6] describes the state-of-the art variance reduction technique for sampling from high-dimensional truncated multivariate normal distributions.

An introduction to reliability models may be found in [15]. For more information on variance reduction in the presence of heavy-tailed distributions, see also [1], [3], [4], and [10].

There is a large literature on estimating the number of SAWs. Although there is currently no known formula for the exact number of SAWs of length n, many approximating methods exist. The most advanced algorithms are the pivot ones. They can handle SAWs of size 10^7; see [8], [28]. For a recent survey, see [41].

A precursor of Algorithm 5.12.1 was given in [26], where the bottleneck elements were identified by estimating gradients of the performance function.

REFERENCES

1. S. Asmussen. Stationary distributions via first passage times. In J. H. Dshalalow, editor, *Advances in Queueing: Theory, Methods and Open Problems*, pages 79–102, New York, 1995. CRC Press.

2. S. Asmussen and P. W. Glynn. *Stochastic Simulation*. Springer-Verlag, New York, 2007.

3. S. Asmussen and D. P. Kroese. Improved algorithms for rare event simulation with heavy tails. *Advances in Applied Probability*, 38(2):545–558, 2006.

4. S. Asmussen, D. P. Kroese, and R. Y. Rubinstein. Heavy tails, importance sampling and cross-entropy. *Stochastic Models*, 21(1):57–76, 2005.

5. S. Asmussen and R. Y. Rubinstein. Complexity properties of steady-state rare-events simulation in queueing models. In J. H. Dshalalow, editor, *Advances in Queueing: Theory, Methods and Open Problems*, pages 429–462, New York, 1995. CRC Press.

6. Z. I. Botev. The normal law under linear restrictions: simulation and estimation via minimax tilting. *Journal of the Royal Statistics Society (B)*, 79:1–24, 2017.

7. H.-P. Chan and T.-L. Lai. A general theory of particle filters in hidden Markov models and some applications. *Annals of Statistics*, 41(6):2877–2904, 2013.

8. N. Clisby. Efficient implementation of the pivot algorithm for self-avoiding walks. *Journal of Statistical Physics*, 140:349–392, 2010.

9. W. G. Cochran. *Sampling Techniques*. John Wiley & Sons, New York, 3rd edition, 1977.

10. P. T. de Boer, D. P. Kroese, and R. Y. Rubinstein. A fast cross-entropy method for estimating buffer overflows in queueing networks. *Management Science*, 50(7):883–895, 2004.

11. A. Doucet, N. de Freitas, and N. Gordon. *Sequential Monte Carlo Methods in Practice*. Springer-Verlag, New York, 2001.

12. T. Elperin, I. B. Gertsbakh, and M. Lomonosov. Estimation of network reliability using graph evolution models. *IEEE Transactions on Reliability*, 40(5):572–581, 1991.

13. G. S. Fishman. *Monte Carlo: Concepts, Algorithms and Applications*. Springer-Verlag, New York, 1996.

14. S. Gal, R. Y. Rubinstein, and A. Ziv. On the optimality and efficiency of common random numbers. *Math. Comput. Simul.*, 26(6):502–512, 1984.

15. I. B. Gertsbakh. *Statistical Reliability Theory*. Marcel Dekker, New York, 1989.

16. M. B. Giles. Multilevel Monte Carlo path simulation. *Operations Research*, 56(8):607–617, 2008.

17. P. Glasserman. *Monte Carlo Methods in Financial Engineering*. Springer-Verlag, New York, 2004.

18. P. Grassberger. Pruned-enriched Rosenbluth method: Simulations of θ polymers of chain length up to 1000000. *Phys. Rev. E*, 56:3682–3693, Sep 1997.

19. D. Gross and C. M. Harris. *Fundamentals of Queueing Theory*. John Wiley & Sons, New York, 2nd edition, 1985.

20. M. Kahn and A. W. Marshall. Methods of reducing sample size in Monte Carlo computations. *Operations Research*, 1:263–278, 1953.

21. J. P. C. Kleijnen. *Statistical Techniques in Simulation, Part 1*. Marcel Dekker, New York, 1974.

22. J. P. C. Kleijnen. Analysis of simulation with common random numbers: A note on Heikes et al. *Simuletter*, 11:7–13, 1976.

23. D. P. Kroese and R. Y. Rubinstein. The transform likelihood ratio method for rare event simulation with heavy tails. *Queueing Systems*, 46:317–351, 2004.

24. D. P. Kroese, T. Taimre, and Z. I. Botev. *Handbook of Monte Carlo Methods*. John Wiley & Sons, 2011.

25. A. M. Law and W. D. Kelton. *Simulation Modeling and Analysis*. McGraw-Hill, New York, 3rd edition, 2000.

26. D. Lieber, R. Y. Rubinstein, and D. Elmakis. Quick estimation of rare events in stochastic networks. *IEEE Transaction on Reliability*, 46:254–265, 1997.

27. J. S. Liu. *Monte Carlo Strategies in Scientific Computing*. Springer-Verlag, New York, 2001.

28. N. N. Madras. *Lectures on Monte Carlo Methods*. American Mathematical Society, 2002.

29. D. L. McLeish. *Monte Carlo Simulation and Finance*. John Wiley & Sons, New York, 2005.

30. M. F. Neuts. *Matrix-Geometric Solutions in Stochastic Models: An Algorithmic Approach*. Dover Publications, New York, 1981.

31. C.-H. Rhee and P. W. Glynn. Unbiased estimation with square root convergence for SDE models. *Operations Research*, 63(5):1026–1043, 2015.

32. C. P. Robert and G. Casella. *Monte Carlo Statistical Methods*. Springer, New York, 2nd edition, 2004.

33. S. M. Ross. *Simulation*. Academic Press, New York, 3rd edition, 2002.

34. R. Y. Rubinstein. *Simulation and the Monte Carlo Method*. John Wiley & Sons, New York, 1981.

35. R. Y. Rubinstein and D. P. Kroese. *The Cross-Entropy Method: A Unified Approach to Combinatorial Optimization, Monte Carlo Simulation and Machine Learning*. Springer-Verlag, New York, 2004.

36. R. Y. Rubinstein and R. Marcus. Efficiency of multivariate control variables in Monte Carlo simulation. *Operations Research*, 33:661–667, 1985.

37. R. Y. Rubinstein and B. Melamed. *Modern Simulation and Modeling*. John Wiley & Sons, New York, 1998.

38. R. Y. Rubinstein and A. Shapiro. *Discrete Event Systems: Sensitivity Analysis and Stochastic Optimization via the Score Function Method*. John Wiley & Sons, New York, 1993.

39. R.Y. Rubinstein, M. Samorodnitsky, and M. Shaked. Antithetic variables, multivariate dependence and simulation of complex stochastic systems. *Management Science*, 31:66–77, 1985.

40. I. M. Sobol. *A Primer for the Monte Carlo Method*. CRC Press, Boca Raton, FL, 1994.

41. E. J. J. van Rensburg. Monte Carlo methods for the self-avoiding walk. *Journal of Physics A: Mathematical and Theoretical*, 42(32):323001, 2009.

42. F. T. Wall and J. J. Erpenbeck. New method for the statistical computation of polymer dimensions. *Journal of Chemical Physics*, 30(3):634–637, 1959.

43. P. Wesseling. *An Introduction to Multigrid Methods*. John Wiley & Sons, 1992.

44. W. Whitt. Bivariate distributions with given marginals. *Annals of Statistics*, 4(6):1280–1289, 1976.

CHAPTER 6

MARKOV CHAIN MONTE CARLO

6.1   INTRODUCTION

In this chapter we present a powerful generic method, called *Markov chain Monte Carlo* (MCMC), for *approximately* generating samples from an arbitrary distribution. This, as we learned in Section 2.5, is typically not an easy task, in particular when $\mathbf{X}$ is a random vector with dependent components. An added advantage of MCMC is that it only requires specification of the target pdf up to a (normalization) constant.

The MCMC method is due to Metropolis et al. [18]. They were motivated by computational problems in statistical physics, and their approach uses the idea of generating a Markov chain whose limiting distribution is equal to the desired target distribution. There are many modifications and enhancement of the original Metropolis [18] algorithm, notably the algorithm introduced by Hastings [11]. Nowadays, any approach that produces an ergodic Markov chain whose stationary distribution is the target distribution is referred to as MCMC or *Markov chain sampling* [20]. The prominent MCMC algorithms are the Metropolis–Hastings and the Gibbs samplers, the latter being particularly useful in Bayesian analysis. Finally, MCMC sampling is the main ingredient in the popular *simulated annealing* technique [1] for discrete and continuous optimization.

The rest of this chapter is organized as follows: In Section 6.2 we present the classic Metropolis–Hastings algorithm, which simulates a Markov chain such that its stationary distribution coincides with the target distribution. An important special case is the *hit-and-run* sampler, discussed in Section 6.3. Section 6.4 deals with the *Gibbs sampler*, where the underlying Markov chain is constructed based

Simulation and the Monte Carlo Method, Third Edition. By R. Y. Rubinstein and D. P. Kroese
Copyright © 2017 John Wiley & Sons, Inc. Published 2017 by John Wiley & Sons, Inc.

on a sequence of conditional distributions. Section 6.5 explains how to sample from distributions arising in the Ising and Potts models, which are extensively used in statistical mechanics, and Section 6.6 deals with applications of MCMC in Bayesian statistics. In Section 6.7 we show that both the Metropolis–Hastings and Gibbs samplers can be viewed as special cases of a general MCMC algorithm and present the slice and reversible jump samplers. Section 6.8 deals with the classic simulated annealing method for finding the global minimum of a multiextremal function, which is based on the MCMC method. Finally, Section 6.9 presents the perfect sampling method, for sampling exactly from a target distribution rather than approximately.

6.2   METROPOLIS–HASTINGS ALGORITHM

The main idea behind the Metropolis–Hastings algorithm is to simulate a Markov chain such that the stationary distribution of this chain coincides with the target distribution.

To motivate the MCMC method, assume that we want to generate a random variable X taking values in $\mathscr{X} = \{1, \ldots, m\}$, according to a target distribution $\{\pi_i\}$, with

$$\pi_i = \frac{b_i}{C}, \quad i \in \mathscr{X}, \tag{6.1}$$

where it is assumed that all $\{b_i\}$ are strictly positive, m is large, and the normalization constant $C = \sum_{i=1}^{m} b_i$ is difficult to calculate. Following Metropolis et al. [18], we construct a Markov chain $\{X_t, t = 0, 1, \ldots\}$ on $\mathscr{X}$ whose evolution relies on an arbitrary transition matrix $\mathbf{Q} = (q_{ij})$ in the following way:

- When $X_t = i$, generate a random variable Y satisfying $\mathbb{P}(Y = j) = q_{ij}$, $j \in \mathscr{X}$. Thus, Y is generated from the m-point distribution given by the i-th row of $\mathbf{Q}$.

- If $Y = j$, let

$$X_{t+1} = \begin{cases} j & \text{with probability } \alpha_{ij} = \min\left\{\frac{\pi_j\, q_{ji}}{\pi_i\, q_{ij}}, 1\right\} = \min\left\{\frac{b_j\, q_{ji}}{b_i\, q_{ij}}, 1\right\}, \\ i & \text{with probability } 1 - \alpha_{ij}. \end{cases}$$

It follows that $\{X_t, t = 0, 1, \ldots\}$ has a one-step transition matrix $\mathbf{P} = (p_{ij})$ given by

$$p_{ij} = \begin{cases} q_{ij}\, \alpha_{ij} & \text{if } i \neq j, \\ 1 - \sum_{k \neq i} q_{ik}\, \alpha_{ik} & \text{if } i = j. \end{cases} \tag{6.2}$$

Now it is easy to check (see Problem 6.1) that, with α_{ij} as above,

$$\pi_i\, p_{ij} = \pi_j\, p_{ji}, \quad i, j \in \mathscr{X}. \tag{6.3}$$

In other words, the *detailed balance equations* (1.38) hold, and hence the Markov chain is time reversible and has stationary probabilities $\{\pi_i\}$. Moreover, this stationary distribution is also the *limiting* distribution if the Markov chain is irreducible and aperiodic. Note that there is no need for the normalization constant C in (6.1) to define the Markov chain.

The extension of the MCMC approach above for generating samples from an arbitrary multidimensional pdf $f(\mathbf{x})$ (instead of π_i) is straightforward. In this case, the nonnegative probability transition function $q(\mathbf{x}, \mathbf{y})$ (taking the place of q_{ij}

above) is often called the *proposal* or *instrumental* function. In viewing this function as a conditional pdf, we can also write $q(\mathbf{y}\,|\,\mathbf{x})$ instead of $q(\mathbf{x}, \mathbf{y})$. The probability $\alpha(\mathbf{x}, \mathbf{y})$ is called the *acceptance probability*. The original Metropolis algorithm [18] was suggested for symmetric proposal functions, that is, for $q(\mathbf{x}, \mathbf{y}) = q(\mathbf{y}, \mathbf{x})$. Hastings modified the original MCMC algorithm to allow nonsymmetric proposal functions. Such an algorithm is called a *Metropolis–Hastings algorithm*. We call the corresponding Markov chain the *Metropolis–Hastings Markov chain*.

In summary, the Metropolis–Hastings algorithm, which, like the acceptance–rejection method, is based on a trial-and-error strategy, is given as follows:

Algorithm 6.2.1: Metropolis–Hastings Algorithm

---

input : Initial state $\mathbf{X}_0$ and sample size N. Target pdf $f(\mathbf{x})$ and proposal function $q(\mathbf{x}, \mathbf{y})$.

output: Markov chain $\mathbf{X}_1, \ldots, \mathbf{X}_N$ approximately distributed according to $f(\mathbf{x})$.

1 **for** $t = 0$ **to** $N - 1$ **do**
2   Generate $\mathbf{Y} \sim q(\mathbf{X}_t, \mathbf{y})$ // draw a proposal
3      $\alpha \leftarrow \min\left\{ \frac{f(\mathbf{Y})\, q(\mathbf{Y}, \mathbf{X}_t)}{f(\mathbf{X}_t)\, q(\mathbf{X}_t, \mathbf{Y})}, 1 \right\}$ // acceptance probability
4   Generate $U \sim \mathsf{U}(0, 1)$
5   **if** $U \leqslant \alpha$ **then**
6         $\mathbf{X}_{t+1} \leftarrow \mathbf{Y}$
7   **else**
8         $\mathbf{X}_{t+1} \leftarrow \mathbf{X}_t$

9 **return** $\mathbf{X}_1, \ldots, \mathbf{X}_N$

The algorithm produces a sequence $\mathbf{X}_1, \mathbf{X}_2, \ldots, \mathbf{X}_N$ of *dependent* random variables, with $\mathbf{X}_t$ approximately distributed according to $f(\mathbf{x})$, for large t.

Since Algorithm 6.2.1 is of the acceptance–rejection type, its efficiency depends on the acceptance probability $\alpha(\mathbf{x}, \mathbf{y})$. Ideally, we would like $q(\mathbf{x}, \mathbf{y})$ to reproduce the desired pdf $f(\mathbf{y})$ as faithfully as possible. A common approach [20] is to first parameterize $q(\mathbf{x}, \mathbf{y})$ as $q(\mathbf{x}, \mathbf{y}; \theta)$ and then use stochastic optimization methods to maximize this with respect to θ. Below we consider several particular choices of $q(\mathbf{x}, \mathbf{y})$.

■ **EXAMPLE 6.1 Independence Sampler**

The simplest Metropolis-type MCMC algorithm is obtained by choosing the proposal function $q(\mathbf{x}, \mathbf{y})$ to be independent of $\mathbf{x}$, that is, $q(\mathbf{x}, \mathbf{y}) = g(\mathbf{y})$ for some pdf $g(\mathbf{y})$. Thus, starting from a previous state $\mathbf{X}$ a candidate state $\mathbf{Y}$ is generated from $g(\mathbf{y})$ and accepted with probability

$$\alpha(\mathbf{X}, \mathbf{Y}) = \min\left\{ \frac{f(\mathbf{Y})\, g(\mathbf{X})}{f(\mathbf{X})\, g(\mathbf{Y})}, 1 \right\}.$$

This procedure is very similar to the original acceptance–rejection methods of Chapter 2, and, as in that method, it is important that the proposal distribution g be close to the target f. Note that, in contrast to the acceptance–rejection method, the independence sampler produces *dependent* samples.

■ **EXAMPLE 6.2   Uniform Sampling**

Being able to sample *uniformly* from some discrete set $\mathscr{Y}$ is important in many applications; see, for example, the algorithms for counting in Chapter 9. A simple general procedure is as follows: Define a *neighborhood* structure on $\mathscr{Y}$. Any neighborhood structure is allowed, as long as the resulting Metropolis–Hastings Markov chain is irreducible and aperiodic. Let $n_{\mathbf{x}}$ be the number of neighbors of a state $\mathbf{x}$. For the proposal distribution, we simply choose each possible neighbor of the current state $\mathbf{x}$ with equal probability. That is, $q(\mathbf{x}, \mathbf{y}) = 1/n_{\mathbf{x}}$. Since the target pdf $f(\mathbf{x})$ here is constant, the acceptance probability is

$$\alpha(\mathbf{x}, \mathbf{y}) = \min\{n_{\mathbf{x}}/n_{\mathbf{y}}, 1\} \ .$$

By construction, the limiting distribution of the Metropolis–Hastings Markov chain is the uniform distribution on $\mathscr{Y}$.

■ **EXAMPLE 6.3   Random Walk Sampler**

In the random walk sampler the proposal state $\mathbf{Y}$, for a given current state $\mathbf{x}$, is given by $\mathbf{Y} = \mathbf{x} + \mathbf{Z}$, where $\mathbf{Z}$ is typically generated from some spherically symmetrical distribution (in the continuous case), such as $\mathsf{N}(\mathbf{0}, \Sigma)$. Note that the proposal function is symmetrical in this case; thus,

$$\alpha(\mathbf{x}, \mathbf{y}) = \min \left\{ \frac{f(\mathbf{y})}{f(\mathbf{x})}, 1 \right\} . \tag{6.4}$$

■ **EXAMPLE 6.4**

Let the random vector $\mathbf{X} = (X_1, X_2)$ have the following two-dimensional pdf:

$$f(\mathbf{x}) = c \, \exp(-(x_1^2 x_2^2 + x_1^2 + x_2^2 - 8x_1 - 8x_2)/2) \ , \tag{6.5}$$

where $c \approx 1/20216.335877$ is a normalization constant. The graph of this density is depicted in Figure 6.1.

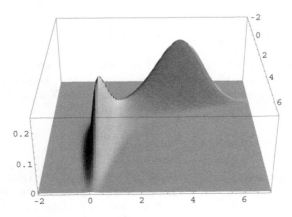

Figure 6.1: The density $f(x_1, x_2)$.

Suppose that we wish to estimate $\ell = \mathbb{E}[X_1]$ via the CMC estimator

$$\widehat{\ell} = \frac{1}{N} \sum_{t=1}^{N} X_{t1} \,,$$

using the random walk sampler to generate a dependent sample $\{\mathbf{X}_t\}$ from $f(\mathbf{x})$. A simple choice for the increment $\mathbf{Z}$ is to draw the components of $\mathbf{Z}$ independently, from a $\mathsf{N}(0, a^2)$ distribution for some $a > 0$. Note that, if a is chosen too small, say less than 0.5, the components of the samples will be strongly positively correlated, which will lead to a large variance for $\widehat{\ell}$. On the other hand, for a too large, say greater than 10, most of the samples will be rejected, leading again to low efficiency. Below we choose a moderate value of a, say $a = 2$. The random walk sampler is now summarized as follows:

Algorithm 6.2.2: Random Walk Sampler

---

1 Initialize $\mathbf{X}_0 \leftarrow (X_{01}, X_{02})$.
2 **for** $t = 0$ **to** $N - 1$ **do**
3 $\quad$ Draw $Z_1, Z_2 \sim \mathsf{N}(0, 1)$ independently.
4 $\quad$ Set $\mathbf{Z} \leftarrow (Z_1, Z_2)$ and $\mathbf{Y} \leftarrow \mathbf{X}_t + 2\,\mathbf{Z}$.
5 $\quad$ Set $\alpha \leftarrow \min\{\frac{f(\mathbf{Y})}{f(\mathbf{X}_t)}, 1\}$.
6 $\quad$ Generate $U \sim \mathsf{U}(0, 1)$.
7 $\quad$ **if** $U \leqslant \alpha$ **then**
8 $\quad\quad$ $\mathbf{X}_{t+1} \leftarrow \mathbf{Y}$
9 $\quad$ **else**
10 $\quad\quad$ $\mathbf{X}_{t+1} \leftarrow \mathbf{X}_t$

We ran this algorithm to produce $N = 10^5$ samples. The last few hundred of these are displayed in the left plot of Figure 6.2. We see that the samples closely follow the contour plot of the pdf, indicating that the correct region has been sampled. This is corroborated by the right plot of Figure 6.2, where we see that the histogram of the x_1 values is close to the true pdf (solid line).

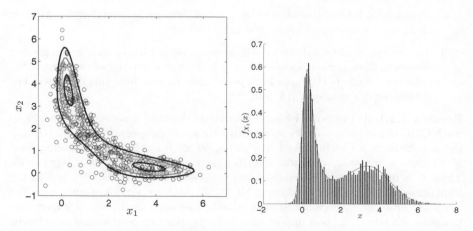

Figure 6.2: The left plot shows some samples of the random walk sampler along with several contour lines of f. The right plot shows the histogram of the x_1 values along with the true density of X_1.

We obtained an estimate $\widehat{\ell} = 1.89$ (the true value is $\mathbb{E}[X_1] \approx 1.85997$). To obtain a CI, we can use (4.21), where $\widetilde{S}$ estimates the asymptotic variance, or employ the batch means method of Section 4.4.2.1. Figure 6.3 displays the estimated (auto)covariance function $\widehat{R}(k)$ for $k = 0, 1, \ldots, 400$. We see that up to about 100 the covariances are nonnegligible. Thus, to estimate the variance of $\widehat{\ell}$, we need to include all nonzero terms in (4.20), not only the variance $R(0)$ of X_1. Summing over the first 400 lags, we obtained an estimate of 10.41 for the asymptotic variance. This gives an estimated relative error for $\widehat{\ell}$ of 0.0185 and an 95% CI of $(1.82, 1.96)$. A similar CI was found when using the batch means method with 500 batches of size 200.

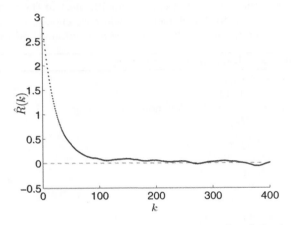

Figure 6.3: The estimated covariance function for the $\{X_{t1}\}$ for lags k up to 400.

While MCMC is a generic method and can be used to generate random samples virtually from any target distribution, regardless of its dimensionality and complexity, potential problems with the MCMC method are:

1. The resulting samples are often highly correlated.

2. Typically, it takes a considerable amount of time until the underlying Markov chain settles down to its steady state.

3. The estimates obtained via MCMC samples often tend to have much greater variances than those obtained from independent sampling of the target distribution. Various attempts have been made to overcome this difficulty. For details see, for example, [14] and [20].

Remark 6.2.1 At this point we must stress that although it is common practice to use MCMC to sample from $f(\mathbf{x})$ in order to estimate any expectation $\ell = \mathbb{E}_f[H(\mathbf{X})]$, the *actual* target for estimating ℓ is $g^*(\mathbf{x}) \propto |H(\mathbf{x})| f(\mathbf{x})$. Namely, sampling from $g^*(\mathbf{x})$ gives a minimum variance estimator (zero variance in the case $H(\mathbf{x}) \geqslant 0$). Thus, it is important to distinguish clearly between using MCMC for generating from some difficult pdf $f(\mathbf{x})$ and using MCMC to estimate a quantity such as ℓ. For the latter problem, much more efficient techniques can be used, such as importance sampling; moreover, a good importance sampling pdf can be obtained adaptively, as with the CE and TLR methods.

6.3 HIT-AND-RUN SAMPLER

The *hit-and-run* sampler, pioneered by Robert Smith [25], is among the first MCMC samplers in the category of *line samplers* [2]. As in the previous section, the objective is to sample from a target distribution $f(\mathbf{x})$ on $\mathscr{X} \subset \mathbb{R}^n$. Line samplers afford the opportunity to reach across the entire feasible region $\mathscr{X}$ in one step.

We first describe the original hit-and-run sampler for generating from a *uniform* distribution on a bounded open region $\mathscr{X}$ of $\mathbb{R}^n$. At each iteration, starting from a current point $\mathbf{x}$, a *direction vector* $\mathbf{d}$ is generated uniformly on the surface of an n-dimensional hypersphere. The intersection of the corresponding bidirectional line (through $\mathbf{x}$) and the enclosing box of $\mathscr{X}$ defines a line segment $\mathscr{L}$. The next point $\mathbf{y}$ is then selected uniformly from the intersection of $\mathscr{L}$ and $\mathscr{X}$.

Figure 6.4 illustrates the hit-and-run algorithm for generating uniformly from the set $\mathscr{X}$ (the gray region), which is bounded by a square. Given the point $\mathbf{x}$ in $\mathscr{X}$, a random direction $\mathbf{d}$ is generated, which defines the line segment $\mathscr{L} = uv$. Then a point $\mathbf{y}$ is chosen uniformly on $\mathscr{M} = \mathscr{L} \cap \mathscr{X}$, for example, by the acceptance–rejection method; that is, one generates a point uniformly on $\mathscr{L}$ and then accepts this point only if it lies in $\mathscr{X}$.

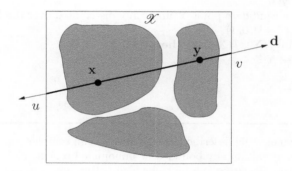

Figure 6.4: Illustration of the hit-and-run algorithm on a square in two dimensions.

Smith [25] showed that hit-and-run asymptotically generates uniformly distributed points over *arbitrary* open regions of $\mathbb{R}^n$. One desirable property of hit-and-run is that it can globally reach any point in the set in one step; that is, there is a strictly positive probability of sampling any neighborhood in the set. This property, coupled with a symmetry property, is important in deriving the limiting distribution. Lovász [15] proved that hit-and-run on a convex body in n dimensions produces an approximately uniformly distributed sample point in polynomial time, $\mathcal{O}(n^3)$, the best-known bound for such a sampling algorithm. He noted that the hit-and-run algorithm appears in practice to offer the most rapid convergence to a uniform distribution [15, 16]. Hit-and-run is unique in that it only takes polynomial time to get out of a corner; in contrast, *ball walk* takes exponential time to get out of a corner [17].

Note that the hit-and-run algorithm described above is a special case of the Metropolis–Hastings Algorithm 6.2.1, where the proposal function $q(\mathbf{x}, \mathbf{y})$ is symmetric and the target $f(\mathbf{x})$ is constant. It follows that each candidate point is accepted with probability 1. To generate from a *general* strictly positive continuous pdf $f(\mathbf{x})$, one can simply modify the uniform hit-and-run algorithm above by accepting the candidate $\mathbf{y}$ with probability

$$\alpha(\mathbf{x}, \mathbf{y}) = \min\{f(\mathbf{y})/f(\mathbf{x}), 1\}, \qquad (6.6)$$

as in Algorithm 6.2.1 (note that $q(\mathbf{y}, \mathbf{x})/q(\mathbf{x}, \mathbf{y})$ equals 1). Thus the general hit-and-run algorithm with the Metropolis acceptance criterion above is summarized as follows [21]:

Algorithm 6.3.1: Hit-and-Run

---

   input : Bounded region $\mathscr{X}$, target pdf f on $\mathscr{X}$, sample size N.
   output: $\mathbf{X}_1, \ldots, \mathbf{X}_N$ approximately distributed according to f.
1 Initialize $\mathbf{X}_0 \in \mathscr{X}$.
2 **for** $t = 0$ to $N - 1$ **do**
3   **repeat**
4   Generate a random direction $\mathbf{d}_t$ according to a uniform distribution on
      the unit n-dimensional hypersphere.
5   Let $\mathscr{M}_t \leftarrow \{\mathbf{x} \in \mathscr{X} : \mathbf{x} = \mathbf{X}_t + \lambda \mathbf{d}_t, \ \lambda \in \mathbb{R}\}$.
6   **until** $\mathscr{M}_t \neq \emptyset$
7   Generate a candidate point $\mathbf{Y}$ uniformly distributed over the line set $\mathscr{M}_t$.
8   Generate $U \sim \mathsf{U}(0, 1)$.
9   **if** $U \leqslant \min\{f(\mathbf{Y})/f(\mathbf{X}_t), 1\}$ **then**
10       $\mathbf{X}_{t+1} \leftarrow \mathbf{Y}$
11   **else**
12       $\mathbf{X}_{t+1} \leftarrow \mathbf{X}_t$

13 **return** $\mathbf{X}_1, \ldots, \mathbf{X}_N$

Chen and Schmeiser [5] describe how the hit-and-run sampler can be generalized to sample from any pdf on any bounded or unbounded region $\mathscr{X}$.

The hit-and-run algorithm can be embedded within an optimization framework to yield two global optimization algorithms: *hide-and-seek* [21] and *improving hit-and-run* [27]. The latter has been applied successfully to practical problems including composite material design and shape optimization, and it has been shown to have polynomial complexity, on average, for a class of quadratic programs. In Section 6.8 we show how to turn an MCMC sampler into an optimization algorithm by using simulated annealing.

6.4 GIBBS SAMPLER

The *Gibbs sampler* (Geman and Geman [7]) uses a somewhat different methodology from the Metropolis–Hastings algorithm and is particularly useful for generating n-dimensional random vectors. The distinguishing feature of the Gibbs sampler is that the underlying Markov chain is constructed, in a deterministic or random fashion, from a sequence of conditional distributions.

Gibbs sampling is advantageous if it is easier to sample from the conditional distributions than from the joint distribution. The essential idea of the Gibbs sampler — updating one part of the previous element while keeping the other parts fixed — is useful in many instances where the state variable is a random variable taking values in a general space, not just in $\mathbb{R}^n$ (see [12]).

Suppose that we wish to sample a random vector $\mathbf{X} = (X_1, \ldots, X_n)$ according to a target pdf $f(\mathbf{x})$. Let $f(x_i \mid x_1, \ldots, x_{i-1}, x_{i+1}, \ldots, x_n)$ represent the conditional pdf of the i-th component, X_i, given the other components $x_1, \ldots, x_{i-1}, x_{i+1}, \ldots, x_n$. The Gibbs sampler is given next.

Algorithm 6.4.1: Gibbs Sampler

input : Initial point $\mathbf{X}_0$, sample size N, and target pdf f.
output: $\mathbf{X}_1, \ldots, \mathbf{X}_N$ approximately distributed according to f.
1 **for** $t = 0$ **to** $N - 1$ **do**
2   Draw Y_1 from the conditional pdf $f(y_1 \mid X_{t,2}, \ldots, X_{t,n})$.
3   **for** $i = 2$ **to** n **do**
4   Draw Y_i from the conditional pdf $f(y_i \mid Y_1, \ldots, Y_{i-1}, X_{t,i+1}, \ldots, X_{t,n})$.
5     $\mathbf{X}_{t+1} \leftarrow \mathbf{Y}$
6 **return** $\mathbf{X}_1, \ldots, \mathbf{X}_N$

Note that in the Gibbs sampler *all* samples are accepted, in contrast to the Metropolis–Hastings algorithm. We will see in Section 6.7 that under mild conditions the limiting distribution of the process $\{\mathbf{X}_t, t = 1, 2, \ldots\}$, generated via the Gibbs sampler, is precisely $f(\mathbf{x})$. Moreover, under some other simple conditions, it can be shown (see [14], [20]) that the convergence to the desired pdf is geometrically fast.

■ **EXAMPLE 6.5** **Example 6.4 (Continued)**

We will show how to sample easily from the pdf f in (6.5) via the Gibbs sampler. We start by writing

$$f(x, y) = c_1(y) \exp\left(-\frac{1 + y^2}{2} \left(x - \frac{4}{1 + y^2} \right)^2 \right),$$

where $c_1(y)$ depends only on y; we see that, conditional on y, X has a normal distribution with expectation $4/(1 + y^2)$ and variance $1/(1 + y^2)$. The conditional distribution of Y given x follows in the same way. The corresponding Gibbs sampler is thus as follows:

Algorithm 6.4.2: Gibbs Sampler for Example 6.4

1 Initialize Y_0.
2 **for** $t = 0$ **to** $N - 1$ **do**
3   Draw $Z_1, Z_2 \sim \mathsf{N}(0, 1)$.
4     $X_{t+1} \leftarrow Z_1/\sqrt{1 + Y_t^2} + 4/(1 + Y_t^2)$
5     $Y_{t+1} \leftarrow Z_2/\sqrt{1 + X_{t+1}^2} + 4/(1 + X_{t+1}^2)$

Remark 6.4.1 (Systematic and Random Gibbs Samplers) Note that Algorithm 6.4.1 presents a *systematic* coordinatewise Gibbs sampler. That is, the vector

X is updated in a deterministic order: $1, 2, \ldots, n, 1, 2, \ldots$ In the *random* coordinatewise Gibbs sampler, the coordinates are chosen randomly, such as by generating them independently from a discrete uniform n-point pdf. In that case the Gibbs sampler can be viewed as an instance of the Metropolis–Hastings sampler, namely with the transition function

$$q(\mathbf{x}, \mathbf{y}) = \frac{1}{n} f(y_i \mid x_1, \ldots, x_{i-1}, x_{i+1}, \ldots, x_n) = \frac{1}{n} \frac{f(\mathbf{y})}{\sum_{y_i} f(\mathbf{y})} ,$$

where $\mathbf{y} = (x_1, \ldots, x_{i-1}, y_i, x_{i+1}, \ldots, x_n)$. Since $\sum_{y_i} f(\mathbf{y})$ can also be written as $\sum_{x_i} f(\mathbf{x})$, we have

$$\varrho(\mathbf{x}, \mathbf{y}) = \frac{f(\mathbf{y})\, q(\mathbf{y}, \mathbf{x})}{f(\mathbf{x})\, q(\mathbf{x}, \mathbf{y})} = \frac{f(\mathbf{y})\, f(\mathbf{x})}{f(\mathbf{x})\, f(\mathbf{y})} = 1 ,$$

so that the acceptance probability $\alpha(\mathbf{x}, \mathbf{y})$ is 1 in this case.

Here is another example of an application of the Gibbs sampler.

■ **EXAMPLE 6.6   Closed Network of Queues in a Product Form**

Consider m customers moving among n queues in a closed queueing network. Denote by $X_i(t)$ the number of customers in queue i, $i = 1, \ldots, n$, and let $\mathbf{X}(t) = (X_1(t), \ldots, X_n(t))$ and $\mathbf{x} = (x_1, \ldots, x_n)$. It is well known [22] that if the limit

$$\lim_{t \to \infty} \mathbb{P}(\mathbf{X}(t) = \mathbf{x}) = \pi(\mathbf{x})$$

exists, then, for exponentially distributed service times, the joint discrete pdf $\pi(\mathbf{x})$ can be written in *product form* as

$$\pi(\mathbf{x}) = C \prod_{i=1}^{n} f_i(x_i), \quad \text{for } \sum_{i=1}^{n} x_i = m , \tag{6.7}$$

where the $\{f_i(x_i),\ x_i \geq 0\}$ are *known* discrete pdfs, and C is a normalization constant. For a concrete example, see Problem 6.11.

The constant C is in general difficult to compute. To proceed, writing $S(\mathbf{x}) = \sum_{i=1}^{n} x_i$ and $\mathscr{X}^* = \{\mathbf{x} : S(\mathbf{x}) = m\}$, we have

$$C^{-1} = \sum_{\mathbf{x} \in \mathscr{X}^*} \prod_{i=1}^{n} f_i(x_i) , \tag{6.8}$$

which requires the evaluation of the product of n pdfs for each $\mathbf{x}$ in the set $\mathscr{X}^*$. This set has a total of $|\mathscr{X}^*| = \binom{m+n-1}{n-1}$ elements (see Problem 6.10), which rapidly grows very large.

We now show how to compute C based on Gibbs sampling. To apply the Gibbs sampler, we need to be able to generate samples from the conditional distribution of X_i given the other components. Note that we only have to generate $X_1, \ldots, X_{n-1}$, since $X_n = m - \sum_{k=1}^{n-1} X_k$. For $i = 1, \ldots, n-1$, we have

$$f(x_i \mid x_1, \ldots, x_{i-1}, x_{i+1}, \ldots, x_{n-1}) \propto f_i(x_i)\, f_n\!\left(m - \sum_{k=1}^{n-1} x_i\right) \tag{6.9}$$

for $x_i \in \{0, 1, \ldots, m - x_1 - \cdots - x_{i-1} - x_{i+1} - \cdots - x_{n-1}\}$. Sampling from these conditional pdfs can often be done efficiently, in particular when the $\{f_i\}$ are members of an exponential family; see also Problem 6.11.

Now that we can sample (approximately) from $\pi(\mathbf{x})$, it is straightforward to estimate the normalization constant C by observing that

$$\mathbb{E}_\pi \left[\frac{1}{\prod_{i=1}^n f_i(X_i)} \right] = \sum_{\mathbf{x} \in \mathscr{X}^*} \frac{1}{\prod_{i=1}^n f_i(x_i)} \, C \prod_{i=1}^n f_i(x_i) = |\mathscr{X}^*| \, C \; .$$

This suggests the following estimator for C, obtained from a random sample $\mathbf{X}_1, \ldots, \mathbf{X}_N$ from π:

$$\widehat{C} = \binom{m+n-1}{n-1}^{-1} \frac{1}{N} \sum_{k=1}^N \prod_{i=1}^n \frac{1}{f_i(X_{ki})} \; ,$$

where X_{ki} is the i-th component of $\mathbf{X}_k$.

6.5 ISING AND POTTS MODELS

6.5.1 Ising Model

The Ising model is one of the most popular and most extensively studied models in statistical mechanics. It describes the interaction of idealized magnets, called *spins*, that are located on a two- or three-dimensional lattice. In the basic two-dimensional case the spins are located on the lattice $\{1, \ldots, n\} \times \{1, \ldots, n\}$, and each of the n^2 sites has four nearest neighbors, possibly including boundary sites, which "wrap around" to the other side of the lattice, creating a so-called *torus*. See Figure 6.5, where the four light gray sites are the neighbors of the dark gray site.

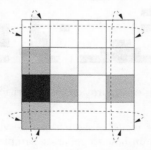

Figure 6.5: The boundary sites wrap around. The neighbors of the dark gray site are the light gray sites.

Let $\{1, \ldots, n^2\}$ be an enumeration of the sites. Each spin can be in one of two states: -1 or 1. Each of the 2^{n^2} *configurations* of spins $\mathbf{s} = (s_1, \ldots, s_{n^2})$ carries an amount of *total energy*

$$E(\mathbf{s}) = -A \sum_{i \leftrightarrow j} s_i \, s_j - B \sum_i s_i \; ,$$

where A and B are constants; in most studies $A = 1$ and $B = 0$, which we will now assume. The quantities $\sum_{i \leftrightarrow j} s_i s_j$ and $\sum_i s_i$ are called the *interaction energy* and *magnetization*, respectively. The notation $\sum_{i \leftrightarrow j}$ indicates that the summation is taken over neighboring pairs (i, j).

In thermal equilibrium the distribution of the spins, say π, follows the Boltzmann law: $\pi(\mathbf{s}) \propto \exp(-E(\mathbf{s})/T)$, where T is a fixed temperature. In other words, we have

$$\pi(\mathbf{s}) = \frac{\mathrm{e}^{\frac{1}{T} \sum_{i \leftrightarrow j} s_i s_j}}{\mathcal{Z}},$$

where $\mathcal{Z}$ is the normalization constant, called the *partition function*. Apart from $\mathcal{Z}$, particular quantities of interest are the *mean energy per spin* $\mathbb{E}_\pi[\sum_{i \leftrightarrow j} S_i S_j / n^2]$ and the *mean magnetization per spin* $\mathbb{E}_\pi[\sum_i S_i / n^2]$. These quantities can be obtained via Monte Carlo simulation, provided that one can sample efficiently from the target distribution π (see below).

In Figure 6.6 a sample from π is given (black $= 1$, white $= -1$) for $n = 30$ at the so-called *critical temperature* $T = 2/\ln(1 + \sqrt{2}) \approx 2.269$.

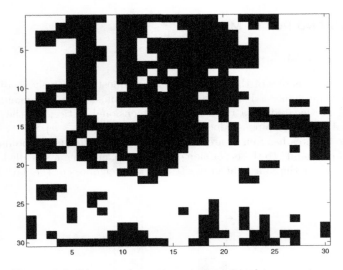

Figure 6.6: Ising configuration at the critical temperature.

We next define the *Potts model* — which can be viewed as a generalization of the Ising model — and explain how to generate samples from this extended model and thus, in particular, how to generate Figure 6.6.

6.5.2   Potts Model

Let $\{1, \dots, J\}$ be an enumeration of spatial positions (sites), and let ψ_{ij} be some symmetrical and positive function relating the sites to each other, for example,

$$\psi_{ij} = \begin{cases} \beta \,(> 0) & \text{if } i \text{ and } j \text{ are neighbors,} \\ 0 & \text{otherwise.} \end{cases} \tag{6.10}$$

Assign to each site i a "color" x_i. Suppose that there are K such colors, labeled $\{1, \ldots, K\}$. Define $\mathbf{x} = (x_1, \ldots, x_J)$, and let $\mathscr{X}$ be the space of such configurations. On $\mathscr{X}$ we define the target pdf $f(\mathbf{x}) \propto \mathrm{e}^{H(\mathbf{x})}$ with

$$H(\mathbf{x}) = \sum_{i<j} \psi_{ij} \, I_{\{x_i = x_j\}} \ .$$

To see that the Ising model is a special case of the Potts model, define $x_i = I_{\{s_i = 1\}}$ and ψ_{ij} as in (6.10), with $\beta = 4/T$. Then

$$\frac{1}{T} \sum_{i \leftrightarrow j} s_i \, s_j = \frac{1}{T} \sum_{i \leftrightarrow j} 2 \left(I_{\{x_i = x_j\}} - \frac{1}{2} \right) = \sum_{i<j} \psi_{ij} \, I_{\{x_i = x_j\}} + \text{const} \ ,$$

so that $\pi(\mathbf{s}) = f(\mathbf{x})$.

Next, we show how to generate a sample from the target pdf $f(\mathbf{x})$. To do so, we define auxiliary random variables $Y_{ij}, 1 \leqslant i < j \leqslant J$, such that conditional on $\mathbf{X} = \mathbf{x}$ the $\{Y_{ij}\}$ are independent, and each Y_{ij} is uniformly distributed on the interval $[0, a_{ij}]$, with $a_{ij} = \exp(\psi_{ij} I_{\{x_i = x_j\}}) \geqslant 1$. In other words, the conditional pdf of $\mathbf{Y} = \{Y_{ij}\}$ given $\mathbf{X} = \mathbf{x}$ is

$$f(\mathbf{y} \,|\, \mathbf{x}) = \prod_{i<j} \frac{I_{\{y_{ij} \leqslant a_{ij}\}}}{a_{ij}} = \prod_{i<j} I_{\{y_{ij} \leqslant a_{ij}\}} \, \mathrm{e}^{-H(\mathbf{x})} \ .$$

The significance of this is that the joint pdf of $\mathbf{X}$ and $\mathbf{Y}$ is now simply

$$f(\mathbf{x}, \mathbf{y}) = f(\mathbf{x}) \, f(\mathbf{y}|\mathbf{x}) \propto \begin{cases} 1 & \text{if } y_{ij} \leqslant a_{ij}, \text{ for all } i < j \ , \\ 0 & \text{otherwise.} \end{cases}$$

In other words, $(\mathbf{X}, \mathbf{Y})$ is *uniformly* distributed. More important, because $f(\mathbf{x} \,|\, \mathbf{y}) \propto f(\mathbf{x}, \mathbf{y})$, we find that $\mathbf{X} \,|\, \mathbf{y}$ is uniformly distributed over the set $\mathscr{A} = \{\mathbf{x} : y_{ij} \leqslant \exp(\psi_{ij} I_{\{x_i = x_j\}})$ for all $i < j\}$. Now, either $y_{ij} \in [0, 1]$ or $y_{ij} \in (1, \mathrm{e}^{\psi_{ij}}]$. In the former case, for any $\mathbf{x} \in \mathscr{A}$, the coordinates x_i and x_j range over all the colors, and by the uniformity, each color is equally likely. But in the latter case, x_i must be equal to x_j. Thus, for a given $\mathbf{y}$, the sites i, j (with $i < j$) for which $y_{ij} > 1$ can be gathered into clusters, and within each such cluster, the sites have identical colors. Moreover, given $\mathbf{y}$, the colors within the clusters are independent and uniformly distributed on $\{1, \ldots, K\}$. The same holds for the colors of the remaining positions, which can be viewed as one-cluster sites.

Hence, we can easily generate both $\mathbf{X} \,|\, \mathbf{y}$ and $\mathbf{Y} \,|\, \mathbf{x}$. As a consequence, we can use the Gibbs sampler to (approximately) sample from $f(\mathbf{x}, \mathbf{y})$; that is, we iteratively sample from $f(\mathbf{x}|\mathbf{y})$ and $f(\mathbf{y}|\mathbf{x})$. Finally, to obtain a sample $\mathbf{X}$ from $f(\mathbf{x})$, we generate $(\mathbf{X}, \mathbf{Y})$ via the Gibbs sampler and simply ignore $\mathbf{Y}$.

To simplify matters further, note that instead of the exact value Y_{ij} it suffices to know only the variable $B_{ij} = I_{\{Y_{ij} \geqslant 1\}}$. Given $\mathbf{X} = \mathbf{x}$, B_{ij} has a $\mathsf{Ber}(1 - \mathrm{e}^{-\psi_{ij}})$ distribution if $x_i = x_j$, and $B_{ij} = 0$ otherwise. This leads to the following so-called Swendsen–Wang algorithm:

Algorithm 6.5.1: Swendsen–Wang

---

1 Given $\{X_i\}$, generate $B_{ij} \sim \mathsf{Ber}(I_{\{X_i = X_j\}}(1 - \mathrm{e}^{-\psi_{ij}}))$ for $1 \leqslant i < j \leqslant J$.

2 Given $\{B_{ij}\}$, generate $X_i, i = 1, \ldots, J$ by clustering all the sites and choosing each cluster color independently and uniformly from $\{1, \ldots, K\}$.

Remark 6.5.1 (Data Augmentation) The idea above of introducing an *auxiliary* variable **y** to make sampling from $f(\mathbf{x})$ easier is also known as *data augmentation*. The composition method described in Section 2.3.3 can be viewed as another example of data augmentation. To illustrate, suppose that we want to sample from the mixture pdf

$$f(x) = \sum_{i=1}^{K} p_i\, f_i(x) \,.$$

Let Y be the discrete random variable taking values in $\{1, \ldots, K\}$ corresponding to the probabilities $\{p_i\}$. The composition method makes it easy to sample from the joint pdf of X and Y: first, draw Y according to $\{p_i\}$ and then sample X conditional on $Y = i$; that is, sample from $f_i(x)$. By simply ignoring Y, we obtain a sample from $f(x)$.

6.6   BAYESIAN STATISTICS

One of the main application areas of the MCMC method is Bayesian statistics. The mainstay of the Bayesian approach is Bayes' rule (1.6), which, in terms of pdfs, can be written as

$$f(\mathbf{y}\,|\,\mathbf{x}) = \frac{f(\mathbf{x}\,|\,\mathbf{y})\, f(\mathbf{y})}{\int f(\mathbf{x}\,|\,\mathbf{y})\, f(\mathbf{y})\, \mathrm{d}\mathbf{y}} \propto f(\mathbf{x}\,|\,\mathbf{y})\, f(\mathbf{y}) \,. \tag{6.11}$$

In other words, for any two random variables $\mathbf{X}$ and $\mathbf{Y}$, the conditional distribution of $\mathbf{Y}$ given $\mathbf{X} = \mathbf{x}$ is proportional to the product of the conditional pdf of $\mathbf{X}$ given $\mathbf{Y} = \mathbf{y}$ and the pdf of $\mathbf{Y}$. Note that instead of writing f_X, f_Y, $f_{X\,|\,Y}$, and $f_{Y\,|\,X}$ in the formula above, we have used the *same letter* f for the pdf of $\mathbf{X}$, $\mathbf{Y}$, and the conditional pdfs. This particular style of notation is typical in Bayesian analysis and can be of great descriptive value, despite its apparent ambiguity. We will use this notation whenever we work in a Bayesian setting.

The significance of (6.11) becomes clear when it is employed in the context of Bayesian parameter estimation, sometimes referred to as *Bayesian learning*. The following example explains the ideas.

■ EXAMPLE 6.7   Coin Flipping and Bayesian Learning

Recall the basic random experiment in Example 1.1 on Page 3, where we toss a biased coin n times. Suppose that the outcomes are $x_1, \ldots, x_n$, with $x_i = 1$ if the i-th toss is heads and $x_i = 0$ otherwise, $i = 1, \ldots, n$. Let p denote the probability of heads. We want to obtain information about p from the data $\mathbf{x} = (x_1, \ldots, x_n)$, for example, construct a CI.

The crucial idea is to summarize the information about p via a probability density $f(p)$. For example, if we know nothing about p, we take $f(p)$ uniformly distributed on the $(0, 1)$ interval, that is, $f(p) = 1, 0 \leqslant p \leqslant 1$. In effect, we treat p as a random variable. Now, obviously, the data $\mathbf{x}$ will affect our knowledge of p, and the way to update this information is to use Bayes' formula:

$$f(p\,|\,\mathbf{x}) \propto f(\mathbf{x}\,|\,p)\, f(p) \,.$$

The density $f(p)$ is called the *prior* density, $f(p\,|\,\mathbf{x})$ is called the *posterior density*, and $f(\mathbf{x}\,|\,p)$ is referred to as the *likelihood*. In our case, given p, the

$\{X_i\}$ are independent and $\mathrm{Ber}(p)$ distributed, so

$$f(\mathbf{x} \mid p) = \prod_{i=1}^{n} p^{x_i}(1-p)^{1-x_i} = p^s (1-p)^{n-s} \,,$$

with $s = x_1 + \cdots + x_n$ representing the total number of successes. Then using a uniform prior $(f(p) = 1)$ we get a posterior pdf

$$f(p \mid \mathbf{x}) = c\, p^s (1-p)^{n-s} \,,$$

which is the pdf of the $\mathsf{Beta}(s+1, n-s+1)$ distribution. The normalization constant is $c = (n+1)\binom{n}{s}$.

A Bayesian CI for p is now formed by taking the appropriate quantiles of the posterior pdf. As an example, suppose that $n = 100$ and $s = 1$. Then, a left one-sided 95% CI for p is $[0, 0.0461]$, where 0.0461 is the 0.95 quantile of the $\mathsf{Beta}(2, 100)$ distribution. To estimate p, we can take the value for which the pdf is maximal, the so-called *mode* of the pdf. In this problem, the mode is 0.01, coinciding with the sample mean. Figure 6.7 gives a plot of the posterior pdf for this problem.

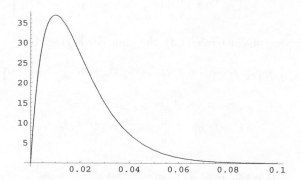

Figure 6.7: Posterior pdf for p, with $n = 100$ and $s = 1$.

Generalizing the previous example, a typical situation where MCMC (in particular, Gibbs sampling) can be used in Bayesian statistics is the following: Suppose that we want to sample from a posterior density $f(\boldsymbol{\theta} \mid \mathbf{x})$, where the data $\mathbf{x}$ are given (fixed) and $\boldsymbol{\theta} = (\theta_1, \ldots, \theta_k)$ is the parameter of interest. Suppose that it is easy to sample from $f(\theta_i \mid \theta_1, \ldots, \theta_{i-1}, \theta_{i+1}, \ldots, \theta_k, \mathbf{x})$ for all i. Then, we can use the Gibbs sampler to obtain a sample $\boldsymbol{\Theta}$ from $f(\boldsymbol{\theta} \mid \mathbf{x})$. The next example, adapted from Gelman et al. [6], illustrates the general idea.

■ EXAMPLE 6.8 Poisson Disruption Problem

Suppose that the random variables $X_1, \ldots, X_n$ describe the number of disasters in n subsequent years. In some random year K the rate of disasters changes from λ_1 to λ_2. Such a K is often called a *change point*. Our prior knowledge of λ_i is summarized by a $\mathsf{Gamma}(a_i, \eta_i)$, where shape parameter a_i is known. In turn, η_i is given by a $\mathsf{Gamma}(b_i, c_i)$ distribution, where both

b_i and c_i are known. Let $\boldsymbol{\lambda} = (\lambda_1, \lambda_2)$ and $\boldsymbol{\eta} = (\eta_1, \eta_2)$. We are given the data $\mathbf{x} = (x_1, \ldots, x_n)$, and the objective is to simulate from the posterior distribution of $\boldsymbol{\theta} = (\lambda_1, \lambda_2, \eta_1, \eta_2, K)$ given $\mathbf{x}$.

For the model we have the following hierarchical structure:

1. K has some discrete pdf $f(K)$ on $1, \ldots, n$.

2. Given K, the $\{\eta_i\}$ are independent and have a $\mathsf{Gamma}(b_i, c_i)$ distribution for $i = 1, 2$.

3. Given K and $\boldsymbol{\eta}$, the $\{\lambda_i\}$ are independent and have a $\mathsf{Gamma}(a_i, \eta_i)$ distribution for $i = 1, 2$.

4. Given $K, \boldsymbol{\eta}$, and $\boldsymbol{\lambda}$, the $\{X_i\}$ are independent and have a $\mathsf{Poi}(\lambda_1)$ distribution for $i = 1, \ldots, K$, and a $\mathsf{Poi}(\lambda_2)$ distribution for $i = K + 1, \ldots, n$.

It follows from point 4. that

$$f(\mathbf{x} \mid \boldsymbol{\lambda}, \boldsymbol{\eta}, K) = \prod_{i=1}^{K} e^{-\lambda_1} \frac{\lambda_1^{x_i}}{x_i!} \prod_{i=K+1}^{n} e^{-\lambda_2} \frac{\lambda_2^{x_i}}{x_i!}$$

$$= e^{-\lambda_1 K} \lambda_1^{\sum_{i=1}^{K} x_i} e^{-\lambda_2(n-K)} \lambda_2^{\sum_{i=K+1}^{n} x_i} \prod_{i=1}^{n} \frac{1}{x_i!} .$$

Moreover, by the product rule (1.4), the joint pdf is given by

$$f(\mathbf{x}, \boldsymbol{\lambda}, \boldsymbol{\eta}, K) \propto f(K) \, e^{-\lambda_1 K} \lambda_1^{\sum_{i=1}^{K} x_i} e^{-\lambda_2(n-K)} \lambda_2^{\sum_{i=K+1}^{n} x_i} \prod_{i=1}^{n} \frac{1}{x_i!}$$

$$\times e^{-\eta_1 \lambda_1} \lambda_1^{a_1-1} \eta_1^{a_1} \times e^{-\eta_2 \lambda_2} \lambda_2^{a_2-1} \eta_2^{a_2}$$

$$\times e^{-c_1 \eta_1} \eta_1^{b_1-1} c_1^{b_1} \times e^{-c_2 \eta_2} \eta_2^{b_2-1} c_2^{b_2} .$$

As a consequence,

$$f(\lambda_1 \mid \lambda_2, \boldsymbol{\eta}, K, \mathbf{x}) \propto e^{-\lambda_1(K+\eta_1)} \lambda_1^{a_1-1+\sum_{i=1}^{K} x_i} .$$

In other words, $(\lambda_1 \mid \lambda_2, \boldsymbol{\eta}, K, \mathbf{x}) \sim \mathsf{Gamma}(a_1 + \sum_{i=1}^{K} x_i, \, K + \eta_1)$. In a similar way, we have

$$(\lambda_2 \mid \lambda_1, \boldsymbol{\eta}, K, \mathbf{x}) \sim \mathsf{Gamma}\left(a_2 + \textstyle\sum_{i=K+1}^{n} x_i, \, n - K + \eta_2\right) ,$$

$$(\eta_1 \mid \boldsymbol{\lambda}, \eta_2, K, \mathbf{x}) \sim \mathsf{Gamma}(a_1 + b_1, \, \lambda_1 + c_1) ,$$

$$(\eta_2 \mid \boldsymbol{\lambda}, \eta_1, K, \mathbf{x}) \sim \mathsf{Gamma}(a_2 + b_2, \, \lambda_2 + c_2) ,$$

$$f(K \mid \boldsymbol{\lambda}, \boldsymbol{\eta}, \mathbf{x}) \propto f(K) \, e^{-K(\lambda_1-\lambda_2)} (\lambda_1/\lambda_2)^{\sum_{i=1}^{K} x_i} .$$

Thus, Gibbs sampling can be used to sample from the posterior pdf $f(\boldsymbol{\lambda}, \boldsymbol{\eta}, K \mid \mathbf{x})$.

6.7 OTHER MARKOV SAMPLERS

There exist many variants of the Metropolis–Hastings and Gibbs samplers. However, all of the known MCMC algorithms can be described via the following framework: Consider a Markov chain $\{(\mathbf{X}_n, \mathbf{Y}_n), n = 0, 1, 2, \ldots\}$ on the set $\mathscr{X} \times \mathscr{Y}$, where

$\mathscr{X}$ is the target set and $\mathscr{Y}$ is an auxiliary set. Let $f(\mathbf{x})$ be the target pdf. Each transition of the Markov chain consists of two parts. The first is $(\mathbf{x}, \tilde{\mathbf{y}}) \to (\mathbf{x}, \mathbf{y})$, according to a transition matrix $\mathbf{Q}$; the second is $(\mathbf{x}, \mathbf{y}) \to (\mathbf{x}', \mathbf{y}')$, according to a transition matrix $\mathbf{R}$. In effect, the transition matrix $\mathbf{P}$ of the Markov chain is given by the product $\mathbf{Q\,R}$. Both steps are illustrated in Figure 6.8 and explained below.

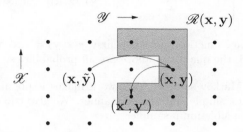

Figure 6.8: Each transition of the Markov chain consists of two steps: the Q-step, followed by the R-step.

The first step, the *Q-step*, changes the $\mathbf{y}$-coordinate but leaves the $\mathbf{x}$-coordinate intact. In particular, $\mathbf{Q}$ is of the form $\mathbf{Q}[(\mathbf{x}, \tilde{\mathbf{y}}), (\mathbf{x}, \mathbf{y})] = \mathbf{Q}_{\mathbf{x}}(\tilde{\mathbf{y}}, \mathbf{y})$, where $\mathbf{Q}_{\mathbf{x}}$ is a transition matrix on $\mathscr{Y}$. Let $q_{\mathbf{x}}$ be a stationary distribution for $\mathbf{Q}_{\mathbf{x}}$, assuming that it exists.

The second step, the *R-step*, is determined by the stationary distribution $q_{\mathbf{x}}$ and the neighborhood structure on the set $\mathscr{X} \times \mathscr{Y}$. Specifically, we define for each point $(\mathbf{x}, \mathbf{y})$ a set of neighbors $\mathscr{R}(\mathbf{x}, \mathbf{y})$ such that *if $(\mathbf{x}', \mathbf{y}')$ is a neighbor of $(\mathbf{x}, \mathbf{y})$ then the converse is also true*; see Figure 6.8, where the shaded area indicates the neighborhood set of $(\mathbf{x}, \mathbf{y})$. The crucial step is now to define the transition matrix $\mathbf{R}$ as

$$\mathbf{R}[(\mathbf{x}, \mathbf{y}), (\mathbf{x}', \mathbf{y}')] = c(\mathbf{x}, \mathbf{y})\, f(\mathbf{x}')\, q_{\mathbf{x}'}(\mathbf{y}') \quad \text{for all} \quad (\mathbf{x}', \mathbf{y}') \in \mathscr{R}(\mathbf{x}, \mathbf{y})\;,$$

where $c(\mathbf{x}, \mathbf{y}) = \sum_{(\mathbf{x}', \mathbf{y}') \in \mathscr{R}(\mathbf{x}, \mathbf{y})} f(\mathbf{x}')\, q_{\mathbf{x}'}(\mathbf{y}')$. Note that $c(\mathbf{x}, \mathbf{y}) = c(\mathbf{x}', \mathbf{y}')$ when $(\mathbf{x}, \mathbf{y})$ and $(\mathbf{x}', \mathbf{y}')$ belong to the same neighborhood set. With this choice of $\mathbf{Q}$ and $\mathbf{R}$ it can be shown (see Problem 6.15) that the Markov chain has a stationary distribution

$$\mu(\mathbf{x}, \mathbf{y}) = f(\mathbf{x})\, q_{\mathbf{x}}(\mathbf{y})\;, \tag{6.12}$$

which is also the limiting distribution, provided that the chain is irreducible and aperiodic. In particular, by ignoring the $\mathbf{y}$-coordinate, we see that the limiting pdf of $\mathbf{X}_n$ is the required target $f(\mathbf{x})$. This leads to the following *generalized Markov sampler* [12]:

Algorithm 6.7.1: Generalized Markov Sampler

---

1 Initialize $(\mathbf{X}_0, \mathbf{Y}_0)$.
2 **for** $t = 0$ **to** $N - 1$ **do**
3 $\quad$ Given $(\mathbf{X}_t, \mathbf{Y}_t)$, generate $\mathbf{Y}$ from $\mathbf{Q}_{\mathbf{x}}(\mathbf{Y}_t, \mathbf{y})$ $\qquad\qquad$ // Q-step
4 $\quad$ Given $\mathbf{Y}$, generate $(\mathbf{X}_{t+1}, \mathbf{Y}_{t+1})$ from $\mathbf{R}[(\mathbf{X}_t, \mathbf{Y}), (\mathbf{x}, \mathbf{y})]$ $\qquad$ // R-step

---

Remark 6.7.1 Denoting $\mathscr{R}^-(\mathbf{x}, \mathbf{y}) = \mathscr{R}(\mathbf{x}, \mathbf{y}) \setminus \{(\mathbf{x}, \mathbf{y})\}$, the sampler can be generalized further (see [12]) by redefining $\mathbf{R}$ as

$$
\mathbf{R}[(\mathbf{x}, \mathbf{y}), (\mathbf{x}', \mathbf{y}')] = \begin{cases} s(\mathbf{x}, \mathbf{y}) \, c(\mathbf{x}, \mathbf{y}) \, f(\mathbf{x}') \, q_{\mathbf{x}'}(\mathbf{y}') & \text{if } (\mathbf{x}', \mathbf{y}') \in \mathscr{R}^-(\mathbf{x}, \mathbf{y}) , \\ 1 - \displaystyle\sum_{(\mathbf{z}, \mathbf{k}) \in \mathscr{R}^-(\mathbf{x}, \mathbf{y})} \mathbf{R}[(\mathbf{x}, \mathbf{y}), (\mathbf{z}, \mathbf{k})] & \text{if } (\mathbf{x}', \mathbf{y}') = (\mathbf{x}, \mathbf{y}) , \end{cases}
$$

$$(6.13)$$

where s is an arbitrary function such that, first, $s(\mathbf{x}, \mathbf{y}) = s(\mathbf{x}', \mathbf{y}')$ for all $(\mathbf{x}', \mathbf{y}') \in \mathscr{R}(\mathbf{x}, \mathbf{y})$ and, second, the quantities above are all probabilities.

The generalized Markov sampler framework makes it possible to obtain many different samplers in a simple and unified manner. We give two examples: the slice sampler and the reversible jump sampler.

6.7.1   Slice Sampler

Suppose that we wish to generate samples from the pdf

$$
f(\mathbf{x}) = b \prod_{k=1}^{m} f_k(\mathbf{x}) , \tag{6.14}
$$

where b is a known or unknown constant and the $\{f_k\}$ are known positive functions — not necessarily densities. We employ Algorithm 6.7.1, where at the Q-step we generate, for a given $\mathbf{X} = \mathbf{x}$, a vector $\mathbf{Y} = (Y_1, \ldots, Y_m)$ by independently drawing each component Y_k from the uniform distribution on $[0, f_k(\mathbf{x})]$. Thus $q_{\mathbf{x}}(\mathbf{y}) = 1/\prod_{k=1}^{m} f_k(\mathbf{x}) = b/f(\mathbf{x})$. Next, we let $\mathscr{R}(\mathbf{x}, \mathbf{y}) = \{(\mathbf{x}', \mathbf{y}) : f_k(\mathbf{x}') \geqslant y_k, \ k = 1, \ldots, m\}$. Then, (since $f(\mathbf{x}') \, q_{\mathbf{x}'}(\mathbf{y}) = b$),

$$
\mathbf{R}[(\mathbf{x}, \mathbf{y}), (\mathbf{x}', \mathbf{y})] = \frac{1}{|\mathscr{R}(\mathbf{x}, \mathbf{y})|} .
$$

In other words, in the R-step, given $\mathbf{x}$ and $\mathbf{y}$, we draw $\mathbf{X}'$ uniformly from the set $\{\mathbf{x}' : f_k(\mathbf{x}') \geqslant y_k, k = 1, \ldots, m\}$. This gives the following *slice sampler*:

Algorithm 6.7.2: Slice Sampler

input : Pdf f of the form (6.14), initial point $\mathbf{X}_0$, and sample size N.
output: $\mathbf{X}_1, \ldots, \mathbf{X}_N$ approximately distributed according to f.

1 **for** $t = 0$ **to** $N - 1$ **do**
2   **for** $k = 1$ **to** m **do**
3   Draw $U_k \sim \mathsf{U}(0, 1)$.
4           $Y_k \leftarrow U_k \, f_k(\mathbf{X}_t)$
5   Draw $\mathbf{X}_{t+1}$ uniformly from the set $\{\mathbf{x} : f_k(\mathbf{x}) \geqslant Y_k, k = 1, \ldots, m\}$.

6 **return** $\mathbf{X}_1, \ldots, \mathbf{X}_N$

■ EXAMPLE 6.9   Slice Sampler

Suppose that we now need to generate a sample from the target pdf

$$f(x) = c \, \frac{x \, e^{-x}}{1+x}, \quad x \geqslant 0 \,,$$

using the slice sampler with $f_1(x) = x/(1+x)$ and $f_2(x) = e^{-x}$.

Suppose that at iteration t, $X_t = z$, and u_1 and u_2 are generated in Lines 2–4. In Line 5, X_{t+1} is drawn uniformly from the set $\{x : f_1(x)/f_1(z) \geqslant u_1, \, f_2(x)/f_2(z) \geqslant u_2\}$, which implies the bounds $x \geqslant \frac{u_1 \, z}{1+z-u_1 \, z}$ and $x \leqslant z - \ln u_2$. Since for $z > 0$ and $0 \leqslant u_1, u_2 \leqslant 1$, the latter bound is larger than the former, the interval to be drawn from in Line 5 is $\left(\frac{u_1 \, z}{1+z-u_1 \, z}, \, z - \ln u_2 \right)$. Figure 6.9 depicts a histogram of $N = 10^5$ samples generated via the slice sampler, along with the true pdf $f(x)$. We see that the two are in close agreement.

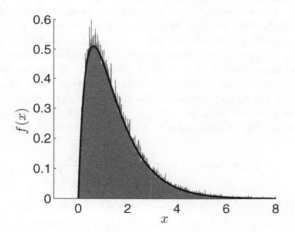

Figure 6.9: True density and histogram of samples produced by the slice sampler.

6.7.2   Reversible Jump Sampler

Reversible jump samplers [9] are useful for sampling from target spaces that contain vectors of different dimensions. This often occurs in Bayesian inference when different models for the data are considered.

■ EXAMPLE 6.10   Regression Data

Let data $y_1, \ldots, y_n$ be the outcomes of independent random variables $\{Y_i\}$ of the form

$$Y_i = \sum_{j=0}^{M} \beta_j \, u_i^j + \varepsilon_i, \qquad \varepsilon_i \sim \mathsf{N}(0,1), \quad i = 1, \ldots, n \,, \tag{6.15}$$

where $u_1, \ldots, u_n$ are known variables, and $M \in \{0, \ldots, M_{\max}\}$ and the parameters $\{\beta_m\}$ are unknown. Let $\mathbf{y} = (y_1, \ldots, y_n)$ and $\boldsymbol{\beta} = (\beta_0, \ldots, \beta_M)$. Taking uniform (i.e., constant) priors for $\{\beta_m\}$ and M, we have the joint pdf

$$f(\mathbf{y}, m, \boldsymbol{\beta}) \propto \exp\left[-\frac{1}{2} \sum_{i=1}^{n} \left(y_i - \sum_{j=0}^{m} \beta_j u_i^j \right)^2 \right]. \tag{6.16}$$

Let $\mathbf{x} = (m, \boldsymbol{\beta})$. Our objective is to draw from the posterior pdf $f(\mathbf{x} \mid \mathbf{y}) = f(m, \boldsymbol{\beta} \mid \mathbf{y})$. This yields information not only about the parameters, but also about which model (expressed by M) is more appropriate. Note here that the dimensionality of $\mathbf{x}$ depends crucially on m, so the standard Gibbs or Metropolis–Hastings sampling is not appropriate.

The reversible jump sampler jumps between spaces of different dimensionalities according to a set of allowed jumps (also called *moves*). In the example above, we could, for instance, allow only jumps between vectors that differ in dimension by at most 1; that is, $\beta_0 \to \beta_0'$, $\beta_0 \to (\beta_0', \beta_1')$, $(\beta_0, \beta_1) \to \beta_0'$, and so on.

To formulate the reversible jump sampler in the generalized Markov sampler framework, we define $\mathscr{Y} = \mathscr{X} \times \mathscr{M}$, where $\mathscr{M}$ is the set of moves, and write a generic element as $(\mathbf{z}, m)$. In the Q-step we take $\mathbf{Q}_{\mathbf{x}}(\cdot, (\mathbf{z}, m)) = p_{\mathbf{x}}(m) \, q_m(\mathbf{x}, \mathbf{z})$. That is, a move of type m is selected according to some discrete pdf $p_{\mathbf{x}}(m)$. For example, the dimension of $\mathbf{x}$ is decreased, increased, or left unchanged. Then a new $\mathbf{z}$ is selected according to some transition function $q_m(\mathbf{x}, \mathbf{z})$. Note that the stationary pdf for the Q-step at $(\mathbf{z}, m)$ then becomes $p_{\mathbf{x}}(m) \, q_m(\mathbf{x}, \mathbf{z})$. The R-step is determined by defining $\mathscr{R}(\mathbf{x}, (\mathbf{z}, m)) = \{(\mathbf{x}, (\mathbf{z}, m)), (\mathbf{z}, (\mathbf{x}, m'))\}$, where m' is the reverse move of m, that is, from $\mathbf{z}$ to $\mathbf{x}$. Then (6.13) reduces to

$$\mathbf{R}[(\mathbf{x}, (\mathbf{z}, m)), (\mathbf{z}, (\mathbf{x}, m'))] = \frac{s(\mathbf{x}, (\mathbf{z}, m))}{1 + 1/\varrho}, \tag{6.17}$$

with $\varrho = \frac{f(\mathbf{z}) \, p_{\mathbf{z}}(m') \, q_{m'}(\mathbf{z}, \mathbf{x})}{f(\mathbf{x}) \, p_{\mathbf{x}}(m) \, q_m(\mathbf{x}, \mathbf{z})}$. Taking $s(\mathbf{x}, (\mathbf{z}, m)) = \min\{1 + \varrho, 1 + 1/\varrho\}$ reduces the right-hand side of (6.17) further to $\min\{\varrho, 1\}$. The transition $(\mathbf{x}, (\mathbf{z}, m)) \to (\mathbf{z}, (\mathbf{x}, m'))$ can thus be interpreted as acceptance of the proposed element $\mathbf{z}$. In effect, $\mathbf{Q}$ is used to propose a new element in accordance with the move m and transition function q, and $\mathbf{R}$ is used to accept or reject it in accordance with the acceptance ratio above. The reversible jump sampler can thus be viewed as a generalization of the Metropolis–Hastings sampler. This gives Algorithm 6.7.3.

Remark 6.7.2 (Dimension Matching) When dealing with continuous random variables, it is important to ensure that the transition densities are properly defined. Suppose that $\dim(\mathbf{x}) = d$ and $\dim(\mathbf{z}) = d' > d$. A possible way to generate a transition $\mathbf{x} \to \mathbf{z}$ is to first draw a $(d' - d)$-dimensional random vector $\mathbf{U}$ according to some density $g(\mathbf{u})$ and then let $\mathbf{z} = \phi(\mathbf{x}, \mathbf{U})$ for some bijection ϕ. This is known as *dimension matching* — the dimension of $(\mathbf{x}, \mathbf{u})$ must match that of $\mathbf{z}$. Note that by (1.20) the transition density is given by $q(\mathbf{x}, \mathbf{z}) = g(\mathbf{u})/|J_{(\mathbf{x}, \mathbf{u})}(\phi)|$, where $|J_{(\mathbf{x}, \mathbf{u})}(\phi)|$ is the absolute value of the determinant of the matrix of Jacobi of ϕ at $(\mathbf{x}, \mathbf{u})$.

Algorithm 6.7.3: Reversible Jump Sampler

---

input : Pdfs $\{p_\mathbf{x}\}$, transition functions $\{q_m\}$, target pdf f, initial point $\mathbf{X}_0$, and sample size N.

output: $\mathbf{X}_1, \ldots, \mathbf{X}_N$ approximately distributed according to f.

1 **for** $t = 0$ **to** $N - 1$ **do**

2 $\quad$ Generate $m \sim p_{\mathbf{X}_t}(m)$.

3 $\quad$ Generate $\mathbf{Z} \sim q_m(\mathbf{X}_t, \mathbf{z})$. Let m' be the reverse move from $\mathbf{Z}$ to $\mathbf{X}_t$.

4 $\quad \alpha \leftarrow \min\left\{ \dfrac{f(\mathbf{Z})\, p_\mathbf{Z}(m')\, q_{m'}(\mathbf{Z}, \mathbf{X}_t)}{f(\mathbf{X}_t)\, p_{\mathbf{X}_t}(m)\, q_m(\mathbf{X}_t, \mathbf{Z})}, 1 \right\}$ $\quad$ // acceptance probability

5

6 $\quad$ Generate $U \sim \mathsf{U}(0, 1)$.

7 $\quad$ **if** $U \leqslant \alpha$ **then**

8 $\quad\quad \mid$ $\mathbf{X}_{t+1} \leftarrow \mathbf{Z}$

9 $\quad$ **else**

10 $\quad\quad \lfloor$ $\mathbf{X}_{t+1} \leftarrow \mathbf{X}_t$

11 **return** $\mathbf{X}_1, \ldots, \mathbf{X}_N$

■ **EXAMPLE 6.11 Example 6.10 (Continued)**

We illustrate the reversible jump sampler using regression data $\mathbf{y} = (y_1, \ldots, y_n)$ of the form (6.15), with $u_i = (i - 1)/20$, $i = 1, \ldots, 101$, $\beta_0 = 1$, $\beta_1 = 0.3$, and $\beta_2 = -0.2$. The data are depicted in Figure 6.10. Although it is obvious that a constant model ($m = 0$) does not fit the data, it is not clear if a linear model ($m = 1$) or a quadratic model ($m = 2$) is more appropriate. To assess the different models, we can run a reversible jump sampler to produce samples from the posterior pdf $f(\mathbf{x} \,|\, \mathbf{y})$, which (up to a normalization constant) is given by the right-hand side of (6.16). A very basic implementation is the following:

Algorithm 6.7.4: Reversible Jump Sampler for Example 6.10

---

1 Initialize $\mathbf{X}_0 = (m', \boldsymbol{\beta}')$

2 **for** $t = 0$ **to** $N - 1$ **do**

3 $\quad$ Generate $m \in \{0, 1, 2\}$ with equal probability.

4 $\quad$ Generate $\boldsymbol{\beta}$ from an $(m + 1)$-dimensional normal pdf g_m with independent components, with means 0 and variances σ^2.

5 $\quad \mathbf{Z} \leftarrow (m, \boldsymbol{\beta})$

6 $\quad \alpha \leftarrow \min\left\{ \dfrac{f(\mathbf{Z} \,|\, \mathbf{y})\, g_{m'}(\boldsymbol{\beta}')}{f(\mathbf{X}_t \,|\, \mathbf{y})\, g_m(\boldsymbol{\beta})}, 1 \right\}$ $\quad$ // acceptance probability

7 $\quad$ Generate $U \sim \mathsf{U}(0, 1)$.

8 $\quad$ **if** $U \leqslant \alpha$ **then**

9 $\quad\quad \mid$ $\mathbf{X}_{t+1} \leftarrow \mathbf{Z}$

10 $\quad$ **else**

11 $\quad\quad \lfloor$ $\mathbf{X}_{t+1} \leftarrow \mathbf{X}_t$

12 $\quad \lfloor$ $(m', \boldsymbol{\beta}') \leftarrow \mathbf{X}_{t+1}$

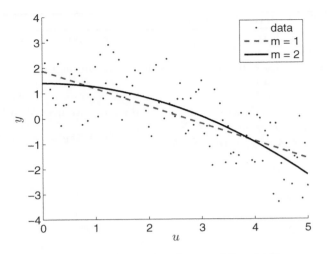

Figure 6.10: Regression data and fitted curves.

The procedure above, with $N = 10^5$ and $\sigma = 2$, produced 22,136 two-dimensional vectors $\boldsymbol{\beta}$ and 77,834 three-dimensional vectors, giving posterior probabilities 0.221 and 0.778 for models 1 and 2, respectively. The posterior probability for the constant model was negligible (0.0003). This indicates that the quadratic model has the best fit. The regression parameters $\boldsymbol{\beta}$ are estimated via the sample means of the $\{\boldsymbol{\beta}_t\}$ for $m_t = 1$ or 2 and are found to be $(1.874, -0.691)$ and $(1.404, -0.011, -0.143)$. The corresponding regression curves are depicted in Figure 6.10.

6.8 SIMULATED ANNEALING

Simulated annealing is a popular optimization technique based on MCMC. This technique uses MCMC sampling to find a mode of a density $f(\mathbf{x})$ (a point $\mathbf{x}^*$ where $f(\mathbf{x})$ is maximal). It involves defining a family of densities of the form $f_T(\mathbf{x}) \propto [f(\mathbf{x})]^{1/T}$, where the parameter T is called the *temperature* of the distribution. MCMC sampling is used to draw a single element $\mathbf{X}^{(k)}$ from f_{T_k} for successively lower temperatures $T_1, T_2, \ldots$. Each element $\mathbf{X}^{(k)}$ is used as the initial element of the next chain. As the temperature is reduced, the distributions become sharply peaked at the global maxima of f. Thus, the $\{\mathbf{X}^{(k)}\}$ converge to a point. They can converge to a local maximum, but this possibility is reduced by careful selection of successive temperatures. The sequence of temperatures, or *annealing schedule*, is therefore critical to the success of the method. A common choice for the annealing schedule is a geometric progression, starting with a specified initial temperature and multiplying by a *cooling factor* in the interval $(0, 1)$ after each iteration.

Simulated annealing can also be applied to nonprobabilistic optimization problems. Given an objective function $S(\mathbf{x})$, a Boltzmann distribution is defined via the density $f(\mathbf{x}) \propto e^{-S(\mathbf{x})}$ or $f(\mathbf{x}) \propto e^{S(\mathbf{x})}$, depending on whether the objective is to minimize or maximize S. Global optima of S are then obtained by searching for the mode of the Boltzmann distribution. We illustrate the method via two worked examples, one based on the Metropolis–Hastings sampler and the other on the Gibbs sampler.

■ **EXAMPLE 6.12   Traveling Salesman Problem**

The traveling salesman problem (TSP) can be formulated as follows: Consider a weighted graph G with n nodes, labeled $1, 2, \ldots, n$. The nodes represent cities, and the edges represent the roads between the cities. Each edge from i to j has weight or cost c_{ij}, representing the length of the road. The problem is to find the shortest *tour* that visits all the cities exactly once except the starting city, which is also the terminating city. An example is given in Figure 6.11, where the bold lines form a possible tour.

Figure 6.11: Find the shortest tour **x** visiting all nodes.

Without loss of generality, we can assume that the graph is *complete* (fully connected) because, if it is not complete, we can always add some costs (distances) equal to $+\infty$. Let $\mathscr{X}$ be the set of all possible tours, and let $S(\mathbf{x})$ the total length of tour $\mathbf{x} \in \mathscr{X}$. We can represent each tour via a *permutation* of $(1, \ldots, n)$. For example, for $n = 4$, the permutation $(1, 3, 2, 4)$ represents the tour $1 \to 3 \to 2 \to 4 \to 1$. Therefore, we will identify a tour with its corresponding permutation. The objective is thus to minimize

$$\min_{\mathbf{x} \in \mathscr{X}} S(\mathbf{x}) \quad = \quad \min_{\mathbf{x} \in \mathscr{X}} \left\{ \sum_{i=1}^{n-1} c_{x_i, x_{i+1}} + c_{x_n, 1} \right\} . \tag{6.18}$$

Note that the number of elements in $\mathscr{X}$ is typically very large, since $|\mathscr{X}| = n!$.

The TSP can be solved via simulated annealing. First, we define the target pdf to be the Boltzmann pdf $f(\mathbf{x}) = c\,e^{-S(\mathbf{x})/T}$. Second, we define a neighborhood structure on the space of permutations $\mathscr{X}$, called *2-opt*. Here the neighbors of an arbitrary permutation $\mathbf{x}$ are found by (1) selecting two different indexes from $\{1, \ldots, n\}$ and (2) reversing the path of $\mathbf{x}$ between those two indexes. For example, if $\mathbf{x} = (1, 2, \ldots, 10)$ and indexes 4 and 7 are selected, then $\mathbf{y} = (1, 2, 3, 7, 6, 5, 4, 8, 9, 10)$; see Figure 6.12. Another example is: if $\mathbf{x} = (6, 7, 2, 8, 3, 9, 10, 5, 4, 1)$ and indexes 6 and 10 are selected, then $\mathbf{y} = (6, 7, 2, 8, 3, 1, 4, 5, 10, 9)$.

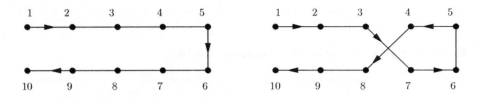

Figure 6.12: Illustration of the 2-opt neighborhood structure.

Third, we apply the Metropolis–Hastings algorithm to sample from the target. We need to supply a transition function $q(\mathbf{x}, \mathbf{y})$ from $\mathbf{x}$ to one of its neighbors. Typically, the two indexes for the 2-opt neighborhood are selected uniformly. This can be done, for example, by drawing a uniform permutation of $(1, \ldots, n)$ (see Section 2.10) and then selecting the first two elements of this permutation. The transition function is here constant: $q(\mathbf{x}, \mathbf{y}) = q(\mathbf{y}, \mathbf{x}) = 1/\binom{n}{2}$. It follows that in this case the acceptance probability is

$$\alpha = \min\left\{\frac{f(\mathbf{y})}{f(\mathbf{x})}, 1\right\} = \begin{cases} 1 & \text{if } S(\mathbf{y}) \leqslant S(\mathbf{x}) \\ e^{-(S(\mathbf{y})-S(\mathbf{x}))/T} & \text{if } S(\mathbf{y}) > S(\mathbf{x}) . \end{cases} \qquad (6.19)$$

As we gradually decrease the temperature T, the Boltzmann distribution becomes more and more concentrated around the global minimizer. Common practice is to decrease the temperature as $T_{t+1} = \beta T_t$ for some $\beta < 1$ close to 1, such as $\beta = 0.99$. This leads to the following generic simulated annealing algorithm with Metropolis–Hastings sampling:

Algorithm 6.8.1: Simulated Annealing: Metropolis–Hastings Sampling

---

input : Objective function S, starting state $\mathbf{X}_0$, initial temperature T_0,
number of iterations N, symmetric proposal function $q(\mathbf{x}, \mathbf{y})$,
constant β.

output: Approximate minimum value of S and corresponding minimizer.

1 **for** $t = 0$ **to** $N - 1$ **do**
2   Generate a new state $\mathbf{Y}$ from the symmetric proposal $q(\mathbf{X}_t, \mathbf{y})$.
3   **if** $S(\mathbf{Y}) < S(\mathbf{X}_t)$ **then**
4        $\mathbf{X}_{t+1} \leftarrow \mathbf{Y}$
5   **else**
6   Draw $U \sim \mathsf{U}(0, 1)$.
7   **if** $U \leqslant e^{-(S(\mathbf{Y})-S(\mathbf{X}_t))/T_t}$ **then**
8           $\mathbf{X}_{t+1} \leftarrow \mathbf{Y}$
9   **else**
10          $\mathbf{X}_{t+1} \leftarrow \mathbf{X}_t$
11     $T_{t+1} \leftarrow \beta T_t$
12 **return** $S(\mathbf{X}_N)$ and $\mathbf{X}_N$

Instead of stopping after a fixed number N of iterations, it is useful to stop when consecutive function values are closer than some distance ε to each other, or when the best found function value has not changed over a fixed number d of iterations.

■ **EXAMPLE 6.13** *n*-Queens Problem

In the *n*-queens problem, the objective is to arrange n queens on a $n \times n$ chessboard in such a way that no queen can capture another queen. An illustration is given in Figure 6.13 for the case $n = 8$. Note that the configuration in Figure 6.13 does not solve the problem. We take $n = 8$ from now on. Note that each row of the chessboard must contain exactly one queen. Denote the position of the queen in the i-th row by x_i; this way each configuration can be represented by a vector $\mathbf{x} = (x_1, \ldots, x_8)$. For example, $\mathbf{x} = (2, 3, 7, 4, 8, 5, 1, 6)$ corresponds to the large configuration in Figure 6.13. Two other examples are given in the same figure. We can now formulate the problem in terms of minimizing a function $S(\mathbf{x})$ that represents the amount of "threat" of the queens. For this we simply add the number of queens *minus* 1, for each column and diagonal that have at least two queens present. For the large configuration in Figure 6.13 there are only two diagonals with two queens, so the score is $(2 - 1) + (2 - 1) = 2$. Note that the minimal S value is 0. One of the optimal solutions is $\mathbf{x}^* = (5, 1, 8, 6, 3, 7, 2, 4)$.

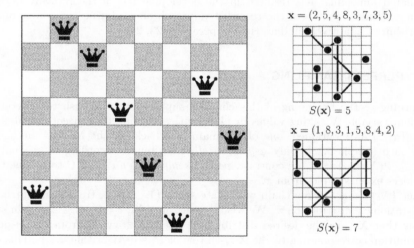

Figure 6.13: Position the eight queens such that no queen can capture another.

We show next how this optimization problem can be solved via simulated annealing using the Gibbs sampler. As in the previous TSP example, each iteration of the algorithm consists of sampling from the Boltzmann pdf $f(\mathbf{x}) = e^{-S(\mathbf{x})/T}$ via the Gibbs sampler, followed by decreasing the temperature. This leads to the following generic simulated annealing algorithm using Gibbs sampling:

Algorithm 6.8.2: Simulated Annealing: Gibbs Sampling

---

1 Initialize $\mathbf{X}_0$.
2 $t \leftarrow 0$
3 **while** $S(\mathbf{X}_t) > 0$ **do**
4   Draw Y_1 from the conditional pdf $f(y_1 \mid X_{t,2}, \ldots, X_{t,n})$.
5   **for** $i = 2$ **to** n **do**
6        $\lfloor$ Draw Y_i from the conditional pdf $f(y_i \mid Y_1, \ldots, Y_{i-1}, X_{t,i+1}, \ldots, X_{t,n})$.
7     $\mathbf{X}_{t+1} \leftarrow \mathbf{Y}$
8     $T_{t+1} \leftarrow \beta T_t$
9     $t \leftarrow t + 1$
10 **return** $\mathbf{X}_t$

Note that in Line 6 each Y_i is drawn from a discrete distribution on $\{1, \ldots, n\}$ with probabilities proportional to $e^{-S(\mathbf{Z}_1)/T_t}, \ldots, e^{-S(\mathbf{Z}_n)/T_t}$, where each $\mathbf{Z}_k$ is equal to the vector $(Y_1, \ldots, Y_{i-1}, k, X_{t,i+1}, \ldots, X_{t,n})$.

Other MCMC samplers can be used in simulated annealing. For example, in the *hide-and-seek* algorithm [21] the general hit-and-run sampler (Section 6.3) is used. Research motivated by the use of hit-and-run and discrete hit-and-run in simulated annealing, has resulted in the development of a theoretically derived cooling schedule that uses the recorded values obtained during the course of the algorithm to adaptively update the temperature [23, 24].

6.9   PERFECT SAMPLING

Returning to the beginning of this chapter, suppose that we wish to generate a random variable X taking values in $\{1, \ldots, m\}$ according to a target distribution $\pi = \{\pi_i\}$. As mentioned, one of the main drawbacks of the MCMC method is that each sample X_t is only *asymptotically* distributed according to π, that is, $\lim_{t \to \infty} \mathbb{P}(X_t = i) = \pi_i$. In contrast, *perfect sampling* is an MCMC technique that produces exact samples from π.

Let $\{X_t\}$ be a Markov chain with state space $\{1, \ldots, m\}$, transition matrix P, and stationary distribution π. We wish to generate $\{X_t, t = 0, -1, -2, \ldots\}$ in such a way that X_0 has the desired distribution. We can draw X_0 from the m-point distribution corresponding to the X_{-1}-th row of P; see Algorithm 2.7.1. This can be done via the IT method, which requires the generation of a random variable $U_0 \sim \mathsf{U}(0, 1)$. Similarly, X_{-1} can be generated from X_{-2} and $U_{-1} \sim \mathsf{U}(0, 1)$. In general, we see that for any negative time $-t$, the random variable X_0 depends on X_{-t} and the independent random variables $U_{-t+1}, \ldots, U_0 \sim \mathsf{U}(0, 1)$.

Next, let us consider m dependent copies of the Markov chain, starting from each of the states $1, \ldots, m$ and using the *same* random numbers $\{U_i\}$ — as in the CRV method. Then, if two paths coincide, or *coalesce*, at some time, from that time on, both paths will be identical. The paths are said to be *coupled*. The main point of the perfect sampling method is that if the chain is ergodic (in particular, if it is aperiodic and irreducible), then *with probability 1 there exists a negative time* $-T$ *such that all m paths will have coalesced before or at time 0*. The situation is illustrated in Figure 6.14.

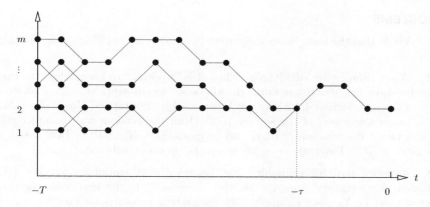

Figure 6.14: All Markov chains have coalesced at time $-\tau$.

Let $\mathbf{U}$ represent the vector of all $U_t, t \leqslant 0$. For each $\mathbf{U}$ we know there exists, with probability 1, a $-T(\mathbf{U}) < 0$ such that by time 0 all m coupled chains defined by $\mathbf{U}$ have coalesced. Moreover, if we start at time $-T$, a *stationary* version of the Markov chain, using again the same $\mathbf{U}$, this stationary chain must, at time $t = 0$, have coalesced with the other ones. Thus, any of the m chains has at time 0 the same distribution as the stationary chain, which is π.

Note that in order to construct T, we do not need to know the whole (infinite vector) $\mathbf{U}$. Instead, we can work backward from $t = 0$ by generating U_{-1} first, and checking if $-T = -1$. If this is not the case, generate U_{-2} and check if $-T = -2$, and so on. This leads to the following algorithm, due to Propp and Wilson [19], called *coupling from the past*:

Algorithm 6.9.1: Coupling from the Past

---

1   Generate $U_0 \sim \mathsf{U}(0,1)$.
2    $\mathbf{U}_0 \leftarrow U_0$
3    $t \leftarrow -1$
4   NotCoalesced $\leftarrow$ **true**
5   **while** NotCoalesced **do**
6   Generate m Markov chains $\{X_i, i = -t, \ldots, 0\}$, starting at t from each of the states $1, \ldots, m$, and using the same random vector $\mathbf{U}_{t+1}$.
7   **if** all chains have coalesced before or at time 0 **then**
8   NotCoalesced $\leftarrow$ **false**
9   **else**
10   Generate $U_t \sim \mathsf{U}(0,1)$.
11          $\mathbf{U}_t \leftarrow (U_t, \mathbf{U}_{t+1})$
12          $t \leftarrow t - 1$

13 **return** X_0

Although perfect sampling seems indeed "perfect" in that it returns an exact sample from the target π rather than an approximate one, practical applications of the technique are, presently, quite limited. Not only is the technique difficult or impossible to use for most continuous simulation systems, it is also much more computationally intensive than simple MCMC.

PROBLEMS

6.1 Verify that the local balance equation (6.3) holds for the Metropolis–Hastings algorithm.

6.2 When running an MCMC algorithm, it is important to know when the transient (or *burn-in*) period has finished; otherwise, steady-state statistical analyses such as those in Section 4.4.2 may not be applicable. In practice, this is often done via a visual inspection of the sample path. Run the random walk sampler with normal target distribution $N(10, 1)$ and proposal $Y \sim N(x, 0.01)$. Take a sample size of $N = 5000$. Determine roughly when the process reaches stationarity.

6.3 A useful tool for examining the behavior of a stationary process $\{X_t\}$ obtained, for example, from an MCMC simulation, is the covariance function $R(t) = \text{Cov}(X_t, X_0)$; see Example 6.4. Estimate the covariance function for the process in Problem 6.2 and plot the results. In Matlab's *signal processing* toolbox, this is implemented under the M-function xcov.m. Try different proposal distributions of the form $N(x, \sigma^2)$ and observe how the covariance function changes.

6.4 Implement the independence sampler with an $\text{Exp}(1)$ target and an $\text{Exp}(\lambda)$ proposal distribution for several values of λ. Similar to the importance sampling situation, things go awry when the sampling distribution gets too far from the target distribution, in this case when $\lambda > 2$. For each run, use a sample size of 10^5 and start with $x = 1$.

    a) For each value $\lambda = 0.2, 1, 2$, and 5, plot a histogram of the data and compare it with the true pdf.

    b) For each value of λ given above, calculate the sample mean and repeat this for 20 independent runs. Make a dotplot of the data (plot them on a line) and notice the differences. Observe that for $\lambda = 5$ most of the sample means are below 1, and thus underestimate the true expectation 1, but a few are significantly greater. Observe also the behavior of the corresponding auto-covariance functions, both between the different λs and, for $\lambda = 5$, within the 20 runs.

6.5 Implement the random walk sampler with an $\text{Exp}(1)$ target distribution, where Z (in the proposal $Y = x + Z$) has a double exponential distribution with parameter λ. Carry out a study similar to that in Problem 6.4 for different values of λ, say $\lambda = 0.1, 1, 5, 20$. Observe that (in this case) the random walk sampler has a more stable behavior than the independence sampler.

6.6 Let $\mathbf{X} = (X, Y)^\top$ be a random column vector with a bivariate normal distribution with expectation vector $\mathbf{0} = (0, 0)^\top$ and covariance matrix

$$\Sigma = \begin{pmatrix} 1 & \varrho \\ \varrho & 1 \end{pmatrix}.$$

    a) Show that $(Y \mid X = x) \sim N(\varrho x, 1 - \varrho^2)$ and $(X \mid Y = y) \sim N(\varrho y, 1 - \varrho^2)$.

    b) Write a systematic Gibbs sampler to draw 10^4 samples from the bivariate distribution $N(\mathbf{0}, \Sigma)$ and plot the data for $\varrho = 0, 0.7$ and 0.9.

6.7 A remarkable feature of the Gibbs sampler is that the conditional distributions in Algorithm 6.4.1 contain sufficient information to generate a sample from the joint distribution. The following result (by Hammersley and Clifford [10]) shows that it is possible to directly express the joint pdf in terms of the conditional pdfs. Namely,

$$f(x,y) = \frac{f_{Y|X}(y|x)}{\int \frac{f_{Y|X}(y|x)}{f_{X|Y}(x|y)}\,dy}\,.$$

Prove this. Generalize this to the n-dimensional case.

6.8   In the Ising model the *expected magnetization per spin* is given by

$$M(T) = \frac{1}{n^2}\,\mathbb{E}_{\pi_T}\left[\sum_i S_i\right],$$

where π_T is the Boltzmann distribution at temperature T. Estimate $M(T)$, for example via the Swendsen–Wang algorithm, for various values of $T \in [0,5]$, and observe that the graph of $M(T)$ changes sharply around the critical temperature $T \approx 2.61$. Take $n = 20$ and use periodic boundaries.

6.9   Run Peter Young's Java applet in

http://physics.ucsc.edu/~peter/java/ising/ising.html

to gain a better understanding of how the Ising model works.

6.10   As in Example 6.6, let $\mathscr{X}^* = \{\mathbf{x} : \sum_{i=1}^n x_i = m,\ x_i \in \{0,\ldots,m\},\ i = 1,\ldots,n\}$. Show that this set has $\binom{m+n-1}{n-1}$ elements.

6.11   In a simple model for a closed queueing network with n queues and m customers, it is assumed that the service times are independent and exponentially distributed, with rate μ_i for queue i, $i = 1,\ldots,n$. After completing service at queue i, the customer moves to queue j with probability p_{ij}. The $\{p_{ij}\}$ are the so-called *routing probabilities*.

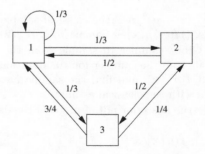

Figure 6.15: A closed queueing network.

It can be shown (e.g., see [13]) that the stationary distribution of the number of customers in the queues is of product form (6.7), with f_i being the pdf of the $\mathsf{G}(1 - y_i/\mu_i)$ distribution; thus $f_i(x_i) \propto (y_i/\mu_i)^{x_i}$. Here the $\{y_i\}$ are constants that are obtained from the following set of *flow balance* equations:

$$y_i = \sum_j y_j\, p_{ji}\,, \quad i = 1,\ldots,n\,, \tag{6.20}$$

which has a one-dimensional solution space. Without loss of generality, y_1 can be set to 1 to obtain a unique solution.

Consider now the specific case of the network depicted in Figure 6.15, with $n = 3$ queues. Suppose that the service rates are $\mu_1 = 2$, $\mu_2 = 1$, and $\mu_3 = 1$. The routing probabilities are given in the figure.

  a) Show that a solution to (6.20) is $(y_1, y_2, y_3) = (1, 10/21, 4/7)$.
  b) For $m = 50$ determine the exact normalization constant C.
  c) Implement the procedure of Example 6.6 to estimate C via MCMC, and compare the estimate for $m = 50$ with the exact value.

6.12   Let $X_1, \ldots, X_n$ be a random sample from the $\mathsf{N}(\mu, \sigma^2)$ distribution. Consider the following Bayesian model:

- $f(\mu, \sigma^2) = 1/\sigma^2$;

- $(\mathbf{x}_i \mid \mu, \sigma) \sim \mathsf{N}(\mu, \sigma^2)$, $i = 1, \ldots, n$ independently.

Note that the prior for (μ, σ^2) is *improper*. That is, it is not a pdf in itself, but if we obstinately apply Bayes' formula, it does yield a proper posterior pdf. In some sense it conveys the least amount of information about μ and σ^2. Let $\mathbf{x} = (x_1, \ldots, x_n)$ represent the data. The posterior pdf is given by

$$f(\mu, \sigma^2 \mid \mathbf{x}) = \left(2\pi\sigma^2\right)^{-n/2} \exp\left\{-\frac{1}{2}\frac{\sum_i (x_i - \mu)^2}{\sigma^2}\right\} \frac{1}{\sigma^2}.$$

We wish to sample from this distribution via the Gibbs sampler.
  a) Show that $(\mu \mid \sigma^2, \mathbf{x}) \sim \mathsf{N}(\bar{x}, \sigma^2/n)$, where $\bar{x}$ is the sample mean.
  b) Prove that

$$f(\sigma^2 \mid \mu, \mathbf{x}) \propto \frac{1}{(\sigma^2)^{n/2+1}} \exp\left(-\frac{n\, V_\mu}{2\, \sigma^2}\right), \tag{6.21}$$

  where $V_\mu = \sum_i (x_i - \mu)^2/n$ is the classical sample variance for known μ. In other words, $(1/\sigma^2 \mid \mu, \mathbf{x}) \sim \mathsf{Gamma}(n/2, nV_\mu/2)$.
  c) Implement a Gibbs sampler to sample from the posterior distribution, taking $n = 100$. Run the sampler for 10^5 iterations. Plot the histograms of $f(\mu \mid \mathbf{x})$ and $f(\sigma^2 \mid \mathbf{x})$, and find the sample means of these posteriors. Compare them with the classical estimates.
  d) Show that the true posterior pdf of μ, given the data, is

$$f(\mu \mid \mathbf{x}) \propto \left((\mu - \bar{x})^2 + V\right)^{-n/2},$$

  where $V = \sum_i (x_i - \bar{x})^2/n$. [Hint: To evaluate the integral

$$f(\mu \mid \mathbf{x}) = \int_0^\infty f(\mu, \sigma^2 \mid \mathbf{x}) \, d\sigma^2$$

  write it first as $(2\pi)^{-n/2}\int_0^\infty t^{n/2-1}\exp(-\frac{1}{2}t c)\,dt$, where $c = nV_\mu$, by applying the change of variable $t = 1/\sigma^2$. Show that the latter integral is proportional to $c^{-n/2}$. Finally, apply the decomposition $V_\mu = (\bar{x} - \mu)^2 + V$.]

6.13   Suppose that $f(\boldsymbol{\theta} \mid \mathbf{x})$ is the posterior pdf for some Bayesian estimation problem. For example, $\boldsymbol{\theta}$ could represent the parameters of a regression model based on the data $\mathbf{x}$. An important use for the posterior pdf is to make predictions about the distribution of other random variables. For example, suppose that the pdf of some random variable $\mathbf{Y}$ depends on $\boldsymbol{\theta}$ via the conditional pdf $f(\mathbf{y} \mid \boldsymbol{\theta})$. The *predictive pdf* of Y given $\mathbf{x}$ is defined as

$$f(\mathbf{y} \mid \mathbf{x}) = \int f(\mathbf{y} \mid \boldsymbol{\theta}) f(\boldsymbol{\theta} \mid \mathbf{x}) \, \mathrm{d}\boldsymbol{\theta} \, ,$$

which can be viewed as the expectation of $f(\mathbf{y} \mid \boldsymbol{\theta})$ under the posterior pdf. Therefore, we can use Monte Carlo simulation to approximate $f(\mathbf{y} \mid \mathbf{x})$ as

$$f(\mathbf{y} \mid \mathbf{x}) \approx \frac{1}{N} \sum_{i=1}^{N} f(\mathbf{y} \mid \boldsymbol{\theta}_i) \, ,$$

where the sample $\{\boldsymbol{\theta}_i, i = 1, \ldots, N\}$ is obtained from $f(\boldsymbol{\theta} \mid \mathbf{x})$ (e.g., via MCMC).

As a concrete application, suppose that the independent measurement data $-0.4326, -1.6656, 0.1253, 0.2877, -1.1465$ come from some $\mathsf{N}(\mu, \sigma^2)$ distribution. Define $\boldsymbol{\theta} = (\mu, \sigma^2)$. Let $Y \sim \mathsf{N}(\mu, \sigma^2)$ be a new measurement. Estimate and draw the predictive pdf $f(y \mid \mathbf{x})$ from a sample $\boldsymbol{\theta}_1, \ldots, \boldsymbol{\theta}_N$ obtained via the Gibbs sampler of Problem 6.12. Take $N = 10{,}000$. Compare this with the "common-sense" Gaussian pdf with expectation $\bar{x}$ (sample mean) and variance s^2 (sample variance).

6.14 In the *zero-inflated Poisson* (ZIP) model, random data $X_1, \ldots, X_n$ are assumed to be of the form $X_i = R_i Y_i$, where the $\{Y_i\}$ have a $\mathsf{Poi}(\lambda)$ distribution and the $\{R_i\}$ have a $\mathsf{Ber}(p)$ distribution, all independent of each other. Given an outcome $\mathbf{x} = (x_1, \ldots, x_n)$, the objective is to estimate both λ and p. Consider the following hierarchical Bayes model:

- $p \sim \mathsf{U}(0, 1)$ (prior for p),
- $(\lambda \mid p) \sim \mathsf{Gamma}(a, b)$ (prior for λ),
- $(r_i \mid p, \lambda) \sim \mathsf{Ber}(p)$ independently   (from the model above),
- $(x_i \mid \mathbf{r}, \lambda, p) \sim \mathsf{Poi}(\lambda r_i)$ independently (from the model above),

where $\mathbf{r} = (r_1, \ldots, r_n)$ and a and b are known parameters. It follows that

$$f(\mathbf{x}, \mathbf{r}, \lambda, p) = \frac{b^a \lambda^{a-1} e^{-b\lambda}}{\Gamma(a)} \prod_{i=1}^{n} \frac{e^{-\lambda r_i} (\lambda r_i)^{x_i}}{x_i!} p^{r_i} (1 - p)^{1 - r_i} \, .$$

We wish to sample from the posterior pdf $f(\lambda, p, \mathbf{r} \mid \mathbf{x})$ using the Gibbs sampler.

    a) Show that
        1. $(\lambda \mid p, \mathbf{r}, \mathbf{x}) \sim \mathsf{Gamma}(a + \sum_i x_i, \; b + \sum_i r_i)$.
        2. $(p \mid \lambda, \mathbf{r}, \mathbf{x}) \sim \mathsf{Beta}(1 + \sum_i r_i, \; n + 1 - \sum_i r_i)$.
        3. $(r_i \mid \lambda, p, \mathbf{x}) \sim \mathsf{Ber}\left(\frac{p\,e^{-\lambda}}{p\,e^{-\lambda} + (1-p) I_{\{x_i=0\}}} \right)$.
    b) Generate a random sample of size $n = 100$ for the ZIP model using parameters $p = 0.3$ and $\lambda = 2$.
    c) Implement the Gibbs sampler, generate a large (dependent) sample from the posterior distribution, and use this to construct 95% Bayesian CIs for p and λ using the data in b). Compare your results with the true values.

6.15   Show that μ in (6.12) satisfies the local balance equations

$$\mu(\mathbf{x},\mathbf{y})\,\mathbf{R}[(\mathbf{x},\mathbf{y}),(\mathbf{x}',\mathbf{y}')] = \mu(\mathbf{x}',\mathbf{y}')\,\mathbf{R}[(\mathbf{x}',\mathbf{y}'),(\mathbf{x},\mathbf{y})]\ .$$

Thus μ is stationary with respect to $\mathbf{R}$, that is, $\mu\mathbf{R} = \mu$. Show that μ is also stationary with respect to $\mathbf{Q}$. Show also that μ is stationary with respect to $\mathbf{P} = \mathbf{QR}$.

6.16   To show that the systematic Gibbs sampler is a special case of the generalized Markov sampler, take $\mathscr{Y}$ to be the set of indexes $\{1,\ldots,n\}$, and define for the Q-step

$$\mathbf{Q}_{\mathbf{x}}(y,y') = \begin{cases} 1 & \text{if } y' = y+1 \text{ or } y' = 1, y = n, \\ 0 & \text{otherwise.} \end{cases}$$

Let the set of possible transitions $\mathscr{R}(\mathbf{x},y)$ be the set of vectors $\{(\mathbf{x}',y)\}$ such that all coordinates of $\mathbf{x}'$ are the same as those of $\mathbf{x}$ except for possibly the y-th coordinate.

  a) Show that the stationary distribution of $\mathbf{Q}_{\mathbf{x}}$ is $q_{\mathbf{x}}(y) = 1/n$, for $y = 1,\ldots,n$.

  b) Show that

$$\mathbf{R}[(\mathbf{x},y),(\mathbf{x}',y)] = \frac{f(\mathbf{x}')}{\displaystyle\sum_{(\mathbf{z},y)\in\mathscr{R}(\mathbf{x},y)} f(\mathbf{z})},\qquad \text{for } (\mathbf{x}',y)\in\mathscr{R}(\mathbf{x},y)\ .$$

  c) Compare with Algorithm 6.4.1.

6.17   Prove that the Metropolis–Hastings algorithm is a special case of the generalized Markov sampler. [Hint: Let the auxiliary set $\mathscr{Y}$ be a copy of the target set $\mathscr{X}$, let $\mathbf{Q}_{\mathbf{x}}$ correspond to the transition function of the Metropolis–Hastings algorithm (i.e., $\mathbf{Q}_{\mathbf{x}}(\cdot,\mathbf{y}) = q(\mathbf{x},\mathbf{y})$), and define $\mathscr{R}(\mathbf{x},\mathbf{y}) = \{(\mathbf{x},\mathbf{y}),(\mathbf{y},\mathbf{x})\}$. Use arguments similar to those for the Markov jump sampler (see (6.17)) to complete the proof.]

6.18   Barker's and Hastings' MCMC algorithms differ from the symmetric Metropolis sampler only in that they define the acceptance ratio $\alpha(\mathbf{x},\mathbf{y})$ to be, respectively, $f(\mathbf{y})/(f(\mathbf{x}) + f(\mathbf{y}))$ and $s(\mathbf{x},\mathbf{y})/(1 + 1/\varrho(\mathbf{x},\mathbf{y}))$ instead of $\min\{f(\mathbf{y})/f(\mathbf{x}),1\}$. Here, $\varrho(\mathbf{x},\mathbf{y}) = f(\mathbf{y})q(\mathbf{y},\mathbf{x})/(f(\mathbf{x})q(\mathbf{x},\mathbf{y}))$ and s is any symmetric function such that $0 \leqslant \alpha(\mathbf{x},\mathbf{y}) \leqslant 1$. Show that both are special cases of the generalized Markov sampler. [Hint: Take $\mathscr{Y} = \mathscr{X}$.]

6.19   Implement the simulated annealing algorithm for the n-queens problem suggested in Example 6.13. How many solutions can you find?

6.20   Implement the Metropolis–Hastings based simulated annealing algorithm for the TSP in Example 6.12. Run the algorithm on some test problems in

  http://www.iwr.uni-heidelberg.de/groups/comopt/software/TSPLIB95/

6.21   Write a simulated annealing algorithm based on the random walk sampler to maximize the function

$$S(x) = \left| \frac{\sin^8(10x) + \cos^5(5x+1)}{x^2 - x + 1} \right|,\qquad x \in \mathbb{R}\ .$$

Use a $N(x, \sigma^2)$ proposal function, given the current state x. Start with $x = 0$. Plot the current best function value against the number of evaluations of S for various values of σ and various annealing schedules. Repeat the experiments several times to assess what works best.

Further Reading

MCMC is one of the principal tools of statistical computing and Bayesian analysis. A comprehensive discussion of MCMC techniques can be found in [20], and practical applications are discussed in [8]. See also [4]. For more details on the use of MCMC in Bayesian analysis, we refer to [6]. A classical reference on simulated annealing is [1]. More general global search algorithms may be found in [26]. An influential paper on stationarity detection in Markov chains, which is closely related to perfect sampling, is [3].

REFERENCES

1. E. H. L. Aarts and J. H. M. Korst. *Simulated Annealing and Boltzmann Machines.* John Wiley & Sons, Chichester, 1989.

2. D. J. Aldous and J. Fill. *Reversible Markov Chains and Random Walks on Graphs.* In preparation. http://www.stat.berkeley.edu /users/aldous/book.html, 2007.

3. S. Asmussen, P. W. Glynn, and H. Thorisson. Stationary detection in the initial transient problem. *ACM Transactions on Modeling and Computer Simulation,* 2(2):130–157, 1992.

4. S. Brooks, A. Gelman, G. Jones, and X.-L. Meng. *Handbook of Markov Chain Monte Carlo.* CRC press, 2011.

5. M.-H. Chen and B. W. Schmeiser. General hit-and-run Monte Carlo sampling for evaluating multidimensional integrals. *Operations Research Letters,* 19(4):161–169, 1996.

6. A. Gelman, J. B. Carlin, H. S. Stern, and D. B. Rubin. *Bayesian Data Analysis.* Chapman & Hall, New York, 2nd edition, 2003.

7. S. Geman and D. Geman. Stochastic relaxation, Gibbs distribution and the Bayesian restoration of images. *IEEE Transactions on PAMI,* 6:721–741, 1984.

8. W.R. Gilks, S. Richardson, and D. J. Spiegelhalter. *Markov Chain Monte Carlo in Practice.* Chapman & Hall, New York, 1996.

9. P. J. Green. Reversible jump Markov chain Monte Carlo computation and Bayesian model determination. *Biometrika,* 82(4):711–732, 1995.

10. J. Hammersley and M. Clifford. Markov fields on finite lattices. Unpublished manuscript, 1970.

11. W. K. Hastings. Monte Carlo sampling methods using Markov chains and their applications. *Biometrika,* 57:92–109, 1970.

12. J. M. Keith, D. P. Kroese, and D. Bryant. A generalized Markov chain sampler. *Methodology and Computing in Applied Probability,* 6(1):29–53, 2004.

13. F. P. Kelly. *Reversibility and Stochastic Networks.* John Wiley & Sons, Chichester, 1979.

14. J. S. Liu. *Monte Carlo Strategies in Scientific Computing.* Springer-Verlag, New York, 2001.

15. L. Lovász. Hit-and-run mixes fast. *Mathematical Programming,* 86:443–461, 1999.

16. L. Lovász and S. S. Vempala. Hit-and-run is fast and fun. Technical report, Microsoft Research, SMS-TR, 2003.

17. L. Lovász and S. Vempala. Hit-and-run from a corner. *SIAM Journal on Computing,* 35(4):985–1005, 2006.

18. M. Metropolis, A. W. Rosenbluth, M. N. Rosenbluth, A. H. Teller, and E. Teller. Equations of state calculations by fast computing machines. *Journal of Chemical Physics,* 21:1087–1092, 1953.

19. J. G. Propp and D. B. Wilson. Exact sampling with coupled Markov chains and applications to statistical mechanics. *Random Structures and Algorithms,* 1 & 2:223–252, 1996.

20. C. P. Robert and G. Casella. *Monte Carlo Statistical Methods.* Springer-Verlag, New York, 2nd edition, 2004.

21. H. E. Romeijn and R. L. Smith. Simulated annealing for constrained global optimization. *Journal of Global Optimization,* 5:101–126, 1994.

22. S. M. Ross. *Simulation.* Academic Press, New York, 3rd edition, 2002.

23. Y. Shen. *Annealing Adaptive Search with Hit-and-Run Sampling Methods for Stochastic Global Optimization Algorithms.* PhD thesis, University of Washington, 2005.

24. Y. Shen, S. Kiatsupaibul, Z. B. Zabinsky, and R. L. Smith. An analytically derived cooling schedule for simulated annealing. *Journal of Global Optimization,* 38(3):333–365, 2007.

25. R. L. Smith. Efficient Monte Carlo procedures for generating points uniformly distributed over bounded regions. *Operations Research,* 32:1296–1308, 1984.

26. Z. B. Zabinsky. *Stochastic Adaptive Search for Global Optimization.* Kluwer Academic Publishers, Dordrecht, 2003.

27. Z. B. Zabinsky, R. L. Smith, J. F. McDonald, H. E. Romeijn, and D. E. Kaufman. Improving hit-and-run for global optimization. *Journal of Global Optimization,* 3:171–192, 1993.

CHAPTER 7

SENSITIVITY ANALYSIS AND MONTE CARLO OPTIMIZATION

7.1 INTRODUCTION

As discussed in Chapter 3, many real-world complex systems in science and engineering can be modeled as *discrete-event systems*. The behavior of such systems is identified via a sequence of discrete events, which causes the system to change from one state to another. Examples include traffic systems, flexible manufacturing systems, computer-communications systems, inventory systems, production lines, coherent lifetime systems, Program Evaluation and Review Technique (PERT) networks, and flow networks. A discrete-event system can be classified as either *static* or *dynamic*. The former are called *discrete-event static systems* (DESS), while the latter are called *discrete-event dynamic systems* (DEDS). The main difference is that DESS do not evolve over time, while DEDS do. The PERT network is a typical example of a DESS, with the sample performance being, e.g., the shortest path in the network. A queueing network, such as the Jackson network in Section 3.4.1, is an example of a DEDS, with the sample performance being, for example, the delay (waiting time of a customer) in the network. In this chapter we will deal mainly with DESS. For a comprehensive study of both DESS and DEDS, the reader is referred to [12], [17], and [21].

Because of their complexity, the performance evaluation of discrete-event systems is usually done by simulation, and it is often associated with the estimation of a response function $\ell(\mathbf{u}) = \mathbb{E}_{\mathbf{u}}[H(\mathbf{X})]$, where the distribution of the sample performance $H(\mathbf{X})$ depends on the control or reference parameter $\mathbf{u} \in \mathscr{V}$. *Sensitivity*

Simulation and the Monte Carlo Method, Third Edition. By R. Y. Rubinstein and D. P. Kroese
Copyright © 2017 John Wiley & Sons, Inc. Published 2017 by John Wiley & Sons, Inc.

analysis is concerned with evaluating sensitivities (gradients, Hessians, etc.) of the response function $\ell(\mathbf{u})$ with respect to parameter vector $\mathbf{u}$, and is based on the score function and Fisher information. It provides guidance for design and operational decisions and plays an important role in selecting system parameters that optimize certain performance measures.

To illustrate, consider the following examples:

1. **Stochastic networks.** Sensitivity analysis can be employed to minimize the mean shortest path in the network with respect, say, to network link parameters, subject to certain constraints. PERT networks and flow networks are common examples. In the former, input and output variables may represent activity durations and minimum project duration, respectively. In the latter, they may represent flow capacities and maximal flow capacities.

2. **Traffic light systems.** The performance measure might be a vehicle's average delay as it proceeds from a given origin to a given destination or the average number of vehicles waiting for a green light at a given intersection. The sensitivity and decision parameters might be the average rate at which vehicles arrive at intersections and the rate of light changes from green to red. Some performance issues of interest are:

   (a) What will the vehicle's average delay be if the interarrival rate at a given intersection increases (decreases), say, by 10–50%? What would be the corresponding impact of adding one or more traffic lights to the system?

   (b) Which parameters are most significant in causing bottlenecks (high congestion in the system), and how can these bottlenecks be prevented or removed most effectively?

   (c) How can the average delay in the system be minimized, subject to certain constraints?

We will distinguish between the so-called *distributional* sensitivity parameters and the *structural* ones. In the former case, we are interested in sensitivities of the expected performance

$$\ell(\mathbf{u}) = \mathbb{E}_{\mathbf{u}}[H(\mathbf{X})] = \int H(\mathbf{x}) f(\mathbf{x}; \mathbf{u}) \, d\mathbf{x} \tag{7.1}$$

with respect to the parameter vector $\mathbf{u}$ of the pdf $f(\mathbf{x}; \mathbf{u})$, while in the latter case, we are interested in sensitivities of the expected performance

$$\ell(\mathbf{u}) = \mathbb{E}[H(\mathbf{X}; \mathbf{u})] = \int H(\mathbf{x}; \mathbf{u}) f(\mathbf{x}) \, d\mathbf{x} \tag{7.2}$$

with respect to the parameter vector $\mathbf{u}$ in the sample performance $H(\mathbf{x}; \mathbf{u})$. As an example, consider a $GI/G/1$ queue. In the first case, $\mathbf{u}$ might be the vector of the inter-arrival and service rates, and in the second case, $\mathbf{u}$ might be the buffer size. Note that often the parameter vector $\mathbf{u}$ includes both the distributional and structural parameters. In such a case we will use the following notation:

$$\ell(\mathbf{u}) = \mathbb{E}_{\mathbf{u}_2}[H(\mathbf{X}; \mathbf{u}_1)] = \int H(\mathbf{x}; \mathbf{u}_1) f(\mathbf{x}; \mathbf{u}_2) \, d\mathbf{x} \,, \tag{7.3}$$

where $\mathbf{u} = (\mathbf{u}_1, \mathbf{u}_2)$. Note that $\ell(\mathbf{u})$ in (7.1) and (7.2) can be considered particular cases of $\ell(\mathbf{u})$ in (7.3), where the corresponding sizes of the vectors $\mathbf{u}_1$ or $\mathbf{u}_2$ equal 0.

■ **EXAMPLE 7.1**

Let $H(\mathbf{X}; u_3, u_4) = \max\{X_1 + u_3, X_2 + u_4\}$, where $\mathbf{X} = (X_1, X_2)$ is a two-dimensional vector with independent components and $X_i \sim f_i(X; u_i)$, $i = 1, 2$. In this example u_1 and u_2 are distributional parameters, and u_3 and u_4 are structural ones.

Consider the following minimization problem using representation (7.3):

$$
\begin{array}{lll}
\text{minimize} & \ell_0(\mathbf{u}) = \mathbb{E}_{\mathbf{u}_1}[H_0(\mathbf{X}; \mathbf{u}_2)], & \mathbf{u} \in \mathscr{V}, \\
(\text{P}_0) \quad \text{subject to}: & \ell_j(\mathbf{u}) = \mathbb{E}_{\mathbf{u}_1}[H_j(\mathbf{X}; \mathbf{u}_2)] \leqslant 0, & j = 1, \dots, k, \\
& \ell_j(\mathbf{u}) = \mathbb{E}_{\mathbf{u}_1}[H_j(\mathbf{X}; \mathbf{u}_2)] = 0, & j = k+1, \dots, M,
\end{array}
\tag{7.4}
$$

where $H_j(\mathbf{X})$ is the j-th sample performance, driven by an input vector $\mathbf{X} \in \mathbb{R}^n$ with pdf $f(\mathbf{x}; \mathbf{u}_1)$, and $\mathbf{u} = (\mathbf{u}_1, \mathbf{u}_2)$ is a decision parameter vector belonging to some parameter set $\mathscr{V} \subset \mathbb{R}^m$.

When the objective function $\ell_0(\mathbf{u})$ and the constraint functions $\ell_j(\mathbf{u})$ are available analytically, (P$_0$) becomes a standard nonlinear programming problem, which can be solved either analytically or numerically by standard nonlinear programming techniques. For example, the Markovian queueing system optimization falls within this domain. Here, however, it will be assumed that the objective function and some of the constraint functions in (P$_0$) are not available analytically (typically due to the complexity of the underlying system), so that one must resort to stochastic optimization methods, particularly Monte Carlo optimization.

The rest of this chapter is organized as follows: Section 7.2 deals with sensitivity analysis of DESS with respect to the distributional parameters. Here we introduce the celebrated *score function* (SF) method. Section 7.3 deals with simulation-based optimization for programs of type (P$_0$) when the expected values $\mathbb{E}_{\mathbf{u}_1}[H_j(\mathbf{X}, \mathbf{u}_2)]$ are replaced by their corresponding sample means. The simulation-based version of (P$_0$) is called the *stochastic counterpart* of the original program (P$_0$). The main emphasis will be placed on the stochastic counterpart of the unconstrained program (P$_0$). Here we show how the stochastic counterpart method can approximate quite efficiently the true unknown optimal solution of the program (P$_0$) using a single simulation. Our results are based on [16, 18, 19], where theoretical foundations of the stochastic counterpart method are established. It is interesting to note that Geyer and Thompson [3] independently discovered the stochastic counterpart method in 1995. They used it to make statistical inference for a particular unconstrained setting of the general program (P$_0$). Section 7.4 presents an introduction to sensitivity analysis and simulation-based optimization of DEDS. Particular emphasis is placed on sensitivity analysis with respect to the distributional parameters of Markov chains using the dynamic version of the SF method. For a comprehensive study of sensitivity analysis and optimization of DEDS, including different types of queueing and inventory models, the reader is referred to [17].

7.2   SCORE FUNCTION METHOD FOR SENSITIVITY ANALYSIS OF DESS

In this section we introduce the celebrated *score function (SF) method* for sensitivity analysis of DESS. The goal of the SF method is to estimate the gradient and higher derivatives of $\ell(\mathbf{u})$ with respect to the distributional parameter vector $\mathbf{u}$, where the expected performance is given (see (7.1)) by

$$\ell(\mathbf{u}) = \mathbb{E}_{\mathbf{u}}[H(\mathbf{X})] ,$$

with $\mathbf{X} \sim f(\mathbf{x}; \mathbf{u})$. As we will see below, the SF approach permits the estimation of *all* sensitivities (gradients, Hessians, etc.) from a *single simulation run* (experiment) for a DESS with tens and quite often with hundreds of parameters. We closely follow [17].

Consider first the case where $\mathbf{u}$ is scalar (denoted therefore u instead of $\mathbf{u}$) and assume that the parameter set $\mathscr{V}$ is an open interval on the real line. Suppose that for all $\mathbf{x}$ the pdf $f(\mathbf{x}; u)$ is continuously differentiable in u and that there exists an integrable function $h(\mathbf{x})$ such that

$$\left| H(\mathbf{x}) \frac{\mathrm{d}f(\mathbf{x}; u)}{\mathrm{d}u} \right| \leqslant h(\mathbf{x}) \tag{7.5}$$

for all $u \in \mathscr{V}$. Then under mild conditions [19] the differentiation and expectation (integration) operators are interchangeable, so that differentiation of $\ell(u)$ yields

$$
\begin{aligned}
\frac{\mathrm{d}\ell(u)}{\mathrm{d}u} &= \frac{\mathrm{d}}{\mathrm{d}u} \int H(\mathbf{x}) f(\mathbf{x}; u) \mathrm{d}\mathbf{x} = \int H(\mathbf{x}) \frac{\mathrm{d}f(\mathbf{x}; u)}{\mathrm{d}u} \, \mathrm{d}\mathbf{x} \\
&= \int H(\mathbf{x}) \frac{\frac{\mathrm{d}f(\mathbf{x}; u)}{\mathrm{d}u}}{f(\mathbf{x}; u)} f(\mathbf{x}; u) \, \mathrm{d}\mathbf{x} = \mathbb{E}_u \left[H(\mathbf{X}) \frac{\mathrm{d} \ln f(\mathbf{X}; u)}{\mathrm{d}u} \right] \\
&= \mathbb{E}_u \left[H(\mathbf{X}) \, \mathcal{S}(u; \mathbf{X}) \right],
\end{aligned}
$$

where

$$\mathcal{S}(u; \mathbf{x}) = \frac{\mathrm{d} \ln f(\mathbf{x}; u)}{\mathrm{d}u}$$

is the *score function* (SF); see also (1.57). This is viewed as a function of u for a given $\mathbf{x}$.

Consider next the multidimensional case. Similar arguments allow us to represent the gradient and the higher order derivatives of $\ell(\mathbf{u})$ in the form

$$\nabla^k \ell(\mathbf{u}) = \mathbb{E}_{\mathbf{u}} \left[H(\mathbf{X}) \, \mathcal{S}^{(k)}(\mathbf{u}; \mathbf{X}) \right], \tag{7.6}$$

where

$$\mathcal{S}^{(k)}(\mathbf{u}; \mathbf{x}) = \frac{\nabla^k f(\mathbf{x}; \mathbf{u})}{f(\mathbf{x}; \mathbf{u})} \tag{7.7}$$

is the *k-th order score function*, $k = 0, 1, 2, \ldots$. In particular, $\mathcal{S}^{(0)}(\mathbf{u}; \mathbf{x}) = 1$ (by definition), $\mathcal{S}^{(1)}(\mathbf{u}; \mathbf{x}) = \mathcal{S}(\mathbf{u}; \mathbf{x}) = \nabla \ln f(\mathbf{x}; \mathbf{u})$, and $\mathcal{S}^{(2)}(\mathbf{u}; \mathbf{x})$ can be represented as

$$
\begin{aligned}
\mathcal{S}^{(2)}(\mathbf{u}; \mathbf{x}) &= \nabla \mathcal{S}(\mathbf{u}; \mathbf{x}) + \mathcal{S}(\mathbf{u}; \mathbf{x}) \, \mathcal{S}(\mathbf{u}; \mathbf{x})^\top \\
&= \nabla^2 \ln f(\mathbf{x}; \mathbf{u}) + \nabla \ln f(\mathbf{x}; \mathbf{u}) \, \nabla \ln f(\mathbf{x}; \mathbf{u})^\top,
\end{aligned}
\tag{7.8}
$$

where $\nabla \ln f(\mathbf{x}; \mathbf{u})^{\top}$ represents that transpose of the column vector $\nabla \ln f(\mathbf{x}; \mathbf{u})$ of partial derivatives of $\ln f(\mathbf{x}; \mathbf{u})$. Note that all partial derivatives are taken with respect to the components of the parameter vector $\mathbf{u}$.

Table 7.1 displays the score functions $\mathcal{S}(\mathbf{u}; x)$ calculated from (7.6) for the commonly used distributions given in Table A.1 in the Appendix. We take $\mathbf{u}$ to be the usual parameters for each distribution. For example, for the $\mathsf{Gamma}(\alpha, \lambda)$ and $\mathsf{N}(\mu, \sigma^2)$ distributions, we take $\mathbf{u} = (\alpha, \lambda)$ and $\mathbf{u} = (\mu, \sigma)$, respectively.

Table 7.1: Score functions for commonly used distributions.

Distribution	$f(x; \mathbf{u})$	$\mathcal{S}(\mathbf{u}; x)$
$\mathsf{Exp}(\lambda)$	$\lambda e^{-\lambda x}$	$\lambda^{-1} - x$
$\mathsf{Gamma}(\alpha, \lambda)$	$\dfrac{\lambda^{\alpha} x^{\alpha-1} e^{-\lambda x}}{\Gamma(\alpha)}$	$\left(\ln(\lambda x) - \frac{\Gamma'(\alpha)}{\Gamma(\alpha)},\ \alpha\lambda^{-1} - x \right)$
$\mathsf{N}(\mu, \sigma^2)$	$\dfrac{1}{\sigma\sqrt{2\pi}} e^{-\frac{1}{2}\left(\frac{x-\mu}{\sigma}\right)^2}$	$\left(\sigma^{-2}(x - \mu),\ -\sigma^{-1} + \sigma^{-3}(x - \mu)^2 \right)$
$\mathsf{Weib}(\alpha, \lambda)$	$\alpha\lambda (\lambda x)^{\alpha-1} e^{-(\lambda x)^{\alpha}}$	$\left(\alpha^{-1} + \ln(\lambda x)[1 - (\lambda x)^{\alpha}],\ \frac{\alpha}{\lambda}[1 - (\lambda x)^{\alpha}] \right)$
$\mathsf{Bin}(n, p)$	$\dbinom{n}{x} p^x (1-p)^{n-x}$	$\dfrac{x - np}{p(1 - p)}$
$\mathsf{Poi}(\lambda)$	$\dfrac{\lambda^x e^{-\lambda}}{x!}$	$\dfrac{x}{\lambda} - 1$
$\mathsf{G}(p)$	$p(1-p)^{x-1}$	$\dfrac{1 - px}{p(1 - p)}$

In general, the quantities $\nabla^k \ell(\mathbf{u})$, $k = 0, 1, \ldots$, are not available analytically, since the response $\ell(\mathbf{u})$ is not available. They can be estimated, however, via simulation as

$$\widehat{\nabla^k \ell}(\mathbf{u}) = \frac{1}{N} \sum_{i=1}^{N} H(\mathbf{X}_i)\, \mathcal{S}^{(k)}(\mathbf{u}; \mathbf{X}_i). \tag{7.9}$$

It is readily seen that the function $\ell(\mathbf{u})$ and *all* the sensitivities $\nabla^k \ell(\mathbf{u})$ can be estimated from a single simulation, since in (7.6) all of them are expressed as expectations with respect to the same pdf, $f(\mathbf{x}; \mathbf{u})$.

The following two toy examples provide more details on the estimation of $\nabla \ell(\mathbf{u})$. Both examples are only for illustration, since $\nabla^k \ell(\mathbf{u})$ is available analytically.

■ **EXAMPLE 7.2**

Let $H(\mathbf{X}) = X$, with $X \sim \mathsf{Ber}(p = u)$, where $u \in [0, 1]$. Using Table 7.1 for the $\mathsf{Bin}(1, p)$ distribution, we easily find that the estimator of $\nabla \ell(u)$ is

$$\widehat{\nabla \ell}(u) = \frac{1}{N} \sum_{i=1}^{N} X_i\, \frac{X_i - u}{u(1 - u)} = \frac{1}{u\,N} \sum_{i=1}^{N} X_i \approx 1, \tag{7.10}$$

where $X_1, \dots, X_N$ is a random sample from $\mathsf{Ber}(u)$. In the second equation we use the fact that $X_i^2 = X_i$. The approximation sign in (7.10) follows from the law of large numbers.

Suppose that $u = \frac{1}{2}$. Suppose also that we took a sample of size $N = 20$ from $\mathsf{Ber}(\frac{1}{2})$ and obtained the following values:

$$\{x_1, \dots, x_{20}\} = \{0, 1, 0, 0, 1, 0, 0, 1, 1, 1, 0, 1, 0, 1, 1, 0, 1, 0, 1, 1\}.$$

From (7.10) we see that the sample derivative is $\widehat{\nabla \ell}(\frac{1}{2}) = 1.1$, while the true one is clearly $\nabla \ell(\frac{1}{2}) = 1$.

■ **EXAMPLE 7.3**

Let $H(\mathbf{X}) = X$, with $X \sim \mathsf{Exp}(\lambda = u)$. This is also a toy example, since $\nabla \ell(u) = -1/u^2$. We see from Table 7.1 that $\mathcal{S}(u; x) = u^{-1} - x$, and therefore

$$\widehat{\nabla \ell}(u) = \frac{1}{N} \sum_{i=1}^{N} X_i \left(u^{-1} - X_i \right) \approx -\frac{1}{u^2} \tag{7.11}$$

is an estimator of $\nabla \ell(u)$, where $X_1, \dots, X_N$ is a random sample from $\mathsf{Exp}(u)$.

■ **EXAMPLE 7.4   Example 7.1 (Continued)**

As before, let $H(\mathbf{X}; u_3, u_4) = \max\{X_1 + u_3, X_2 + u_4\}$, where $\mathbf{X} = (X_1, X_2)$ is a two-dimensional vector with independent components and $X_i \sim f_i(X, u_i)$, $i = 1, 2$. Suppose we are interested in estimating $\nabla \ell(\mathbf{u}_1)$ with respect to the distributional parameter vector $\mathbf{u}_1 = (u_1, u_2)$. We have

$$\widehat{\nabla \ell}(\mathbf{u}_1) = \frac{1}{N} \sum_{i=1}^{N} H(\mathbf{X}_i; u_3, u_4) \, \mathcal{S}(\mathbf{u}_1; \mathbf{X}_i),$$

where $\mathcal{S}(\mathbf{u}_1; \mathbf{X}_i)$ is the column vector $(\mathcal{S}(u_1; X_{1i}), \ \mathcal{S}(u_2; X_{2i}))^{\top}$.

Next, we will apply the importance sampling technique to estimate the sensitivities $\nabla^k \ell(\mathbf{u}) = \mathbb{E}_{\mathbf{u}}[H(\mathbf{X}) \, \mathcal{S}^{(k)}(\mathbf{u}; \mathbf{X})]$ simultaneously for several values of $\mathbf{u}$. To this end, let $g(\mathbf{x})$ be the importance sampling density, and assume, as usual, that the support of $g(\mathbf{x})$ contains the support of $H(\mathbf{x}) f(\mathbf{x}; \mathbf{u})$ for all $\mathbf{u} \in \mathcal{V}$. Then $\nabla^k \ell(\mathbf{u})$ can be written as

$$\nabla^k \ell(\mathbf{u}) = \mathbb{E}_g[H(\mathbf{X}) \, \mathcal{S}^{(k)}(\mathbf{u}; \mathbf{X}) \, W(\mathbf{X}; \mathbf{u})], \tag{7.12}$$

where

$$W(\mathbf{x}; \mathbf{u}) = \frac{f(\mathbf{x}; \mathbf{u})}{g(\mathbf{x})} \tag{7.13}$$

is the likelihood ratio of $f(\mathbf{x}; \mathbf{u})$ and $g(\mathbf{x})$. The likelihood ratio estimator of $\nabla^k \ell(\mathbf{u})$ can be written as

$$\widehat{\nabla^k \ell}(\mathbf{u}) = \frac{1}{N} \sum_{i=1}^{N} H(\mathbf{X}_i) \, \mathcal{S}^{(k)}(\mathbf{u}; \mathbf{X}_i) \, W(\mathbf{X}_i; \mathbf{u}), \tag{7.14}$$

where $\mathbf{X}_1, \ldots, \mathbf{X}_N$ is a random sample from $g(\mathbf{x})$. Note that $\widehat{\nabla^k \ell}(\mathbf{u})$ is an unbiased estimator of $\nabla^k \ell(\mathbf{u})$ for *all* $\mathbf{u}$. This means that, by varying $\mathbf{u}$ and keeping g fixed, we can, in principle, estimate unbiasedly the whole *response surface* $\{\nabla^k \ell(\mathbf{u}), \mathbf{u} \in \mathcal{V}\}$ from a *single simulation*. Often the importance sampling distribution is chosen in the *same* class of distributions as the original one. That is, $g(\mathbf{x}) = f(\mathbf{x}; \mathbf{v})$ for some $\mathbf{v} \in \mathcal{V}$. If not stated otherwise, we assume from now on that $g(\mathbf{x}) = f(\mathbf{x}; \mathbf{v})$; that is, we assume that the importance sampling pdf lies in the same parametric family as the original pdf $f(\mathbf{x}; \mathbf{u})$. If we denote the likelihood ratio estimator of $\ell(\mathbf{u})$ for a given $\mathbf{v}$ by $\widehat{\ell}(\mathbf{u}; \mathbf{v})$,

$$\widehat{\ell}(\mathbf{u}; \mathbf{v}) = \frac{1}{N} \sum_{i=1}^{N} H(\mathbf{X}_i) W(\mathbf{X}_i; \mathbf{u}, \mathbf{v}) , \tag{7.15}$$

with $W(\mathbf{x}; \mathbf{u}, \mathbf{v}) = f(\mathbf{x}; \mathbf{u})/f(\mathbf{x}; \mathbf{v})$, and likewise denote the estimators in (7.14) by $\widehat{\nabla^k \ell}(\mathbf{u}; \mathbf{v})$, then (see Problem 7.4)

$$\widehat{\nabla^k \ell}(\mathbf{u}; \mathbf{v}) = \nabla^k \widehat{\ell}(\mathbf{u}; \mathbf{v}) = \frac{1}{N} \sum_{i=1}^{N} H(\mathbf{X}_i) \, \mathcal{S}^{(k)}(\mathbf{u}; \mathbf{X}_i) \, W(\mathbf{X}_i; \mathbf{u}, \mathbf{v}) . \tag{7.16}$$

Thus the estimators of sensitivities are simply the sensitivities of the estimators.

Next, we apply importance sampling to the two toy examples 7.2 and 7.3, and show how to estimate $\nabla^k \ell(\mathbf{u})$ simultaneously for different values of $\mathbf{u}$ using a single simulation from the importance sampling pdf $f(\mathbf{x}; \mathbf{v})$.

■ **EXAMPLE 7.5 Example 7.2 (Continued)**

Consider again the Bernoulli toy example, with $H(\mathbf{X}) = X$ and $X \sim \mathsf{Ber}(u)$. Suppose that the importance sampling distribution is $\mathsf{Ber}(v)$, that is,

$$g(x) = f(x; v) = v^x (1-v)^{1-x} , \quad x = 0, 1 .$$

Using importance sampling, we can write $\nabla^k \ell(u)$ as

$$\nabla^k \ell(u) = \mathbb{E}_v \left[X \frac{u^X (1-u)^{1-X}}{v^X (1-v)^{1-X}} \, \mathcal{S}^{(k)}(u; X) \right] ,$$

where $X \sim \mathsf{Ber}(v)$. Recall that for $\mathsf{Bin}(1, u)$ we have $\mathcal{S}(u; x) = \frac{x-u}{u(1-u)}$. The corresponding likelihood ratio estimator of $\nabla^k \ell(u)$ is

$$\begin{aligned} \widehat{\nabla^k \ell}(u; v) &= \frac{1}{N} \sum_{i=1}^{N} X_i \frac{u^{X_i} (1-u)^{1-X_i}}{v^{X_i} (1-v)^{1-X_i}} \, \mathcal{S}^{(k)}(u; X_i) \\ &= \frac{u}{v} \frac{1}{N} \sum_{i=1}^{N} X_i \, \mathcal{S}^{(k)}(u; X_i) , \end{aligned} \tag{7.17}$$

where $X_1, \ldots, X_N$ is a random sample from $\mathsf{Ber}(v)$. In the second equation we have used the fact that X_i is either 0 or 1. For $k = 0$ we readily obtain

$$\widehat{\ell}(u; v) = \frac{u}{v} \frac{1}{N} \sum_{i=1}^{N} X_i ,$$

which also follows directly from (7.15), and for $k = 1$ we have

$$\widehat{\nabla \ell}(u; v) = \frac{u}{v} \frac{1}{N} \sum_{i=1}^{N} X_i \frac{X_i - u}{u(1-u)} = \frac{1}{v} \frac{1}{N} \sum_{i=1}^{N} X_i , \tag{7.18}$$

which is the derivative of $\widehat{\ell}(u; v)$, as observed in (7.16). Note that in the special case where $v = u$, the likelihood ratio estimators $\widehat{\ell}(u; u)$ and $\widehat{\nabla \ell}(u; u)$ reduce to the CMC estimator $\frac{1}{N} \sum_{i=1}^{N} X_i$ (sample mean) and the earlier derived score function estimator (7.10), respectively. As a simple illustration, we take a sample from $\mathsf{Ber}(v = 1/2)$ of size $N = 20$ and obtain

$$\{x_1, \ldots, x_{20}\} = \{0, 1, 0, 0, 1, 0, 0, 1, 1, 1, 0, 1, 0, 1, 1, 0, 1, 0, 1, 1\} \,.$$

Suppose that, using this sample, we wish to estimate the quantities $\ell(u) = \mathbb{E}_u[X]$ and $\nabla \ell(u)$ *simultaneously* for $u = 1/4$ and $u = 1/10$. We readily obtain

$$\widehat{\ell}(u = 1/4; v = 1/2) = \frac{1/4}{1/2} \frac{11}{20} = \frac{11}{40} \,,$$

$$\widehat{\ell}(u = 1/10; v = 1/2) = \frac{1/10}{1/2} \frac{11}{20} = \frac{11}{100} \,,$$

and $\widehat{\nabla \ell}(u; v) = 11/10$ for both $u = 1/4$ and $1/10$.

■ **EXAMPLE 7.6**   **Example 7.3 (Continued)**

Let us consider the estimation of $\nabla^k \ell(u)$ simultaneously for several values of u in the second toy example, where $H(X) = X$ and $X \sim \mathsf{Exp}(u)$. Selecting the importance sampling distribution as

$$g(x) = f(x; v) = v \, e^{-vx}, \quad x > 0$$

for some $v > 0$, and using (7.14), we can express $\nabla^k \ell(u)$ as

$$\nabla^k \ell(u) = \mathbb{E}_v \left[X \frac{u \, e^{-uX}}{v \, e^{-vX}} S^{(k)}(u; X) \right] \,,$$

where $X \sim \mathsf{Exp}(v)$ and (see Table 7.1) $S(u; x) = \frac{1 - ux}{u}$. The sample average estimator of $\nabla^k \ell(u)$ (see (7.14)) is

$$\widehat{\nabla^k \ell}(u; v) = \frac{1}{N} \sum_{i=1}^{N} X_i \frac{u \, e^{-uX_i}}{v \, e^{-vX_i}} S^{(k)}(u; X_i) \,, \tag{7.19}$$

where $X_1, \ldots, X_N$ is a random sample from $\mathsf{Exp}(v)$. For $k = 0$, we have

$$\widehat{\ell}(u; v) = \frac{1}{N} \sum_{i=1}^{N} X_i \frac{u \, e^{-uX_i}}{v \, e^{-vX_i}} \approx \frac{1}{u} \,,$$

and for $k = 1$ we obtain

$$\widehat{\nabla \ell}(u; v) = \frac{1}{N} \sum_{i=1}^{N} X_i \frac{u \, e^{-uX_i}}{v \, e^{-vX_i}} \frac{1 - uX_i}{u} \approx -\frac{1}{u^2} \,, \tag{7.20}$$

which is the derivative of $\widehat{\ell}(u; v)$, as observed in (7.16). Note that in the particular case where $v = u$, the importance sampling estimators, $\widehat{\ell}(u; u)$ and $\widehat{\nabla \ell}(u; u)$, reduce to the sample mean (CMC estimator) and the SF estimator (7.11), respectively.

For a given importance sampling pdf $f(\mathbf{x}; \mathbf{v})$, the algorithm for estimating the sensitivities $\nabla^k \ell(\mathbf{u})$, $k = 0, 1, \ldots$, for *multiple values* $\mathbf{u}$ *from a single simulation run* is given next.

Algorithm 7.2.1: Score Function Method with Importance Sampling

---

1 Generate a sample $\mathbf{X}_1, \ldots, \mathbf{X}_N$ from the importance sampling pdf $f(\mathbf{x}; \mathbf{v})$, which must be chosen in advance.

2 Calculate the sample performance $H(\mathbf{X}_i)$ and the scores $\mathcal{S}^{(k)}(\mathbf{u}; \mathbf{X}_i)$, $i = 1, \ldots, N$, for the desired parameter value(s) $\mathbf{u}$.

3 Calculate $\widehat{\nabla^k \ell}(\mathbf{u}; \mathbf{v})$ according to (7.16).

From Algorithm 7.2.1 it follows that in order to estimate the sensitivities $\nabla^k \ell(\mathbf{u})$, $k = 1, 2, \ldots$, all we need is to apply formula (7.16), which involves calculation of the performance $H(\mathbf{X}_i)$ and estimation of the scores $\mathcal{S}^{(k)}(\mathbf{u}; \mathbf{X}_i)$ based on a sample $\mathbf{X}_1, \ldots, \mathbf{X}_N$ obtained from the importance sampling pdf $f(\mathbf{x}; \mathbf{v})$.

Confidence regions for $\nabla^k \ell(\mathbf{u})$ can be obtained by standard statistical techniques. In particular (e.g., see [19] and Section 1.11), $N^{1/2} \left[\widehat{\nabla^k \ell}(\mathbf{u}; \mathbf{v}) - \nabla^k \ell(\mathbf{u}) \right]$ converges to a multivariate normal random vector with mean zero and covariance matrix

$$\mathrm{Cov}_{\mathbf{v}}(H \, \mathcal{S}^{(k)} \, W) = \mathbb{E}_{\mathbf{v}} \left[H^2 \, W^2 \, \mathcal{S}^{(k)} \mathcal{S}^{(k) \, T} \right] - [\nabla^k \ell(\mathbf{u})][\nabla^k \ell(\mathbf{u})]^\top, \qquad (7.21)$$

using the abbreviations $H = H(\mathbf{X})$, $\mathcal{S}^{(k)} = \mathcal{S}^{(k)}(\mathbf{u}; \mathbf{x})$ and $W = W(\mathbf{X}; \mathbf{u}, \mathbf{v})$. From this point on, we will use these abbreviations when convenient, abbreviating $\mathcal{S}^{(1)}$ further to $\mathcal{S}$.

In particular, in the case $k = 0$, the variance of $\widehat{\ell}(\mathbf{u}; \mathbf{v})$, under the importance sampling density $f(\mathbf{x}; \mathbf{v})$, can be written as

$$\mathrm{Var}\left(\widehat{\ell}(\mathbf{u}; \mathbf{v}) \right) = \mathbb{E}_{\mathbf{v}} \left[H^2 \, W^2 \right] - \ell^2(\mathbf{u}) . \qquad (7.22)$$

The crucial issue clearly is how to choose a good importance sampling pdf that ensures low-variance estimates of $\ell(\mathbf{u})$ and $\nabla \ell(\mathbf{u})$. As we will see, this is not a simple task. We start with the variance of $\widehat{\ell}(\mathbf{u}; \mathbf{v})$. We will show that for exponential families of the form (A.9) it can be derived explicitly. Specifically, with $\boldsymbol{\theta}$ taking the role of $\mathbf{u}$ and $\boldsymbol{\eta}$ the role of $\mathbf{v}$, we have

$$
\begin{aligned}
\mathbb{E}_{\boldsymbol{\eta}} \left[H^2(\mathbf{X}) W^2(\mathbf{X}; \boldsymbol{\theta}, \boldsymbol{\eta}) \right] &= \mathbb{E}_{\boldsymbol{\theta}} \left[H^2(\mathbf{X}) W(\mathbf{X}; \boldsymbol{\theta}, \boldsymbol{\eta}) \right] \\
&= \int H^2(\mathbf{x}) \frac{c(\boldsymbol{\theta})}{c(\boldsymbol{\eta})} \, \mathrm{e}^{(\boldsymbol{\theta} - \boldsymbol{\eta}) \cdot \mathbf{t}(\mathbf{x})} \, c(\boldsymbol{\theta}) \, \mathrm{e}^{\boldsymbol{\theta} \cdot \mathbf{t}(\mathbf{x})} \, h(\mathbf{x}) \, \mathrm{d}\mathbf{x} \\
&= \frac{c^2(\boldsymbol{\theta})}{c(\boldsymbol{\eta})} \int H^2(\mathbf{x}) \, \mathrm{e}^{(2\boldsymbol{\theta} - \boldsymbol{\eta}) \cdot \mathbf{t}(\mathbf{x})} \, h(\mathbf{x}) \, \mathrm{d}\mathbf{x} \\
&= \frac{c^2(\boldsymbol{\theta})}{c(\boldsymbol{\eta}) \, c(2\boldsymbol{\theta} - \boldsymbol{\eta})} \, \mathbb{E}_{2\boldsymbol{\theta} - \boldsymbol{\eta}} \left[H^2(\mathbf{X}) \right] \\
&= \mathbb{E}_{\boldsymbol{\eta}} \left[W^2(\mathbf{X}; \boldsymbol{\theta}, \boldsymbol{\eta}) \right] \, \mathbb{E}_{2\boldsymbol{\theta} - \boldsymbol{\eta}} \left[H^2(\mathbf{X}) \right] . \qquad (7.23)
\end{aligned}
$$

Note that $\mathbb{E}_{\boldsymbol{\eta}} \left[W^2(\mathbf{X}; \boldsymbol{\theta}, \boldsymbol{\eta}) \right] = \mathbb{E}_{\boldsymbol{\theta}} \left[W(\mathbf{X}; \boldsymbol{\theta}, \boldsymbol{\eta}) \right]$.

Table 7.2 shows the expectations $\mathbb{E}_v\left[W^2(X;u,v)\right]$ for common exponential families in Tables A.1 and 7.1. Note that in Table 7.2 only *one* parameter is changed, the one denoted by u and changed to v. The values of $\mathbb{E}_v\left[W^2(X;u,v)\right]$ were calculated via (A.9) and (7.23). In doing so, we first reparameterized the distribution in terms of (A.9), with $\theta = \psi(u)$ and $\eta = \psi(v)$, and then calculated

$$\mathbb{E}_\eta\left[W^2(X;\theta,\eta)\right] = \frac{c^2(\theta)}{c(\eta)\,c(2\theta - \eta)} \; . \tag{7.24}$$

Last, we substituted u and v back in order to obtain the desired $\mathbb{E}_v\left[W^2(X;u,v)\right]$.

Table 7.2: $\mathbb{E}_v[W^2]$ for commonly used distributions.

Distribution	$f(x;u)$	$\theta = \psi(u)$	$c(\theta)$	$\mathbb{E}_v[W^2(X;u,v)]$
Gamma(α,u)	$\dfrac{u^\alpha x^{\alpha-1}e^{-ux}}{\Gamma(\alpha)}$	$-u$	$\dfrac{(-\theta)^\alpha}{\Gamma(\alpha)}$	$\left(\dfrac{u^2}{v(2u-v)}\right)^\alpha$
N(u,σ^2)	$\dfrac{1}{\sigma\sqrt{2\pi}}e^{-\frac{1}{2}\left(\frac{x-u}{\sigma}\right)^2}$	$\dfrac{u}{\sigma^2}$	$\dfrac{e^{-\frac{1}{2}\theta^2\sigma^2}}{\sigma\sqrt{2\pi}}$	$e^{\left(\frac{u-v}{\sigma}\right)^2}$
Weib(α,u)	$\alpha u\,(ux)^{\alpha-1}e^{-(ux)^\alpha}$	$-u^\alpha$	$\alpha\,\theta$	$\dfrac{(u/v)^{2\alpha}}{2\,(u/v)^\alpha - 1}$
Bin(n,u)	$\dbinom{n}{x}u^x(1-u)^{n-x}$	$\ln\left(\dfrac{u}{1-u}\right)$	$(1+e^\theta)^{-n}$	$\left(\dfrac{u^2 - 2uv + v}{(1-u)v}\right)^n$
Poi(u)	$\dfrac{u^x e^{-u}}{x!}$	$\ln u$	e^{-e^θ}	$e^{\left(\frac{(u-v)^2}{v}\right)}$
G(u)	$u(1-u)^{x-1}$	$\ln(1-u)$	$1-e^\theta$	$\dfrac{u^2(v-1)}{v(u^2 - 2u + v)}$

Let us consider, first, the Gamma(α,u) pdf. It readily follows that in order for the estimator $\widehat{\ell}(u;v)$ to be meaningful $(\mathrm{Var}(\widehat{\ell}(u;v)) < \infty)$, one has to ensure that $2u - v > 0$, $(v < 2\,u)$; otherwise, W will "blow up" the variance of the importance sampling estimator $\widehat{\ell}(u;v)$. A more careful analysis [19] (see also Proposition A.4.2 in the Appendix) indicates that in this case v should be chosen smaller than u (instead of smaller than $2u$) because the optimal importance sampling pdf $f(x;v^*)$ has a "fatter" tail than the original pdf $f(x;u)$. A similar result holds for the estimators of $\nabla^k \ell(\mathbf{u})$ and for other exponential families.

Let us consider, next, the multidimensional case $\mathbf{X} = (X_1, \ldots, X_n)$. Assume for concreteness that the $\{X_i\}$ are independent and $X_i \sim \mathsf{Exp}(u_i)$. It is not difficult to derive (see Problem 7.3) that in this case

$$\mathbb{E}_\mathbf{v}[W^2] = \mathbb{E}_\mathbf{u}[W] = \prod_{k=1}^{n}\frac{1}{1-\delta_k^2} \; , \tag{7.25}$$

where $\delta_k = (u_k - v_k)/u_k$, $k = 1, \ldots, n$ is the *relative perturbation* in u_k. For the special case where δ_k does not depend on k, say, $\delta_k = \delta$, $k = 1, \ldots, n$, we obtain

$$\mathrm{Var}_v(HW) = (1 - \delta^2)^{-n}\,\mathbb{E}_{2u-v}\left[H^2\right] - \ell^2 \; . \tag{7.26}$$

It should be noted that for fixed δ (even with $v < 2u$, which corresponds to $\delta < 1$), the variance of HW increases exponentially in n. For small values of δ, the first term on the right-hand side of (7.26) can be approximated by

$$(1 - \delta^2)^{-n} = \exp\left\{-n \ln\left(1 - \delta^2\right)\right\} \approx \exp\left\{n\delta^2\right\} ,$$

since for small x, $\ln(1 + x) \approx x$. In fact, for the variance of HW to be manageably small, the value $n\delta^2$ should not be too large. That is, as n increases, δ^2 should satisfy

$$\delta^2 = \mathcal{O}(n^{-1}) . \tag{7.27}$$

As is shown in [19] an assumption similar to (7.27) applies for rather general distributions and, in particular, for the exponential family.

Formula (7.27) is associated with the so-called *trust region*, that is, the region where the likelihood ratio estimator $\widehat{\nabla^k \ell}(\mathbf{u}; \mathbf{v})$ can be trusted to give a reasonably good approximation of $\nabla^k \ell(\mathbf{u})$. As an illustration, consider the case where $u_i = u$, $v_i = v$ for all i, and $n = 100$. In this case, the estimator $\widehat{\ell}(u; v)$ performs reasonably well for δ not exceeding 0.1, that is, when the relative perturbation in u is within 10%. For larger relative perturbations, the term $\mathbb{E}_v[W^2]$ "blows up" the variance of the estimators. Similar results also hold for the derivatives of $\ell(u)$.

The (negative) results concerning the unstable behavior of the likelihood ratio W and the rapid decrease of the trust region with the dimensionality n (see (7.27)) do not leave much room for importance sampling to be used for estimating $\nabla^k \ell(\mathbf{u})$, $k \geqslant 0$, in high dimensions. For such problems we therefore suggest the use of the score function estimators given in (7.9) (those that do not contain the likelihood ratio term W) as estimators of the true $\nabla^k \ell(\mathbf{u})$. For low-dimensional problems, say $n \leqslant 10$, the importance sampling estimator (7.14) for $\nabla^k \ell(\mathbf{u})$ could still be used, provided that the trust region is properly chosen, such as when the relative perturbation δ from the original parameter vector $\mathbf{u}$ does not exceed 10–20%. Even in this case, in order to prevent the degeneration of the importance sampling estimates, it is crucial to choose the reference parameter vector $\mathbf{v}$ such that the associated importance sampling pdf $f(\mathbf{x}; \mathbf{v})$ has a "fatter" tail than the original pdf $f(\mathbf{x}; \mathbf{u})$; see further Section A.4 of the Appendix.

7.3   SIMULATION-BASED OPTIMIZATION OF DESS

For the optimization program (P_0) in (7.4), let us suppose that the objective function

$$\ell_0(\mathbf{u}) = \mathbb{E}_{\mathbf{u}_1}[H_0(\mathbf{X}; \mathbf{u}_2)]$$

and some of the constraint functions

$$\ell_j(\mathbf{u}) = \mathbb{E}_{\mathbf{u}_1}[H_j(\mathbf{X}; \mathbf{u}_2)]$$

are not available in analytical form, so that in order to solve (P_0) we must resort to simulation-based optimization. This approach involves using the sample average versions, $\widehat{\ell}_0(\mathbf{u})$ and $\widehat{\ell}_j(\mathbf{u})$ instead of $\ell_0(\mathbf{u})$ and $\ell_j(\mathbf{u})$, respectively. Recall that the parameter vector $\mathbf{u} = (\mathbf{u}_1, \mathbf{u}_2)$ can have distributional and structural components.

Let us consider, first, a general treatment of simulation-based programs of type (P_0), with an emphasis on how to estimate the optimal solution $\mathbf{u}^*$ of the program

(P_0) using a *single simulation* run. Assume that we are given a random sample $\mathbf{X}_1, \mathbf{X}_2, \ldots, \mathbf{X}_N$ from the pdf $f(\mathbf{x}; \mathbf{u}_1)$ and consider the following two cases.

Case A. Either of the following holds true:

1. It is too expensive to store long samples $\mathbf{X}_1, \mathbf{X}_2, \ldots, \mathbf{X}_N$ and the associated sequences $\{\widehat{\ell}_j(\mathbf{u})\}$.

2. The sample performance, $\widehat{\ell}_j(\mathbf{u})$, cannot be computed simultaneously for different values of $\mathbf{u}$. However, we are allowed to set the control vector, $\mathbf{u}$, at any desired value $\mathbf{u}^{(t)}$ and then compute the random variables $\widehat{\ell}_j(\mathbf{u}^{(t)})$ and (quite often) the associated derivatives (gradients) $\widehat{\nabla \ell}_j(\mathbf{u})$, at $\mathbf{u} = \mathbf{u}^{(t)}$.

Case B. Both of the following hold true:

1. It is easy to compute and store the whole sample, $\mathbf{X}_1, \mathbf{X}_2, \ldots, \mathbf{X}_N$.

2. Given a sample $\mathbf{X}_1, \mathbf{X}_2, \ldots, \mathbf{X}_N$, it is easy to compute the sample performance $\widehat{\ell}_j(\mathbf{u})$ for any desired value $\mathbf{u}$.

The main difference between Case A and Case B is that the former is associated with an *on-line* optimization approach, called *stochastic approximation*, while the latter is associated with an *off-line* optimization approach, called *stochastic counterpart optimization* or *sample average approximation*. For references on stochastic approximation and the stochastic counterpart method, we refer the reader to [11] and [19], respectively.

The following two subsections deal separately with the stochastic approximation and the stochastic counterpart methods.

7.3.1 Stochastic Approximation

Stochastic approximation originated with Robbins and Monro [14] and Kiefer and Wolfowitz [8]. Kiefer and Wolfowitz dealt with on-line minimization of *smooth convex* problems of the form

$$\min_{\mathbf{u}} \ell(\mathbf{u}), \quad \mathbf{u} \in \mathcal{V}, \tag{7.28}$$

and they assumed the feasible set $\mathcal{V}$ to be convex and that at any fixed-in-advance point $\mathbf{u} \in \mathcal{V}$ an estimate $\widehat{\nabla \ell}(\mathbf{u})$ of the true gradient $\nabla \ell(\mathbf{u})$ can be computed. Here we will apply their stochastic approximation in the context of simulation-based optimization.

The stochastic approximation method iterates in $\mathbf{u}$, by way of the following recursive formula:

$$\mathbf{u}^{(t+1)} = \Pi_{\mathcal{V}}(\mathbf{u}^{(t)} - \beta_t \widehat{\nabla \ell}(\mathbf{u}^{(t)})), \tag{7.29}$$

where $\beta_1, \beta_2, \ldots$ is a sequence of positive step sizes and $\Pi_{\mathcal{V}}$ denotes the projection onto the set $\mathcal{V}$, that is, $\Pi_{\mathcal{V}}(\mathbf{u})$ is the point in $\mathcal{V}$ closest to $\mathbf{u}$. The projection $\Pi_{\mathcal{V}}$ is needed in order to enforce feasibility of the generated points $\{\mathbf{u}^{(t)}\}$. If the problem is unconstrained, that is, the feasible set $\mathcal{V}$ coincides with the whole space, then this projection is the identity mapping and can be omitted from (7.29).

It is readily seen that (7.29) represents a *gradient descent procedure* in which the exact gradients are replaced by their estimates. Indeed, if the exact value $\nabla \ell(\mathbf{u}^{(t)})$ of the gradient were available, then $-\nabla \ell(\mathbf{u}^{(t)})$ would give the direction of steepest descent at the point $\mathbf{u}^{(t)}$. This would guarantee that if $\nabla \ell(\mathbf{u}^{(t)}) \neq 0$, then moving along this direction the value of the objective function decreases, that is, $\ell(\mathbf{u}^{(t)} - \beta \nabla \ell(\mathbf{u}^{(t)})) < \ell(\mathbf{u}^{(t)})$ for $\beta > 0$ small enough. The iterative procedure (7.29) mimics this idea by using estimates of the gradients instead of the actual ones. Note further that a new random sample $\mathbf{X}_1, \ldots, \mathbf{X}_N$ should be generated to calculate each $\widehat{\nabla \ell}(\mathbf{u}^{(t)})$, $t = 1, 2, \ldots$.

We will now present several alternative estimators $\widehat{\nabla \ell}(\mathbf{u})$ of $\nabla \ell(\mathbf{u})$ regarding the model in Example 7.1.

■ EXAMPLE 7.7   Example 7.1 (Continued)

As before, let $H(\mathbf{X}; u_3, u_4) = \max\{X_1 + u_3, X_2 + u_4\}$, where $\mathbf{X} = (X_1, X_2)$ is a two-dimensional vector with independent components, and $X_i \sim f_i(X; u_i)$, $i = 1, 2$. Here we are interested in estimating the four-dimensional vector $\nabla \ell(\mathbf{u})$, where $\ell(\mathbf{u}) = \mathbb{E}_{\mathbf{u}_1}[H(\mathbf{X}; \mathbf{u}_2)]$, $\mathbf{u} = (\mathbf{u}_1, \mathbf{u}_2) = (u_1, u_2, u_3, u_4)$, with respect to both the distributional parameter vector $\mathbf{u}_1 = (u_1, u_2)$, and the structural vector $\mathbf{u}_2 = (u_3, u_4)$.

We will consider three alternative estimators for $\nabla \ell(\mathbf{u})$. These estimators are called (a) *direct*, (b) *inverse-transform*, and (c) *push-out* estimators. More details on these estimators and their various applications are given in [17].

(a) *The direct estimator of* $\nabla \ell(\mathbf{u})$. We have

$$\ell(\mathbf{u}) = \mathbb{E}_{\mathbf{u}_1}[H(\mathbf{X}; \mathbf{u}_2)], \tag{7.30}$$

$$\frac{\partial \ell(\mathbf{u})}{\partial u_1} = \mathbb{E}_{\mathbf{u}_1}[H(\mathbf{X}; \mathbf{u}_2) \nabla \ln f_1(X_1; u_1)], \tag{7.31}$$

$$\frac{\partial \ell(\mathbf{u})}{\partial u_3} = \mathbb{E}_{\mathbf{u}_1}\left[\frac{\partial H(\mathbf{X}; \mathbf{u}_2)}{\partial u_3}\right], \tag{7.32}$$

and similarly for $\partial \ell(\mathbf{u})/\partial u_2$ and $\partial \ell(\mathbf{u})/\partial u_4$. Here

$$\frac{\partial H(\mathbf{X}; \mathbf{u}_2)}{\partial u_3} = \begin{cases} 1, & \text{if } X_1 + u_3 > X_2 + u_4, \\ 0, & \text{otherwise}, \end{cases} \tag{7.33}$$

and similarly for $\partial H(\mathbf{X}; \mathbf{u}_2)/\partial u_4$. The sample estimators of $\partial \ell(\mathbf{u})/\partial u_i$, $i = 1, \ldots, 4$, can be obtained *directly* from their expected-value counterparts — hence the name *direct estimators*. For example, the estimator of $\partial \ell(\mathbf{u})/\partial u_3$ can be written as

$$\widehat{\nabla \ell}_3^{(1)}(\mathbf{u}) = \frac{1}{N} \sum_{i=1}^{N} \frac{\partial H(\mathbf{X}_i; \mathbf{u}_2)}{\partial u_3}, \tag{7.34}$$

where $\mathbf{X}_1, \ldots, \mathbf{X}_N$ is a sample from $f(\mathbf{x}; \mathbf{u}_1) = f_1(x_1; u_1) f_2(x_2; u_2)$, and similarly for the remaining estimators $\widehat{\nabla \ell}_i^{(1)}(\mathbf{u})$ of $\partial \ell(\mathbf{u})/\partial u_i$, $i = 1, 2, 4$.

(b) *The inverse-transform estimator of* $\nabla \ell(\mathbf{u})$. Using the inverse transformations $X_i = F_i^{-1}(Z_i; u_i)$, where $Z_i \sim \mathsf{U}(0, 1)$, $i = 1, 2$, we can write $H(\mathbf{X}; \mathbf{u}_2)$ alternatively as

$$\breve{H}(\mathbf{Z}; \mathbf{u}) = \max\{F_1^{-1}(Z_1; u_1) + u_3, F_2^{-1}(Z_2; u_2) + u_4\},$$

where $\mathbf{Z} = (Z_1, Z_1)$. The expected performance $\ell(\mathbf{u})$ and the gradient $\nabla\ell(\mathbf{u})$ can be now written as

$$\ell(\mathbf{u}) = \mathbb{E}_{\mathsf{U}}[\breve{H}(\mathbf{Z}; \mathbf{u})]$$

and

$$\nabla\ell(\mathbf{u}) = \mathbb{E}_{\mathsf{U}}[\nabla\breve{H}(\mathbf{Z}; \mathbf{u})] \,,$$

respectively. Here U denotes the uniform distribution. It is readily seen that in the inverse-transform setting all four parameters u_1, u_2, u_3, u_4 become *structural* parameters. The estimator of $\nabla\ell(\mathbf{u})$ based on the inverse-transform method, denoted as $\widehat{\nabla\ell}^{(2)}(\mathbf{u})$, is therefore

$$\widehat{\nabla\ell}^{(2)}(\mathbf{u}) = \frac{1}{N}\sum_{i=1}^{N}\nabla\breve{H}(\mathbf{Z}_i; \mathbf{u}) \,, \tag{7.35}$$

where the partial derivatives of $\breve{H}(\mathbf{z}; \mathbf{u})$ can be obtained similarly to (7.33). Note that the first-order derivatives of $\breve{H}(\mathbf{z}; \mathbf{u})$ are piecewise-continuous functions with discontinuities at points for which $F_1^{-1}(z_2; u_1) + u_2 = F_2^{-1}(z_2; u_2) + u_4$.

(c) *The push-out estimator of* $\nabla\ell(\mathbf{u})$. Define the following two random variables: $\widetilde{X}_1 = X_1 + u_3$ and $\widetilde{X}_2 = X_2 + u_4$. Now, the original sample performance $H(\mathbf{X}; u_3, u_4) = \max\{X_1 + u_3, X_2 + u_4\}$ and the expected value $\ell(\mathbf{u})$ can be written as $\widetilde{H}(\widetilde{\mathbf{X}}) = \max\{\widetilde{X}_1, \widetilde{X}_2\}$ and as

$$\ell(\mathbf{u}) = \mathbb{E}_{\widetilde{f}}[\widetilde{H}(\widetilde{\mathbf{X}})] = \mathbb{E}_{\widetilde{f}}[\max\{\widetilde{X}_1, \widetilde{X}_2\}] \,, \tag{7.36}$$

respectively. Here $\widetilde{f}$ is the pdf of $\widetilde{\mathbf{X}}$; thus, $\widetilde{f}(\mathbf{x}; \mathbf{u}) = f_1(x_1 - u_3; u_1) f_2(x_2 - u_4; u_2) = \widetilde{f}_1(x; u_1, u_3) \widetilde{f}_2(x; u_2, u_4)$. In this case we say that the original structural parameters u_3 and u_4 in $H(\cdot)$ are "pushed out" into the pdf $\widetilde{f}$.

As an example, suppose that $X_j \sim \mathsf{Exp}(u_j)$, $j = 1, 2$. Then the cdf $\widetilde{F}_1(x)$ of $\widetilde{X}_1 = X_1 + u_3$ and the cdf $\widetilde{F}_2(x)$ of $\widetilde{X}_2 = X_2 + u_4$ can be written, respectively, as

$$\widetilde{F}_1(x) = P(\widetilde{X}_1 \leqslant x) = \mathbb{P}(X_1 \leqslant x - u_3) = F_1(x - u_3)$$

and

$$\widetilde{F}_2(x) = \mathbb{P}(\widetilde{X}_2 \leqslant x - u_4) = F_2(x - u_4) \,.$$

Clearly,

$$\widetilde{f}_1(x; u_1, u_3) = \begin{cases} u_1\,e^{-u_1(x-u_3)}, & x \geqslant u_3 \,, \\ 0 & \text{otherwise,} \end{cases} \tag{7.37}$$

and

$$\widetilde{f}_2(x; u_2, u_4) = \begin{cases} u_2\,e^{-u_2(x-u_4)}, & x \geqslant u_4 \\ 0 & \text{otherwise.} \end{cases} \tag{7.38}$$

Because in representation (7.36) all four parameters $u_1, \ldots, u_4$ are *distributional*, in order to estimate $\nabla^k\ell(\mathbf{u})$, we can apply the SF method. In particular, we can estimate the gradient

$$\nabla\ell(\mathbf{u}) = \mathbb{E}_{\widetilde{f}}[\max\{\widetilde{X}_1, \widetilde{X}_2\}\,\nabla\ln\widetilde{f}(\widetilde{\mathbf{X}}; \mathbf{u})]$$

as

$$\widehat{\nabla \ell}^{(3)}(\mathbf{u}) = \frac{1}{N} \sum_{i=1}^{N} \max\{\widetilde{X}_{1i}, \widetilde{X}_{2i}\} \nabla \ln \widetilde{f}(\widetilde{\mathbf{X}}_i; \mathbf{u}) . \tag{7.39}$$

Recall that $\partial H(\mathbf{X}; \mathbf{u}_1)/\partial u_1$ and $\partial H(\mathbf{X}; u_1)/\partial u_2$ are piecewise-constant functions (see (7.33)) and that $\partial^2 H(\mathbf{X}; \mathbf{u}_1)/\partial u_1^2$ and $\partial^2 H(\mathbf{X}; \mathbf{u}_1)/\partial u_2^2$ vanish almost everywhere. Consequently, the associated second-order derivatives cannot be interchanged with the expectation operator in (7.30). Yet, the transformed function $\ell(\mathbf{u})$ in (7.36) and its sample version $\widehat{\nabla \ell}^{(3)}(\mathbf{u})$ are both *differentiable* in $\mathbf{u}$ everywhere, provided that $f_1(x_1 - u_3; u_1)$ and $f_2(x_2 - u_4; u_2)$ are smooth. Thus, subject to smoothness of $\widetilde{f}(\mathbf{x}; \mathbf{u})$, the push-out technique *smoothes out* the original *non-smooth* performance measure $H(\cdot)$.

Let us return to stochastic optimization. As we will see in Theorem 7.3.1 below, starting from some fixed initial value $\mathbf{u}^{(1)}$, and under some reasonable assumptions, the sequence $\{\mathbf{u}^{(t)}\}$ converges asymptotically in t to the minimizer $\mathbf{u}^*$ of the objective function $\ell(\mathbf{u})$ over $\mathcal{V}$. Typically, in order to guarantee the convergence, the following two conditions are imposed on the step sizes: (a) $\sum_{t=1}^{\infty} \beta_t = \infty$, and (b) $\sum_{t=1}^{\infty} \beta_t^2 < \infty$. For example, take $\beta_t \equiv c/t$ for some $c > 0$. There are many theorems on the convergence and rate of convergence for stochastic optimization. One of the simplest, from [11], is presented next.

Theorem 7.3.1 *Assume that ℓ is smooth and strictly convex, that is,*

$$\ell(\mathbf{u} + \mathbf{h}) \geqslant \ell(\mathbf{u}) + [\nabla \ell(\mathbf{u})]^\top \mathbf{h} + \frac{\beta}{2} \mathbf{h}^\top \mathbf{h}, \quad \beta > 0 . \tag{7.40}$$

Assume further that the errors in the stochastic gradient vector $\widehat{\nabla \ell}(\mathbf{u})$ have a bounded second moment, that is,

$$\mathbb{E}\left[\|\widehat{\nabla \ell}(\mathbf{u}) - \nabla \ell(\mathbf{u})\|^2\right] \leqslant C^2 < \infty .$$

Then, for an arbitrary (deterministic) positive sequence $\{\beta_t\}$ such that

$$\sum_{t=1}^{\infty} \beta_t = \infty, \quad \sum_{t=1}^{\infty} \beta_t^2 < \infty ,$$

the vector $\mathbf{u}^{(t)}$ converges asymptotically to $\mathbf{u}^*$ (the minimizer of $\ell(\mathbf{u})$) in the sense of mean square. If, moreover,

$$\beta_t = c/t ,$$

with an appropriate constant c (whether a given c is appropriate depends only on β in (7.40)), then for all t we have the following bounds:

$$\mathbb{E}\left[\|\mathbf{u}^{(t)} - \mathbf{u}^*\|^2\right] \leqslant \frac{A(\beta, c)}{t} \|\mathbf{u}^{(1)} - \mathbf{u}^*\|^2$$

and

$$\mathbb{E}\left[\ell(\mathbf{u}^{(t)}) - \ell(\mathbf{u}^*)\right] \leqslant \mathcal{O}(1/t) ,$$

where $A(\beta, c)$ is some constant depending on β and c.

The attractive features of the stochastic approximation method are its simplicity and ease of implementation in those cases in which the projection $\Pi_{\mathscr{Y}}(\cdot)$ can be easily computed. However, it also has severe shortcomings. The crucial question in its implementation is the choice of the step sizes $\{\beta_t\}$. Small step sizes slow the progress toward the optimum, and large step sizes cause the iterations to "zigzag". Also a few wrong steps in the beginning of the procedure may require many iterations to correct. For instance, the algorithm is extremely sensitive to the choice of the constant c in the step size rule $\beta_t = c/t$. Therefore, various step size rules have been suggested in which the step sizes are chosen adaptively. An easy way to balance the need for both small and large step sizes, is to use a sequence $\{\beta_t\}$ of the form

$$\beta_t = \frac{a}{(t+1+A)^\alpha},$$

for $A \geqslant 0$ and $0 \leqslant a \leqslant 1$, as suggested in Spall [21, Section 4.4]. Another drawback of the stochastic approximation method is that it lacks good stopping criteria and often has difficulties handling even relatively simple linear constraints.

Last, we would like to stress that even in cases where the direct, push-out, or inverse-transform estimators are not applicable (e.g., when H or f are not known or are difficult to evaluate), it is still possible to obtain estimates of the gradient of $\ell(\mathbf{u})$ via simulation, by using difference estimators. In particular, the *central difference estimator* of the i-th component of $\nabla \ell(\mathbf{u})$, that is, $\partial \ell(\mathbf{u})/\partial u_i$, is given by

$$\frac{\widehat{\ell}(\mathbf{u} + \mathbf{e}_i \, \delta/2) - \widehat{\ell}(\mathbf{u} - \mathbf{e}_i \, \delta/2)}{\delta},$$

where $\mathbf{e}_i$ denotes the i-th unit vector, and $\widehat{\ell}(\mathbf{u} + \mathbf{e}_i \, \delta/2)$ and $\widehat{\ell}(\mathbf{u} - \mathbf{e}_i \, \delta/2)$ can be any estimators of $\ell(\mathbf{u} + \mathbf{e}_i \, \delta/2)$ and $\ell(\mathbf{u} - \mathbf{e}_i \, \delta/2)$, respectively. The difference parameter δ should be small enough to reduce the bias of the estimator (which is of order $\mathcal{O}(\delta^2)$, see [2, Page 209]), but large enough to keep the variance of the estimator small.

It is important to use *common random variables* (CRVs) in the implementation, as this reduces the variance of the difference estimator; see Section 5.2. For the case where $\ell(\mathbf{u}) = \mathbb{E}[h(\mathbf{U}; \mathbf{u})]$, with $\mathbf{U} \sim \mathsf{U}(0,1)^d$, for some function h, this leads to the following algorithm:

Algorithm 7.3.1: Central Difference Estimation with CRVs

---

1 Generate $\mathbf{U}_1, \ldots, \mathbf{U}_N \sim \mathsf{U}(0,1)^d$.
2 Let $L_k \leftarrow h(\mathbf{U}_k; \mathbf{u} - \mathbf{e}_i \, \delta/2)$ and $R_k \leftarrow h(\mathbf{U}_k; \mathbf{u} + \mathbf{e}_i \, \delta/2)$, $k = 1, \ldots, N$.
3 Compute the sample covariance matrix corresponding to the pairs $\{(L_k, R_k)\}$:

$$C = \begin{pmatrix} \frac{1}{N-1} \sum_{k=1}^{N} (L_k - \bar{L})^2 & \frac{1}{N-1} \sum_{k=1}^{N} (L_k - \bar{L})(R_k - \bar{R}) \\ \frac{1}{N-1} \sum_{k=1}^{N} (L_k - \bar{L})(R_k - \bar{R}) & \frac{1}{N-1} \sum_{k=1}^{N} (R_k - \bar{R})^2 \end{pmatrix}.$$

4 Estimate $\partial \ell(\mathbf{u})/\partial \theta_i$ via the central difference estimator $\dfrac{\bar{R} - \bar{L}}{\delta}$, with an estimated variance of

$$\mathrm{SE} = \frac{C_{1,1} + C_{2,2} - 2\,C_{1,2}}{\delta^2 \, N}.$$

7.3.2 Stochastic Counterpart Method

The underlying idea in the stochastic counterpart approach is to replace all the expected value functions in the deterministic program (P_0) by their sample average equivalents and then solve the latter by standard mathematical programming techniques. The resultant optimal solution provides an estimator of the corresponding true optimal solution of the original program (P_0).

If not stated otherwise, we will consider here the unconstrained program

$$\min_{\mathbf{u} \in \mathscr{V}} \ell(\mathbf{u}) = \min_{\mathbf{u} \in \mathscr{V}} \mathbb{E}_{\mathbf{u}_1}[H(\mathbf{X}; \mathbf{u}_2)] . \tag{7.41}$$

The general constrained program (P_0) is treated in [19].

Assume that $\mathscr{V}$ is an open set and that $\ell(\mathbf{u})$ is continuously differentiable on $\mathscr{V}$. Then, by the first-order necessary conditions, the gradient of $\ell(\mathbf{u})$ at an optimal solution point, $\mathbf{u}^*$, must vanish. Consequently, the optimal solution $\mathbf{u}^*$ can be found by solving the equation system

$$\nabla \ell(\mathbf{u}) = \mathbf{0} , \quad \mathbf{u} \in \mathscr{V} . \tag{7.42}$$

Using the importance sampling pdf $f(\mathbf{x}; \mathbf{v}_1)$, we can write the stochastic counterpart of (7.41) as

$$\min_{\mathbf{u} \in \mathscr{V}} \widehat{\ell}(\mathbf{u}; \mathbf{v}_1) = \min_{\mathbf{u} \in \mathscr{V}} \frac{1}{N} \sum_{i=1}^{N} H(\mathbf{X}_i; \mathbf{u}_2) W(\mathbf{X}_i; \mathbf{u}_1, \mathbf{v}_1) , \tag{7.43}$$

where $\mathbf{X}_1, \ldots, \mathbf{X}_N$ is a random sample from the importance sampling pdf $f(\mathbf{x}; \mathbf{v}_1)$ and

$$W(\mathbf{x}; \mathbf{u}_1, \mathbf{v}_1) = \frac{f(\mathbf{x}; \mathbf{u}_1)}{f(\mathbf{x}; \mathbf{v}_1)} .$$

Assuming further that $\widehat{\ell}(\mathbf{u}; \mathbf{v}_1)$ is continuously differentiable on $\mathscr{V}$, we find the optimal solution of (7.43) by solving the equation system

$$\widehat{\nabla \ell}(\mathbf{u}; \mathbf{v}_1) = \nabla \widehat{\ell}(\mathbf{u}; \mathbf{v}_1) = \mathbf{0} , \quad \mathbf{u} \in \mathscr{V} , \tag{7.44}$$

which itself may be viewed as a stochastic counterpart of the deterministic system (7.42). Thus we simply take the gradient of the likelihood ratio estimator $\widehat{\ell}(\mathbf{u}; \mathbf{v}_1)$ as an estimator for the gradient of ℓ at $\mathbf{u}$; see also (7.16).

Note that (7.44) can be written as

$$\nabla \widehat{\ell}(\mathbf{u}; \mathbf{v}_1) = \frac{1}{N} \sum_{i=1}^{N} \{H(\mathbf{X}_i; \mathbf{u}_2) \nabla \ln f(\mathbf{X}_i; \mathbf{u}_1) W(\mathbf{X}_i; \mathbf{u}_1, \mathbf{v}_1)$$
$$+ \nabla H(\mathbf{X}_i; \mathbf{u}_2) W(\mathbf{X}_i; \mathbf{u}_1, \mathbf{v}_1)\} = 0, \quad \mathbf{u} \in \mathscr{V}. \tag{7.45}$$

Recall that we can view the above problem as the sample average approximation of the true (or expected value) problem (7.41). The function $\widehat{\ell}(\mathbf{u}; \mathbf{v}_1)$ is random in the sense that it depends on the corresponding sample $\mathbf{X}_1, \ldots, \mathbf{X}_N$. However, note that once the sample is generated, $\widehat{\ell}(\mathbf{u}; \mathbf{v}_1)$ becomes a deterministic function whose values and derivatives can be computed for a given value of the argument $\mathbf{u}$. Consequently, the problem (7.43) becomes a deterministic optimization problem, and one can solve it with an appropriate deterministic optimization algorithm. For example, in the unconstrained case it can be solved by using, say, the steepest descent method, that is,

$$\mathbf{u}^{(t+1)} = \Pi_{\mathscr{V}}(\mathbf{u}^{(t)} - \beta_t \nabla \widehat{\ell}(\mathbf{u}^{(t)}; \mathbf{v}_1)) , \tag{7.46}$$

where the step size β_t is obtained by a line search, for example,

$$\beta_t \equiv \underset{\beta}{\operatorname{argmin}}\{\widehat{\ell}(\mathbf{u}^{(t)} - \beta \nabla \widehat{\ell}(\mathbf{u}^{(t)}; \mathbf{v}_1); \mathbf{v}_1)\} ,$$

and, as before, $\Pi_{\mathscr{V}}$ denotes the projection onto the set $\mathscr{V}$. Note that this procedure is different from the stochastic approximation method (7.29) in three respects:

1. The step sizes β_t are calculated by a line search instead of being defined a priori.

2. The same sample $\mathbf{X}_1, \ldots, \mathbf{X}_N$ is used for all $\nabla \widehat{\ell}(\mathbf{u}^{(t)}; \mathbf{v}_1)$.

3. Typically, a reasonably large sample size N is used in (7.46) as compared to stochastic optimization in (7.29).

Next, we consider the particular case of the program (7.41) where $\mathbf{u}$ is a distributional decision parameter vector, that is, we consider the following program:

$$\min_{\mathbf{u} \in \mathscr{V}} \ell(\mathbf{u}) = \min_{\mathbf{u} \in \mathscr{V}} \mathbb{E}_{\mathbf{u}}[H(\mathbf{X})] . \tag{7.47}$$

To estimate the optimal solution $\mathbf{u}^*$ of the program (7.47), we will use the score function method. In this case the program (7.43) reduces to

$$\min_{\mathbf{u} \in \mathscr{V}} \widehat{\ell}(\mathbf{u}; \mathbf{v}) = \min_{\mathbf{u} \in \mathscr{V}} \frac{1}{N} \sum_{i=1}^{N} H(\mathbf{X}_i) W(\mathbf{X}_i; \mathbf{u}, \mathbf{v}) , \tag{7.48}$$

where $\mathbf{X}_1, \ldots, \mathbf{X}_N$ is a random sample from the importance sampling pdf $f(\mathbf{x}; \mathbf{v})$, and

$$W(\mathbf{x}; \mathbf{u}, \mathbf{v}) = \frac{f(\mathbf{x}; \mathbf{u})}{f(\mathbf{x}; \mathbf{v})} .$$

By analogy to (7.45), we have

$$\frac{1}{N} \sum_{i=1}^{N} H(\mathbf{X}_i) \nabla \ln f(\mathbf{X}_i; \mathbf{u}) W(\mathbf{X}_i; \mathbf{u}, \mathbf{v}) = 0, \quad \mathbf{u} \in \mathscr{V}. \tag{7.49}$$

Remark 7.3.1 The size of the trust region is an important consideration when solving the stochastic counterpart (7.48) based on likelihood ratios. In particular, the following two requirements must hold:

(a) The reference parameter vector $\mathbf{v}$ must be chosen carefully so that that the importance sampling pdf $f(\mathbf{x}; \mathbf{v})$ has a "fatter" tail than the original pdf $f(\mathbf{x}; \mathbf{u})$ (see also Section A.4 of the appendix); otherwise, the likelihood ratio $W(\mathbf{x}; \mathbf{u}, \mathbf{v})$ is likely to degenerate.

(b) The trust region $\mathscr{V}$ of all possible values of $\mathbf{v}$ should be decided in advance so that $\mathscr{V}$ is not too wide. In particular, it should satisfy (7.27). If the region $\mathscr{V}$

is too wide, the likelihood ratio term $W(\mathbf{x}; \mathbf{u}, \mathbf{v})$ will "blow up" the variance of the estimate of (7.48). In this occurs, alternative methods (not involving likelihood ratio terms), like the steepest descent method (7.46), should be used. Common alternatives are the steepest descent and the stochastic approximation methods.

■ EXAMPLE 7.8 Stochastic Shortest Path

Consider again the stochastic shortest path problem in Example 5.15 on Page 158. Let the components $\{X_i\}$ be independent and $X_i \sim \mathsf{Exp}(u_i^{-1})$, $i = 1, \ldots, 5$, with $\mathbf{u} = (1, 2, u, 4, 5)$, for some $u > 0$. Our objective now is to solve the following program:

$$\min_{u \in \mathscr{V}} \ell(u) = \min_{u \in \mathscr{V}}[3.1 - \mathbb{E}_u[S(\mathbf{X})] + 0.1\,u]\,, \qquad (7.50)$$

where $\mathscr{V} = \{u : 1 \leqslant u \leqslant 4\}$, $S(\mathbf{X}) = \min\{X_1 + X_4,\ X_1 + X_3 + X_5,\ X_2 + X_3 + X_4,\ X_2 + X_5\}$ denotes the length of the shortest path. The function $\ell(u)$ is difficult to evaluate exactly but can be estimated easily via Monte Carlo simulation, yielding a "noisy" function $\widehat{\ell}(u)$. Figure 7.1 displays estimates and confidence intervals for various values of u, using for each estimate a sample size of $N = 50{,}000$. The minimizer, say u^*, seems to lie in the interval [1,2].

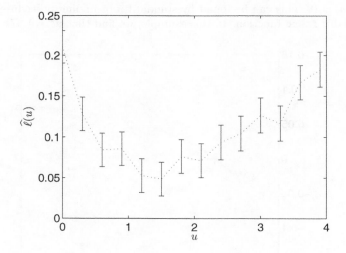

Figure 7.1: Minimize $\ell(u)$ with respect to u. Estimates and 95% confidence intervals indicate that the minimum is attained between 1 and 2.

To find u^* via the stochastic counterpart method, we can proceed as follows. First, the derivative of $\ell(u)$ satisfies

$$\nabla \ell(u) = \frac{1}{10} - \mathbb{E}_v[S(\mathbf{X})\,\mathcal{S}(u; \mathbf{X})\,W(\mathbf{x}; u, v)]$$

$$= \frac{1}{10} - \mathbb{E}_v\left[S(\mathbf{X})\frac{X_3 - u}{u^2}\frac{v}{u}\,\mathrm{e}^{-X_3(u^{-1} - v^{-1})}\right]\,,$$

where the score function $S(u; \mathbf{X})$ is given by

$$S(u; \mathbf{X}) = \frac{\partial}{\partial u} \ln \left(u^{-1} e^{-X_3/u} \right) = \frac{X_3 - u}{u^2} . \tag{7.51}$$

Hence, the stochastic counterpart of $\nabla \ell(u) = 0$ is

$$\widehat{\nabla \ell}(u; v) = \frac{1}{10} - \frac{v}{N} \sum_{i=1}^{N} S(\mathbf{X}_i) \frac{X_{3i} - u}{u^3} e^{-X_{3i}(u^{-1} - v^{-1})} = 0 ,$$

with $\mathbf{X}_1, \ldots, \mathbf{X}_N$ simulated under parameter v. The choice of v here is *crucial*. In [17] the following method is proposed for choosing a "good" parameter v. We start by imposing some constraints on v, that is, $v_1 \leqslant v \leqslant v_2$, reflecting our prior knowledge about the optimal u^* (lying between v_1 and v_2). In our case, we could take $v_1 = 1$ and $v_2 = 4$, for example, since $\mathcal{V} = \{u : \leqslant u \leqslant 4\}$. Next, we take v from this interval such that the tail of the corresponding distribution is as fat as possible to ensure that the likelihood ratio behaves well. In this case, it means taking $v = 4$.

Figure 7.2 shows the graph of $\widehat{\nabla \ell}(u; v)$ as a function of u (solid line). It was obtained from a sample of size $N = 500,000$ using $v = 4$. Recall that, by definition, $\nabla \ell(u^*) = 0$. As an estimate for u^* we take the point $\widehat{u^*}$ such that $\widehat{\nabla \ell}(\widehat{u^*}; v) = 0$. This can be found by standard root-finding procedures, such as Matlab's `fzero` function. In this example, we find that $\widehat{u^*} = 1.37$.

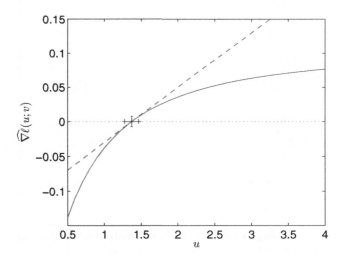

Figure 7.2: Estimate and 95% confidence interval for u^*. The estimate is found as the root of $\widehat{\nabla \ell}(u; v)$.

We can proceed to describe how to construct a confidence interval for u^* using a confidence interval for $\nabla \ell(u^*)$. For ease of notation, we write $g(u)$ for $\nabla \ell(u)$ and $\widehat{g}(u)$ for its estimate, $\widehat{\nabla \ell}(u; v)$, and we assume that both $g(u)$ and

$\widehat{g}(u)$ are monotonically increasing. Since $\widehat{g}(u)$ is of the form

$$\widehat{g}(u) = \frac{1}{10} - \frac{1}{N}\sum_{i=1}^{N} Z_i \,,$$

where $\{Z_i\}$ are iid random variables, an approximate $(1-\alpha)$ confidence interval for $g(u^*) = 0$ is $(-C, C)$, with $C = z_{1-\alpha/2}\, S_Z/\sqrt{N}$. Here S_Z is the sample standard deviation of the $\{Z_i\}$ and $z_{1-\alpha/2}$ is the $1 - \alpha/2$ quantile of the standard normal distribution. As a consequence, $(g^{-1}(-C), g^{-1}(C))$ is an approximate $(1 - \alpha)$ confidence interval for u^*. For small C we have $g^{-1}(C) \approx u^* + C/g'(u^*)$, where g' is the derivative of g, that is, the second derivative of ℓ. The latter is given by

$$g'(u) = \nabla^2\ell(u) = -\mathbb{E}_v\left[S(\mathbf{X})\,\frac{2u^2 - 4uX_3 + X_3^2}{u^4}\frac{v}{u}\,e^{-X_3(u^{-1}-v^{-1})}\right]$$

and it can be estimated via its stochastic counterpart, using the same sample as we used to obtain $\widehat{g}(u)$. Indeed, the estimate of $g'(u)$ is simply the derivative of $\widehat{g}$ at u. Thus, an approximate $(1 - \alpha)$ confidence interval for u^* is $\widehat{u^*} \pm C/\widehat{g'}(\widehat{u^*})$. This is illustrated in Figure 7.2, where the dashed line corresponds to the tangent line to $\widehat{g}(u)$ at the point $(\widehat{u}, 0)$, and 95% confidence intervals for $g(\widehat{u})$ and u^* are plotted vertically and horizontally, respectively. The particular values for these confidence intervals were found to be $(-0.0075, 0.0075)$ and $(1.28, 1.46)$.

Last, it is important to choose the parameter v under which the simulation is carried out greater than u^*. This is shown in Figure 7.3, where 10 replications of $\widehat{g}(u)$ are plotted for cases $v = 0.5$ and $v = 4$.

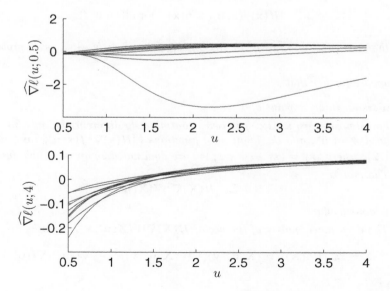

Figure 7.3: Ten replications of $\widehat{\nabla\ell}(u; v)$ are simulated under $v = 0.5$ and $v = 4$.

Note that in the first case the estimates of $g(u) = \widehat{\nabla \ell}(u; v)$ fluctuate widely, whereas in the second case they remain stable. As a consequence, u^* cannot be reliably estimated under $v = 0.5$. For $v = 4$ no such problems occur. All this is in accordance with the general principle that the importance sampling distribution should have heavier tails than the target distribution. Specifically, under $v = 4$ the pdf of X_3 has heavier tails than under $v = u^*$, whereas the opposite is true for $v = 0.5$.

In general, let $\widehat{\ell}^*$ and $\widehat{\mathbf{u}}^*$ denote the optimal objective value and the optimal solution of the sample average problem (7.48), respectively. By the law of large numbers, $\widehat{\ell}(\mathbf{u}; \mathbf{v})$ converges to $\ell(\mathbf{u})$ with probability 1 (w.p.1) as $N \to \infty$. As shown in [19] under mild additional conditions, $\widehat{\ell}^*$ and $\widehat{\mathbf{u}}^*$ converge w.p.1 to their corresponding optimal objective value and to the optimal solution of the true problem (7.47), respectively. That is, $\widehat{\ell}^*$ and $\widehat{\mathbf{u}}^*$ are consistent estimators of their true counterparts ℓ^* and $\mathbf{u}^*$, respectively. Moreover, [19] establishes a central limit theorem and valid confidence regions for the tuple $(\ell^*, \mathbf{u}^*)$. The following theorem summarizes the basic statistical properties of $\widehat{\mathbf{u}}^*$ for the unconstrained program formulation. Additional discussion, including proofs for both the unconstrained and constrained programs, can be found in [19].

Theorem 7.3.2 *Let $\mathbf{u}^*$ be a unique minimizer of $\ell(\mathbf{u})$ over $\mathscr{V}$.*

A. *Suppose that*

1. *The set $\mathscr{V}$ is compact.*
2. *For almost every $\mathbf{x}$, the function $f(\mathbf{x}; \cdot)$ is continuous on $\mathscr{V}$.*
3. *The family of functions $\{|H(\mathbf{x})f(\mathbf{x}; \mathbf{u})|, \mathbf{u} \in \mathscr{V}\}$ is dominated by an integrable function $h(\mathbf{x})$, that is,*

$$|H(\mathbf{x})f(\mathbf{x}; \mathbf{u})| \leqslant h(\mathbf{x}) \quad \text{for all } \mathbf{u} \in \mathscr{V}.$$

Then the optimal solution $\widehat{\mathbf{u}}^*$ of (7.48) converges to $\mathbf{u}^*$ as $N \to \infty$, with probability one.

B. *Suppose further that*

1. *$\mathbf{u}^*$ is an interior point of $\mathscr{V}$.*
2. *For almost every $\mathbf{x}, f(\mathbf{x}; \cdot)$ is twice continuously differentiable in a neighborhood $\mathscr{U}$ of $\mathbf{u}^*$, and the families of functions $\{\|H(\mathbf{x})\nabla^k f(\mathbf{x}; \mathbf{u})\| : \mathbf{u} \in \mathscr{U}, k = 1, 2\}$, where $\|\mathbf{x}\| = (x_1^2 + \cdots + x_n^2)^{\frac{1}{2}}$, are dominated by an integrable function.*
3. *The matrix*

$$B = \mathbb{E}_{\mathbf{v}}\left[H(\mathbf{X})\nabla^2 W(\mathbf{X}; \mathbf{u}^*, \mathbf{v})\right] \tag{7.52}$$

   is nonsingular.
4. *The covariance matrix of the vector $H(\mathbf{X})\nabla W(\mathbf{X}; \mathbf{u}^*, \mathbf{v})$, given by*

$$\Sigma = \mathbb{E}_{\mathbf{v}}\left[H^2(\mathbf{X})\nabla W(\mathbf{X}; \mathbf{u}^*, \mathbf{v})(\nabla W(\mathbf{X}; \mathbf{u}^*, \mathbf{v}))^{\top}\right] - \nabla\ell(\mathbf{u}^*)(\nabla\ell(\mathbf{u}^*))^{\top},$$

   exists.

Then the random vector $N^{1/2}(\widehat{\mathbf{u}}^* - \mathbf{u}^*)$ converges in distribution to a normal random vector with zero mean and covariance matrix

$$B^{-1}\Sigma B^{-1}. \tag{7.53}$$

The asymptotic efficiency of the estimator $N^{1/2}(\widehat{\mathbf{u}^*} - \mathbf{u}^*)$ is controlled by the covariance matrix given in (7.53). Under the assumptions of Theorem 7.3.2, this covariance matrix can be consistently estimated by $\widehat{B}^{-1}\widehat{\Sigma}\widehat{B}^{-1}$, where

$$\widehat{B} = \frac{1}{N} \sum_{i=1}^{N} H(\mathbf{X}_i) \nabla^2 W(\mathbf{X}_i; \widehat{\mathbf{u}^*}, \mathbf{v}) \tag{7.54}$$

and

$$\widehat{\Sigma} = \frac{1}{N} \sum_{i=1}^{N} H^2(\mathbf{X}_i) \nabla W(\mathbf{X}_i; \widehat{\mathbf{u}}, \mathbf{v})(\nabla W(\mathbf{X}_i; \widehat{\mathbf{u}^*}, \mathbf{v}))^\top - \widehat{\nabla \ell}(\widehat{\mathbf{u}^*}; \mathbf{v})(\widehat{\nabla \ell}(\widehat{\mathbf{u}^*}; \mathbf{v}))^\top \tag{7.55}$$

are consistent estimators of the matrices B and Σ, respectively. Observe that these matrices can be estimated from the same sample $\{\mathbf{X}_1, \ldots, \mathbf{X}_N\}$ simultaneously with the estimator $\widehat{\mathbf{u}^*}$. Observe also that the matrix B coincides with the Hessian matrix $\nabla^2 \ell(\mathbf{u}^*)$ and is, therefore, symmetric and independent of the choice of the importance sampling parameter vector $\mathbf{v}$.

Although Theorem 7.3.2 was formulated for the distributional case only, similar arguments [19] apply to the stochastic counterpart (7.43), involving both distributional and structural parameter vectors $\mathbf{u}_1$ and $\mathbf{u}_2$, respectively.

The statistical inference for the estimators $\widehat{\ell^*}$ and $\widehat{\mathbf{u}^*}$ allows the construction of stopping rules, validation analysis, and error bounds for the obtained solutions. In particular, it is shown in Shapiro [20] that if the function $\ell(\mathbf{u})$ is twice differentiable, then the above stochastic counterpart method produces estimators that converge to an optimal solution of the true problem at the same asymptotic rate as the stochastic approximation method, provided that the stochastic approximation method is applied with the *asymptotically optimal* step sizes. Moreover, it is shown in Kleywegt, Shapiro, and Homem de Mello [10] that if the underlying probability distribution is discrete and $\ell(\mathbf{u})$ is piecewise linear and convex, then w.p.1 the stochastic counterpart method (also called the *sample path method*) provides an exact optimal solution. For a recent survey on simulation-based optimization see Kleywegt and Shapiro [9].

The following example deals with unconstrained minimization of $\ell(\mathbf{u})$, where $\mathbf{u} = (\mathbf{u}_1, \mathbf{u}_2)$ and therefore contains both distributional and structural parameter vectors.

■ EXAMPLE 7.9 Examples 7.1 and 7.7 (Continued)

Consider minimization of the function

$$\ell(\mathbf{u}) = \mathbb{E}_{\mathbf{u}_1}[H(\mathbf{X}; \mathbf{u}_2)] + \mathbf{b}^\top \mathbf{u},$$

where

$$H(\mathbf{X}; u_3, u_4) = \max\{X_1 + u_3, X_2 + u_4\}, \tag{7.56}$$

$\mathbf{u} = (\mathbf{u}_1, \mathbf{u}_2)$, $\mathbf{u}_1 = (u_1, u_2)$, $\mathbf{u}_2 = (u_3, u_4)$, $\mathbf{X} = (X_1, X_2)$ is a two-dimensional vector with independent components, $X_i \sim f_i(x; u_i)$, $i = 1, 2$, with $X_i \sim \mathsf{Exp}(u_i)$, and $\mathbf{b} = (b_1, \ldots, b_4)$ is a cost vector.

To find the estimate of the optimal solution $\mathbf{u}^*$, we will use, as in Example 7.7, the direct, inverse-transform and push-out estimators of $\nabla \ell(\mathbf{u})$. In particular, we will define a system of nonlinear equations of type (7.44), which is generated by the corresponding direct, inverse-transform, and push-out estimators of $\nabla \ell(\mathbf{u})$. Note that each such estimator will be associated with a proper likelihood ratio function $W(\cdot)$.

(a) *The direct estimator of* $\nabla \ell(\mathbf{u})$. In this case

$$W(\mathbf{X}; \mathbf{u}_1, \mathbf{v}_1) = \frac{f_1(X_1; u_1) f_2(X_2; u_2)}{f_1(X_1; v_1) f_2(X_2; v_2)},$$

where $\mathbf{X} \sim f_1(x_1; v_1) f_2(x_2; v_2)$ and $\mathbf{v}_1 = (v_1, v_2)$. Using the likelihood ratio term given above, we can rewrite formulas (7.31) and (7.32) as

$$\frac{\partial \ell(\mathbf{u})}{\partial u_1} = \mathbb{E}_{\mathbf{v}_1} \left[H(\mathbf{X}; \mathbf{u}_2) W(\mathbf{X}; \mathbf{u}_1, \mathbf{v}_1) \nabla \ln f_1(X_1; u_1) \right] + b_1 \qquad (7.57)$$

and

$$\frac{\partial \ell(\mathbf{u})}{\partial u_3} = \mathbb{E}_{\mathbf{v}_1} \left[\frac{\partial H(\mathbf{X}; \mathbf{u}_2)}{\partial u_3} W(\mathbf{X}; \mathbf{u}_1, \mathbf{v}_1) \right] + b_3 , \qquad (7.58)$$

respectively, and similarly $\partial \ell(\mathbf{u})/\partial u_2$ and $\partial \ell(\mathbf{u})/\partial u_4$. By analogy to (7.34), the importance sampling estimator of $\partial \ell(\mathbf{u})/\partial u_3$ can be written as

$$\widehat{\nabla \ell}_3^{(1)}(\mathbf{u}; \mathbf{v}_1) = \frac{1}{N} \sum_{i=1}^{N} \frac{\partial H(\mathbf{X}_i; \mathbf{u}_2)}{\partial u_3} W(\mathbf{X}_i; \mathbf{u}_1, \mathbf{v}_1) + b_3 , \qquad (7.59)$$

where $\mathbf{X}_1, \ldots, \mathbf{X}_N$ is a random sample from $f(\mathbf{x}; \mathbf{v}_1) = f_1(x_1; v_1) f_2(x_2; v_2)$, and similarly for the remaining importance sampling estimators $\widehat{\nabla \ell}_i^{(1)}(\mathbf{u}; \mathbf{v}_1)$ of $\partial \ell(\mathbf{u})/\partial u_i$, $i = 1, 2, 4$. With this at hand, the estimate of the optimal solution $\mathbf{u}^*$ can be obtained from the solution of the following four-dimensional system of nonlinear equations:

$$\widehat{\nabla \ell}^{(1)}(\mathbf{u}) = \mathbf{0}, \quad \mathbf{u} \in \mathscr{V}, \qquad (7.60)$$

where $\widehat{\nabla \ell}^{(1)} = (\widehat{\nabla \ell}_1^{(1)}, \ldots, \widehat{\nabla \ell}_4^{(1)})$.

(b) *The inverse-transform estimator of* $\nabla \ell(\mathbf{u})$. Taking (7.35) into account, the estimate of the optimal solution $\mathbf{u}^*$ can be obtained by solving, by analogy to (7.60), the following four-dimensional system of nonlinear equations:

$$\widehat{\nabla \ell}^{(2)}(\mathbf{u}) = \mathbf{0}, \quad \mathbf{u} \in \mathscr{V} . \qquad (7.61)$$

Here, as before,

$$\widehat{\nabla \ell}^{(2)}(\mathbf{u}) = \frac{1}{N} \sum_{i=1}^{N} \nabla \breve{H}(\mathbf{Z}_i; \mathbf{u}) + \mathbf{b} ,$$

and $\mathbf{Z}_1, \ldots, \mathbf{Z}_N$ is a random sample from the two-dimensional uniform pdf with independent components, that is, $\mathbf{Z} = (Z_1, Z_2)$ and $Z_j \sim \mathsf{U}(0, 1)$, $j = 1, 2$. Alternatively, we could estimate $\mathbf{u}^*$ using the ITLR method. In this case, by analogy to (7.61), the four-dimensional system of nonlinear equations can be written as

$$\widetilde{\nabla \ell}^{(2)}(\mathbf{u}) = \mathbf{0}, \quad \mathbf{u} \in \mathscr{V} , \qquad (7.62)$$

with

$$\widetilde{\nabla\ell}^{(2)}(\mathbf{u}) = \frac{1}{N} \sum_{i=1}^{N} \nabla \breve{H}(\mathbf{X}_i; \mathbf{u}) W(\mathbf{X}_i; \boldsymbol{\theta}) + \mathbf{b} \,,$$

and

$$W(\mathbf{X}_i; \boldsymbol{\theta}) = \frac{1}{h_1(X_{1i}; \theta_1) \, h_2(X_{2i}; \theta_2)},$$

where $\boldsymbol{\theta} = (\theta_1, \theta_2)$, $\mathbf{X} = (X_1, X_2) \sim h_1(x_1; \theta_1) h_2(x_2; \theta_2)$, and, for example, $h_i(x; \theta_i) = \theta_i x^{\theta_i - 1}$, $i = 1, 2$; that is, $h_i(\cdot)$ is a Beta pdf.

(c) *The push-out estimator of* $\nabla\ell(\mathbf{u})$. Using (7.39), the estimate of the optimal solution $\mathbf{u}^*$ can be obtained from the solution of the following four-dimensional system of nonlinear equations:

$$\widehat{\nabla\ell}^{(3)}(\mathbf{u}; \mathbf{v}) = \mathbf{0}, \quad \mathbf{u} \in \mathscr{V}, \tag{7.63}$$

where

$$\widehat{\nabla\ell}^{(3)}(\mathbf{u}; \mathbf{v}) = \frac{1}{N} \sum_{i=1}^{N} \max\{\widetilde{X}_{1i}, \widetilde{X}_{2i}\} \, W(\widetilde{\mathbf{X}}_i; \mathbf{u}, \mathbf{v}) \nabla \ln \widetilde{f}(\widetilde{\mathbf{X}}_i; \mathbf{u}) + \mathbf{b} \,,$$

$$W(\widetilde{\mathbf{X}}_i; \mathbf{u}, \mathbf{v}) = \frac{f_1(X_1 - u_3; u_1) \, f_2(X_2 - u_4; u_2)}{f_1(X_1 - v_3; v_1) \, f_2(X_2 - v_4; v_2)}$$

and $\widetilde{\mathbf{X}}_i \sim \widetilde{f}(\mathbf{x}) = f_1(x_1 - v_3; v_1) f_2(x_2 - v_4; v_2)$.

Let us return finally to the stochastic counterpart of the general program (P_0). From the foregoing discussion, it can be written as

$$
\begin{aligned}
&\text{minimize} && \widehat{\ell}_0(\mathbf{u}; \mathbf{v}_1), && \mathbf{u} \in \mathscr{V}, \\
(\widehat{P}_N) \quad &\text{subject to:} && \widehat{\ell}_j(\mathbf{u}; \mathbf{v}_1) \leqslant 0, && j = 1, \ldots, k, \\
& && \widehat{\ell}_j(\mathbf{u}; \mathbf{v}_1) = 0, && j = k + 1, \ldots, M,
\end{aligned}
\tag{7.64}
$$

with

$$\widehat{\ell}_j(\mathbf{u}; \mathbf{v}_1) = \frac{1}{N} \sum_{i=1}^{N} H_j(\mathbf{X}_i; \mathbf{u}_2) \, W(\mathbf{X}_i; \mathbf{u}_1, \mathbf{v}_1), \quad j = 0, 1, \ldots, M, \tag{7.65}$$

where $\mathbf{X}_1, \ldots, \mathbf{X}_N$ is a random sample from the importance sampling pdf $f(\mathbf{x}; \mathbf{v}_1)$, and the $\{\widehat{\ell}_j(\mathbf{u}; \mathbf{v}_1)\}$ are viewed as functions of $\mathbf{u}$ rather than as estimators for a fixed $\mathbf{u}$.

Note again that once the sample $\mathbf{X}_1, \ldots, \mathbf{X}_N$ is generated, the functions $\widehat{\ell}_j(\mathbf{u}; \mathbf{v}_1)$, $j = 0, \ldots, M$, become *explicitly* determined via the functions $H_j(\mathbf{X}_i; \mathbf{u}_2)$ and $W(\mathbf{X}_i; \mathbf{u}_1, \mathbf{v}_1)$. Let us assume that the corresponding gradients $\nabla\widehat{\ell}_j(\mathbf{u}; \mathbf{v}_1)$ can be calculated, for any $\mathbf{u}$, from a single simulation run, so that we can solve the optimization problem $(\widehat{P}_N)$ by standard methods of mathematical programming. The resultant optimal function value and the optimal decision vector of the program $(\widehat{P}_N)$ provide estimators of the optimal values ℓ^* and $\mathbf{u}^*$, respectively, of the original one (P_0). It is important to understand that what makes this approach feasible is the fact that once the sample $\mathbf{X}_1, \ldots, \mathbf{X}_N$ is generated, the functions $\widehat{\ell}_j(\mathbf{u})$, $j = 0, \ldots, M$ become known explicitly, provided that the sample functions $\{H_j(\mathbf{X}; \mathbf{u}_2)\}$ are explicitly available for any $\mathbf{u}_2$. Recall that if $H_j(\mathbf{X}; \mathbf{u}_2)$ is available only for some fixed in advance $\mathbf{u}_2$, rather than simultaneously for all values $\mathbf{u}_2$, then stochastic approximation algorithms can be applied instead of the stochastic counterpart method. Note that in the case where the $\{H_j(\cdot)\}$ do not depend on $\mathbf{u}_2$, one can solve the program $(\widehat{P}_N)$ (from a single simulation run) using the SF method, provided that the trust region of the program $(\widehat{P}_N)$ does not exceed the one defined in (7.27). If this is not the case, iterative gradient-type methods need to be used, since they do not involve likelihood ratios.

The algorithm for estimating the optimal solution, $\mathbf{u}^*$, of the program (P_0) via the stochastic counterpart $(\widehat{P}_N)$ can be written as follows:

Algorithm 7.3.2: Estimation of $\mathbf{u}^*$

---

1 Generate a random sample $\mathbf{X}_1, \ldots, \mathbf{X}_N$ from $f(\mathbf{x}; \mathbf{v}_1)$.
2 Calculate the functions $H_j(\mathbf{X}_i; \mathbf{u}_2)$, $j = 0, \ldots, M$, $i = 1, \ldots, N$ via simulation.
3 Solve the program $(\widehat{P}_N)$ by standard mathematical programming methods.
4 Return the resultant optimal solution, $\widehat{\mathbf{u}}^*$ of $(\widehat{P}_N)$, as an estimate of $\mathbf{u}^*$.

The third step of Algorithm 7.3.2 typically calls for iterative numerical procedures, which may require, in turn, calculation of the functions $\widehat{\ell}_j(\mathbf{u})$, $j = 0, \ldots, M$, and their gradients (and possibly Hessians), for multiple values of the parameter vector $\mathbf{u}$. Our extensive simulation studies for typical DESS with sizes up to 100 decision variables show that the optimal solution $\widehat{\mathbf{u}}^*$ of the program $(\widehat{P}_N)$ constitutes a reliable estimator of the true optimal solution, $\mathbf{u}^*$, provided that the program $(\widehat{P}_N)$ is convex (see [19] and the Appendix), the trust region is not too large, and the sample size N is quite large (on the order of 1000 or more).

7.4 SENSITIVITY ANALYSIS OF DEDS

Let $X_1, X_2, \ldots$ be an input sequence of m-dimensional random vectors driving an output process $\{H_t, t = 0, 1, 2, \ldots\}$. That is, $H_t = H_t(\mathbf{X}_t)$ for some function H_t, where the vector $\mathbf{X}_t = (X_1, X_2, \ldots, X_t)$ represents the history of the input process up to time t. Let the pdf of $\mathbf{X}_t$ be given by $f_t(\mathbf{x}_t; \mathbf{u})$, which depends on some parameter vector $\mathbf{u}$. Assume that $\{H_t\}$ is a regenerative process with a regenerative cycle of length τ. Typical examples are an ergodic Markov chain and the waiting time process in the $GI/G/1$ system. In both cases (see Section 4.4.2.2) the expected steady-state performance, $\ell(\mathbf{u})$, can be written as

$$\ell(\mathbf{u}) = \frac{\mathbb{E}_{\mathbf{u}}[R]}{\mathbb{E}_{\mathbf{u}}[\tau]} = \frac{\mathbb{E}_{\mathbf{u}}\left[\sum_{t=1}^{\tau} H_t(\mathbf{X}_t)\right]}{\mathbb{E}_{\mathbf{u}}[\tau]} , \tag{7.66}$$

where R is the reward during a cycle. As for static models, we show here how to estimate from a *single simulation run* the performance $\ell(\mathbf{u})$, and the derivatives $\nabla^k \ell(\mathbf{u})$, $k = 1, 2 \ldots$, for different values of $\mathbf{u}$.

Consider first the estimation of $\ell_R(\mathbf{u}) = \mathbb{E}_{\mathbf{u}}[R]$ when the $\{X_i\}$ are iid with pdf $f(x; \mathbf{u})$; thus, $f_t(\mathbf{x}_t) = \prod_{i=1}^{t} f(x_i)$. Let $g(x)$ be any importance sampling pdf, and let $g_t(\mathbf{x}_t) = \prod_{i=1}^t g(x_i)$. It will be shown that $\ell_R(\mathbf{u})$ can be represented as

$$\ell_R(\mathbf{u}) = \mathbb{E}_g\left[\sum_{t=1}^{\tau} H_t(\mathbf{X}_t)\, W_t(\mathbf{X}_t; \mathbf{u})\right], \tag{7.67}$$

where $\mathbf{X}_t \sim g_t(\mathbf{x}_t)$ and $W_t(\mathbf{X}_t; \mathbf{u}) = f_t(\mathbf{x}_t; \mathbf{u})/g_t(\mathbf{x}_t) = \prod_{j=1}^t f(X_j; \mathbf{u})/g(X_j)$. To proceed, we write

$$\sum_{t=1}^{\tau} H_t = \sum_{t=1}^{\infty} H_t\, I_{\{\tau \geqslant t\}}. \tag{7.68}$$

Since $\tau = \tau(\mathbf{X}_t)$ is completely determined by $\mathbf{X}_t$, the indicator $I_{\{\tau \geqslant t\}}$ can be viewed as a function of $\mathbf{x}_t$; we write $I_{\{\tau \geqslant t\}}(\mathbf{x}_t)$. Accordingly, the expectation of $H_t\, I_{\{\tau \geqslant t\}}$ is

$$\begin{aligned}
\mathbb{E}_{\mathbf{u}}[H_t\, I_{\{\tau \geqslant t\}}] &= \int H_t(\mathbf{x}_t)\, I_{\{\tau \geqslant t\}}(\mathbf{x}_t)\, f_t(\mathbf{x}_t; \mathbf{u})\, \mathrm{d}\mathbf{x}_t \\
&= \int H_t(\mathbf{x}_t)\, I_{\{\tau \geqslant t\}}(\mathbf{x}_t)\, W_t(\mathbf{x}_t; \mathbf{u})\, g_t(\mathbf{x}_t)\, \mathrm{d}\mathbf{x}_t \\
&= \mathbb{E}_g[H_t(\mathbf{X}_t)\, I_{\{\tau \geqslant t\}}(\mathbf{X}_t)\, W_t(\mathbf{X}_t; \mathbf{u})].
\end{aligned} \tag{7.69}$$

The result (7.67) follows by combining (7.68) and (7.69). For the special case where $H_t \equiv 1$, (7.67) reduces to

$$\mathbb{E}_{\mathbf{u}}[\tau] = \mathbb{E}_g\left[\sum_{t=1}^{\tau} W_t\right],$$

abbreviating $W_t(\mathbf{X}_t; \mathbf{u})$ to W_t. Derivatives of (7.67) can be presented in a similar form. In particular, under standard regularity conditions ensuring the interchangeability of the differentiation and the expectation operators, one can write

$$\nabla^k \ell_R(\mathbf{u}) = \mathbb{E}_g\left[\sum_{t=1}^{\tau} H_t \nabla^k W_t\right] = \mathbb{E}_g\left[\sum_{t=1}^{\tau} H_t\, \mathcal{S}_t^{(k)}\, W_t\right], \tag{7.70}$$

where $\mathcal{S}_t^{(k)}$ is the k-th order score function corresponding to $f_t(\mathbf{x}_t; \mathbf{u})$, as in (7.7).

Now let $\{X_{11}, \ldots, X_{\tau_1 1}, \ldots, X_{1N}, \ldots, X_{\tau_N N}\}$ be a sample of N regenerative cycles from the pdf $g(x)$. Then, using (7.70), we can estimate $\nabla^k \ell_R(\mathbf{u})$, $k = 0, 1, \ldots$ from a *single simulation run* as

$$\widehat{\nabla^k \ell_R}(\mathbf{u}) = \frac{1}{N}\sum_{i=1}^{N}\sum_{t=1}^{\tau_i} H_{ti}\, \mathcal{S}_{ti}^{(k)}\, W_{ti}, \tag{7.71}$$

where $W_{ti} = \prod_{j=1}^t \frac{f(X_{ji}; \mathbf{u})}{g(X_{ji})}$ and $X_{ji} \sim g(x)$. Notice that here $\widehat{\nabla^k \ell_R}(\mathbf{u}) = \nabla^k \widehat{\ell_R}(\mathbf{u})$. For the special case where $g(\mathbf{x}) = f(\mathbf{x}; \mathbf{u})$, that is, when using the original pdf $f(\mathbf{x}; \mathbf{u})$, one has

$$\nabla^k \ell_R(\mathbf{u}) = \mathbb{E}_{\mathbf{u}}\left[\sum_{t=1}^{\tau} H_t\, \mathcal{S}_t^{(k)}\right]. \tag{7.72}$$

For $k = 1$, writing $\mathcal{S}_t$ for $\mathcal{S}_t^{(1)}$, the score function process $\{\mathcal{S}_t\}$ is given by

$$\mathcal{S}_t = \sum_{j=1}^{t} \nabla \ln f(X_j; \mathbf{u}). \tag{7.73}$$

■ **EXAMPLE 7.10**

Let $X \sim \mathsf{G}(p)$. That is, $f(x; p) = p\,(1-p)^{x-1}$, $x = 1, 2, \ldots$. Then (see also Table 7.1)

$$\mathcal{S}_t = \frac{\partial}{\partial p} \ln f_t(\mathbf{X}_t; p) = \frac{t - p \sum_{j=1}^{t} X_j}{p(1-p)} \; .$$

■ **EXAMPLE 7.11**

Let $X \sim \mathsf{Gamma}(\alpha, \lambda)$. That is, $f(x; \lambda, \alpha) = \frac{\lambda^{\alpha} x^{\alpha-1} e^{-\lambda x}}{\Gamma(\alpha)}$ for $x > 0$. Suppose we are interested in the sensitivities with respect to λ. Then

$$\mathcal{S}_t = \frac{\partial}{\partial \lambda} \ln f_t(\mathbf{X}_t; \lambda, \alpha) = t\,\alpha\,\lambda^{-1} - \sum_{i=1}^{t} X_i \; .$$

Let us return now to the estimation of $\ell(\mathbf{u}) = \mathbb{E}_{\mathbf{u}}[R]/\mathbb{E}_{\mathbf{u}}[\tau]$ and its sensitivities. In view of (7.70) and the fact that $\tau = \sum_{t=1}^{\tau} 1$ can be viewed as a special case of (7.67), with $H_t = 1$, one can write $\ell(\mathbf{u})$ as

$$\ell(\mathbf{u}) = \frac{\mathbb{E}_g[\sum_{t=1}^{\tau} H_t\, W_t]}{\mathbb{E}_g[\sum_{t=1}^{\tau} W_t]} \tag{7.74}$$

and by direct differentiation of (7.74) write $\nabla \ell(\mathbf{u})$ as

$$\nabla \ell(\mathbf{u}) = \frac{\mathbb{E}_g[\sum_{t=1}^{\tau} H_t\, \nabla W_t]}{\mathbb{E}_g[\sum_{t=1}^{\tau} W_t]} - \frac{\mathbb{E}_g[\sum_{t=1}^{\tau} H_t\, W_t]}{\mathbb{E}_g[\sum_{t=1}^{\tau} W_t]} \cdot \frac{\mathbb{E}_g[\sum_{t=1}^{\tau} \nabla W_t]}{\mathbb{E}_g[\sum_{t=1}^{\tau} W_t]} \tag{7.75}$$

(observe that $W_t = W_t(\mathbf{X}_t, \mathbf{u})$ is a function of $\mathbf{u}$ but $H_t = H_t(\mathbf{X}_t)$ is not). Observe also that above, $\nabla W_t = W_t\, \mathcal{S}_t$. Higher-order partial derivatives with respect to parameters of interest can then be obtained from (7.75). Utilizing (7.74) and (7.75), one can estimate $\ell(\mathbf{u})$ and $\nabla \ell(\mathbf{u})$, for all $\mathbf{u}$, as

$$\widehat{\ell}(\mathbf{u}) = \frac{\sum_{i=1}^{N} \sum_{t=1}^{\tau_i} H_{ti}\, W_{ti}}{\sum_{i=1}^{N} \sum_{t=1}^{\tau_i} W_{ti}} \tag{7.76}$$

and

$$\widehat{\nabla \ell}(\mathbf{u}) = \frac{\sum_{i=1}^{N} \sum_{t=1}^{\tau_i} H_{ti} W_{ti} \mathcal{S}_{ti}}{\sum_{i=1}^{N} \sum_{t=1}^{\tau_i} W_{ti}} - \frac{\sum_{i=1}^{N} \sum_{t=1}^{\tau_i} H_{ti} W_{ti}}{\sum_{i=1}^{N} \sum_{t=1}^{\tau_i} W_{ti}} \cdot \frac{\sum_{i=1}^{N} \sum_{t=1}^{\tau_i} W_{ti} \mathcal{S}_{ti}}{\sum_{i=1}^{N} \sum_{t=1}^{\tau_i} W_{ti}} \;, \tag{7.77}$$

respectively, and similarly for higher-order derivatives. Notice again that in this case, $\widehat{\nabla \ell}(\mathbf{u}) = \nabla \widehat{\ell}(\mathbf{u})$. The algorithm for estimating the gradient $\nabla \ell(\mathbf{u})$ at different values of $\mathbf{u}$ using a single simulation run can be written as follows:

Algorithm 7.4.1: $\nabla \ell(\mathbf{u})$ Estimation

---

1 Generate a random sample $\{X_1, \ldots, X_T\}$, $T = \sum_{i=1}^{N} \tau_i$, from $g(x)$.
2 Generate the output processes $\{H_t\}$ and $\{\nabla W_t\} = \{W_t \mathcal{S}_t\}$.
3 Calculate $\widehat{\nabla \ell}(\mathbf{u})$ from (7.77).

Confidence intervals (regions) for the sensitivities $\nabla^k \ell(\mathbf{u})$, $k = 0, 1$, utilizing the SF estimators $\nabla^k \widehat{\ell}(\mathbf{u})$, $k = 0, 1$, can be derived analogously to those for the standard regenerative estimator of Chapter 4 and are left as an exercise.

■ **EXAMPLE 7.12 Waiting Time**

The waiting time process in a $GI/G/1$ queue is driven by sequences of inter-arrival times $\{A_t\}$ and service times $\{S_t\}$ via the Lindley equation

$$H_t = \max\{H_{t-1} + S_t - A_t, 0\}, \quad t = 1, 2, \ldots, \tag{7.78}$$

with $H_0 = 0$; see (4.33) and Problem 5.3. Writing $X_t = (S_t, A_t)$, the $\{X_t, t = 1, 2, \ldots\}$ are iid. The process $\{H_t, t = 0, 1, \ldots\}$ is a regenerative process, which regenerates every time $H_t = 0$. Let $\tau > 0$ denote the first such time, and let H denote the steady-state waiting time. We wish to estimate the steady-state performance

$$\ell = \mathbb{E}[H] = \frac{\mathbb{E}\left[\sum_{t=1}^{\tau} H_t\right]}{\mathbb{E}[\tau]} .$$

Consider, for instance, the case where $S \sim \mathsf{Exp}(\mu)$, $A \sim \mathsf{Exp}(\lambda)$, and S and A are independent. Thus, H is the steady-state waiting time in the $M/M/1$ queue, and $\mathbb{E}[H] = \lambda/(\mu(\mu - \lambda))$ for $\mu > \lambda$; see, for example, [6]. Suppose we carry out the simulation using the service rate $\tilde{\mu}$ and wish to estimate $\ell(\mu) = \mathbb{E}[H]$ for different values of μ using the same simulation run. Let $(S_1, A_1), \ldots, (S_\tau, A_\tau)$ denote the service and inter-arrival times in the first cycle, respectively. Then, for the first cycle

$$W_t = W_{t-1} \frac{\mu \, e^{-\mu S_t}}{\tilde{\mu} \, e^{-\tilde{\mu} S_t}}, \quad t = 1, 2, \ldots, \tau \quad (W_0 = 1) ,$$

$$\mathcal{S}_t = \mathcal{S}_{t-1} + \frac{1}{\mu} - S_t, \quad t = 1, 2, \ldots, \tau \quad (\mathcal{S}_0 = 0) ,$$

and H_t is as given in (7.78). From these, the sums $\sum_{t=1}^{\tau} H_t W_t$, $\sum_{t=1}^{\tau} W_t$, $\sum_{t=1}^{\tau} W_t \mathcal{S}_t$, and $\sum_{t=1}^{\tau} H_t W_t \mathcal{S}_t$ can be computed. Repeating this for the subsequent cycles, we can estimate $\ell(\mu)$ and $\nabla\ell(\mu)$ from (7.76) and (7.77), respectively. Figure 7.4 displays the estimates and true values for $1.5 \leqslant \mu \leqslant 5.5$ using a single simulation run of $N = 10^5$ cycles. The simulation was carried out under the service rate $\tilde{\mu} = 2$ and arrival rate $\lambda = 1$. We see that both $\ell(\mu)$ and $\nabla\ell(\mu)$ are estimated accurately over the whole range. Note that for $\mu < 2$ the confidence interval for $\ell(\mu)$ grows rapidly wider. The estimation should not be extended much below $\mu = 1.5$, as the importance sampling will break down, resulting in unreliable estimates.

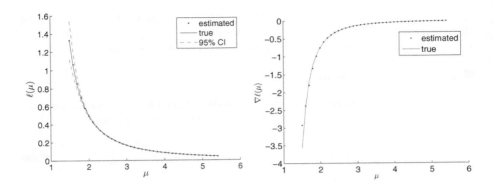

Figure 7.4: Estimated and true values for the expected steady-state waiting time and its derivative as a function of μ.

Although (7.76) and (7.77) were derived for the case where the $\{X_i\}$ are iid, much of the theory can be readily modified to deal with the dependent case. As an example, consider the case where $X_1, X_2, \ldots$ form an ergodic Markov chain and R is of the form

$$R = \sum_{t=1}^{\tau} c_{X_{t-1}, X_t} \, , \tag{7.79}$$

where c_{ij} is the cost of going from state i to j and R represents the cost accrued in a cycle of length τ. Let $\mathbf{P} = (p_{ij})$ be the one-step transition matrix of the Markov chain. Following reasoning similar to that for (7.67) and defining $H_t = c_{X_{t-1}, X_t}$, we see that

$$\mathbb{E}_{\mathbf{P}}[R] = \mathbb{E}_{\widetilde{\mathbf{P}}} \left[\sum_{t=1}^{\tau} H_t W_t \right],$$

where $\widetilde{\mathbf{P}} = (\widetilde{p}_{ij})$ is another transition matrix, and

$$W_t = W_t(\mathbf{X}_t; \mathbf{P}, \widetilde{\mathbf{P}}) = \prod_{k=1}^{t} \frac{p_{X_{k-1}, X_k}}{\widetilde{p}_{X_{k-1}, X_k}}$$

is the likelihood ratio. The pdf of $\mathbf{X}_t$ is given by

$$f_t(\mathbf{x}_t; \mathbf{P}) = \prod_{k=1}^{t} p_{X_{k-1}, X_k} \, .$$

The score function can again be obtained by taking the derivative of the logarithm of the pdf. Since, $\mathbb{E}_{\mathbf{P}}[\tau] = \mathbb{E}_{\widetilde{\mathbf{P}}}[\sum_{t=1}^{\tau} W_t]$, the long-run average cost $\ell(\mathbf{P}) = \mathbb{E}_{\mathbf{P}}[R]/\mathbb{E}_{\mathbf{P}}[\tau]$ can be estimated via (7.76) — and its derivatives by (7.77) — simultaneously for various $\mathbf{P}$ using a single simulation run under $\widetilde{\mathbf{P}}$.

■ **EXAMPLE 7.13   Markov Chain: Example 4.9 (Continued)**

Consider again the two-state Markov chain with transition matrix $\mathbf{P} = (p_{ij})$ and cost matrix C given by

$$\mathbf{P} = \begin{pmatrix} p_1 & 1 - p_1 \\ p_2 & 1 - p_2 \end{pmatrix} = (\mathbf{p} \quad 1 - \mathbf{p})$$

and

$$C = \begin{pmatrix} 0 & 1 \\ 2 & 3 \end{pmatrix},$$

respectively, where $\mathbf{p}$ denotes the vector $(p_1, p_2)^\top$. Our goal is to estimate $\ell(\mathbf{p})$ and $\nabla \ell(\mathbf{p})$ using (7.76) and (7.77) for various $\mathbf{p}$ from a single simulation run under $\widetilde{\mathbf{p}} = (\frac{1}{2}, \frac{1}{5})^\top$. Assume, as in Example 4.9 on Page 124, that starting from state 1, we obtain the sample trajectory $(x_0, x_1, x_2, \ldots, x_{10}) = (1, 2, 2, 2, 1, 2, 1, 1, 2, 2, 1)$, which has four cycles with lengths $\tau_1 = 4$, $\tau_2 = 2$, $\tau_3 = 1$, $\tau_4 = 3$, and corresponding transition probabilities $(p_{12}, p_{22}, p_{22}, p_{21}); (p_{12}, p_{21}); (p_{11}); (p_{12}, p_{22}, p_{21})$. The cost in the first cycle is given by (7.79). We consider the cases (1) $\mathbf{p} = \widetilde{\mathbf{p}} = (\frac{1}{2}, \frac{1}{5})^\top$ and (2) $\mathbf{p} = (\frac{1}{5}, \frac{1}{2})^\top$. The transition matrices for the two cases are

$$\widetilde{\mathbf{P}} = \begin{pmatrix} \frac{1}{2} & \frac{1}{2} \\ \frac{1}{5} & \frac{4}{5} \end{pmatrix} \quad \text{and} \quad \mathbf{P} = \begin{pmatrix} \frac{1}{5} & \frac{4}{5} \\ \frac{1}{2} & \frac{1}{2} \end{pmatrix}.$$

Note that the first case pertains to the nominal Markov chain.

In the first cycle, costs $H_{11} = 1, H_{21} = 3, H_{31} = 3$, and $H_{41} = 2$ are incurred. The likelihood ratios under case (2) are $W_{11} = \frac{p_{12}}{\widetilde{p}_{12}} = 8/5$, $W_{21} = W_{11} \frac{p_{22}}{\widetilde{p}_{22}} = 1$, $W_{31} = W_{21} \frac{p_{22}}{\widetilde{p}_{22}} = \frac{5}{8}$, and $W_{41} = W_{31} \frac{p_{21}}{\widetilde{p}_{21}} = \frac{25}{16}$, while in case (1) they are all 1. Next, we derive the score functions (in the first cycle) with respect to p_1 and p_2. Note that

$$f_4(\mathbf{x}_4; \mathbf{p}) = p_{12} \, p_{22}^2 \, p_{21} = (1 - p_1)(1 - p_2)^2 \, p_2 \, .$$

It follows that in case (2),

$$\frac{\partial}{\partial p_1} \ln f_4(\mathbf{x}_4; \mathbf{p}) = \frac{-1}{1 - p_1} = -\frac{5}{4}$$

and

$$\frac{\partial}{\partial p_2} \ln f_4(\mathbf{x}_4; \mathbf{p}) = \frac{-2}{1 - p_2} + \frac{1}{p_2} = -2 \, ,$$

so that the score function at time $t = 4$ in the first cycle is given by $\mathcal{S}_{41} = (-\frac{5}{4}, -2)$. Similarly, $\mathcal{S}_{31} = (-\frac{5}{4}, -4)$, $\mathcal{S}_{21} = (-\frac{5}{4}, -2)$, and $\mathcal{S}_{11} = (-\frac{5}{4}, 0)$. The quantities for the other cycles are derived in the same way, and the results are summarized in Table 7.3.

Table 7.3: Summary of costs, likelihood ratios, and score functions.

i	H_{ti}	W_{ti} (case 2)	S_{ti} (case 1)	S_{ti} (case 2)
1	1, 3, 3, 2	$\frac{8}{5}, 1, \frac{5}{8}, \frac{25}{16}$	$(-2,0), (-2,-\frac{5}{4}), (-2,-\frac{5}{2}), (-2,\frac{5}{2})$	$(-\frac{5}{4},0), (-\frac{5}{4},-2), (-\frac{5}{4},-4), (-\frac{5}{4},-2)$
2	1, 2	$\frac{8}{5}, 4$	$(-2,0),(-2,5)$	$(-\frac{5}{4},0),(-\frac{5}{4},2)$
3	0	$\frac{2}{5}$	$(2,0)$	$(5,0)$
4	1, 3, 2	$\frac{8}{5}, 1, \frac{5}{2}$	$(-2,0), (-2,-\frac{5}{4}), (-2,\frac{15}{4})$	$(-\frac{5}{4},0), (-\frac{5}{4},-2), (-\frac{5}{4},0)$

By substituting these values in (7.76) and (7.77), the reader can verify that $\widehat{\ell}(\tilde{\mathbf{p}}) = 1.8$, $\widehat{\ell}(\mathbf{p}) \approx 1.81$, $\widehat{\nabla\ell}(\tilde{\mathbf{p}}) = (-0.52, -0.875)$, and $\widehat{\nabla\ell}(\mathbf{p}) \approx (0.22, -1.23)$.

PROBLEMS

7.1   Consider the unconstrained minimization program

$$\min_u \ell(u) = \min_u \left\{ \mathbb{E}_u[X] + \frac{b}{u} \right\}, \quad u \in (0,1), \tag{7.80}$$

where $X \sim \mathsf{Ber}(u)$.

a) Show that the stochastic counterpart of $\nabla\ell(u) = 0$ can be written (see (7.18)) as

$$\widehat{\nabla\ell}(u) = \nabla\widehat{\ell}(u; v) = \frac{1}{v}\frac{1}{N}\sum_{i=1}^{N} X_i - \frac{b}{u^2} = 0, \tag{7.81}$$

where $X_1, \ldots, X_N$ is a random sample from $\mathsf{Ber}(v)$.

b) Assume that the sample $\{0,1,0,0,1,0,0,1,1,1,0,1,0,1,1,0,1,0,1,1\}$ was generated from $\mathsf{Ber}(v = 1/2)$. Show that the optimal solution u^* is estimated as

$$\widehat{u^*} = \left(\frac{b}{1.1} \right)^{1/2}.$$

7.2   Consider the unconstrained minimization program

$$\min_u \ell(u) = \min_u \{\mathbb{E}_u[X] + bu\}, \quad u \in (0.5, 2.0), \tag{7.82}$$

where $X \sim \mathsf{Exp}(u)$. Show that the stochastic counterpart of $\nabla\ell(u) = -\frac{1}{u^2} + b = 0$ can be written (see (7.20)) as

$$\nabla\widehat{\ell}(u; v) = \frac{1}{N}\sum_{i=1}^{N} X_i \frac{e^{-uX_i}(1 - u\,X_i)}{v\,e^{-vX_i}} + b = 0, \tag{7.83}$$

where $X_1, \ldots, X_n$ is a random sample from $\mathsf{Exp}(v)$.

7.3   Prove (7.25).

7.4   Show that $\nabla^k W(\mathbf{x}; \mathbf{u}, \mathbf{v}) = S^{(k)}(\mathbf{u}; \mathbf{x}) W(\mathbf{x}; \mathbf{u}, \mathbf{v})$ and hence prove (7.16).

7.5   Let $X_i \sim \mathsf{N}(u_i, \sigma_i^2)$, $i = 1, \ldots, n$ be independent random variables. Here we are interested in sensitivities with respect to $\mathbf{u} = (u_1, \ldots, u_n)$ only. Show that, for $i = 1, \ldots, n$,

$$\mathbb{E}_{\mathbf{v}}[W^2] = \exp \left(\sum_{i=1}^{n} \frac{(u_i - v_i)^2}{\sigma_i^2} \right)$$

and

$$[\mathcal{S}^{(1)}(\mathbf{u}; \mathbf{X})]_i = \sigma_i^{-2}(x_i - u_i) .$$

7.6   Let the components X_i, $i = 1, \ldots, n$, of a random vector $\mathbf{X}$ be independent, and distributed according to the exponential family

$$f_i(x_i; \mathbf{u}_i) = c_i(\mathbf{u}_i) \, e^{b_i(\mathbf{u}_i) t_i(x_i)} \, h_i(x_i) ,$$

where $b_i(\mathbf{u}_i)$, $t_i(x_i)$, and $h_i(x_i)$ are real-valued functions and $c_i(\mathbf{u}_i)$ is normalization constant. The corresponding pdf of $\mathbf{X}$ is given by

$$f(\mathbf{x}; \mathbf{u}) = c(\mathbf{u}) \exp \left(\sum_{i=1}^{n} b_i(\mathbf{u}_i) t_i(x_i) \right) h(\mathbf{x}) ,$$

where $\mathbf{u} = (\mathbf{u}_1^\top, \ldots, \mathbf{u}_n^\top)$, $c(\mathbf{u}) = \prod_{i=1}^{n} c_i(\mathbf{u}_i)$, and $h(\mathbf{x}) = \prod_{i=1}^{n} h_i(x_i)$.

a)   Show that $\mathrm{Var}_{\mathbf{v}}(HW) = \frac{c(\mathbf{u})^2}{c(\mathbf{v}) \, c(\mathbf{w})} \, \mathbb{E}_{\mathbf{w}}[H^2] - \ell(\mathbf{u})^2$, where $\mathbf{w}$ is determined by $b_i(\mathbf{w}_i) = 2 b_i(\mathbf{u}_i) - b_i(\mathbf{v}_i)$, $i = 1, \ldots, n$.

b)   Show that

$$\mathbb{E}_{\mathbf{v}}[H^2 W^2] = \mathbb{E}_{\mathbf{v}}[W^2] \, \mathbb{E}_{\mathbf{w}}[H^2] .$$

7.7   Consider the exponential pdf $f(x; u) = u \exp(-ux)$. Show that if $H(x)$ is a monotonically increasing function, then the expected performance $\ell(u) = \mathbb{E}_u[H(X)]$ is a monotonically decreasing convex function of $u \in (0, \infty)$.

7.8   Let $X \sim \mathsf{N}(u, \sigma^2)$. Suppose that σ is known and fixed. For a given u, consider the function

$$\mathcal{L}(v) = \mathbb{E}_v[H^2 W^2] .$$

a)   Show that if $\mathbb{E}_u[H^2] < \infty$ for all $u \in \mathbb{R}$, then $\mathcal{L}(v)$ is convex and continuous on $\mathbb{R}$. Show further that if, additionally, $\mathbb{E}_{u_n}[H^2] > 0$ for any u, then $\mathcal{L}(v)$ has a unique minimizer, v^*, over $\mathbb{R}$.

b)   Show that if $H^2(x)$ is monotonically increasing on $\mathbb{R}$, then $v^* > u$.

7.9   Let $X \sim \mathsf{N}(u, \sigma^2)$. Suppose that u is known, and consider the parameter σ. Note that the resulting exponential family is not of canonical form (A.9). However, parameterizing it by $\theta = \sigma^{-2}$ transforms it into canonical form, with $t(x) = -(x - u)^2/2$ and $c(\theta) = (2\pi)^{-1/2} \theta^{1/2}$.

a)   Show that

$$\mathbb{E}_\eta[W^2] = \frac{\theta}{\eta^{1/2}(2\theta - \eta)^{1/2}} ,$$

provided that $0 < \eta < 2\theta$.

b)   Show that, for a given θ, the function

$$\mathcal{L}(\eta) = \mathbb{E}_\eta[H^2 W^2]$$

has a unique minimizer, η^*, on the interval $(0, 2\theta)$, provided that the expectation, $\mathbb{E}_\eta[H^2]$, is finite for all $\eta \in (0, 2\theta)$ and does not tend to 0 as η approaches 0 or 2θ. (Notice that this implies that the corresponding optimal value, $\sigma^* = \eta^{*-1/2}$, of the reference parameter, σ, is also unique.)

c) Show that if $H^2(x)$ is strictly convex on $\mathbb{R}$, then $\eta^* < \theta$. (Notice that this implies that $\sigma^* > \sigma$.)

7.10 Consider the performance

$$H(X_1, X_2; u_3, u_4) = \min\{\max(X_1, u_3), \max(X_2, u_4)\} ,$$

where X_1 and X_2 have continuous densities $f(x_1; u_1)$ and $f(x_2; u_2)$, respectively. If we let $Y_1 = \max(X_1, u_3)$ and $Y_2 = \max(X_2, u_4)$ and write the performance as $\min(Y_1, Y_2)$, then Y_1 and Y_2 would take values u_3 and u_4 with nonzero probability. Hence the random vector $\mathbf{Y} = (Y_1, Y_2)$ would not have a density function at point (u_3, u_4), since its distribution is a mixture of continuous and discrete ones. Consequently, the push-out method would fail in its current form. To overcome this difficulty, we carry out a transformation. We first write Y_1 and Y_2 as

$$Y_1 = u_3 \max\left(\frac{X_1}{u_3}, 1\right), \quad Y_2 = u_4 \max\left(\frac{X_2}{u_4}, 1\right)$$

and then replace $\mathbf{X} = (X_1, X_2)$ by the random vector $\widetilde{\mathbf{X}} = (\widetilde{X}_1, \widetilde{X}_2)$, where

$$\widetilde{X}_1 = \max\left(\frac{X_1}{u_3}, 1\right)$$

and

$$\widetilde{X}_2 = \max\left(\frac{X_2}{u_4}, 1\right).$$

Prove that the density of the random vector $(\widetilde{X}_1, \widetilde{X}_2)$ is differentiable with respect to the variables (u_3, u_4), provided that both $\widetilde{X}_1$ and $\widetilde{X}_2$ are greater than 1.

7.11   Delta method. Let $\mathbf{X} = (X_1, \ldots, X_n)$ and $\mathbf{Y} = (Y_1, \ldots, Y_m)$ be random (column) vectors, with $\mathbf{Y} = \mathbf{g}(\mathbf{X})$ for some mapping $\mathbf{g}$ from $\mathbb{R}^n$ to $\mathbb{R}^m$. Let $\Sigma_{\mathbf{X}}$ and $\Sigma_{\mathbf{Y}}$ denote the corresponding covariance matrices. Suppose that $\mathbf{X}$ is close to its mean $\boldsymbol{\mu}$. A first-order Taylor expansion of $\mathbf{g}$ around $\boldsymbol{\mu}$ gives

$$\mathbf{Y} \approx \mathbf{g}(\boldsymbol{\mu}) + J_{\boldsymbol{\mu}}(\mathbf{g})(\mathbf{X} - \boldsymbol{\mu}),$$

where $J_{\boldsymbol{\mu}}(\mathbf{g})$ is the matrix of Jacobi of $\mathbf{g}$ (the matrix whose (i, j)-th entry is the partial derivative $\partial g_i / \partial x_j$) evaluated at $\boldsymbol{\mu}$. Show that, as a consequence,

$$\Sigma_{\mathbf{Y}} \approx J_{\boldsymbol{\mu}}(\mathbf{g}) \Sigma_{\mathbf{X}} J_{\boldsymbol{\mu}}(\mathbf{g})^\top .$$

This is called the *delta method* in statistics.

Further Reading

The SF method in the simulation context has been discovered and rediscovered independently, starting in the late 1960s. The earlier work on SF appeared in

Aleksandrov, Sysoyev, and Shemeneva [1] in 1968 and Rubinstein [15] in 1969. Motivated by the pioneering paper of Ho, Eyler, and Chien [7] on *infinitesimal perturbation analysis* (IPA) in 1979, the SF method was rediscovered at the end of the 1980s by Glynn [5] in 1990 and independently in 1989 by Reiman and Weiss [13], who called it the *likelihood ratio method*. Since then, both the IPA and SF methods have evolved and have now reached maturity; see Glasserman [4], Pflug [12], Rubinstein and Shapiro [19], and Spall [21].

To the best of our knowledge, the stochastic counterpart method in the simulation context was first suggested by Rubinstein in his PhD thesis [15]. It was applied there to estimate the optimal parameters in a complex simulation-based optimization model. It was shown numerically that the *off-line* stochastic counterpart method produces better estimates than the standard *on-line* stochastic approximation. For some later work on the stochastic counterpart method and stochastic approximation, see [16]. Alexander Shapiro should be credited for developing theoretical foundations for stochastic programs and, in particular, for the stochastic counterpart method. For relevant references, see Shapiro's elegant paper [20] and also [18, 19]. As mentioned, Geyer and Thompson [3] independently discovered the stochastic counterpart method in the early 1990s, and used it to make statistical inference in a particular unconstrained setting.

REFERENCES

1. V. M Aleksandrov, V. I. Sysoyev, and V. V. Shemeneva. Stochastic optimization. *Engineering Cybernetics*, 5:11–16, 1968.

2. S. Asmussen and P. W. Glynn. *Stochastic Simulation*. Springer-Verlag, New York, 2007.

3. C. J. Geyer and E. A. Thompson. Annealing Markov chain Monte-Carlo with applications to ancestral inference. *Journal of the American Statistical Association*, 90:909–920, 1995.

4. P. Glasserman. *Gradient Estimation via Perturbation Analysis*. Kluwer, Norwell, MA, 1991.

5. P. W. Glynn. Likelihood ratio gradient estimation for stochastic systems. *Communications of the ACM*, 33(10):75–84, 1990.

6. D. Gross and C. M. Harris. *Fundamentals of Queueing Theory*. John Wiley & Sons, New York, 2nd edition, 1985.

7. Y. C. Ho, M. A. Eyler, and T. T. Chien. A gradient technique for general buffer storage design in a serial production line. *International Journal on Production Research*, 17(6):557–580, 1979.

8. J. Kiefer and J. Wolfowitz. Stochastic estimation of the maximum of regression function. *Annals of Mathematical Statistics*, 23:462–466, 1952.

9. A. J. Kleywegt and A. Shapiro. Stochastic optimization. In G. Salvendy, editor, *Handbook of Industrial Engineering*, pages 2625–2650, New York, 2001. John Wiley & Sons.

10. A. J. Kleywegt, A. Shapiro, and T. Homem de Mello. The sample average approximation method for stochastic discrete optimization. *SIAM Journal on Optimization*, 12:479–502, 2001.

11. H. J. Kushner and D. S. Clark. *Stochastic Approximation Methods for Constrained and Unconstrained Systems.* Springer-Verlag, New York, 1978.

12. G. Ch. Pflug. *Optimization of Stochastic Models.* Kluwer, Boston, 1996.

13. M. I. Reiman and A. Weiss. Sensitivity analysis for simulations via likelihood ratios. *Operations Research*, 37(5):830–844, 1989.

14. H. Robbins and S. Monro. Stochastic approximation methods. *Annals of Mathematical Statistics*, 22:400–407, 1951.

15. R. Y. Rubinstein. *Some Problems in Monte Carlo Optimization.* PhD thesis, University of Riga, Latvia, 1969. (In Russian).

16. R. Y. Rubinstein. *Monte Carlo Optimization Simulation and Sensitivity of Queueing Network.* John Wiley & Sons, New York, 1986.

17. R. Y. Rubinstein and B. Melamed. *Modern Simulation and Modeling.* John Wiley & Sons, New York, 1998.

18. R. Y. Rubinstein and A. Shapiro. Optimization of static simulation models by the score function method. *Mathematics and Computers in Simulation*, 32:373–392, 1990.

19. R. Y. Rubinstein and A. Shapiro. *Discrete Event Systems: Sensitivity Analysis and Stochastic Optimization via the Score Function Method.* John Wiley & Sons, New York, 1993.

20. A. Shapiro. Simulation based optimization: Convergence analysis and statistical inference. *Stochastic Models*, 12:425–454, 1996.

21. J. C. Spall. *Introduction to Stochastic Search and Optimization: Estimation, Simulation, and Control.* John Wiley & Sons, New York, 2003.

CHAPTER 8

CROSS-ENTROPY METHOD

8.1 INTRODUCTION

The *cross-entropy* (CE) method [45] is a relatively new Monte Carlo technique for both estimation and optimization. In the estimation setting, the CE method provides an adaptive way to find the optimal importance sampling distribution for quite general problems. By formulating an optimization problem as an estimation problem, the CE method becomes a general and powerful stochastic search algorithm. The method is based on a simple iterative procedure where each iteration contains two phases: (1) generate a random data sample (trajectories, vectors, etc.) according to a specified mechanism; (2) update the parameters of the random mechanism on the basis of the data in order to produce a better sample in the next iteration.

The CE method has its origins in an adaptive algorithm for rare-event estimation based on *variance minimization* (VM) [39]. This procedure was soon modified [40] to an adaptive algorithm for both rare-event estimation and combinatorial optimization, where the original VM program was replaced by a similar CE minimization program. In this chapter we present a general introduction to the CE method. For a comprehensive treatment, we refer the reader to [45].

The rest of this chapter is organized as follows: Section 8.2 presents a general CE algorithm for the estimation of rare-event probabilities, and Section 8.3 introduces a slight modification of this algorithm for solving combinatorial optimization problems. In Sections 8.4–8.6 we discuss the applications of the CE method to several problems, such as the max-cut problem and the TSP, and provide supportive numerical results on the performance of the algorithm. In Sections 8.7 and 8.8, we show how the CE method can deal with continuous and noisy optimization problems, respectively. Last, in Section 8.9, we introduce a nonparametric version of the CE method, called the *MinxEnt* method.

8.2 ESTIMATION OF RARE-EVENT PROBABILITIES

In this section we apply the CE method in the context of efficient estimation of small probabilities. Consider, in particular, the estimation of

$$\ell = \mathbb{P}_{\mathbf{u}}(S(\mathbf{X}) \geq \gamma) = \mathbb{E}_{\mathbf{u}}\left[I_{\{S(\mathbf{X}) \geq \gamma\}}\right] \tag{8.1}$$

for some fixed level γ. Here $S(\mathbf{X})$ is the sample performance, $\mathbf{X}$ is a random vector with pdf $f(\cdot; \mathbf{u})$ belonging to some parametric family $\{f(\cdot; \mathbf{v}), \mathbf{v} \in \mathscr{V}\}$, and $\{S(\mathbf{X}) \geq \gamma\}$ is assumed to be a rare event. We can estimate ℓ using the likelihood ratio estimator (see also (5.58))

$$\widehat{\ell} = \frac{1}{N} \sum_{k=1}^{N} I_{\{S(\mathbf{X}_k) \geq \gamma\}} W(\mathbf{X}_k; \mathbf{u}, \mathbf{v}), \tag{8.2}$$

where $\mathbf{X}_1, \ldots, \mathbf{X}_N$ is a random sample from $f(\mathbf{x}; \mathbf{v})$ and $W(\mathbf{X}_k; \mathbf{u}, \mathbf{v}) = f(\mathbf{X}_k; \mathbf{u})/f(\mathbf{X}_k; \mathbf{v})$ is the likelihood ratio.

■ **EXAMPLE 8.1 Stochastic Shortest Path**

Let us return to Example 5.15 (see also Example 5.1), where the objective is to efficiently estimate the probability ℓ that the shortest path from node A to node B in the network of Figure 8.1 has a length of at least γ. The random lengths $X_1, \ldots, X_5$ of the links are assumed to be independent and exponentially distributed with means $u_1, \ldots, u_5$, respectively.

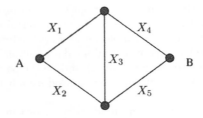

Figure 8.1: Shortest path from A to B.

We define $\mathbf{X} = (X_1, \ldots, X_5)$, $\mathbf{u} = (u_1, \ldots, u_5)$, and

$$S(\mathbf{X}) = \min\{X_1 + X_4, \ X_1 + X_3 + X_5, \ X_2 + X_5, \ X_2 + X_3 + X_4\},$$

so that we can cast the problem in the framework of (8.1). As explained in Example 5.15, we can estimate (8.1) via (8.2) by drawing $X_1, \ldots, X_5$ independently from exponential distributions that are possibly *different* from the original ones. That is, $X_i \sim \mathsf{Exp}(v_i^{-1})$ instead of $X_i \sim \mathsf{Exp}(u_i^{-1})$, $i = 1, \ldots, 5$. The corresponding likelihood ratio was given in (5.72).

The challenge is how to select a vector $\mathbf{v} = (v_1, \ldots, v_5)$ that gives the most accurate estimate of ℓ for a given simulation effort. In the toy Example 5.15, this was achieved by first choosing the trial vector $\mathbf{w}$ equal to $\mathbf{u}$ and then applying the CE updating formula (5.68), possibly iterating the latter. This approach was possible because the event $\{S(\mathbf{x}) \geqslant \gamma\}$ was *not rare*. However, for the current problem, (5.68) cannot be applied directly, since for rare events it returns, with high probability, the indeterminate expression $\frac{0}{0}$. To overcome this difficulty, we will use a different approach to selecting a good $\mathbf{v}$ by adopting a *two-stage* procedure where *both the level γ and the reference parameters $\mathbf{v}$ are updated*. One of the strengths of the CE method for rare-event simulation is that it provides a fast way to estimate accurately the optimal parameter vector $\mathbf{v}^*$.

Returning to the general situation, recall from Section 5.7 that for estimation problems of the form (8.1) the ideal (zero variance) importance sampling density is given by

$$g^*(\mathbf{x}) = \frac{f(\mathbf{x}; \mathbf{u}) \, I_{\{S(\mathbf{x}) \geqslant \gamma\}}}{\ell} \,,$$

which is the conditional pdf of $\mathbf{X}$ given $S(\mathbf{X}) \geqslant \gamma$. The idea behind the CE method is to get as close as possible to the optimal importance sampling distribution by using the Kullback–Leibler CE distance as a measure of closeness. Using a parametric class of densities $\{f(\mathbf{x}; \mathbf{v}), \mathbf{v} \in \mathscr{V}\}$, we see (from (5.60)) that the optimal reference parameter $\mathbf{v}^*$ is given by

$$\mathbf{v}^* = \operatorname*{argmax}_{\mathbf{v} \in \mathscr{V}} \mathbb{E}_{\mathbf{u}}[I_{\{S(\mathbf{X}) \geqslant \gamma\}} \ln f(\mathbf{X}; \mathbf{v})] \,. \tag{8.3}$$

We can, in principle, estimate $\mathbf{v}^*$ as

$$\operatorname*{argmax}_{\mathbf{v} \in \mathscr{V}} \frac{1}{N} \sum_{k=1}^{N} I_{\{S(\mathbf{X}_k) \geqslant \gamma\}} \ln f(\mathbf{X}_k; \mathbf{v}) \,, \tag{8.4}$$

with $\mathbf{X}_1, \ldots, \mathbf{X}_N \sim f(\cdot; \mathbf{u})$ — that is, using the stochastic counterpart of (8.3). However, as mentioned in Example 8.1, this is void of meaning if $\{S(\mathbf{X}) \geqslant \gamma\}$ is a rare event under $f(\cdot; \mathbf{u})$, since then most likely all indicators in the sum above will be zero.

To circumvent this problem, we will use a multilevel approach where we generate a sequence of reference parameters $\{\mathbf{v}_t, \, t \geqslant 0\}$ and a sequence of levels $\{\gamma_t, \, t \geqslant 1\}$ while iterating in both γ_t and $\mathbf{v}_t$. Our ultimate goal is to have $\mathbf{v}_t$ close to $\mathbf{v}^*$ after some number of iterations and to use $\mathbf{v}_t$ in the importance sampling density $f(\cdot; \mathbf{v}_t)$ to estimate ℓ.

We start with $\mathbf{v}_0 = \mathbf{u}$. Let ϱ be a not too small number, say $10^{-2} \leqslant \varrho \leqslant 10^{-1}$. In the first iteration, we choose $\mathbf{v}_1$ to be the optimal parameter for estimating $\mathbb{P}_{\mathbf{u}}(S(\mathbf{X}) \geqslant \gamma_1)$, where γ_1 is the $(1 - \varrho)$-quantile of $S(\mathbf{X})$. That is, γ_1 is the largest real number for which

$$\mathbb{P}_{\mathbf{u}}(S(\mathbf{X}) \geqslant \gamma_1) \geqslant \varrho \,.$$

Thus, if we simulate under $\mathbf{u}$, then level γ_1 is reached with a reasonably high probability of around ϱ. This enables us to estimate both γ_1 and $\mathbf{v}_1$ via Monte

Carlo simulation. Namely we can estimate γ_1 from a random sample $\mathbf{X}_1, \ldots, \mathbf{X}_N$ from $f(\cdot; \mathbf{u})$ as follows: Calculate the performances $S(\mathbf{X}_i)$ for all i, and order them from smallest to largest, that is, $S_{(1)} \leqslant \ldots \leqslant S_{(N)}$. Then γ_1 is estimated via the sample $(1 - \varrho)$-quantile $\widehat{\gamma}_1 = S_{(\lceil (1-\varrho)N \rceil)}$, where $\lceil a \rceil$ denotes the smallest integer larger than or equal to a (the so-called *ceiling* of a). The reference parameter $\mathbf{v}_1$ can be estimated via (8.4), replacing γ with the estimate of γ_1. Note that we can use here the *same* sample for estimating both $\mathbf{v}_1$ and γ_1. This means that $\mathbf{v}_1$ is estimated on the basis of the $\lceil \varrho N \rceil$ best samples, that is, the samples $\mathbf{X}_i$ for which $S(\mathbf{X}_i)$ is greater than or equal to $\widehat{\gamma}_1$. These form the *elite samples* in the first iteration; let N^e denote the number of elite samples.

In the subsequent iterations, we repeat these steps. Thus we have the following two updating phases, starting from $\mathbf{v}_0 = \widehat{\mathbf{v}}_0 = \mathbf{u}$:

1. **Adaptive updating of γ_t.** For a fixed $\mathbf{v}_{t-1}$, let γ_t be the $(1 - \varrho)$-quantile of $S(\mathbf{X})$ under $\mathbf{v}_{t-1}$. To estimate γ_t, draw a random sample $\mathbf{X}_1, \ldots, \mathbf{X}_N$ from $f(\cdot; \widehat{\mathbf{v}}_{t-1})$ and evaluate the sample $(1 - \varrho)$-quantile $\widehat{\gamma}_t$.

2. **Adaptive updating of $\mathbf{v}_t$.** For fixed γ_t and $\mathbf{v}_{t-1}$, derive $\mathbf{v}_t$ as

$$\mathbf{v}_t = \underset{\mathbf{v} \in \mathscr{V}}{\operatorname{argmax}} \, \mathbb{E}_{\mathbf{v}_{t-1}} \left[I_{\{S(\mathbf{X}) \geqslant \gamma_t\}} W(\mathbf{X}; \mathbf{u}, \mathbf{v}_{t-1}) \ln f(\mathbf{X}; \mathbf{v}) \right] . \qquad (8.5)$$

The stochastic counterpart of (8.5) is as follows: for fixed $\widehat{\gamma}_t$ and $\widehat{\mathbf{v}}_{t-1}$, derive $\widehat{\mathbf{v}}_t$ as the solution

$$\widehat{\mathbf{v}}_t = \underset{\mathbf{v} \in \mathscr{V}}{\operatorname{argmax}} \, \frac{1}{N} \sum_{\mathbf{X}_k \in \mathcal{E}_t} W(\mathbf{X}_k; \mathbf{u}, \widehat{\mathbf{v}}_{t-1}) \ln f(\mathbf{X}_k; \mathbf{v}) , \qquad (8.6)$$

where $\mathcal{E}_t$ is the *set of elite samples* in the t-th iteration, that is, the samples $\mathbf{X}_k$ for which $S(\mathbf{X}_k) \geqslant \widehat{\gamma}_t$.

The procedure terminates when at some iteration T a level $\widehat{\gamma}_T$ is reached that is at least γ, and thus the original value of γ can be used without our getting too few samples. We then reset $\widehat{\gamma}_T$ to γ, reset the corresponding elite set, and deliver the final reference parameter $\widehat{\mathbf{v}}^*$, again using (8.6). This $\widehat{\mathbf{v}}^*$ is then used in (8.2) to estimate ℓ.

Algorithm 8.2.1: Main CE Algorithm for Rare-Event Estimation

---

1 Initialize $\widehat{\mathbf{v}}_0 \leftarrow \mathbf{u}$, $N^e \leftarrow \lceil (1 - \varrho)N \rceil$, $t \leftarrow 0$.
2 Continue $\leftarrow$ **true**
3 while Continue **do**
4 $\quad$ $t \leftarrow t + 1$
5 $\quad$ Generate a random sample $\mathbf{X}_1, \ldots, \mathbf{X}_N$ from the density $f(\cdot; \widehat{\mathbf{v}}_{t-1})$.
6 $\quad$ Calculate the performances $S(\mathbf{X}_1), \ldots, S(\mathbf{X}_N)$.
7 $\quad$ Order the performances from smallest to largest: $S_{(1)} \leqslant \ldots \leqslant S_{(N)}$.
8 $\quad$ Let $\widehat{\gamma}_t \leftarrow S_{(N^e)}$.
9 $\quad$ **if** $\widehat{\gamma}_t > \gamma$ **then**
10 $\quad\quad$ $\widehat{\gamma}_t \leftarrow \gamma$
11 $\quad\quad$ Continue $\leftarrow$ **false**
12 $\quad$ Use the *same* sample $\mathbf{X}_1, \ldots, \mathbf{X}_N$ to solve the stochastic program (8.6).
13 Generate a sample $\mathbf{X}_1, \ldots, \mathbf{X}_{N_1}$ according to the pdf $f(\cdot; \widehat{\mathbf{v}}_T)$, where T is the final iteration number, t, and estimate ℓ via (8.2).

Remark 8.2.1 In typical applications the sample size N in Line 5 can be chosen smaller than the final sample size N_1 in Line 13.

Note that Algorithm 8.2.1 breaks down the complex problem of estimating the very small probability ℓ into a sequence of simple problems, generating a sequence of pairs $\{(\widehat{\gamma}_t, \widehat{\mathbf{v}}_t)\}$, depending on ϱ, which is called the *rarity parameter*. Convergence of Algorithm 8.2.1 is discussed in [45]. Other convergence proofs for the CE method may be found in [34] and [13].

Remark 8.2.2 (Maximum Likelihood Estimator) Optimization problems of the form (8.6) appear frequently in statistics. In particular, if the W term is omitted — which will turn out to be important in CE optimization — we can recast (8.6) as

$$\widehat{\mathbf{v}}_t = \underset{\mathbf{v}}{\operatorname{argmax}} \prod_{\mathbf{X}_k \in \mathcal{E}_t} f(\mathbf{X}_k; \mathbf{v}) ,$$

where the product is the joint density of the elite samples. Consequently, $\widehat{\mathbf{v}}_t$ is chosen such that the joint density of the elite samples is maximized. Viewed as a function of the parameter $\mathbf{v}$, rather than of the data $\{\mathcal{E}_t\}$, this joint density is called the *likelihood*. In other words, $\widehat{\mathbf{v}}_t$ is the *maximum likelihood estimator* (it maximizes the likelihood) of $\mathbf{v}$ based on the elite samples. When the W term is present, the form of the updating formula remains similar. Recall from Section 5.7 that for exponential families the updating rules for $\widehat{\mathbf{v}}_t$ can be obtained analytically; see also Section A.3 of the Appendix.

To gain a better understanding of the CE algorithm, we also present its *deterministic* version.

Algorithm 8.2.2: Deterministic Version of the CE Algorithm

---

1 Initialize $\widehat{\mathbf{v}}_0 \leftarrow \mathbf{u}$, $N^{\mathrm{e}} \leftarrow \lceil (1 - \varrho)N \rceil$, $t \leftarrow 0$.
2 Continue $\leftarrow$ **true**
3 **while** Continue **do**
4     $t \leftarrow t + 1$
5   Calculate γ_t as

$$\gamma_t \leftarrow \max \left\{ s \, : \, \mathbb{P}_{\mathbf{v}_{t-1}}(S(\mathbf{X}) \geqslant s) \geqslant \varrho \right\} . \tag{8.7}$$

   if $\gamma_t > \gamma$ **then**
6        $\gamma_t \leftarrow \gamma$
7   Continue $\leftarrow$ **false**
8   Calculate $\mathbf{v}_t$ (see (8.5)) as
      $\mathbf{v}_t \leftarrow \underset{\mathbf{v}}{\operatorname{argmax}} \mathbb{E}_{\mathbf{v}_{t-1}} \left[I_{\{S(\mathbf{X}) \geqslant \gamma_t\}} W(\mathbf{X}; \mathbf{u}, \mathbf{v}_{t-1}) \ln f(\mathbf{X}; \mathbf{v}) \right] . \tag{8.8}$

Note that, when compared with Algorithm 8.2.1, Line 13 is redundant in Algorithm 8.2.2.

To provide further insight into Algorithm 8.2.1, we will apply it step by step in a number of toy examples.

■ **EXAMPLE 8.2 Exponential Distribution**

Let us revisit Examples 5.9, 5.11, and 5.13, where the goal was to estimate, via Monte Carlo simulation, the probability $\ell = \mathbb{P}_u(X \geqslant \gamma)$, with $X \sim \mathsf{Exp}(u^{-1})$.

Suppose that γ is large in comparison with u, so that $\ell = e^{-\gamma/u}$ is a rare-event probability. The updating formula for $\widehat{v}_t$ in (8.6) follows from the optimization of

$$\sum_{X_k \in \mathcal{E}_t} W_k \ln\left(v^{-1} e^{-X_k/v}\right) = -\sum_{X_k \in \mathcal{E}_t} W_k \ln v - \sum_{X_k \in \mathcal{E}_t} W_k \frac{X_k}{v} ,$$

where $W_k = e^{-X_k(u^{-1} - v^{-1})} v/u$. To find the maximum of the right-hand side, we take derivatives and equate the result to 0:

$$-\sum_{X_k \in \mathcal{E}_t} \frac{W_k}{v} + \sum_{X_k \in \mathcal{E}_t} \frac{W_k X_k}{v^2} = 0 .$$

Solving this for v yields $\widehat{v}_t$. Thus,

$$\widehat{v}_t = \frac{\sum_{X_k \in \mathcal{E}_t} W_k X_k}{\sum_{X_k \in \mathcal{E}_t} W_k} . \tag{8.9}$$

In other words, $\widehat{v}_t$ is simply the sample mean of the elite samples weighted by the likelihood ratios. Note that, without the weights $\{W_k\}$, we would simply have the maximum likelihood estimator of v for the $\mathsf{Exp}(v^{-1})$ distribution based on the elite samples described in Remark 8.2.2. Note further that the updating formula (8.9) follows directly from (5.68). Similarly, the *deterministic* updating formula (8.8) gives

$$v_t = \frac{\mathbb{E}_{\mathbf{u}}\left[I_{\{X \geqslant \gamma_t\}} X\right]}{\mathbb{E}_{\mathbf{u}}\left[I_{\{X \geqslant \gamma_t\}}\right]} = \mathbb{E}_{\mathbf{u}}[X \mid X \geqslant \gamma_t] = u + \gamma_t ,$$

where γ_t is the $(1 - \varrho)$-quantile of the $\mathsf{Exp}(v_{t-1}^{-1})$ distribution. Thus, $\gamma_t = -v_{t-1} \ln \varrho$.

Assume, for concreteness, that $u = 1$ and $\gamma = 32$, which corresponds to $\ell \approx 1.27 \cdot 10^{-14}$. Table 8.1 shows the evolution of $\widehat{\gamma}_t$ and $\widehat{v}_t$ for $\varrho = 0.05$ using sample size $N = 1000$. Note that iteration $t = 0$ corresponds to the original exponential pdf with expectation $u = 1$, while iterations $t = 1, 2, 3$, correspond to exponential pdfs with expectations $\widehat{v}_t$, $t = 1, 2, 3$, respectively. Figure 8.2 illustrates the iterative procedure. We see that Algorithm 8.2.1 requires three iterations to reach the final level $\widehat{\gamma}_3 = 32$. In the third iteration the lowest value of the elite samples, $S_{(N^e)}$, turns out to be greater than 32, so that in the final Line 12 of the algorithm we use $\widehat{\gamma}_3 = \gamma = 32$ instead. The corresponding reference parameter $\widehat{v}_3$ is found to be 32.82. Note that both parameters $\widehat{\gamma}_t$ and $\widehat{v}_t$ increase gradually, each time "blowing up" the tail of the exponential pdf.

The final step of Algorithm 8.2.1 now invokes the likelihood ratio estimator (8.2) to estimate ℓ, and uses a sample size N_1 that is typically larger than N.

Table 8.1: Evolution of $\widehat{\gamma}_t$ and $\widehat{v}_t$ for $\varrho = 0.05$ with $\gamma = 32$ and $N = 1000$ samples.

t	$\widehat{\gamma}_t$	$\widehat{v}_t$
0	–	1
1	2.91	3.86
2	11.47	12.46
3	32	32.82

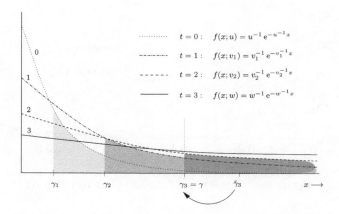

Figure 8.2: A three-level realization of Algorithm 8.2.1. Each shaded region has area ϱ.

■ EXAMPLE 8.3 Degeneracy

When γ is the maximum of $S(\mathbf{x})$, no "overshooting" of γ in Algorithm 8.2.1 occurs and therefore γ_t does not need to be reset. In such cases the sampling pdf may *degenerate* toward the *atomic* pdf that has all its mass concentrated at the points $\mathbf{x}$ where $S(\mathbf{x})$ is maximal. As an example, suppose we use a $\mathsf{Beta}(v, 1), v \geqslant 1$ family of importance sampling distributions, with nominal parameter $u = 1$ (corresponding to the uniform distribution), and take $S(X) = X$ and $\gamma = 1$. We find the updating formula for v from the optimization of

$$\sum_{X_k \in \mathscr{E}_k} W_k \ln(v X_k^{v-1}) = \sum_{X_k \in \mathscr{E}_k} W_k \ln v + \sum_{X_k \in \mathscr{E}_k} W_k (v - 1) \ln X_k,$$

with $W_k = 1/(v X_k^{v-1})$. Hence,

$$\widehat{v}_t = \frac{\sum_{X_k \in \mathscr{E}_k} W_k}{-\sum_{X_k \in \mathscr{E}_k} W_k \ln X_k}.$$

Table 8.2 and Figure 8.3 show the evolution of parameters in the CE algorithm using $\varrho = 0.8$ and $N = 1000$. We see that $\widehat{\gamma}_t$ rapidly increases to γ and that the sampling pdf degenerates to the atomic density with mass at 1.

Table 8.2: The evolution of $\widehat{\gamma}_t$ and $\widehat{v}_t$ for the $\mathsf{Beta}(v, 1)$ example, with $\varrho = 0.8$ and $\gamma = 1$, using $N = 1000$ samples.

t	$\widehat{\gamma}_t$	$\widehat{v}_t$	t	$\widehat{\gamma}_t$	$\widehat{v}_t$
0	–	1	5	0.896	31.2
1	0.207	1.7	6	0.949	74.3
2	0.360	3.1	7	0.979	168.4
3	0.596	6.4	8	0.990	396.5
4	0.784	14.5	9	0.996	907.7

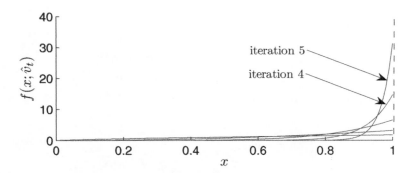

Figure 8.3: Degeneracy of the sampling distribution.

■ **EXAMPLE 8.4   Coin Flipping**

Consider the experiment where we flip n fair coins. We can describe this experiment via n independent Bernoulli random variables, $X_1, \ldots, X_n$, each with success parameter $1/2$. We write $\mathbf{X} = (X_1, \ldots, X_n) \sim \mathsf{Ber}(\mathbf{u})$, where $\mathbf{u} = (1/2, \ldots, 1/2)$ is the vector of success probabilities. Note that the range of $\mathbf{X}$ (the set of possible values it can take) contains 2^n elements. Suppose that we are interested in estimating $\ell = \mathbb{P}_{\mathbf{u}}(S(\mathbf{X}) \geqslant \gamma)$, with $S(\mathbf{X}) = \sum_{k=1}^n X_k$. We want to employ importance sampling using $\mathbf{X} \sim \mathsf{Ber}(\mathbf{p})$ for a possibly different parameter vector $\mathbf{p} = (p_1, \ldots, p_n)$. Consider two cases: (a) $\gamma = (n+1)/2$ (with n odd) and (b) $\gamma = n$. It is readily seen that for both cases (a) and (b) the optimal importance sampling parameter vector is $\mathbf{p}^* = (1/2, \ldots, 1/2)$ and $\mathbf{p}^* = (1, \ldots, 1)$, respectively. The corresponding probabilities are $\ell = \frac{1}{2}$ and $\ell = \frac{1}{2^n}$, respectively. Note that in the first case ℓ is not a rare-event probability, but it is so for the second case (provided that n is large). Note also that in the second case $\mathsf{Ber}(\mathbf{p}^*)$ corresponds to a *degenerated* distribution that places all probability mass at the point $(1, 1, \ldots, 1)$.

Since $\{\mathsf{Ber}(\mathbf{p})\}$ forms an exponential family that is parameterized by the mean, it immediately follows from (5.68) that the updating formula for $\mathbf{p}$ in Algorithm 8.2.1 at the t-th iteration coincides with (8.9) and is given by

$$\widehat{p}_{t,i} = \frac{\sum_{\mathbf{X}_k \in \mathcal{E}_t} W_k X_{ki}}{\sum_{\mathbf{X}_k \mathcal{E}_t} W_k}, \quad i = 1, \ldots, n, \tag{8.10}$$

where X_{ki} is the i-th component of the k-th sample vector $\mathbf{X}_k \sim \mathsf{Ber}(\widehat{\mathbf{p}}_{t-1})$, and W_k is the corresponding likelihood ratio:

$$W_k = \prod_{i=1}^n q_i^{X_{ki}} r_i^{1-X_{ki}},$$

with $q_i = p_{0,i}/\widehat{p}_{t-1,i}$ and $r_i = (1 - p_{0,i})/(1 - \widehat{p}_{t-1,i})$, $i = 1, \ldots, n$. Thus, the i-th probability is updated as a weighted average of the number of 1s in the i-th position over all vectors in the elite sample.

As we will see below, this simple coin flipping example can shed light on how rare-events estimation is connected with combinatorial optimization.

Remark 8.2.3 It is important to note that if we employ the deterministic CE Algorithm 8.2.2 to any rare-event-type problem where the underlying distributions have finite supports *without fixing γ in advance*, it will iterate until it reaches some γ, denoted as γ_* (not necessarily the true optimal γ^*), and then stop. The corresponding importance sampling pdf $f(\mathbf{x}; \mathbf{v}_*)$ will be *degenerated*. For the coin flipping example given above, we will typically have in case (b) that $\gamma_* = \gamma^* = n$. The main Algorithm 8.2.1 behaves similarly, but in a stochastic rather than a deterministic sense. More precisely, for pdfs with finite supports and γ not fixed in advance, it will generate a tuple $(\widehat{\gamma}_T, \widehat{\mathbf{v}}_T)$ with $f(\mathbf{x}; \widehat{\mathbf{v}}_T)$ corresponding again typically to a degenerate pdf. This property of Algorithms 8.2.2 and 8.2.1 will be of crucial importance when we come to combinatorial optimization problems in the next section. As mentioned, a combinatorial optimization problem can be viewed as a rare-event estimation problem in the sense that its optimal importance sampling pdf $f(\mathbf{x}; \mathbf{v}^*)$ is degenerated and coincides with the pdf generated by the deterministic rare-event Algorithm 8.2.2, provided that it keeps iterating in γ without fixing it in advance.

In the next example, we illustrate the behavior of Algorithm 8.2.1 when applied to a typical static simulation problem of estimating $\ell = \mathbb{P}(S(\mathbf{X}) \geqslant \gamma)$. Note that the likelihood ratio estimator $\widehat{\ell}$ of ℓ in (8.2) is of the form $\widehat{\ell} = N^{-1} \sum_{k=1}^{n} Z_k$. We measure the efficiency of the estimator by its relative error (RE), which (recall (4.6)) is defined as

$$\mathrm{RE} = \mathrm{Var}(\widehat{\ell})^{1/2} / \mathbb{E}[\widehat{\ell}]$$

and which is estimated by $S/(\widehat{\ell}\sqrt{N})$, with

$$S^2 = N^{-1} \sum_{k=1}^{N} Z_k^2 - (\widehat{\ell})^2$$

being the sample variance of the $\{Z_i\}$. Assuming asymptotic normality of the estimator, the confidence intervals now follow in a standard way. For example, a 95% relative confidence interval for ℓ is given by

$$\widehat{\ell} \pm 1.96\,\widehat{\ell}\,\mathrm{RE}\ .$$

■ EXAMPLE 8.5 Stochastic Shortest Path: Example 8.1 (Continued)

Consider again the stochastic shortest path graph of Figure 8.1. Let us take the same nominal parameter vector $\mathbf{u}$ as in Example 5.15, that is, $\mathbf{u} = (1, 1, 0.3, 0.2, 0.1)$, and estimate the probability ℓ that the minimum path length is greater than $\gamma = 6$. Note that in Example 5.15 $\gamma = 1.5$ is used, which gives rise to an event that is not rare.

Crude Monte Carlo (CMC), with 10^8 samples — a very large simulation effort — gave an estimate $8.01 \cdot 10^{-6}$ with an estimated relative error of 0.035.

To apply Algorithm 8.2.1 to this problem, we need to establish the updating rule for the reference parameter $\mathbf{v} = (v_1, \ldots, v_5)$. Since the components $X_1, \ldots, X_5$ are independent and form an exponential family parameterized by the mean, this updating formula follows immediately from (5.68), that is,

$$\widehat{v}_{t,i} = \frac{\sum_{k=1}^{N} I_{\{S(\mathbf{X}_k) \geqslant \widehat{\gamma}_t\}} W(\mathbf{X}_k; \mathbf{u}, \widehat{\mathbf{v}}_{t-1})\, X_{ki}}{\sum_{k=1}^{N} I_{\{S(\mathbf{X}_k) \geqslant \widehat{\gamma}_t\}} W(\mathbf{X}_k; \mathbf{u}, \widehat{\mathbf{v}}_{t-1})}, \quad i = 1, \ldots, 5\,, \tag{8.11}$$

with $W(\mathbf{X}; \mathbf{u}, \mathbf{v})$ given in (5.72).

We take in all our experiments with Algorithm 8.2.1 the rarity parameter $\varrho = 0.1$, the sample size in Line 5 of the algorithm $N = 10^3$, and for the final sample size $N_1 = 10^5$. Table 8.3 displays the results of Lines 1–12 of the CE algorithm. We see that after five iterations level $\gamma = 6$ is reached.

Table 8.3: Convergence of the sequence $\{(\widehat{\gamma}_t, \widehat{\mathbf{v}}_t)\}$.

t	$\widehat{\gamma}_t$	$\widehat{\mathbf{v}}_t$				
0		1.0000	1.0000	0.3000	0.2000	0.1000
1	1.1656	1.9805	2.0078	0.3256	0.2487	0.1249
2	2.1545	2.8575	3.0006	0.2554	0.2122	0.0908
3	3.1116	3.7813	4.0858	0.3017	0.1963	0.0764
4	4.6290	5.2803	5.6542	0.2510	0.1951	0.0588
5	6.0000	6.7950	6.7094	0.2882	0.1512	0.1360

Using the estimated optimal parameter vector of $\widehat{\mathbf{v}}_5 = (6.7950, 6.7094, 0.2882, 0.1512, 0.1360)$, we get in Line 13 of the CE algorithm an estimate of $7.85 \cdot 10^{-6}$, with an estimated relative error of 0.035 — the same as for the CMC method with 10^8 samples. However, whereas the CMC method required more than an hour of computation time, the CE algorithm was finished in only one second, with a Matlab implementation on a 1500 MHz computer. We see that with a minimal amount of work, we have achieved a dramatic reduction of the simulation effort.

Table 8.4 presents the performance of Algorithm 8.2.1 for the stochastic shortest path model presented above, where instead of the exponential random variables we used $\mathsf{Weib}(\alpha_i, \lambda_i)$ random variables, with $\alpha_i = 0.2$ and $\lambda_i = u_i^{-1}$, $i = 1, \ldots, 5$, where the $\{u_i\}$ are the same as before, that is, $\mathbf{u} = (1, 1, 0.3, 0.2, 0.1)$.

Table 8.4: Evolution of $\widehat{\mathbf{v}}_t$ for estimating the optimal parameter $\mathbf{v}^*$ with the TLR method and $\alpha = 0.2$. The estimated probability is $\widehat{\ell} = 3.30 \cdot 10^{-6}$, RE = 0.03.

t	$\widehat{\gamma}_t$	$\widehat{v}_{1t}$	$\widehat{v}_{2t}$	$\widehat{v}_{3t}$	$\widehat{v}_{4t}$	$\widehat{v}_{5t}$
0		1	1	1	1	1
1	3.633	2.0367	2.1279	0.9389	1.3834	1.4624
2	100.0	3.2690	3.3981	1.1454	1.3674	1.2939
3	805.3	4.8085	4.7221	0.9660	1.1143	0.9244
4	5720	6.6789	6.7252	0.6979	0.9749	1.0118
5	10000	7.5876	7.8139	1.0720	1.3152	1.2252

The Weibull distribution with shape parameter α less than 1 is an example of a *heavy-tailed* distribution. We use the TLR method (see Section 5.11) to estimate ℓ for $\gamma = 10,000$. Specifically, we first write (see (5.101)) $X_k = u_k Z_k^{1/\alpha}$, with $Z_k \sim \mathsf{Exp}(1)$, and then use importance sampling on the $\{Z_k\}$, changing the mean of Z_k from 1 to v_k, $k = 1, \ldots, 5$. The corresponding updating formula is again of the form (8.11), namely,

$$\widehat{v}_{t,i} = \frac{\sum_{k=1}^{N} I_{\{\widetilde{S}(\mathbf{Z}_k) \geqslant \widehat{\gamma}_t\}} \widetilde{W}(\mathbf{Z}_k; \mathbf{1}, \widehat{\mathbf{v}}_{t-1}) \, Z_{ki}}{\sum_{k=1}^{N} I_{\{\widetilde{S}(\mathbf{Z}_k) \geqslant \widehat{\gamma}_t\}} \widetilde{W}(\mathbf{Z}_k; \mathbf{1}, \widehat{\mathbf{v}}_{t-1})}, \quad i = 1, \ldots, 5 \,,$$

with $\widetilde{W}(\mathbf{Z}; \mathbf{1}, \mathbf{v})$ the likelihood ratio, and $\widetilde{S}(\mathbf{Z}) = S(\mathbf{X})$. Note that the "nominal" parameter here is $\mathbf{1} = (1, 1, 1, 1, 1)$, rather than $(1, 1, 0.3, 0.2, 0.1)$. Instead of using the TLR method, one could use the standard CE method here, where the components are sampled from $\{\mathsf{Weib}(\alpha, v_i^{-1})\}$ and the $\{v_i\}$ are updated adaptively. One would obtain results similar (estimate and relative error) to those for the TLR case. The TLR is a convenient and quite general tool for importance sampling simulation, but it does not provide additional variance reduction; see also Exercises 8.5 and 8.6.

8.2.1 Root-Finding Problem

Many applications require one to estimate, for given ℓ, the *root*, γ, of the nonlinear equation

$$\mathbb{P}_{\mathbf{u}}(S(\mathbf{X}) \geqslant \gamma) = \mathbb{E}_{\mathbf{u}}[I_{\{S(\mathbf{X}) \geqslant \gamma\}}] = \ell \tag{8.12}$$

rather than estimate ℓ itself. We call such a problem a *root-finding* problem.

An estimate of γ in (8.12) based on the sample equivalent of $\mathbb{E}_{\mathbf{u}}[I_{\{S(\mathbf{X}) \geqslant \gamma\}}]$ can be obtained, for example, via stochastic approximation; see Chapter 7 and [46]. Alternatively, we can obtain γ using the CE method. The aim is to find a good trial vector $\widehat{\mathbf{v}}_T$ such that γ can be estimated as the smallest number $\widehat{\gamma}$ such that

$$\frac{1}{N_1} \sum_{k=1}^{N_1} I_{\{S(\mathbf{X}_k) \geqslant \widehat{\gamma}\}} W(\mathbf{X}_k; \mathbf{u}, \widehat{\mathbf{v}}_T) \leqslant \ell \,. \tag{8.13}$$

In particular our main Algorithm 8.2.1 can be modified as follows.

Algorithm 8.2.3: Root-Finding Algorithm

---

1 Initialize $\widehat{\mathbf{v}}_0 \leftarrow \mathbf{u}$, $N^{\mathrm{e}} \leftarrow \lceil (1 - \varrho) N \rceil$, $t \leftarrow 0$.
2 Continue $\leftarrow$ **true**
3 **while** Continue **do**
4 $\quad$ $t \leftarrow t + 1$
5 $\quad$ Generate a random sample $\mathbf{X}_1, \ldots, \mathbf{X}_N$ from the density $f(\cdot; \widehat{\mathbf{v}}_{t-1})$.
6 $\quad$ Calculate the performances $S(\mathbf{X}_1), \ldots, S(\mathbf{X}_N)$.
7 $\quad$ Order the performances from smallest to largest: $S_{(1)} \leqslant \ldots \leqslant S_{(N)}$.
8 $\quad$ Let $\widehat{\gamma}_t \leftarrow S_{(N^{\mathrm{e}})}$.
9 $\quad$ Let $\widehat{\ell}_t \leftarrow \frac{1}{N} \sum_{k=1}^{N} I_{\{S(\mathbf{X}_k) \geqslant \widehat{\gamma}_t\}} W(\mathbf{X}_k; \mathbf{u}, \widehat{\mathbf{v}}_{t-1})$.
10 $\quad$ **if** $\widehat{\ell}_t < \ell$ **then**
11 $\quad\quad$ $\widehat{\ell}_t \leftarrow \ell$
12 $\quad\quad$ Continue $\leftarrow$ **false**
13 $\quad$ Determine $\widehat{\mathbf{v}}_t$ via (8.6) using the *same* sample $\mathbf{X}_1, \ldots, \mathbf{X}_N$.
14 Estimate γ via (8.13) using a sample $\mathbf{X}_1, \ldots, \mathbf{X}_{N_1} \sim f(\cdot; \widehat{\mathbf{v}}_T)$, where T is the final iteration number, t.

8.2.2   Screening Method for Rare Events

Here we show how the screening method, introduced in Section 5.12, works for estimating rare-event probabilities of the form $\ell = \mathbb{P}_{\mathbf{u}}(S(\mathbf{X}) \geq \gamma) = \mathbb{E}_{\mathbf{u}}[I_{\{S(\mathbf{X}) \geq \gamma\}}]$, where we assume, as in Section 5.12, that the components of $\mathbf{X}$ are independent, that each component is distributed according to a one-dimensional exponential family parameterized by the mean, and that $S(\mathbf{x})$ (and hence $H(\mathbf{x}) = I_{\{S(\mathbf{x}) \geq \gamma\}}$) is monotonically increasing in each component of $\mathbf{x}$. In particular, we will present a modification of the two-stage Algorithm 5.12.1.

As in Algorithm 5.12.1, the main idea of the first stage of our modified algorithm is to identify the bottleneck parameters *without involving the likelihood ratio*. One might wonder how this could be possible given the fact that the estimation of the rare-event probability ℓ is essentially based on likelihood ratios. The trick is to execute the first stage (the screening part) by replacing γ with some γ_0 such that $\ell_0 = \mathbb{P}_{\mathbf{u}}(S(\mathbf{X}) \geq \gamma_0)$ *is not* a rare-event probability, say $10^{-2} \leq \ell_0 \leq 10^{-1}$. As soon as γ_0 is determined, the execution of the first stage is similar to the one in Algorithm 5.12.1. It reduces to finding the estimator, say $\widehat{\mathbf{v}}_0$, of the optimal parameter vector $\mathbf{v}_0^*$ obtained from (8.4), where γ is replaced by γ_0. Note that (8.4) does not contain the likelihood ratio term $W(\mathbf{X}; \mathbf{u}, \mathbf{w})$. It is important to note again that the components of $\mathbf{v}_0^*$ are at least as large as the corresponding elements of $\mathbf{u}$, and thus we can classify the bottleneck and nonbottleneck parameters according to the relative perturbation $\delta_i = \frac{\widehat{v}_i - u_i}{u_i}$, $i = 1, \ldots, n$, which is the core of the screening algorithm.

Below we present the modified version of the two-stage screening CE-SCR Algorithm 5.12.1 suitable for rare events. We use the same notation as in Section 5.12.

Algorithm 8.2.4: Two-Stage Screening Algorithm for Rare Events

---

input   : Sample size N, performance function S, level γ, tolerance δ, number of repetitions d.

output: Estimator $\widehat{\ell}_B$ of $\mathbb{P}(S(\mathbf{X}) \geq \gamma)$.

1 Initialize $B \leftarrow \{1, \ldots, n\}$.
2 Generate a sample $\mathbf{X}_1, \ldots, \mathbf{X}_N$ from $f(\mathbf{x}; \mathbf{u})$ and compute $\widehat{\gamma}_0$ as the $(1 - \varrho)$-sample quantile of the sample performances $\{S(\mathbf{X}_i)\}$.
3 **for** $t = 1$ **to** d **do**
4 $\quad$ Generate a sample $\mathbf{X}_1, \ldots, \mathbf{X}_N$ from $f(\mathbf{x}; \mathbf{u})$.
5 $\quad$ **for** $i = 1$ **to** n **do**
6 $\qquad$ $\widehat{v}_i \leftarrow \dfrac{\sum_{k=1}^{N} I_{\{S(\mathbf{X}_k) \geq \widehat{\gamma}_0\}} X_{ki}}{\sum_{k=1}^{N} I_{\{S(\mathbf{X}_k) \geq \widehat{\gamma}_0\}}}$
7 $\qquad$ $\delta_i \leftarrow \dfrac{\widehat{v}_i - u_i}{u_i}$ $\qquad$ // calculate the relative perturbation
8 $\qquad$ **if** $\delta_i < \delta$ **then**
9 $\qquad\quad$ $\widehat{v}_i = u_i$
10 $\qquad\quad$ $B \leftarrow B \setminus \{i\}$ $\qquad$ // remove i from the bottleneck set

11 Generate a sample $\mathbf{X}_1, \ldots, \mathbf{X}_N$ from $f(\mathbf{x}; \widehat{\mathbf{v}})$
12 $\widehat{\ell}_B \leftarrow \dfrac{1}{N} \sum_{k=1}^{N} I_{\{S(\mathbf{X}_k) \geq \widehat{\gamma}_0\}} W_B(\mathbf{X}_{kB}; \widehat{\mathbf{v}}_B)$
13 **return** $\widehat{\ell}_B$

8.2.2.1 Numerical Results Next, we present numerical studies with Algorithm 8.2.4 for the $m \times n$ bridge system in Figure 5.7 on page 176. We are interested in estimating the rare-event probability $\ell = \mathbb{P}(S(\mathbf{X}) \geqslant \gamma)$ that the length $S(\mathbf{X})$ of the shortest path through the graph is greater than or equal to γ, where

$$S(\mathbf{X}) = \min\{Y_{11} + \cdots + Y_{1n}, \ldots, Y_{m1} + \cdots + Y_{mn}\}$$

and the Y_{ij} are defined in (5.109). Note that the operator "max" in (5.110) is replaced by "min" here. The random variables X_{ijk} are assumed to be $\mathsf{Exp}(u_{ijk})$ distributed. As in the numerical example in Section 5.12.0.1, the $\{X_{ijk}\}$ are *not* parameterized by the mean here, so one needs to take instead $1/u_{ijk}$ and $1/\hat{v}_{ijk}$ to compare the relative perturbations as described above. For the same reason, the parameter values corresponding to the bottleneck elements should be *smaller* than those for the nonbottleneck ones. As in Section 5.12, we purposely select (in advance) some elements of our model to be bottlenecks.

Table 8.5 presents the performance of Algorithm 8.2.4 for the 2×2 model with eight bottlenecks, using $\delta = 0.1$, $\gamma = 6$ and the sample sizes $N = 50,000$ and $N_1 = 500,000$. In particular, we set the bottleneck parameters $u_{111}, u_{112}, u_{121}, u_{122}, u_{211}, u_{212}, u_{221}, u_{222}$ to 1 and the remaining 12 elements to 4.

Table 8.5: Performance of Algorithm 8.2.4 for the 2×2 model. We set $\delta = 0.1$, $\gamma = 6$, $N = 50,000$, $N_1 = 500,000$.

	CE	VM	CE-SCR	VM-SCR
Mean $\widehat{\ell}$	2.92E-8	2.96E-8	2.88E-8	2.81E-8
Max $\widehat{\ell}$	3.93E-8	3.69E-8	3.56E-8	3.29E-8
Min $\widehat{\ell}$	2.46E-8	2.65E-8	2.54E-8	2.45E-8
RE	0.166	0.102	0.109	0.077
CPU	6.03	9.31	6.56	9.12

From the results of Table 8.5 it follows that, for this relatively small model, both CE and VM perform similarly to their screening counterparts. We will see further on that as the complexity of the model increases, VM-SCR outperforms its three alternatives and, in particular, CE-SCR.

Table 8.6 presents the typical dynamics for detecting the bottleneck parameters at the first stage of Algorithm 8.2.4 for the 2×2 model above with 20 parameters, 8 of which are bottlenecks. Similar to Table 5.3, in Table 8.6 the 0s and 1s indicate which parameters are detected as nonbottleneck and bottleneck ones, respectively, and t denotes the iteration number at the first stage of the algorithm.

Table 8.6: Typical dynamics for detecting the bottleneck parameters at the first stage of Algorithm 8.2.4.

t	u_{111}	u_{112}	u_{113}	u_{114}	u_{115}	u_{121}	u_{122}	u_{123}	u_{124}	u_{125}
0	1	1	1	1	1	1	1	1	1	1
1	1	1	1	0	0	1	1	1	0	0
2	1	1	0	0	0	1	1	0	0	0
3	1	1	0	0	0	1	1	0	0	0
4	1	1	0	0	0	1	1	0	0	0
5	1	1	0	0	0	1	1	0	0	0

t	u_{211}	u_{212}	u_{213}	u_{214}	u_{215}	u_{221}	u_{222}	u_{223}	u_{224}	u_{225}
0	1	1	1	1	1	1	1	1	1	1
1	1	1	1	0	0	1	1	0	1	1
2	1	1	1	0	0	1	1	0	1	1
3	1	1	0	0	0	1	1	0	0	0
4	1	1	0	0	0	1	1	0	0	0
5	1	1	0	0	0	1	1	0	0	0

It is readily seen that, after the first iteration, we have 13 bottleneck parameters and, after the second one, 11 bottleneck parameters; after the third iteration, the process stabilizes, delivering the 8 true bottleneck parameters.

Table 8.7 presents a typical evolution of the sequence $\{(\widehat{\gamma}_t, \widehat{\mathbf{v}}_t)\}$ for the elements of the 2×2 model given above for the VM and VM-SCR methods. We see in this table that the bottleneck elements decrease more than three times, while the nonbottleneck elements fluctuate around their nominal values 4.

Table 8.7: Typical evolution of the sequence $\{\widehat{\mathbf{v}}_t\}$ for the VM and VM-SCR methods.

	VM					VM-SCR				
t	$\widehat{v}_{111}$	$\widehat{v}_{112}$	$\widehat{v}_{113}$	$\widehat{v}_{114}$	$\widehat{v}_{115}$	$\widehat{v}_{111}$	$\widehat{v}_{112}$	$\widehat{v}_{113}$	$\widehat{v}_{114}$	$\widehat{v}_{115}$
0	1.000	1.000	4.000	4.000	4.000	1.000	1.000	4	4	4
1	0.759	0.771	3.944	3.719	3.839	0.760	0.771	4	4	4
2	0.635	0.613	3.940	3.681	3.734	0.638	0.605	4	4	4
3	0.524	0.517	4.060	3.297	3.608	0.506	0.491	4	4	4
4	0.443	0.415	3.370	3.353	3.909	0.486	0.447	4	4	4
5	0.334	0.332	3.689	3.965	4.250	0.402	0.371	4	4	4
6	0.378	0.365	3.827	3.167	4.188	0.348	0.317	4	4	4
7	0.357	0.358	3.881	4.235	4.929	0.375	0.347	4	4	4
8	0.285	0.271	4.011	2.982	4.194	0.285	0.298	4	4	4
9	0.287	0.301	3.249	2.879	3.409	0.288	0.254	4	4	4

We conclude with a larger model, consisting of a 3×10 model in which $u_{111}, u_{112},$ $u_{211}, u_{212}, u_{311},$ and u_{312} are chosen as bottleneck parameters and are set to 1, while the remaining parameters are set to 4. Table 8.8 presents the performance of Algorithm 8.2.4 for this model using $\delta = 0.1$, $\gamma = 6$, and $N = N_1 = 400,000$. In this case, both CE and VM find the true six bottlenecks. Note that VM-SCR is the most accurate of the three alternatives and that CE underestimates ℓ. Thus, for this relatively large model, CE without screening is affected by the degeneracy of the likelihood ratio, presenting a product of 150 terms.

Table 8.8: Performance of Algorithm 8.2.4 for the 3×10 model with six bottlenecks. We set $\delta = 0.1$, $\gamma = 6$, $N = 400,000$, $N_1 = 400,000$.

	CE	VM	CE-SCR	VM-SCR
Mean $\widehat{\ell}$	2.44E-8	5.34E-8	5.28E-8	5.17E-8
Max $\widehat{\ell}$	5.82E-8	7.18E-8	8.34E-8	6.93E-8
Min $\widehat{\ell}$	4.14E-15	2.76E-8	2.74E-8	4.32E-8
RE	1.05	0.28	0.33	0.15
CPU	247	482	303	531

8.2.3 CE Method Combined with Sampling from the Zero-Variance Distribution

In Algorithm 8.2.1 a general procedure is described for estimating the optimal CE parameter $\mathbf{v}^*$ using a *multilevel approach*. However, as observed in [10], such an approach may not always be necessary or desirable. We next describe how to estimate $\mathbf{v}^*$ directly from g^* without a multilevel approach. We assume that one can easily sample (approximately) from g^*. For example, we can use any of the Markov chain samplers described in Chapter 6 to simulate from g^*. Let $\mathbf{X}_1, \dots, \mathbf{X}_N$ be approximately distributed according to g^*; then we can estimate $\mathbf{v}^*$ via $\widehat{\mathbf{v}}^* = \operatorname{argmax}_{\mathbf{v}} \sum_{k=1}^{N} \ln f(\mathbf{X}_k; \mathbf{v})$. Thus the CE program reduces to a standard maximum likelihood optimization problem. Once $\widehat{\mathbf{v}}^*$ is computed, we use the importance sampling estimator

$$\widehat{\ell} = \frac{1}{N_1} \sum_{k=1}^{N_1} \frac{f(\mathbf{X}_k; \mathbf{u})}{f(\mathbf{X}_k; \widehat{\mathbf{v}}^*)} I_{\{S(\mathbf{X}_k) \geqslant \gamma\}}, \quad \mathbf{X}_1, \dots, \mathbf{X}_{N_1} \sim f(\mathbf{x}; \widehat{\mathbf{v}}^*) \qquad (8.14)$$

to estimate ℓ. This motivates the following single-level CE algorithm:

Algorithm 8.2.5: Single-Level CE Method for Rare-Event Estimation

---
input : Performance function S, level γ, sample sizes N and N_1.
output: Estimator $\widehat{\ell}$ of the rare-event probability $\ell = \mathbb{P}(S(\mathbf{X}) \geqslant \gamma)$.
1 Run a Markov chain sampler to generate $\mathbf{X}_1, \dots, \mathbf{X}_N$ approximately distributed according to $g^*(\mathbf{x}) \propto f(\mathbf{x}) I_{\{S(\mathbf{x}) \geqslant \gamma\}}$.
2 Compute $\widehat{\mathbf{v}}^*$ by solving the maximum likelihood optimization problem

$$\widehat{\mathbf{v}}^* = \operatorname*{argmax}_{\mathbf{v}} \sum_{k=1}^{N} \ln f(\mathbf{X}_k; \mathbf{v}) .$$

3 Given $\widehat{\mathbf{v}}^*$, deliver the importance sampling estimator $\widehat{\ell}$ in (8.14).

8.3   CE METHOD FOR OPTIMIZATION

In this section we explain how the CE method works for optimization. Suppose that our task is to maximize a function $S(\mathbf{x})$ over some set $\mathscr{X}$. Let us denote the maximum by γ^*; thus

$$\gamma^* = \max_{\mathbf{x} \in \mathscr{X}} S(\mathbf{x}) \,. \tag{8.15}$$

The problem is called a *discrete* or *continuous* optimization problem based on whether $\mathscr{X}$ is discrete or continuous. An optimization problem involving both discrete and continuous variables is called a *mixed* optimization problem. A discrete optimization problem is sometimes called a *combinatorial optimization problem*, which is the main focus of this section.

The CE method takes a novel approach to optimization problems by casting the original problem (8.15) into an *estimation problem of rare-event probabilities*. By doing so, the CE method aims to locate an optimal parametric sampling distribution, that is, a probability distribution on $\mathscr{X}$, rather than locating the optimal solution directly. To this end, we define a collection of indicator functions $\{I_{\{S(\mathbf{x}) \geqslant \gamma\}}\}$ on $\mathscr{X}$ for various levels $\gamma \in \mathbb{R}$. Next, let $\{f(\cdot\,; \mathbf{v}), \mathbf{v} \in \mathscr{V}\}$ be a family of probability densities on $\mathscr{X}$ parameterized by a real-valued parameter vector $\mathbf{v}$. For a fixed $\mathbf{u} \in \mathscr{V}$ we associate with (8.15) the problem of estimating the rare-event probability

$$\ell(\gamma) = \mathbb{P}_{\mathbf{u}}(S(\mathbf{X}) \geqslant \gamma) = \mathbb{E}_{\mathbf{u}}\left[I_{\{S(\mathbf{X}) \geqslant \gamma\}}\right], \tag{8.16}$$

where $\mathbb{P}_{\mathbf{u}}$ is the probability measure under which the random state $\mathbf{X}$ has a discrete pdf $f(\cdot\,; \mathbf{u})$ and $\mathbb{E}_{\mathbf{u}}$ denotes the corresponding expectation operator. We call the estimation problem (8.16) the *associated stochastic problem*.

It is crucial to understand that one of the main goals of CE in optimization is to generate a sequence of pdfs $f(\cdot\,; \widehat{\mathbf{v}}_0), f(\cdot\,; \widehat{\mathbf{v}}_1), \ldots$, converging to a degenerate measure (Dirac measure) that assigns all probability mass to a single state $\mathbf{x}_T$, for which, by definition, the function value is either optimal or very close to it.

As soon as the associated stochastic problem is defined, we approximate the optimal solution, say $\mathbf{x}^*$, of (8.15) by applying Algorithm 8.2.1 for rare-event estimation, but without fixing γ in advance. It is plausible that if $\widehat{\gamma}^*$ is close to γ^*, then $f(\cdot\,; \widehat{\mathbf{v}}_T)$ assigns most of its probability mass close to $\mathbf{x}^*$. Thus any $\mathbf{X}$ drawn from this distribution can be used as an approximation to the optimal solution $\mathbf{x}^*$ and the corresponding function value as an approximation to the true optimal γ^* in (8.15).

To provide some more insight into the relation between combinatorial optimization and rare-event estimation, we revisit the coin flipping problem of Example 8.4, but from an optimization rather than an estimation perspective. This will serve as a preview to our discussion of *all real combinatorial optimization problems*, such as the maximal cut problem and the TSP considered in the next section. Even though the sample function $S(\mathbf{X})$ and the trajectory generation algorithm will differ from the toy example below, the updating of the sequence $\{(\gamma_t, \mathbf{v}_t)\}$ will always be determined by the *same* principles.

■ **EXAMPLE 8.6   Flipping n Coins: Example 8.4 Continued**

Suppose that we want to maximize

$$S(\mathbf{x}) = \sum_{i=1}^{n} x_i \,,$$

where $x_i = 0$ or 1 for all $i = 1, \ldots, n$. Clearly, the optimal solution to (8.15) is $\mathbf{x}^* = (1, \ldots, 1)$. The simplest way to put the deterministic program (8.15) into a stochastic framework is to associate with each component x_i, $i = 1, \ldots, n$ a Bernoulli random variable X_i, $i = 1, \ldots, n$. For simplicity, we can assume that all $\{X_i\}$ are independent and that each component i has success probability $1/2$. By doing so, we turn the associated stochastic problem (8.16) into a rare-event estimation problem. Taking into account that there is a single solution $\mathbf{x}^* = (1, \ldots, 1)$, using the CMC method, we obtain $\ell(\gamma^*) = 1/|\mathscr{X}|$, where $|\mathscr{X}| = 2^n$, which for large n is a very small probability. Instead of estimating $\ell(\gamma)$ via CMC, we can estimate it via importance sampling using $X_i \sim \mathsf{Ber}(p_i)$, $i = 1, \ldots, n$.

The next step is, clearly, to apply Algorithm 8.2.1 to (8.16) without fixing γ in advance. As mentioned in Remark 8.2.3, CE Algorithm 8.2.1 should be viewed as the stochastic counterpart of the deterministic CE Algorithm 8.2.2, and the latter will iterate until it reaches a local maximum. We thus obtain a sequence $\{\widehat{\gamma}_t\}$ that converges to a local or global maximum, which can be taken as an estimate for the true optimal solution γ^*.

In summary, in order to solve a combinatorial optimization problem, we will employ the CE Algorithm 8.2.1 for rare-event estimation without fixing γ in advance. By doing so, we can treat the CE algorithm for optimization as a modified version of Algorithm 8.2.1. In particular, by analogy to Algorithm 8.2.1, we choose a not very small number ϱ, say $\varrho = 10^{-2}$, initialize the parameter vector $\mathbf{u}$ by setting $\mathbf{v}_0 = \mathbf{u}$, and proceed as follows:

1. **Adaptive updating of $\boldsymbol{\gamma}_t$.** For a fixed $\mathbf{v}_{t-1}$, let γ_t be the $(1 - \varrho)$-quantile of $S(\mathbf{X})$ under $\mathbf{v}_{t-1}$. As before, an estimator $\widehat{\gamma}_t$ of γ_t can be obtained by drawing a random sample $\mathbf{X}_1, \ldots, \mathbf{X}_N$ from $f(\cdot; \mathbf{v}_{t-1})$ and then evaluating the sample $(1 - \varrho)$-quantile of the performances as

$$\widehat{\gamma}_t = S_{(\lceil (1-\varrho)N \rceil)} \,. \tag{8.17}$$

2. **Adaptive updating of $\mathbf{v}_t$.** For fixed γ_t and $\mathbf{v}_{t-1}$, obtain $\mathbf{v}_t$ as the solution of the program

$$\max_{\mathbf{v}} D(\mathbf{v}) = \max_{\mathbf{v}} \mathbb{E}_{\mathbf{v}_{t-1}} [I_{\{S(\mathbf{X}) \geqslant \gamma_t\}} \ln f(\mathbf{X}; \mathbf{v})] \,. \tag{8.18}$$

The stochastic counterpart of (8.18) is then as follows: for fixed $\widehat{\gamma}_t$ and $\widehat{\mathbf{v}}_{t-1}$, obtain $\widehat{\mathbf{v}}_t$ as the solution of the following program:

$$\max_{\mathbf{v}} \widehat{D}(\mathbf{v}) = \max_{\mathbf{v}} \frac{1}{N} \sum_{k=1}^{N} I_{\{S(\mathbf{X}_k) \geqslant \widehat{\gamma}_t\}} \ln f(\mathbf{X}_k; \mathbf{v}) \,. \tag{8.19}$$

It is important to observe that, in contrast to (8.5) and (8.6) (for the rare-event setting), (8.18) and (8.19) *do not contain the likelihood ratio terms W*. The reason is that in the rare-event setting the initial (nominal) parameter $\mathbf{u}$ is specified in advance and is an essential part of the estimation problem. In contrast, the initial reference vector $\mathbf{u}$ in the associated stochastic problem is quite arbitrary. In effect, by dropping the W term, we can efficiently estimate at each iteration t the CE optimal reference parameter vector $\mathbf{v}_t$ for the rare-event probability $\mathbb{P}_{\mathbf{v}_t}(S(\mathbf{X}) \geqslant \gamma_t) \geq \mathbb{P}_{\mathbf{v}_{t-1}}(S(\mathbf{X}) \geqslant \gamma_t)$, even for high-dimensional problems.

Remark 8.3.1 (Smoothed Updating) Instead of updating the parameter vector $\mathbf{v}$ directly via the solution of (8.19), we use the *smoothed* version

$$\widehat{\mathbf{v}}_t = \alpha \widetilde{\mathbf{v}}_t + (1 - \alpha)\widehat{\mathbf{v}}_{t-1}, \tag{8.20}$$

where $\widetilde{\mathbf{v}}_t$ is the parameter vector obtained from the solution of (8.19) and α is called the *smoothing parameter*, and typically $0.7 < \alpha \leqslant 1$. Clearly, for $\alpha = 1$ we have our original updating rule. The reason for using the smoothed (8.20) instead of the original updating rule is twofold: (a) to smooth out the values of $\widehat{\mathbf{v}}_t$ and (b) to reduce the probability that some component $\widehat{v}_{t,i}$ of $\widehat{\mathbf{v}}_t$ will be 0 or 1 at the first few iterations. This is particularly important when $\widehat{\mathbf{v}}_t$ is a vector or matrix of *probabilities*. Note that for $0 < \alpha \leqslant 1$, we always have $\widehat{v}_{t,i} > 0$, while for $\alpha = 1$, we might have (even at the first iterations) $\widehat{v}_{t,i} = 0$ or $\widehat{v}_{t,i} = 1$ for some indexes i. As result, the algorithm will converge to a wrong solution.

Thus, the main CE optimization algorithm, which includes smoothed updating of parameter vector $\mathbf{v}$ and which presents a slight modification of Algorithm 8.2.1 can be summarized as follows.

Algorithm 8.3.1: Main CE Algorithm for Optimization

---

1 Initialize $\widehat{\mathbf{v}}_0 \leftarrow \widehat{\mathbf{v}}_0$.

2 **repeat**

3      $t \leftarrow t + 1$

4   Generate a sample $\mathbf{X}_1, \ldots, \mathbf{X}_N$ from the density $f(\cdot; \widehat{\mathbf{v}}_{t-1})$ and compute the sample $(1 - \varrho)$-quantile $\widehat{\gamma}_t$ of the performances according to (8.17).

5   Use the *same* sample $\mathbf{X}_1, \ldots, \mathbf{X}_N$ and solve the stochastic program (8.19). Denote the solution by $\widetilde{\mathbf{v}}_t$.

6   Apply (8.20) to smooth out the vector $\widetilde{\mathbf{v}}_t$.

7 **until** *a stopping criterion is met.*

Remark 8.3.2 (Minimization) When $S(\mathbf{x})$ is to be *minimized* instead of maximized, we simply change the inequalities "$\geqslant$" to "$\leqslant$" and take the ϱ-quantile instead of the $(1 - \varrho)$-quantile. Alternatively, we can just maximize $-S(\mathbf{x})$.

As a stopping criterion, we can use, for example: if for some $t \geqslant d$, say $d = 5$,

$$\widehat{\gamma}_t = \widehat{\gamma}_{t-1} = \cdots = \widehat{\gamma}_{t-d}, \tag{8.21}$$

then stop. As an alternative estimate for γ^*, we can consider

$$\widetilde{\gamma}_T = \max_{0 \leqslant s \leqslant T} \widehat{\gamma}_s. \tag{8.22}$$

Note that the initial vector $\hat{\mathbf{v}}_0$, the sample size N, the stopping parameter d, and the number ϱ have to be specified in advance, but the rest of the algorithm is "self-tuning". Note also that, by analogy to the simulated annealing algorithm, γ_t may be viewed as the "annealing temperature". In contrast to simulated annealing, where the cooling scheme is chosen in advance, in the CE algorithm it is updated adaptively.

■ **EXAMPLE 8.7 Flipping Coins: Example 8.6 (Continued)**

In this case, the random vector $\mathbf{X} = (X_1, \ldots, X_n) \sim \mathsf{Ber}(\mathbf{p})$ and the parameter vector $\mathbf{v}$ is $\mathbf{p}$. Consequently, the pdf is

$$f(\mathbf{X}; \mathbf{p}) = \prod_{i=1}^{n} p_i^{X_i} (1 - p_i)^{1-X_i} \, ,$$

and since each X_i can only be 0 or 1,

$$\frac{\partial}{\partial p_i} \ln f(\mathbf{X}; \mathbf{p}) = \frac{X_i}{p_i} - \frac{1 - X_i}{1 - p_i} = \frac{1}{(1 - p_i)p_i}(X_i - p_i) \, .$$

Now we can find the optimal parameter vector $\mathbf{p}$ of (8.19) by setting the first derivatives with respect to p_i equal to zero for $i = 1, \ldots, n$, that is,

$$\frac{\partial}{\partial p_i} \sum_{k=1}^{N} I_{\{S(\mathbf{X}_k) \geqslant \gamma\}} \ln f(\mathbf{X}_k; \mathbf{p}) = \frac{1}{(1 - p_i)\, p_i} \sum_{i=1}^{N} I_{\{S(\mathbf{X}_k) \geqslant \gamma\}} \, (X_{ki} - p_i) = 0 \, .$$

Thus, we obtain

$$p_i = \frac{\sum_{k=1}^{N} I_{\{S(\mathbf{X}_k) \geqslant \gamma\}} X_{ki}}{\sum_{k=1}^{N} I_{\{S(\mathbf{X}_k) \geqslant \gamma\}}} \, , \tag{8.23}$$

which gives the same updating formula as (8.10) *except for the W term*. Recall that the updating formula (8.23) holds for all one-dimensional exponential families that are parameterized by the mean; see (5.68). Note also that the parameters are simply updated via their maximum likelihood estimators, using only the elite samples; see Remark 8.2.2.

Algorithm 8.3.1 can, in principle, be applied to any discrete and continuous optimization problem. However, for each problem two essential actions need to be taken:

1. We need to specify how the samples are generated. In other words, we need to specify the family of densities $\{f(\cdot; \mathbf{v})\}$.

2. We need to update the parameter vector $\mathbf{v}$ based on CE minimization program (8.19), which is the *same* for all optimization problems.

In general, there are many ways to generate samples from $\mathscr{X}$, and it is not always immediately clear which method will yield better results or easier updating formulas.

Remark 8.3.3 (Parameter Selection) The choice of the sample size N and the rarity parameter ϱ depends on the size of the problem and the number of parameters in the associated stochastic problem. Typical choices are $\varrho = 0.1$ or $\varrho = 0.01$, and $N = cK$, where K is the number of parameters that need to be estimated/updated and c is a constant between 1 and 10.

By analogy to Algorithm 8.2.2, we also present the deterministic version of Algorithm 8.3.1, which will be used below.

Algorithm 8.3.2: Deterministic CE Algorithm for Optimization

---

1 Choose some $\mathbf{v}_0$.
2 **repeat**
3   Calculate γ_t as
$$\gamma_t \leftarrow \max \left\{ s \,:\, \mathbb{P}_{\mathbf{v}_{t-1}}(S(\mathbf{X}) \geqslant s) \geqslant \varrho \right\} . \tag{8.24}$$
4   Calculate $\mathbf{v}_t$ as
$$\mathbf{v}_t \leftarrow \underset{\mathbf{v}}{\mathrm{argmax}} \, \mathbb{E}_{\mathbf{v}_{t-1}} \left[I_{\{S(\mathbf{X}) \geqslant \gamma_t\}} \ln f(\mathbf{X}; \mathbf{v}) \right] . \tag{8.25}$$
5 **until** $\gamma_t = \gamma_{t-1} = \cdots = \gamma_{t-d}$ `// for, say, d = 5`

Remark 8.3.4 Note that instead of the CE distance we could minimize the variance of the estimator, as discussed in Section 5.7. As mentioned, the main reason for using CE is that for exponential families the parameters can be updated analytically, rather than numerically as for the VM procedure.

Below we present several applications of the CE method to combinatorial optimization, namely the max-cut, the bipartition, and the TSP. We demonstrate numerically the efficiency of the CE method and its fast convergence for several case studies. For additional applications of CE, see [45] and the list of references at the end of this chapter.

8.4 MAX-CUT PROBLEM

The maximal cut or *max-cut* problem can be formulated as follows: Given a graph $G = G(V, E)$ with a set of nodes $V = \{1, \ldots, n\}$ and a set of edges E between the nodes, partition the nodes of the graph into two arbitrary subsets V_1 and V_2 such that the sum of the weights (costs) c_{ij} of the edges going from one subset to the other is maximized. Note that some of the c_{ij} may be 0 — indicating that there is actually no edge from i to j.

As an example, consider the graph in Figure 8.4, with corresponding cost matrix $C = (c_{ij})$ given by

$$C = \begin{pmatrix} 0 & 2 & 2 & 5 & 0 \\ 2 & 0 & 1 & 0 & 3 \\ 2 & 1 & 0 & 4 & 2 \\ 5 & 0 & 4 & 0 & 1 \\ 0 & 3 & 2 & 1 & 0 \end{pmatrix} . \tag{8.26}$$

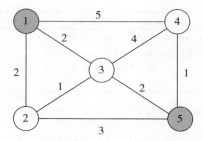

Figure 8.4: A six-node network with the cut $\{\{1,5\},\{2,3,4\}\}$.

Here the cut $\{\{1,5\},\{2,3,4\}\}$ has cost

$$c_{12} + c_{13} + c_{14} + c_{52} + c_{53} + c_{54} = 2 + 2 + 5 + 3 + 2 + 1 = 15 \ .$$

A cut can be conveniently represented via its corresponding *cut vector* $\mathbf{x} = (x_1, \ldots, x_n)$, where $x_i = 1$ if node i belongs to same partition as 1 and 0 otherwise. For example, the cut in Figure 8.4 can be represented via the cut vector $(1, 0, 0, 0, 1)$. For each cut vector $\mathbf{x}$, let $\{V_1(\mathbf{x}), V_2(\mathbf{x})\}$ be the partition of V induced by $\mathbf{x}$, such that $V_1(\mathbf{x})$ contains the set of indexes $\{i : x_i = 1\}$. If not stated otherwise, we set $x_1 = 1 \in V_1$.

Let $\mathscr{X}$ be the set of all cut vectors $\mathbf{x} = (1, x_2, \ldots, x_n)$, and let $S(\mathbf{x})$ be the corresponding cost of the cut. Then

$$S(\mathbf{x}) = \sum_{i \in V_1(\mathbf{x}),\, j \in V_2(\mathbf{x})} c_{ij} \ . \tag{8.27}$$

It is readily seen that the total number of cut vectors is

$$|\mathscr{X}| = 2^{n-1}. \tag{8.28}$$

We will assume below that the graph is *undirected*. Note that for a *directed* graph the cost of a cut $\{V_1, V_2\}$ includes the cost of the edges both from V_1 to V_2 and from V_2 to V_1. In this case, the cost corresponding to a cut vector $\mathbf{x}$ is therefore

$$S(\mathbf{x}) = \sum_{i \in V_1(\mathbf{x}),\, j \in V_2(\mathbf{x})} (c_{ij} + c_{ji}) \ . \tag{8.29}$$

Next, we generate random cuts and update of the corresponding parameters using the CE Algorithm 8.3.1. The most natural and easiest way to generate the cut vectors is to let $X_2, \ldots, X_n$ be independent Bernoulli random variables with success probabilities $p_2, \ldots, p_n$.

Algorithm 8.4.1: Random Cuts Generation

---

1 Generate an n-dimensional random vector $\mathbf{X} = (X_1, \ldots, X_n)$ from $\mathsf{Ber}(\mathbf{p})$ with independent components, where $\mathbf{p} = (1, p_2, \ldots, p_n)$.
2 Construct the partition $\{V_1(\mathbf{X}), V_2(\mathbf{X})\}$ of V and calculate the performance $S(\mathbf{X})$ as in (8.27).

The updating formulas for $\widehat{p}_{t,i}$ are the same as for the toy Example 8.7 and are given in (8.23).

The following toy example illustrates, step by step, the workings of the deterministic CE Algorithm 8.3.2. The small size of the problem allows us to make all calculations analytically, that is, using directly the updating rules (8.24) and (8.25) rather than their stochastic counterparts.

■ **EXAMPLE 8.8   Illustration of Algorithm 8.3.2**

Consider the five-node graph presented in Figure 8.4. The 16 possible cut vectors (see (8.28)) and the corresponding cut values are given in Table 8.9.

Table 8.9: Possible cut vectors of Example 8.8.

$\mathbf{X}$	V_1	V_2	$S(\mathbf{X})$
(1,0,0,0,0)	$\{1\}$	$\{2,3,4,5\}$	9
(1,1,0,0,0)	$\{1,2\}$	$\{3,4,5\}$	11
(1,0,1,0,0)	$\{1,3\}$	$\{2,4,5\}$	14
(1,0,0,1,0)	$\{1,4\}$	$\{2,3,5\}$	9
(1,0,0,0,1)	$\{1,5\}$	$\{2,3,4\}$	15
(1,1,1,0,0)	$\{1,2,3\}$	$\{4,5\}$	14
(1,1,0,1,0)	$\{1,2,4\}$	$\{3,5\}$	11
(1,1,0,0,1)	$\{1,2,5\}$	$\{3,4\}$	11
(1,0,1,1,0)	$\{1,3,4\}$	$\{2,5\}$	6
(1,0,1,0,1)	$\{1,3,5\}$	$\{2,4\}$	16
(1,0,0,1,1)	$\{1,4,5\}$	$\{2,3\}$	13
(1,1,1,1,0)	$\{1,2,3,4\}$	$\{5\}$	6
(1,1,1,0,1)	$\{1,2,3,5\}$	$\{4\}$	10
(1,1,0,1,1)	$\{1,2,4,5\}$	$\{3\}$	9
(1,0,1,1,1)	$\{1,3,4,5\}$	$\{2\}$	6
(1,1,1,1,1)	$\{1,2,3,4,5\}$	$\emptyset$	0

Clearly, in this case the optimal cut vector is $\mathbf{x}^* = (1,0,1,0,1)$ with $S(\mathbf{x}^*) = \gamma^* = 16$.

We will show next that in the deterministic Algorithm 8.3.2, adapted to the max-cut problem, the parameter vectors $\mathbf{p}_0, \mathbf{p}_1, \ldots$ converge to the optimal $\mathbf{p}^* = (1,0,1,0,1)$ after two iterations, provided that $\varrho = 10^{-1}$ and $\mathbf{p}_0 = (1, 1/2, 1/2, 1/2, 1/2)$.

Iteration 1

In the first step of the first iteration, we have to determine γ_1 from

$$\gamma_t = \max \left\{ \gamma \text{ s.t. } \mathbb{E}_{\mathbf{p}_{t-1}}[I_{\{S(\mathbf{X}) \geqslant \gamma\}}] \geqslant 0.1 \right\} . \qquad (8.30)$$

It is readily seen that under the parameter vector $\mathbf{p}_0$, $S(\mathbf{X})$ takes values in $\{0, 6, 9, 10, 11, 13, 14, 15, 16\}$ with probabilities $\{1/16, 3/16, 3/16, 1/16, 3/16, 1/16, 2/16, 1/16, 1/16\}$. Hence we find $\gamma_1 = 15$. In the second step, we need to solve

$$\mathbf{p}_t = \operatorname*{argmax}_{\mathbf{p}} \mathbb{E}_{\mathbf{p}_{t-1}} \left[I_{\{S(\mathbf{X}) \geqslant \gamma_t\}} \ln f(\mathbf{X}; \mathbf{p}) \right] , \qquad (8.31)$$

which has the solution

$$p_{t,i} = \frac{\mathbb{E}_{\mathbf{p}_{t-1}}[I_{\{S(\mathbf{X}) \geqslant \gamma_t\}} X_i]}{\mathbb{E}_{\mathbf{p}_{t-1}}[I_{\{S(\mathbf{X}) \geqslant \gamma_t\}}]} .$$

There are only two vectors $\mathbf{x}$ for which $S(\mathbf{x}) \geqslant 15$, namely, $(1, 0, 0, 0, 1)$ and $(1, 0, 1, 0, 1)$, and both have probability $1/16$ under $\mathbf{p}_0$. Thus,

$$p_{1,i} = \begin{cases} \dfrac{2/16}{2/16} = 1 & \text{for } i = 1, 5, \\ \dfrac{1/16}{2/16} = \dfrac{1}{2} & \text{for } i = 3, \\ \dfrac{0}{2/16} = 0 & \text{for } i = 4, 2. \end{cases}$$

Iteration 2

In the second iteration $S(\mathbf{X})$ is 15 or 16 with probability $1/2$. Applying again (8.30) and (8.31) yields the optimal $\gamma_2 = 16$ and the optimal $\mathbf{p}_2 = (1, 0, 1, 0, 1)$, respectively.

Remark 8.4.1 (Alternative Stopping Rule) Note that the stopping rule (8.21), which is based on convergence of the sequence $\{\widehat{\gamma}_t\}$ to γ^*, stops Algorithm 8.3.1 when the sequence $\{\widehat{\gamma}_t\}$ does not change. An alternative stopping rule is to stop when the sequence $\{\widehat{\mathbf{p}}_t\}$ is very close to a degenerated one, for example if $\min\{\widehat{p}_i, 1 - \widehat{p}_i\} < \varepsilon$ for all i, where ε is some small number.

The code in Table 8.5 gives a simple Matlab implementation of the CE algorithm for the max-cut problem, with cost matrix (8.26). Note that, although the max-cut examples presented here are of relatively small size, basically the *same* CE program can be used to tackle max-cut problems of much higher dimension, comprising hundreds or thousands of nodes.

```
global C;
C = [ 0   2   2   5   0;                    % cost matrix
      2   0   1   0   3;
      2   1   0   4   2;
      5   0   4   0   1;
      0   3   2   1   0];
m = 5; N = 100; Ne = 10; eps = 10^-3; p = 1/2*ones(1,m); p(1) = 1;
while max(min(p,1-p)) > eps
    x = (rand(N,m) < ones(N,1)*p);          % generate cut vectors
    SX = S(x);
    sortSX = sortrows([x SX], m+1);
    p = mean(sortSX(N-Ne+1:N, 1:m))         % update the parameters
end

function perf = S(x)                        % performance function
global C;
N = size(x,1);
for i=1:N
    V1 = find(x(i,:));                      % {V1,V2} is the partition
    V2 = find(~x(i,:));
perf(i,1) = sum(sum(C(V1,V2)));             % size of the cut
end
```

Figure 8.5: Matlab CE program to solve the max-cut problem with cost matrix (8.26).

■ **EXAMPLE 8.9   Maximal Cuts for the Dodecahedron Graph**

To further illustrate the behavior of the CE algorithm for the max-cut problem, consider the so-called *dodecahedron graph* in Figure 8.6. Suppose that all edges have cost 1. We wish to partition the node set into two subsets (color the nodes black and white) such that the cost across the cut, given by (8.27), is maximized. Although this problem exhibits a lot of symmetry, it is not clear beforehand what the solution(s) should be.

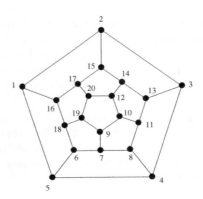

Figure 8.6: Dodecahedron graph.

The performance of the CE algorithm is depicted in Figure 8.7 using $N = 200$ and $\varrho = 0.1$.

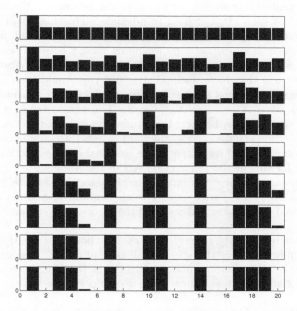

Figure 8.7: Evolution of the CE algorithm for the dodecahedron max-cut problem.

Observe that the probability vector $\widehat{\mathbf{p}}_t$ quickly (eight iterations) converges to a degenerate vector — corresponding (for this particular case) to the solution $\mathbf{x}^* = (1, 0, 1, 1, 0, 0, 1, 0, 0, 1, 1, 0, 0, 1, 0, 0, 1, 1, 1, 0)$. Thus, $V_1^* = \{1, 3, 4, 7, 10, 11, 14, 17, 18, 19\}$. This required around 1600 function evaluations, as compared to $2^{19} - 1 \approx 5 \cdot 10^5$ if all cut vectors were to be enumerated. The maximal value is 24. It is interesting that, because of the symmetry, there are in fact many optimal solutions. We found that during each run the CE algorithm "focuses" on one (not always the same) of the solutions.

Max-Cut Problem with r Partitions

We can readily extend the max-cut procedure to the case where the node set V is partitioned into $r > 2$ subsets $\{V_1, \ldots, V_r\}$ such that the sum of the total weights of all edges going from subset V_a to subset V_b, $a, b = 1, \ldots, r$, $(a < b)$ is maximized. Thus, for each partition $\{V_1, \ldots, V_r\}$, the value of the objective function is

$$\sum_{a=1}^{r} \sum_{b=a+1}^{r} \sum_{i \in V_a, \, j \in V_b} c_{ij} \, .$$

In this case, we can follow the basic steps of Algorithm 8.3.1 using independent r-point distributions, instead of independent Bernoulli distributions, and update the probabilities as

$$\widehat{p}_{t,ij} = \frac{\sum_{\mathbf{X}_k \in \mathscr{E}_t} I_{\{X_{ki}=j\}}}{|\mathscr{E}_t|} \, . \tag{8.32}$$

8.5 PARTITION PROBLEM

The partition problem is similar to the max-cut problem. The only difference is that the *size* of each class is *fixed* in advance. This has implications for the trajectory generation. Consider, for example, a partition problem in which V has to be partitioned into two *equal* sets, assuming that n is even. We could simply use Algorithm 8.4.1 for the random cut generation, that is, generate $\mathbf{X} \sim \text{Ber}(\mathbf{p})$ and *reject* partitions that have unequal size, but this would be highly inefficient. We can speed up this method by drawing directly from the *conditional* distribution of $\mathbf{X} \sim \text{Ber}(\mathbf{p})$ given $X_1 + \cdots + X_n = n/2$. The parameter $\mathbf{p}$ is then updated in exactly the same way as before. Unfortunately, generating from a conditional Bernoulli distribution is not as straightforward as generating independent Bernoulli random variables. A useful technique is the so-called *drafting* method. We provide computer code for this method in Section A.2 of the Appendix.

As an alternative, we describe next a simple algorithm for the generation of a random bipartition $\{V_1, V_2\}$ with exactly m elements in V_1 and $n - m$ elements in V_2 that works well in practice. Extension of the algorithm to r-partition generation is simple.

The algorithm requires the generation of random permutations $\Pi = (\Pi_1, \ldots, \Pi_n)$ of $(1, \ldots, n)$, uniformly over the space of all permutations. This can be done via Algorithm 2.10.2. We demonstrate our algorithm first for a five-node network, assuming $m = 2$ and $m - n = 3$ for a given vector $\mathbf{p} = (p_1, \ldots, p_5)$.

■ **EXAMPLE 8.10 Generating a Bi-Partition for $m = 2$ and $n = 5$**

1. Generate a random permutation $\Pi = (\Pi_1, \ldots, \Pi_5)$ of $(1, \ldots, 5)$, uniformly over the space of all 5! permutations. Let $(\pi_1, \ldots, \pi_5)$ be a particular outcome, for example, $(\pi_1, \ldots, \pi_5) = (3, 5, 1, 2, 4)$. This means that we will draw independent Bernoulli random variables in the following order: $\text{Ber}(p_3)$, $\text{Ber}(p_5)$, $\text{Ber}(p_1)$, ….

2. Given $\Pi = (\pi_1, \ldots, \pi_5)$ and the vector $\mathbf{p} = (p_1, \ldots, p_5)$, generate independent Bernoulli random variables $X_{\pi_1}, X_{\pi_2}, \ldots$ from $\text{Ber}(p_{\pi_1})$, $\text{Ber}(p_{\pi_2})$, …, respectively, *until either exactly* $m = 2$ unities or $n - m = 3$ zeros are generated. Note that, in general, the number of samples is a random variable with the range from $\min\{m, n - m\}$ to n. Assume, for concreteness, that the first four independent Bernoulli samples (from $\text{Ber}(p_3), \text{Ber}(p_5), \text{Ber}(p_1), \text{Ber}(p_2)$ given above) result in the following outcome $(0, 0, 1, 0)$. Since we have already generated three 0s, we can set $X_4 \equiv 1$ and deliver $\{V_1(\mathbf{X}), V_2(\mathbf{X})\} = \{(1, 4), (2, 3, 5)\}$ as the desired partition.

3. If in the previous step $m = 2$ unities are generated, set the remaining three elements to 0; if instead three 0s are generated, set the remaining two elements to 1 and deliver $\mathbf{X} = (X_1, \ldots, X_n)$ as the final partition vector. Construct the partition $\{V_1(\mathbf{X}), V_2(\mathbf{X})\}$ of V.

With this example in hand, the random partition generation algorithm can be written as follows:

Algorithm 8.5.1: Random Partition Generation Algorithm

---

1 Generate a random permutation $\Pi = (\Pi_1, \ldots, \Pi_n)$ of $(1, \ldots, n)$ uniformly over the space of all $n!$ permutations.

2 Given $\Pi = (\pi_1, \ldots, \pi_n)$, independently generate Bernoulli random variables $X_{\pi_1}, X_{\pi_2}, \ldots$ from $\mathsf{Ber}(p_{\pi_1}), \mathsf{Ber}(p_{\pi_2}), \ldots$, respectively, *until m 1s or $n - m$ 0s* are generated.

3 If in the previous step m 1s are generated, set the remaining elements to 0; if, on the other hand, $n - m$ 0s are generated, set the remaining elements to 1s. Deliver $\mathbf{X} = (X_1, \ldots, X_n)$ as the final partition vector.

4 Construct the partition $\{V_1(\mathbf{X}), V_2(\mathbf{X})\}$ of V and calculate the performance $S(\mathbf{X})$ according to (8.27).

We take the updating formula for the reference vector **p** *exactly the same* as in (8.10).

8.5.1 Empirical Computational Complexity

Finally, we come to consider the computational complexity of Algorithm 8.3.1 for the max-cut and the partition problems, which can be defined as

$$\kappa_n = T_n(N_n G_n + U_n) . \tag{8.33}$$

Here T_n is the total number of iterations needed before Algorithm 8.3.1 stops; N_n is the sample size, that is, the total number of maximal cuts and partitions generated at each iteration; G_n is the cost of generating the random Bernoulli vectors of size n for Algorithm 8.3.1; $U_n = \mathcal{O}(N_n n^2)$ is the cost of updating the tuple $(\widehat{\gamma}_t, \widehat{\mathbf{p}}_t)$. The last follows from the fact that computing $S(\mathbf{X})$ in (8.27) is a $\mathcal{O}(n^2)$ operation.

For the model in (8.62) we found empirically that $T_n = \mathcal{O}(\ln n)$, provided that $100 \leqslant n \leqslant 1000$. For the max-cut problem, considering that we take $n \leqslant N_n \leqslant 10\,n$ and that G_n is $\mathcal{O}(n)$, we obtain $\kappa_n = \mathcal{O}(n^3 \ln n)$. In our experiments, the complexity we observed was more like

$$\kappa_n = \mathcal{O}(n \ln n) .$$

The partition problem has similar computational characteristics. It is important to note that these empirical complexity results are solely for the model with the cost matrix (8.62).

8.6 TRAVELING SALESMAN PROBLEM

The CE method can also be applied to solve the traveling salesman problem (TSP). Recall (see Example 6.12 for a more detailed formulation) that the objective is to find the shortest tour through all the nodes in a graph G. As in Example 6.12, we assume that the graph is complete and that each tour is represented as a permutation $\mathbf{x} = (x_1, \ldots, x_n)$ of $(1, \ldots, n)$. Without loss of generality, we can set $x_1 = 1$, so that the set of all possible tours $\mathscr{X}$ has cardinality $|\mathscr{X}| = (n-1)!$. Let $S(\mathbf{x})$ be the total length of tour $\mathbf{x} \in \mathscr{X}$, and let $C = (c_{ij})$ be the cost matrix. Our goal is thus to solve

$$\min_{\mathbf{x} \in \mathscr{X}} S(\mathbf{x}) = \min_{\mathbf{x} \in \mathscr{X}} \left\{ \sum_{i=1}^{n-1} c_{x_i, x_{i+1}} + c_{x_n, 1} \right\} . \tag{8.34}$$

In order to apply the CE algorithm, we need to specify a parameterized random mechanism to generate the random tours. As mentioned, the updating formulas for the parameters follow, as always, from CE minimization.

An easy way to explain how the tours are generated and how the parameters are updated is to relate (8.34) to an *equivalent* minimization problem. Let

$$\widetilde{\mathscr{X}} = \{(x_1, \ldots, x_n) \,:\, x_1 = 1, \quad x_i \in \{1, \ldots, n\}\,, \quad i = 2, \ldots, n\} \tag{8.35}$$

be the set of vectors that correspond to tours that start in 1 and can visit the same city more than once. Note that $|\widetilde{\mathscr{X}}| = n^{n-1}$ and $\mathscr{X} \subset \widetilde{\mathscr{X}}$. When $n = 4$, we could have, for example, $\mathbf{x} = (1, 3, 1, 3) \in \widetilde{\mathscr{X}}$, corresponding to the *path* (not tour) $1 \to 3 \to 1 \to 3 \to 1$. Define the function $\widetilde{S}$ on $\widetilde{\mathscr{X}}$ by $\widetilde{S}(\mathbf{x}) = S(\mathbf{x})$, if $\mathbf{x} \in \mathscr{X}$ and $\widetilde{S}(\mathbf{x}) = \infty$ otherwise. Then, obviously, (8.34) is equivalent to the minimization problem

$$\text{minimize } \widetilde{S}(\mathbf{x}) \text{ over } \mathbf{x} \in \widetilde{\mathscr{X}}\,. \tag{8.36}$$

A simple method to generate a random path $\mathbf{X} = (X_1, \ldots, X_n)$ in $\widetilde{\mathscr{X}}$ is to use a Markov chain on the graph G, starting at node 1 and stopping after n steps. Let $\mathbf{P} = (p_{ij})$ denote the one-step transition matrix of this Markov chain. We assume that the diagonal elements of $\mathbf{P}$ are 0 and that all other elements of $\mathbf{P}$ are strictly positive, but otherwise $\mathbf{P}$ is a general $n \times n$ stochastic matrix.

The pdf $f(\cdot; \mathbf{P})$ of $\mathbf{X}$ is thus parameterized by the matrix $\mathbf{P}$, and its logarithm is given by

$$\ln f(\mathbf{x}; \mathbf{P}) = \sum_{r=1}^{n} \sum_{i,j} I_{\{\mathbf{x} \in \widetilde{\mathscr{X}}_{ij}(r)\}} \ln p_{ij}\,,$$

where $\widetilde{\mathscr{X}}_{ij}(r)$ is the set of all paths in $\widetilde{\mathscr{X}}$ for which the r-th transition is from node i to j. The updating rules for this modified optimization problem follow from (8.18), with $\{S(\mathbf{X}_i) \geqslant \gamma_t\}$ replaced with $\{\widetilde{S}(\mathbf{X}_i) \leqslant \gamma_t\}$, under the condition that the rows of $\mathbf{P}$ sum up to 1. Using Lagrange multipliers $u_1, \ldots, u_n$, we obtain the maximization problem

$$\max_{\mathbf{P}} \min_{u_1, \ldots, u_n} \left\{ \mathbb{E}_{\mathbf{P}}\left[I_{\{\widetilde{S}(\mathbf{X}) \leqslant \gamma\}} \ln f(\mathbf{X}; \mathbf{P}) \right] + \sum_{i=1}^{n} u_i \left(\sum_{j=1}^{n} p_{ij} - 1 \right) \right\}\,. \tag{8.37}$$

Differentiating the expression within braces above with respect to p_{ij} yields, for all $j = 1, \ldots, n$,

$$\frac{\mathbb{E}_{\mathbf{P}}\left[I_{\{\widetilde{S}(\mathbf{X}) \leqslant \gamma\}} \sum_{r=1}^{n} I_{\{\mathbf{X} \in \widetilde{\mathscr{X}}_{ij}(r)\}} \right]}{p_{ij}} + u_i = 0\,. \tag{8.38}$$

Summing over $j = 1, \ldots, n$ gives $\mathbb{E}_{\mathbf{P}}\left[I_{\{\widetilde{S}(\mathbf{X}) \leqslant \gamma\}} \sum_{r=1}^{n} I_{\{\mathbf{X} \in \widetilde{\mathscr{X}}_i(r)\}} \right] = -u_i$, where $\widetilde{\mathscr{X}}_i(r)$ is the set of paths for which the r-th transition starts from node i. It follows that the optimal p_{ij} is given by

$$p_{ij} = \frac{\mathbb{E}_{\mathbf{P}}\left[I_{\{\widetilde{S}(\mathbf{X}) \leqslant \gamma\}} \sum_{r=1}^{n} I_{\{\mathbf{X} \in \widetilde{\mathscr{X}}_{ij}(r)\}} \right]}{\mathbb{E}_{\mathbf{P}}\left[I_{\{\widetilde{S}(\mathbf{X}) \leqslant \gamma\}} \sum_{r=1}^{n} I_{\{\mathbf{X} \in \widetilde{\mathscr{X}}_i(r)\}} \right]}\,. \tag{8.39}$$

The corresponding estimator is

$$
\widehat{p}_{ij} = \frac{\displaystyle\sum_{k=1}^{N} I_{\{\widetilde{S}(\mathbf{X}_k) \leqslant \gamma\}} \sum_{r=1}^{n} I_{\{\mathbf{X}_k \in \widetilde{\mathscr{X}}_{ij}(r)\}}}{\displaystyle\sum_{k=1}^{N} I_{\{\widetilde{S}(\mathbf{X}_k) \leqslant \gamma\}} \sum_{r=1}^{n} I_{\{\mathbf{X}_k \in \widetilde{\mathscr{X}}_{i}(r)\}}} .
\tag{8.40}
$$

This has a very simple interpretation. To update p_{ij}, we simply take the fraction of times in which the transition from i to j occurs, taking into account only those paths that have a total length less than or equal to γ.

This is how one could, *in principle*, carry out the sample generation and parameter updating for problem (8.36): generate paths via a Markov process with transition matrix $\mathbf{P}$ and use the updating formula (8.40). However, *in practice*, we would never generate the tours this way, since most paths would visit cities (other than 1) more than once, and therefore their $\widetilde{S}$ values would be ∞ — that is, most of the paths would not constitute tours. In order to avoid the generation of irrelevant paths, we proceed as follows:

Algorithm 8.6.1: Trajectory Generation Using Node Transitions

---

1 Define $\mathbf{P}^{(1)} = \mathbf{P}$ and $X_1 = 1$.
2 **for** $t = 1$ **to** $n - 1$ **do**
3   Obtain $\mathbf{P}^{(t+1)}$ from $\mathbf{P}^{(t)}$ by first setting the X_t-th column of $\mathbf{P}^{(t)}$ to 0 and then normalizing the rows to sum up to 1.
4   Generate X_{t+1} from the distribution formed by the X_t-th row of $\mathbf{P}^{(t)}$.

A fast implementation of the algorithm above, due to Slava Vaisman, is given by the following procedure, which has complexity $\mathcal{O}(n^2)$. Here i is the currently visited node, and $(b_1, \ldots, b_n)$ is used to keep track of which states have been visited: $b_i = 1$ if node i has already been visited and 0 otherwise.

Algorithm 8.6.2: Fast Generation of Trajectories.

---

1 Let $b_1 \leftarrow 1$ and $b_j \leftarrow 0$ for all $j \neq 1$.
2 $X_1 \leftarrow 1, i \leftarrow 1$
3 **for** $t = 1$ **to** n **do**
4   Generate $U \sim \mathsf{U}(0,1)$ and let $R \leftarrow U \sum_{j=1}^{n} (1 - b_j) \, p_{ij}$.
5   Let sum $\leftarrow 0$ and $j \leftarrow 0$.
6   **while** sum $< R$ **do**
7         $j \leftarrow j + 1$
8   **if** $b_j = 0$ **then** sum $\leftarrow$ sum $+ p_{ij}$
9   Set $X_t \leftarrow j$, $b_j \leftarrow 1$ and $i \leftarrow j$.

It is important to realize that the updating formula for p_{ij} remains the same. By using Algorithm 8.6.1, we are merely *speeding up* our naive trajectory generation by only generating *tours*. As a consequence, each trajectory will visit each city once, and transitions from i to j can at most occur once. It follows that

$$
\widetilde{\mathscr{X}}_{ij}(r) = \widetilde{\mathscr{X}}_{i}(r) = \emptyset, \quad \text{for } r \geqslant 2 ,
$$

so that the updating formula for p_{ij} can be written as

$$\widehat{p}_{ij} = \frac{\sum_{k=1}^{N} I_{\{S(\mathbf{X}_k) \leqslant \gamma\}} I_{\{\mathbf{X}_k \in \mathscr{X}_{ij}\}}}{\sum_{k=1}^{N} I_{\{S(\mathbf{X}_k) \leqslant \gamma\}}} , \qquad (8.41)$$

where $\mathscr{X}_{ij}$ is the set of tours in which the transition from i to j is made. This has the same "natural" interpretation discussed for (8.40).

For the initial matrix $\widehat{\mathbf{P}}_0$, one could simply take all off-diagonal elements equal to $1/(n-1)$, provided that all cities are connected.

Note that ϱ and α should be chosen as in Remark 8.3.3, and the sample size for TSP should be $N = c\,n^2$, with $c > 1$, say $c = 5$.

■ **EXAMPLE 8.11   TSP on Hammersley Points**

To shed further light on the CE method applied to the TSP, consider a shortest (in Euclidean distance sense) tour through a set of *Hammersley points*. These form an example of *low-discrepancy* sequences that cover a d-dimensional unit cube in a pseudo-random but orderly way. To find the 2^5 two-dimensional Hammersley points of order 5, construct first the x-coordinates by taking all binary fractions $x = 0.x_1 x_2 \ldots x_5$. Then let the corresponding y coordinate be obtained from x by reversing the binary digits. For example, if $x = 0.11000$ (binary), which is $x = 1/2 + 1/4 = 3/4$ (decimal), then $y = 0.00011$ (binary), which is $y = 3/32$ (decimal). The Hammersley points, in order of increasing y, are thus

$\{(0,0), (16,1), (8,2), (24,3), (4,4), (20,5), (12,6), (28,7), (2,8), (18,9), (10,10),$
$(26,11), (6,12), (22,13), (14,14), (30,15), (1,16), (17,17), (9,18), (25,19), (5,20),$
$(21,21), (13,22), (29,23), (3,24), (19,25), (11,26), (27,27), (7,28), (23,29),$
$(15,30), (31,31)\}/32 .$

Table 8.10 and Figure 8.8 show the behavior of the CE algorithm applied to the Hammersley TSP. In particular, Table 8.10 depicts the progression of $\widehat{\gamma}_t$ and S_t^b, which denote the largest of the elite values in iteration t and the best value encountered so far, respectively. Similarly, Figure 8.8 shows the evolution of the transition matrices $\mathbf{P}_t$. Here the initial elements $p_{0,ij}$, $i \neq j$ are all set to $1/(n-1) = 1/31$; the diagonal elements are 0. We used a sample size of $N = 5\,n^2 = 5120$, rarity parameter $\varrho = 0.03$, and smoothing parameter $\alpha = 0.7$. The algorithm was stopped when no improvement in $\widehat{\gamma}_t$ during three consecutive iterations was observed.

Table 8.10: Progression of the CE algorithm for the Hammersley TSP.

t	S_t^b	$\widehat{\gamma}_t$	t	S_t^b	$\widehat{\gamma}_t$
1	11.0996	13.2284	16	5.95643	6.43456
2	10.0336	11.8518	17	5.89489	6.31772
3	9.2346	10.7385	18	5.83683	6.22153
4	8.27044	9.89423	19	5.78224	6.18498
5	7.93992	9.18102	20	5.78224	6.1044
6	7.54475	8.70609	21	5.78224	6.0983
7	7.32622	8.27284	22	5.78224	6.06036
8	6.63646	7.94316	23	5.78224	6.00794
9	6.63646	7.71491	24	5.78224	5.91265
10	6.61916	7.48252	25	5.78224	5.86394
11	6.43016	7.25513	26	5.78224	5.86394
12	6.20255	7.07624	27	5.78224	5.83645
13	6.14147	6.95727	28	5.78224	5.83645
14	6.12181	6.76876	29	5.78224	5.83645
15	6.02328	6.58972			

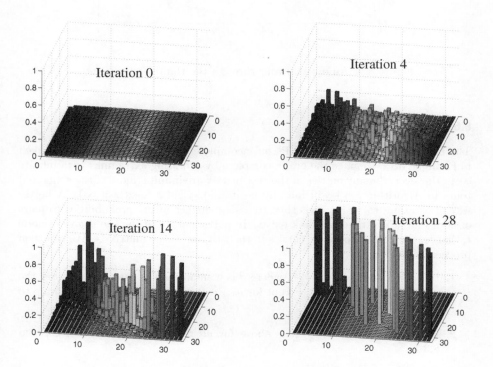

Figure 8.8: Evolution of $\mathbf{P}_t$ in the CE algorithm for the Hammersley TSP.

The optimal tour length for the Hammersley problem is $\gamma^* = 5.78224$ (rounded), which coincides with $\widehat{\gamma}_{29}$ found in Table 8.10. A corresponding solution (optimal tour) is (1, 5, 9, 17, 13, 11, 15, 18, 22, 26, 23, 19, 21, 25, 29, 27, 31, 30, 32, 28, 24, 20, 16, 8, 12, 14, 10, 6, 4, 2, 7, 3), depicted in Figure 8.9. There are several other optimal tours (see Problem 8.14), but all exhibit a straight line through the points $(10,10)/32$, $(14,14)/32$, $(17,17)/32$, and $(21,21)/32$.

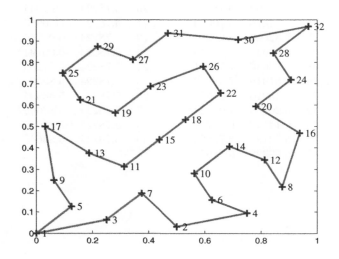

Figure 8.9: An optimal tour through the Hammersley points.

8.6.1 Incomplete Graphs

The easiest way to deal with TSPs on incomplete graphs is, as already remarked in Example 6.12, to make the graph complete by acquiring extra links with infinite cost. However, if many entries in the cost matrix are infinite, most of the generated tours in Algorithm 8.6.1 will initially be invalid (yield a length of ∞). A better way of choosing $\mathbf{P}_0 = (p_{0,ij})$ is then to assign smaller initial probabilities to pairs of nodes for which no direct link exists. In particular, let d_i be the *degree* of node i, that is, the number of finite entries in the i-th row of the matrix C. We can then proceed as follows:

1. If $c_{ij} = \infty$, set $p_{0,ij}$ to $\frac{1-\delta}{d_i}$, where δ is a small number, say $\delta = 0.1$. Set the remaining elements to ε, except for $p_{0,ii} = 0$. Since the rows of $\mathbf{P}_0$ sum up to 1, we have $\varepsilon = \frac{\delta}{n-d_i-1}$.

2. Keep the $p_{0,ij} = \varepsilon = \frac{\delta}{n-d_i-1}$ above for *all iterations* of the CE Algorithm 8.3.1.

Since δ is the sum of all $p_{t,ij}$ corresponding to the ∞ elements in the i-th row of C, and since all such $p_{t,ij}$ are equal to each other (ε), we can generate a transition from each state i using only a $(d_i + 1)$-point distribution rather than the n-point distribution formed by the i-th row of $\widehat{\mathbf{P}}_t$. Indeed, if we *relabel* the elements of this row such that the first d_i entries correspond to existing links, while the next $n - d_i - 1$ correspond to nonexisting links, then we obtain the following faster procedure for generating transitions:

Algorithm 8.6.3: A Fast Procedure for Generating Transitions

---

1 Generate a random variable $U \sim \mathsf{U}(0, 1)$.
2 **if** $U \leqslant 1 - \delta$ **then**
3   Generate the next transition from the discrete d_i-point pdf with
    probabilities $p_{t,ij}/(1 - \delta)$, $j = 1, \ldots, d_i$.
4 **else**
5   Generate the next transition by drawing a discrete random variable Z
    uniformly distributed over the points $d_i + 1, \ldots, n - 1$ (recall that these
    points correspond to the ∞ elements in the i-th row of C).

It is important to note that the small elements of $\mathbf{P}_0$ corresponding to infinities in matrix C should be kept the same from iteration to iteration rather than being updated. By doing so, one obtains considerable speedup in trajectory generation.

8.6.2 Node Placement

We now present an alternative algorithm for trajectory generation due to Margolin [33] called the node *placement algorithm*. In contrast to Algorithm 8.6.1, which generates *transitions* from node to node (based on the transition matrix $P = (p_{ij})$), in Algorithm 8.6.4 below, a similar matrix

$$\mathbf{P} = \begin{pmatrix} p_{(1,1)} & p_{(1,2)} & \cdots & p_{(1,n)} \\ p_{(2,1)} & p_{(2,2)} & \cdots & p_{(2,n)} \\ \vdots & \vdots & \vdots & \vdots \\ p_{(n,1)} & p_{(n,2)} & \cdots & p_{(n,n)} \end{pmatrix} \tag{8.42}$$

generates *node placements*. Specifically, $p_{(i,j)}$ corresponds to the probability of node i being visited at the j-th place in a tour of n cities. In other words, $p_{(i,j)}$ can be viewed as the probability that city (node) i is "arranged" to be visited at the j-th place in a tour of n cities. More formally, a *node placement vector* is a vector $\mathbf{y} = (y_1, \ldots, y_n)$ such that y_i denotes the place of node i in the tour $\mathbf{x} = (x_1, \ldots, x_n)$. The precise meaning is given by the correspondence

$$y_i = j \quad \Longleftrightarrow \quad x_j = i \,, \tag{8.43}$$

for all $i, j \in \{1, \ldots, n\}$. For example, the node placement vector $\mathbf{y} = (3, 4, 2, 6, 5, 1)$ in a six-node network defines uniquely the tour $\mathbf{x} = (6, 3, 1, 2, 5, 4)$. The performance of each node placement $\mathbf{y}$ can be defined as $\overline{S}(\mathbf{y}) = S(\mathbf{x})$, where $\mathbf{x}$ is the unique path corresponding to $\mathbf{y}$.

Algorithm 8.6.4: Trajectory Generation Using Node Placements

---

1 Define $\mathbf{P}^{(1)} \leftarrow \mathbf{P}$. Let $k \leftarrow 1$.
2 for $t = 1$ **to** n **do**
3 | Generate Y_t from the distribution formed by the t-th row of $\mathbf{P}^{(t)}$.
4 | Obtain the matrix $\mathbf{P}^{(t+1)}$ from $\mathbf{P}^{(t)}$ by first setting the Y_t-th column of
  | $\mathbf{P}^{(t)}$ to 0 and then normalizing the rows to sum up to 1.
5 Determine the tour by (8.43) and evaluate the length of the tour by (8.34).

It is readily seen that the updating formula for $p_{(i,j)}$ is now

$$\widehat{p}_{(i,j)} = \frac{\displaystyle\sum_{k=1}^{N} I_{\{\overline{S}(\mathbf{Y}_k) \leqslant \gamma\}} I_{\{Y_{ki}=j\}}}{\displaystyle\sum_{k=1}^{N} I_{\{\overline{S}(\mathbf{Y}_k) \leqslant \gamma\}}} . \tag{8.44}$$

Our simulation results with the TSP and other problems do not indicate clear superiority of either Algorithm 8.6.1 or Algorithm 8.6.4 in terms of the efficiency (speed and accuracy) of the main CE Algorithm 8.3.1.

8.6.3   Case Studies

To illustrate the accuracy and robustness of the CE algorithm, we applied the algorithm to a number of benchmark problems from the TSP library

http://www.iwr.uni-heidelberg.de/groups/comopt/software/TSPLIB95/tsp/

In all cases the *same* set of CE parameters were chosen: $\varrho = 0.03$, $\alpha = 0.7$, $N = 5\,n^2$; and we use the stopping rule (8.21) with the parameter $d = 3$.

Table 8.11 presents the performance of Algorithm 8.3.1 for a selection of *symmetric* TSPs from this library. To study the variability in the solutions, each problem was repeated 10 times. In the table, *min*, *mean*, and *max* denote the smallest (i.e., best), average, and largest of the 10 estimates for the optimal value. The true optimal value is denoted by γ^*.

The average CPU time in seconds and the average number of iterations are given in the last two columns. The size of the problem (number of nodes) is indicated in its name. For example, st70 has $n = 70$ nodes. Similar case studies for the *asymmetric case* can be found in Table 2.5 of [45].

At this end, note that CE is ideally suitable for parallel computation, since parallel computing speeds up the process by almost a factor of r, where r is the number of parallel processors.

One might wonder why the CE Algorithm 8.2.1, with such simple updating rules and quite arbitrary parameters α and ϱ, performs so nicely for combinatorial optimization problems. A possible explanation is that the objective function S for combinatorial optimization problems is typically close to being additive; see, for example, the objective function S for the TSP problem in (8.34). For other optimization problems (e.g., optimizing complex multiextremal continuous functions), one needs to make a more careful and more conservative choice of the parameters α and ϱ.

Table 8.11: Case studies for the TSP.

File	γ^*	Min	Mean	Max	CPU	$\bar{T}$
burma14	3323	3323	3325.6	3336	0.14	12.4
ulysses16	6859	6859	6864	6870	0.21	14.1
ulysses22	7013	7013	7028.9	7069	1.18	22.1
bayg29	1610	1610	1628.6	1648	4.00	28.2
bays29	2020	2020	2030.9	2045	3.83	27.1
dantzig42	699	706	717.7	736	19.25	38.4
eil51	426	428	433.9	437	65.0	63.35
berlin52	7542	7618	7794	8169	64.55	59.9
st70	675	716	744.1	765	267.5	83.7
eil76	538	540	543.5	547	467.3	109.0
pr76	108159	109882	112791	117017	375.3	88.9

8.7 CONTINUOUS OPTIMIZATION

We will briefly discuss how the CE method can be applied to solve continuous optimization problems. Let $S(\mathbf{x})$ be a real-valued function on $\mathbb{R}^n$. To maximize the function via CE, we need to specify a family of parameterized distributions to generate samples in $\mathbb{R}^n$. This family must include, at least in the limiting case, the degenerate distribution that puts all its probability mass on an optimal solution. A simple choice is to use a multivariate normal distribution, parameterized by a mean vector $\boldsymbol{\mu} = (\mu_1, \ldots, \mu_n)$ and a covariance matrix Σ. When the covariance matrix is chosen to be diagonal — that is, the components of $\mathbf{X}$ are independent — the CE updating formulas become particularly easy. In particular, denoting $\{\mu_i\}$ and $\{\sigma_i\}$ the means and standard deviations of the components, the updating formulas are (see Problem 8.18)

$$\widehat{\mu}_{t,i} = \frac{\sum_{\mathbf{X}_k \in \mathscr{E}_t} X_{ki}}{|\mathscr{E}_t|} \qquad i = 1, \ldots, n, \tag{8.45}$$

and

$$\widehat{\sigma}_{t,i} = \sqrt{\frac{\sum_{\mathbf{X}_k \in \mathscr{E}_t} (X_{ki} - \widehat{\mu}_{t,i})^2}{|\mathscr{E}_t|}}, \qquad i = 1, \ldots, n, \tag{8.46}$$

where X_{ki} is the i-th component of $\mathbf{X}_k$ and $\mathbf{X}_1, \ldots, \mathbf{X}_n$ is a random sample from $\mathsf{N}(\widehat{\boldsymbol{\mu}}_{t-1}, \widehat{\Sigma}_{t-1})$. In other words, the means and standard deviations are simply updated via the corresponding maximum likelihood estimators based on the elite samples $\mathscr{E}_t = \{\mathbf{X}_k : S(\mathbf{X}_k) \geqslant \widehat{\gamma}_t\}$.

■ **EXAMPLE 8.12   The Peaks Function**

Matlab's `peaks` function,

$$S(\mathbf{x}) = 3\,(1 - x_1)^2\,\exp(-(x_1^2) - (x_2 + 1)^2) - 10\,(x_1/5 - x_1^3 - x_2^5)\,\exp(-x_1^2 - x_2^2)$$
$$- 1/3\,\exp(-(x_1 + 1)^2 - x_2^2)\,,$$

has various local maxima. In Section A.5 of the Appendix, a simple Matlab implementation of CE Algorithm 8.3.1 is given for finding the global maximum of this function, which is approximately $\gamma^* = 8.10621359$ and is attained at $\mathbf{x}^* = (-0.0093151, 1.581363)$. The choice of the initial value for $\boldsymbol{\mu}$ is not important, but the initial standard deviations should be chosen large enough to ensure initially a close to uniform sampling of the region of interest. The CE algorithm is stopped when all standard deviations of the sampling distribution are less than some small ε.

Figure 8.10 gives the evolution of the worst and best of the elite samples, that is, $\widehat{\gamma}_t$ and S_t^*, for each iteration t. We see that the values quickly converge to the optimal value γ^*.

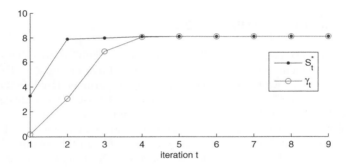

Figure 8.10: Evolution of the CE algorithm for the `peaks` function.

Remark 8.7.1 (Injection) When using the CE method to solve practical optimization problems with many constraints and many local optima, it is sometimes necessary to prevent the sampling distribution from shrinking too quickly. A simple but effective approach is the following *injection* method [6]. Let S_t^* denote the best performance found at the t-th iteration, and (in the normal case) let σ_t^* denote the largest standard deviation at the t-th iteration. If σ_t^* is sufficiently small and $|S_t^* - S_{t-1}^*|$ is also small, then add some small value to each standard deviation, for example, a constant δ or the value $c\,|S_t^* - S_{t-1}^*|$, for some fixed δ and c. When using CE with injection, a possible stopping criterion is to stop after a fixed number of injections.

8.8   NOISY OPTIMIZATION

One of the distinguishing features of the CE Algorithm 8.3.1 is that it can easily handle noisy optimization problems, that is, when the objective function $S(\mathbf{x})$ is corrupted with noise. We denote such a noisy function by $\widehat{S}(\mathbf{x})$. We assume that for each $\mathbf{x}$ we can readily obtain an outcome of $\widehat{S}(\mathbf{x})$, for example, via generation of some additional random vector $\mathbf{Y}$, whose distribution may depend on $\mathbf{x}$.

A classical example of noisy optimization is simulation-based optimization [46]. A typical instance is the *buffer allocation problem*, where the objective is to allocate n buffer spaces among the $m - 1$ "niches" (storage areas) between m machines in a serial production line so as to optimize some performance measure, such as the steady-state throughput. This performance measure is typically not available analytically and thus must be estimated via simulation. A detailed description of the buffer allocation problem, and of how CE can be used to solve this problem, is given in [45].

Another example is the noisy TSP, where, say, the cost matrix (c_{ij}), denoted now by $\mathbf{Y} = (Y_{ij})$, is random. Think of Y_{ij} as the random time to travel from city i to city j. The total cost of a tour $\mathbf{x} = (x_1, \ldots, x_n)$ is given by

$$\widehat{S}(\mathbf{x}) = \sum_{i=1}^{n-1} Y_{x_i, x_{i+1}} + Y_{x_n, x_1} \, . \tag{8.47}$$

We assume that $\mathbb{E}[Y_{ij}] = c_{ij}$.

The main CE optimization Algorithm 8.3.1 for deterministic functions $S(\mathbf{x})$ is also valid for noisy ones $\widehat{S}(\mathbf{x})$. Extensive numerical studies [45] with the noisy version of Algorithm 8.3.1 show that it works nicely because, during the course of the optimization, it *filters out* efficiently the noise component from $\widehat{S}(\mathbf{x})$. However, to get reliable estimates of the optimal solution of combinatorial optimization problems, one is required to increase the sample size N by a factor 2 to 5 in each iteration of Algorithm 8.3.1. Clearly, this factor increases with the "power" of the noise.

■ EXAMPLE 8.13   Noisy TSP

Suppose that in the first test case of Table 8.11, `burma14`, some uniform noise is added to the cost matrix. In particular, suppose that the cost of traveling from i to j is given by $Y_{ij} \sim \mathsf{U}(c_{ij} - 8, c_{ij} + 8)$, where c_{ij} is the cost for the deterministic case. The expected cost is thus $\mathbb{E}[Y_{ij}] = c_{ij}$, and the total cost $\widehat{S}(\mathbf{x})$ of a tour $\mathbf{x}$ is given by (8.47). The CE algorithm for optimizing the unknown $S(\mathbf{x}) = \mathbb{E}[\widehat{S}(\mathbf{x})]$ remains exactly the same as in the deterministic case, except that $S(\mathbf{x})$ is replaced with $\widehat{S}(\mathbf{x})$ and a different stopping criterion than (8.21) needs to be employed. A simple rule is to stop when the transition probabilities $\widehat{p}_{t,ij}$ satisfy $\min(\widehat{p}_{t,ij}, 1 - \widehat{p}_{t,ij}) < \varepsilon$ for *all* i and j, similar to Remark 8.4.1. We repeated the experiment 10 times, taking a sample size twice as large as for the deterministic case, that is, $N = 10 \cdot n^2$. For the stopping criterion given above we took $\varepsilon = 0.02$. The other parameters remained the same as those described in Section 8.6.3. CE found the optimal solution eight times, which is comparable to its performance in the deterministic case.

Figure 8.11 displays the evolution of the worst performance of the elite samples $(\widehat{\gamma}_t)$ for both the deterministic and noisy case denoted by $\widehat{\gamma}_{1t}$ and $\widehat{\gamma}_{2t}$, respectively.

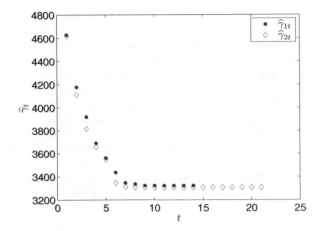

Figure 8.11: Evolution of the worst of the elite samples for a deterministic and noisy TSP.

We see in both cases a similar rapid drop in the level $\widehat{\gamma}_t$. It is important notice that even though here the algorithm in both the deterministic and noisy cases converges to the optimal solution, the $\{\widehat{\gamma}_{2t}\}$ for the noisy case do not converge to $\gamma^* = 3323$, in contrast to the $\{\widehat{\gamma}_{1t}\}$ for the deterministic case. This is because the latter estimates eventually the $(1 - \varrho)$-quantile of the deterministic $S(\mathbf{x}^*)$, whereas the former estimates the $(1 - \varrho)$-quantile of $\widehat{S}(\mathbf{x}^*)$, which is random. To estimate $S(\mathbf{x}^*)$ in the noisy case, one needs to take the sample average of $\widehat{S}(\mathbf{x}_T)$, where $\mathbf{x}_T$ is the solution found at the final iteration.

8.9   MINXENT METHOD

In the standard CE method for rare-event simulation, the importance sampling density for estimating $\ell = \mathbb{P}_f(S(\mathbf{X}) \geqslant \gamma)$ is restricted to some parametric family, say $\{f(\cdot; \mathbf{v}), \mathbf{v} \in \mathcal{V}\}$, and the optimal density $f(\cdot; \mathbf{v}^*)$ is found as the solution to the *parametric* CE minimization program (8.3). In contrast to CE, we present below a *nonparametric* method called the *MinxEnt* method. The idea is to minimize the CE distance to g^* over *all* pdfs rather than over $\{f(\cdot; \mathbf{v}), \mathbf{v} \in \mathcal{V}\}$. However, the program $\min_g \mathcal{D}(g, g^*)$ is void of meaning, since the minimum (zero) is attained at the unknown $g = g^*$. A more useful approach is to first specify a *prior* density h, which conveys the available information on the "target" g^*, and then choose the "instrumental" pdf g as close as possible to h, subject to certain *constraints* on g. If no prior information on the target g^* is known, the prior h is simply taken to be a constant, corresponding to the uniform distribution (continuous or discrete). This leads to the following minimization framework [3], [4], and [22]:

$$
\text{(P}_0) \begin{cases} \min_g \mathcal{D}(g,h) = \min_g \int \ln \frac{g(\mathbf{x})}{h(\mathbf{x})} \, g(\mathbf{x}) \, d\mathbf{x} = \min_g \mathbb{E}_g \left[\ln \frac{g(\mathbf{X})}{h(\mathbf{X})} \right] \\[2mm] \text{s.t.} \quad \int S_i(\mathbf{x}) \, g(\mathbf{x}) \, d\mathbf{x} = \mathbb{E}_g[S_i(\mathbf{X})] = \gamma_i, \quad i = 1, \dots, m \,, \\[2mm] \int g(\mathbf{x}) \, d\mathbf{x} = 1 \,. \end{cases} \tag{8.48}
$$

Here g and h are n-dimensional pdfs, $S_i(\mathbf{x})$, $i = 1, \dots, m$, are given functions, and $\mathbf{x}$ is an n-dimensional vector. The program (P$_0$) is Kullback's *minimum cross-entropy* (MinxEnt) program. Note that this is a *convex* functional optimization problem, because the objective function is a convex function of g, and the constraints are affine in g.

If the prior h is constant, then $\mathcal{D}(g,h) = \int g(\mathbf{x}) \ln g(\mathbf{x}) \, d\mathbf{x} + \text{constant}$, so that the minimization of $\mathcal{D}(g,h)$ in (P$_0$) can be replaced with the *maximization* of

$$
\mathcal{H}(g) = - \int g(\mathbf{x}) \ln g(\mathbf{x}) \, d\mathbf{x} = -\mathbb{E}_g[\ln g(\mathbf{X})] \,, \tag{8.49}
$$

where $\mathcal{H}(g)$ is the *Shannon entropy*; see (1.46). (Here we use a different notation to emphasize the dependence on g.) The corresponding program is Jaynes' *MaxEnt* program [20]. Note that the former minimizes the Kullback–Leibler cross-entropy, while the latter maximizes the Shannon entropy [22].

In typical counting and combinatorial optimization problems h is chosen as an n-dimensional pdf with uniformly distributed marginals. For example, in the SAT counting problem, we assume that each component of the n-dimensional random vector $\mathbf{X}$ is $\mathsf{Ber}(1/2)$ distributed. While estimating rare events in stochastic models, like queueing models, h is the fixed underlying pdf. For example, in the $M/M/1$ queue h would be a two-dimensional pdf with independent marginals, where the first marginal is the interarrival $\mathsf{Exp}(\lambda)$ pdf and the second is the service $\mathsf{Exp}(\mu)$ pdf.

The MinxEnt program, which presents a constrained functional optimization problem, can be solved via Lagrange multipliers. The solution for the discrete case is derived in Example 1.19 on page 39. A similar solution can be derived, for example , via calculus of variations [3], for the general case. In particular, the solution of the MinxEnt problem is given [22] by

$$
g(\mathbf{x}) = \frac{h(\mathbf{x}) \exp \left\{ \sum_{i=1}^m \lambda_i \, S_i(\mathbf{x}) \right\}}{\mathbb{E}_h \left[\exp \left\{ \sum_{i=1}^m \lambda_i \, S_i(\mathbf{X}) \right\} \right]} \,, \tag{8.50}
$$

where λ_i, $i = 1, \dots, m$, are obtained from the solution of the following system of equations:

$$
\frac{\mathbb{E}_h \left[S_i(\mathbf{X}) \exp \left\{ \sum_{j=1}^m \lambda_j \, S_j(\mathbf{X}) \right\} \right]}{\mathbb{E}_h \left[\exp \left\{ \sum_{j=1}^m \lambda_j \, S_j(\mathbf{X}) \right\} \right]} = \gamma_i, \quad i = 1, \dots, m \,, \tag{8.51}
$$

where $\mathbf{X} \sim h(\mathbf{x})$.

Note that $g(\mathbf{x})$ can be written as

$$
g(\mathbf{x}) = C(\boldsymbol{\lambda}) \, h(\mathbf{x}) \, \exp \left\{ \sum_{i=1}^m \lambda_i \, S_i(\mathbf{x}) \right\}, \tag{8.52}
$$

where

$$C^{-1}(\boldsymbol{\lambda}) = \mathbb{E}_h \left[\exp \left\{ \sum_{i=1}^{m} \lambda_i \, S_i(\mathbf{X}) \right\} \right] \tag{8.53}$$

is the normalization constant. Note also that $g(\mathbf{x})$ is a density function; in particular, $g(\mathbf{x}) \geq 0$.

Consider the MinxEnt program (P_0) with a single constraint, that is,

$$\begin{cases} \min_g \mathcal{D}(g, h) = \min_g \mathbb{E}_g \left[\ln \frac{g(X)}{h(X)} \right] \\ \text{s.t.} \quad \mathbb{E}_g[S(X)] = \gamma , \\ \quad \int g(x) \, \mathrm{d}x = 1 . \end{cases} \tag{8.54}$$

In this case (8.50) and (8.51) reduce to

$$g(\mathbf{x}) = \frac{h(\mathbf{x}) \, \exp\{\lambda \, S(\mathbf{x})\}}{\mathbb{E}_h \left[\exp\{\lambda \, S(\mathbf{X})\} \right]} \tag{8.55}$$

and

$$\frac{\mathbb{E}_h \left[S(\mathbf{X}) \exp \left\{ \lambda \, S(\mathbf{X}) \right\} \right]}{\mathbb{E}_h \left[\exp \left\{ \lambda \, S(\mathbf{X}) \right\} \right]} = \gamma , \tag{8.56}$$

respectively.

In the particular case where $S(\mathbf{x})$, $\mathbf{x} = (x_1, \ldots, x_n)$ is a coordinatewise separable function, that is,

$$S(\mathbf{x}) = \sum_{i=1}^{n} S_i(x_i) , \tag{8.57}$$

and the components X_i, $i = 1, \ldots, n$ of the random vector $\mathbf{X}$ are independent under $h(\mathbf{x}) = h(x_1) \cdots h(x_n)$, the joint pdf $g(\mathbf{x})$ in (8.55) reduces to the *product of marginal pdfs*. In particular, the i-th component of $g(\mathbf{x})$ can be written as

$$g_i(x) = \frac{h_i(x) \exp\{\lambda S_i(x)\}}{\mathbb{E}_{h_i} \left[\exp \left\{ \lambda S_i(x) \right\} \right]}, \quad i = 1, \ldots, n . \tag{8.58}$$

Remark 8.9.1 (The MinxEnt Program with Inequality Constraints) It is not difficult to extend the MinxEnt program to contain *inequality* constraints. Suppose that the following M inequality constraints are added to the MinxEnt program (8.48):

$$\mathbb{E}_g[S_i(\mathbf{X})] \geqslant \gamma_i, \qquad i = m + 1, \ldots, m + M .$$

The solution of this MinxEnt program is given by

$$g(\mathbf{x}) = \frac{h(\mathbf{x}) \exp \left\{ \sum_{i=1}^{m+M} \lambda_i \, S_i(\mathbf{x}) \right\}}{\mathbb{E}_h \left[\exp \left\{ \sum_{i=1}^{m+M} \lambda_i \, S_i(\mathbf{X}) \right\} \right]} , \tag{8.59}$$

where the Lagrange multipliers $\lambda_1, \ldots, \lambda_{m+M}$ are the solutions to the dual convex optimization problem

$$\max_{\boldsymbol{\lambda}, \beta} \quad \sum_{i=1}^{m+M} \lambda_i \, \gamma_i - \beta - \mathbb{E}_h \left[\exp \left(-1 - \beta + \sum_{i=1}^{m+M} \lambda_i S_i(\mathbf{x}) \right) \right]$$

$$\text{subject to:} \quad \lambda_i \geqslant 0, \quad i = m + 1, \ldots, m + M.$$

Thus, only the Lagrange multipliers corresponding to an inequality must be constrained from below by zero. Similar to (1.80), this can be solved in two steps, where β can be determined explicitly as a normalization constant but the $\{\lambda_i\}$ have to be determined numerically.

In the special case of a single inequality constraint (i.e., $m = 0$ and $M = 1$), the dual program can be solved directly (see also Problem 8.30), yielding the following solution:

$$\lambda = \begin{cases} 0 & \text{if } \mathbb{E}_h[S(\mathbf{X})] \geqslant \gamma \,, \\ \lambda^* & \text{if } \mathbb{E}_h[S(\mathbf{X})] < \gamma \,, \end{cases}$$

where λ^* is obtained from (8.56). That is, if $\mathbb{E}_h[S(\mathbf{X})] < \gamma$, then the inequality MinxEnt solution agrees with the equality MinxEnt solution; otherwise, the optimal sampling pdf remains the prior h.

Remark 8.9.2 It is well known [14] that the optimal solution of the single-dimensional single-constrained MinxEnt program (8.54) coincides with the celebrated optimal *exponential change of measure* (ECM). Note that normally in a multidimensional ECM one twists each component separately, using possibly different twisting parameters. In contrast, the optimal solution to the MinxEnt program (see (8.58)) is parameterized by a *single-dimensional* parameter λ.

If not otherwise stated, we consider below only the single-constrained case (8.54). As in the standard CE method we could also use a *multilevel* approach, and apply a sequence of instrumentals $\{g_t\}$ and levels $\{\gamma_t\}$. Starting with $g_0 = f$ and always taking prior $h = f$, we determine γ_t and g_t as follows:

1. Update γ_t as
$$\gamma_t = \mathbb{E}_{g_t}[S(\mathbf{X}) \mid S(\mathbf{X}) \geqslant q_t] \,,$$
   where q_t is the $(1 - \varrho)$-quantile of $S(\mathbf{X})$ under g_{t-1}.

2. Update g_t as the solution to the above MinxEnt program for level γ_t rather than γ.

The updating formula for γ_t is based on the constraint $\mathbb{E}_g[S(\mathbf{X})] = \gamma$ in the MinxEnt program. However, instead of simply updating as $\gamma_t = \mathbb{E}_{g_{t-1}}[S(\mathbf{X})]$, we take the expectation of $S(\mathbf{X})$ with respect to g_{t-1} *conditional* on $S(\mathbf{X})$ being greater than its $(1 - \varrho)$ quantile, here denoted as q_t. In contrast, in the standard CE method the level γ_t is simply updated as q_t.

Note that each g_t is completely determined by its Lagrange multiplier, say λ_t, which is the solution to (8.56) with γ_t instead of γ. In practice, both γ_t and λ_t have to be replaced by their stochastic counterparts $\widehat{\gamma}_t$ and $\widehat{\lambda}_t$, respectively. Specifically, γ_t can be estimated from a random sample $\mathbf{X}_1, \ldots, \mathbf{X}_N$ of g_{t-1} as the average of the $N^e = \lceil (1 - \varrho)N \rceil$ elite sample performances:

$$\widehat{\gamma}_t = \frac{\sum_{i=N-N^e+1}^{N} S_{(i)}}{N^e} \,, \tag{8.60}$$

where $S_{(i)}$ denotes the i-th order-statistics of the sequence $S(\mathbf{X}_1), \ldots, S(\mathbf{X}_N)$. And λ_t can be estimated by solving, with respect to λ, the stochastic counterpart of (8.56), which is

$$\frac{\sum_{k=1}^{N} e^{\lambda S(\mathbf{X}_k)} S(\mathbf{X}_k)}{\sum_{k=1}^{N} e^{\lambda S(\mathbf{X}_k)}} = \widehat{\gamma}_t \,. \tag{8.61}$$

PROBLEMS

8.1   In Example 8.2, show that the true CE-optimal parameter for estimating $\mathbb{P}(X \geqslant 32)$ is given by $v^* = 33$.

8.2   Write a CE program to reproduce Table 8.1 in Example 8.2. Use the final reference parameter $\widehat{v}_3$ to estimate ℓ via importance sampling, using a sample size of $N_1 = 10^6$. Estimate the relative error and give an approximate 95% confidence interval. Check if the true value of ℓ is contained in this interval.

8.3   In Example 8.2, calculate the exact relative error for the importance sampling estimator $\widehat{\ell}$ when using the CE optimal parameter $v^* = 33$ and compare it with the one estimated in Problem 8.2. How many samples are required to estimate ℓ with the same relative error, using CMC?

8.4   Implement the CE Algorithm 8.2.1 for the stochastic shortest path problem in Example 8.5 and reproduce Table 8.3.

8.5   Slightly modify the program used in Problem 8.4 to allow Weibull-distributed lengths. Reproduce Table 8.4 and make a new table for $\alpha = 5$ and $\gamma = 2$ (the other parameters remain the same).

8.6   Make a table similar to Table 8.4 by employing the standard CE method. That is, take $\mathsf{Weib}(\alpha, v_i^{-1})$ as the importance sampling distribution for the i-th component and update the $\{v_i\}$ via (8.6).

8.7   Consider again the stochastic shortest path problem in Example 8.5, but now with nominal parameter $\mathbf{u} = (0.25, 0.5, 0.1, 0.3, 0.2)$. Implement the root-finding Algorithm 8.2.3 to estimate for which level γ the probability ℓ is equal to 10^{-5}. Also, give a 95% confidence interval for γ, for example, using the bootstrap method.

8.8   Suppose $\mathbf{X} = (X_1, \ldots, X_n) \sim \mathsf{Ber}(\mathbf{p})$, with $\mathbf{p} = (p_1, \ldots, p_n)$, meaning that $X_i \sim \mathsf{Ber}(p_i)$, $i = 1, \ldots, n$ independent. Define $S(\mathbf{X}) = X_1 + \cdots + X_n$. We wish to estimate $\ell = \mathbb{P}(S(\mathbf{X}) \geqslant \gamma)$ via importance sampling, using vector $\mathbf{q} = (q_1, \ldots, q_n)$ instead of $\mathbf{p}$. To determine a good importance sampling vector $\mathbf{q}$, we could employ the multilevel CE Algorithm 8.2.1 or the single-level CE Algorithm 8.2.5. The latter requires an approximate sample from the zero-variance pdf $g^*(\mathbf{x}) \propto f(\mathbf{x}; \mathbf{p}) I_{\{S(\mathbf{x}) \geqslant \gamma\}}$, where $f(\mathbf{x}; \mathbf{p})$ is the nominal pdf. In the numerical questions below take $n = 80$, $\gamma = 48$, and $p_i = p = 0.1$, $i = 1, \ldots, n$.

    a)   Implement Algorithm 8.2.1 using parameters $\varrho = 0.01$ and $N = 10,000$. Note that the CE updating formula is given in (8.10). Plot the empirical distribution function of the parameters $\{q_i\}$ found in the final iteration.

    b)   Implement a systematic Gibbs sampler to generate a dependent sample $\mathbf{X}_1, \ldots, \mathbf{X}_N$ from the conditional distribution of $\mathbf{X}$ given $S(\mathbf{X}) \geqslant \gamma$. Take $N = 10,000$. Use this sample to obtain

$$q_i = \frac{\sum_{k=1}^{N} X_{ki}}{N}, \quad i = 1, \ldots, n \;.$$

    Plot the empirical distribution function of the $\{q_i\}$ and compare with question **a)**.

    c)   Compare the importance sampling estimators using the parameters $\{q_i\}$ found in questions **a)** and **b)**. Take a sample size $N_1 = 10^6$.

8.9   Adapt the cost matrix in the max-cut program of Table 8.5 and apply it to the dodecahedron max-cut problem in Example 8.9. Produce various optimal solutions and find out how many of these exist in total, disregarding the fivefold symmetry.

8.10   Consider the following symmetric cost matrix for the max-cut problem:

$$C = \begin{pmatrix} Z_{11} & B_{12} \\ B_{21} & Z_{22} \end{pmatrix}, \tag{8.62}$$

where Z_{11} is an $m \times m$ $(m < n)$ symmetric matrix in which all the upper-diagonal elements are generated from a $U(a, b)$ distribution (and all the lower diagonal elements follow by symmetry), Z_{22} is an $(n - m) \times (n - m)$ symmetric matrix that is generated in the same way as Z_{11}, and all the other elements are c, apart from the diagonal elements, which are 0.

  a) Show that if $c > b(n - m)/m$, the optimal cut is given by $V^* = \{\{1, \ldots, m\}, \{m + 1, \ldots, n\}\}$.
  b) Show that the optimal value of the cut is $\gamma^* = c\, m\, (n - m)$.
  c) Implement and run the CE algorithm on this synthetic max-cut problem for a network with $n = 400$ nodes, with $m = 200$. Generate Z_{11} and Z_{22} from the $U(0, 1)$ distribution and take $c = 1$. For the CE parameters take $N = 1000$ and $\varrho = 0.1$. List for each iteration the best and worst of the elite samples and the Euclidean distance $\|\widehat{\mathbf{p}}_t - \mathbf{p}^*\| = \sqrt{(\widehat{p}_{t,i} - p_i^*)^2}$ as a measure of how close the reference vector is to the optimal reference vector $\mathbf{p}^* = (1, 1, \ldots, 1, 0, 0, \ldots, 0)$.

8.11   Consider a TSP with cost matrix $C = (c_{ij})$ defined by $c_{i,i+1} = 1$ for all $i = 1, 2, \ldots, n - 1$, and $c_{n,1} = 1$, while the remaining elements $c_{ij} \sim U(a, b)$, $j \neq i + 1$, $1 < a < b$, and $c_{ii} = 0$.

  a) Verify that the optimal permutation/tour is given by $\mathbf{x}^* = (1, 2, 3, \ldots, n)$, with minimal value $\gamma^* = n$.
  b) Implement a CE algorithm to solve an instance of this TSP for the case $n = 30$ and make a table of the performance, listing the best and worst of the elite samples at each iteration, as well as

$$p_t^{mm} = \min_{1 \leqslant i \leqslant n} \max_{1 \leqslant j \leqslant n} \widehat{p}_{t,ij},$$

$t = 1, 2, \ldots,$ which corresponds to the min max value of the elements of the matrix $\widehat{\mathbf{P}}_t$ at iteration t. Use $d = 3$, $\varrho = 0.01$, $N = 4500$, and $\alpha = 0.7$. Also, keep track of the overall best solution.

8.12   Run Algorithm 8.3.1 on the data from the URL

`http://www.iwr.uni-heidelberg.de/groups/comopt/software/TSPLIB95/atsp/`

and obtain a table similar to Table 8.11.

8.13   Select a TSP of your choice. Verify the following statements about the choice of CE parameters:

  a) After reducing ϱ or increasing α, the convergence is faster but we can be trapped in a local minimum.
  b) After reducing ϱ, we need to decrease simultaneously α, and vice versa, in order to avoid convergence to a local minimum.

    c) In increasing the sample size N, we can simultaneously reduce ϱ or (and) increase α.

8.14 Find out how many optimal solutions there are for the Hammersley TSP in Example 8.11.

8.15 Consider a complete graph with n nodes. With each edge from node i to j there is an associated cost c_{ij}. In the *longest path problem* the objective is to find the longest self-avoiding path from a certain *source* node to a *sink* node.

    a) Assuming the source node is 1 and the sink node is n, formulate the longest path problem similar to the TSP. (The main difference is that the paths in the longest path problem can have different lengths.)

    b) Specify a path generation mechanism and the corresponding CE updating rules.

    c) Implement a CE algorithm for the longest path problem and apply it to a test problem.

8.16 Write a CE program that solves the eight-queens problem using the same configuration representation $\mathbf{X} = (X_1, \ldots, X_8)$ as in Example 6.13. A straightforward way to generate the configurations is to draw each X_i independently from a probability vector $(p_{i1}, \ldots, p_{i8})$, $i = 1, \ldots, 8$. Take $N = 500$, $\alpha = 0.7$, and $\varrho = 0.1$.

8.17 In the *permutation flow shop problem* (PFSP), n jobs have to be processed (in the same order) on m machines. The objective is to find the permutation of jobs that will minimize the *makespan*, that is, the time at which the last job is completed on machine m. Let $t(i, j)$ be the processing time for job i on machine j, and let $\mathbf{x} = (x_1, x_2, \ldots, x_n)$ be a job permutation. Then the completion time $C(x_i, j)$ for job i on machine j can be calculated as follows:

$$
\begin{aligned}
C(x_1, 1) &= t(x_1, 1) , \\
C(x_i, 1) &= C(x_{i-1}, 1) + t(x_i, 1), \forall i = 2, \ldots, n , \\
C(x_1, j) &= C(x_1, j - 1) + t(x_1, j), \forall j = 2, \ldots, m , \\
C(x_i, j) &= \max\{C(x_{i-1}, j), C(x_i, j - 1)\} + t(x_i, j) , \\
&\quad \text{for all } i = 2, \ldots, n; \quad j = 2, \ldots, m .
\end{aligned}
$$

The objective is to minimize $S(\mathbf{x}) = C(x_n, m)$. The trajectory generation for the PFSP is similar to that of the TSP.

    a) Implement a CE algorithm to solve this problem.

    b) Run the algorithm for a benchmark problem from the Internet, for example `http://ina2.eivd.ch/Collaborateurs/etd/problemes.dir/ordonnancement.dir/ordonnancement.html`.

8.18 Verify the updating formulas (8.45) and (8.46).

8.19 Plot Matlab's `peaks` function and verify that it has three local maxima.

8.20 Use the CE program in Section A.5 of the Appendix to maximize the function $S(x) = e^{-(x-2)^2} + 0.8 e^{-(x+2)^2}$. Examine the convergence of the algorithm by plotting in the same figure the sequence of normal sampling densities.

8.21   Use the CE method to minimize the *trigonometric function*

$$S(\mathbf{x}) = 1 + \sum_{i=1}^{n} 8 \sin^2(\eta(x_i - x_i^*)^2) + 6 \sin^2(2\eta(x_i - x_i^*)^2) + \mu(x_i - x_i^*)^2 , \quad (8.63)$$

with $\eta = 7$, $\mu = 1$, and $x_i^* = 0.9, i = 1, \ldots, n$. The global minimum $\gamma^* = 1$ is attained at $\mathbf{x}^* = (0.9, \ldots, 0.9)$. Display the graph and density plot of this function and give a table for the evolution of the algorithm.

8.22   A well-known test case in continuous optimization is the *Rosenbrock* function (in n dimensions):

$$S(\mathbf{x}) = \sum_{i=1}^{n-1} 100 \, (x_{i+1} - x_i^2)^2 + (x_i - 1)^2 . \quad (8.64)$$

The function has a global minimum $\gamma^* = 0$, attained at $\mathbf{x}^* = (1, 1, \ldots, 1)$. Implement a CE algorithm to minimize this function for dimensions $n = 2, 5, 10$, and 20. Observe how injection (Remark 8.7.1) affects the accuracy and speed of the algorithm.

8.23   Suppose that $\mathscr{X}$ in (8.15) is a (possibly nonlinear) region defined by the following system of inequalities:

$$G_i(\mathbf{x}) \leqslant 0, \quad i = 1, \ldots, L . \quad (8.65)$$

The *proportional penalty* approach to constrained optimization is to modify the objective function as follows:

$$\widetilde{S}(\mathbf{x}) = S(\mathbf{x}) + \sum_{i=1}^{L} P_i(\mathbf{x}) , \quad (8.66)$$

where $P_i(\mathbf{x}) = C_i \max(G_i(\mathbf{x}), 0)$ and $C_i > 0$ measures the importance (cost) of the i-th penalty. It is clear that as soon as the constrained problem (8.15), (8.65) is reduced to the unconstrained one (8.15) — using (8.66) instead of S — we can again apply Algorithm 8.3.1.

Apply the proportional penalty approach to the constrained minimization of the Rosenbrock function of dimension 10 for the constraints below. List for each case the minimal value obtained by the CE algorithm (with injection, if necessary) and the CPU time. In all experiments, use $\varepsilon = 10^{-3}$ for the stopping criterion (stop if all standard deviations are less than ε) and $C = 1000$. Repeat the experiments 10 times to check if indeed a global minimum is found.

a) $\sum_{j=1}^{10} x_j \leqslant -8$

b) $\sum_{j=1}^{10} x_j \geqslant 15$

c) $\sum_{j=1}^{10} x_j \leqslant -8, \quad \sum_{j=1}^{10} x_j^2 \geqslant 15$

d) $\sum_{j=1}^{10} x_j \geqslant 15, \quad \sum_{j=1}^{10} x_j^2 \leqslant 22.5$

8.24   Use the CE method to minimize the function

$$S(\mathbf{x}) = 1000 - x_1^2 - 2x_2^2 - x_3^2 - x_1 x_2 - x_1 x_3 ,$$

subject to the constraints $x_j \geqslant 0, j = 1, 2, 3$, and

$$8x_1 + 14x_2 + 7x_3 - 56 = 0 \,,$$
$$x_1^2 + x_2^2 + x_3^2 - 25 = 0 \,.$$

First, eliminate two of the variables by expressing x_2 and x_3 in terms of x_1. Note that this gives *two* different expressions for the pair (x_2, x_3). In the CE algorithm, generate the samples $\mathbf{X}$ by first drawing X_1 according to a truncated normal distribution on $[0, 5]$. Then choose either the first or the second expression for (X_2, X_3) with equal probability. Verify that the optimal solution is approximately $\mathbf{x}^* = (3.51, 0.217, 3.55)$, with $S(\mathbf{x}^*) = 961.7$. Give the solution and the optimal value in seven significant digits.

8.25   Add $\mathsf{U}(-0.1, 0.1)$, $\mathsf{N}(0, 0.01)$, and $\mathsf{N}(0, 1)$ noise to the objective function in Problem 8.20. Formulate an appropriate stopping criterion, for example, based on $\widehat{\sigma}_t$. For each case, observe how $\widehat{\gamma}_t$, $\widehat{\mu}_t$, and $\widehat{\sigma}_t$ behave.

8.26   Add $\mathsf{N}(0, 1)$ noise to the Matlab `peaks` function and apply the CE algorithm to find the global maximum. Display the contour plot and the path followed by the mean vectors $\{\widehat{\boldsymbol{\mu}}_t\}$, starting with $\widehat{\boldsymbol{\mu}}_0 = (1.3, -2.7)$ and using $N = 200$ and $\varrho = 0.1$. Stop when all standard deviations are less than $\varepsilon = 10^{-3}$. In a separate plot, display the evolution of the worst and best of the elite samples ($\widehat{\gamma}_t$ and S_t^*) at each iteration of the CE algorithm. In addition, evaluate and plot the noisy objective function in $\widehat{\boldsymbol{\mu}}_t$ for each iteration. Observe that, in contrast to the deterministic case, the $\{\widehat{\gamma}_t\}$ and $\{S_t^*\}$ do not converge to γ^* because of the noise, but eventually $S(\widehat{\boldsymbol{\mu}}_t)$ fluctuates around the optimum γ^*. More important, observe that the means $\{\widehat{\boldsymbol{\mu}}_t\}$ do converge to the optimal $\mathbf{x}^*$.

8.27   Select a particular instance (cost matrix) of the synthetic TSP in Problem 8.11. Make this TSP *noisy* by defining the random cost Y_{ij} from i to j in (8.47) to be $\mathsf{Exp}(c_{ij}^{-1})$ distributed. Apply the CE Algorithm 8.3.1 to the noisy problem and compare the results with those in the deterministic case. Display the evolution of the algorithm in a graph, plotting the maximum distance, $\max_{i,j} |\widehat{p}_{t,ij} - p_{ij}^*|$, as a function of t.

8.28   Let $X_1, \ldots, X_n$ be independent random variables, each with marginal pdf f. Suppose that we wish to estimate $\ell = \mathbb{P}_f(X_1 + \cdots + X_n \geqslant \gamma)$ using MinxEnt. For the prior pdf, we could choose $h(\mathbf{x}) = f(x_1)f(x_2) \cdots f(x_n)$, that is, the joint pdf. We consider only a single constraint in the MinxEnt program, namely, $S(\mathbf{x}) = x_1 + \cdots + x_n$. As in (8.55), the solution to this program is given by

$$g(\mathbf{x}) = c\,h(\mathbf{x})\,e^{\lambda S(\mathbf{x})} = c \prod_{j=1}^n e^{\lambda x_j} f(x_j) \,,$$

where $c = 1/\mathbb{E}_h[e^{\lambda S(\mathbf{X})}] = (\mathbb{E}_f[e^{\lambda X}])^{-n}$ is a normalization constant and λ satisfies (8.56). Show that the new marginal pdfs are obtained from the old ones by an *exponential twist*, with twisting parameter $-\lambda$; see also (A.13).

8.29   Problem 8.28 can be generalized to the case where $S(\mathbf{x})$ is a coordinatewise separable function, as in (8.57), and the components $\{X_i\}$ are independent under the prior pdf $h(\mathbf{x})$. Show that also in this case the components under the optimal MinxEnt pdf $g(\mathbf{x})$ are independent and determine the marginal pdfs.

8.30 Write the Lagrangian dual problem for the MinxEnt problem with constraints discussed in Remark 8.9.1.

Further Reading

The CE method was pioneered in [39] as an adaptive algorithm for estimating probabilities of rare events in complex stochastic networks. Originally it was based on variance minimization. It was soon realized [40, 41] that the same technique (using CE rather than VM) could be used not only for estimation but also for optimization purposes.

A gentle tutorial on the CE method is given in [16] and a more comprehensive treatment can be found in [45]. In 2005 a whole volume (134) of the *Annals of Operations Research* was devoted to the CE method. The CE home page, featuring many links, articles, references, tutorials, and computer programs on CE, can be found at http://www.cemethod.org.

The CE method has applications in many areas, including buffer allocation [1], queueing models of telecommunication systems [15, 17], control and navigation [18], signal detection [31], DNA sequence alignment [23], scheduling and vehicle routing [11], reinforcement learning [32, 35], project management [12] and heavy-tail distributions [2], [27]. Applications to more classical combinatorial optimization problems are given in [41], [42], and [43]. The continuous counterpart is discussed in [26], and applications to clustering analysis are given in [6] and [28]. Various CE estimation and noisy optimization problems for reliability systems and network design can be found in [19], [24], [25], [36], [37], and [38]. Convergence issues are discussed in [13], [34], and Section 3.5 of [45]. More recent references may be found in [8] and [29].

An approach closely related to CE is the *probability collectives* work of Dr. David Wolpert and his collaborators. This approach uses information theory as a bridge to relate game theory, statistical physics, and distributed optimization; see, for example, [48, 49].

Since the pioneering work of Shannon [47] and Kullback [30], the relationship between statistics and information theory has become a fertile area of research. The work of Kapur and Kesavan, such as [21, 22], has provided great impetus to the study of entropic principles in statistics. Rubinstein [44] introduced the idea of updating the probability vector for combinatorial optimization problems and rare events using the marginals of the MinxEnt distribution. For some fundamental contributions to MinxEnt, see [3, 4]. In [5, 9, 7] a powerful generalization and unification of the ideas behind the MinxEnt and CE methods is presented under the name *generalized cross-entropy* (GCE).

REFERENCES

1. G. Alon, D. P. Kroese, T. Raviv, and R. Y. Rubinstein. Application of the cross-entropy method to the buffer allocation problem in a simulation-based environment. *Annals of Operations Research*, 134:137–151, 2005.

2. S. Asmussen, D. P. Kroese, and R. Y. Rubinstein. Heavy tails, importance sampling and cross-entropy. *Stochastic Models*, 21(1):57–76, 2005.

3. A. Ben-Tal, D. E. Brown, and R. L. Smith. Relative entropy and the convergence of the posterior and empirical distributions under incomplete and conflicting information. Manuscript, University of Michigan, 1988.

4. A. Ben-Tal and M. Teboulle. Penalty functions and duality in stochastic programming via ϕ divergence functionals. *Mathematics of Operations Research*, 12:224–240, 1987.

5. Z. I. Botev. *Stochastic Methods for Optimization and Machine Learning.* ePrintsUQ, http://eprint.uq.edu.au/archive/00003377/, BSc (Hons) Thesis, Department of Mathematics, School of Physical Sciences, The University of Queensland, 2005.

6. Z. I. Botev and D. P. Kroese. Global likelihood optimization via the cross-entropy method, with an application to mixture models. In R. G. Ingalls, M. D. Rossetti, J. S. Smith, and B. A. Peters, editors, *Proceedings of the 2004 Winter Simulation Conference*, pages 529–535, Washington, DC, December 2004.

7. Z. I. Botev and D. P. Kroese. The generalized cross entropy method, with applications to probability density estimation. *Methodology and Computing in Applied Probability*, 13(1):1–27, 2011.

8. Z. I. Botev, D. P. Kroese, R.Y. Rubinstein, and P. LEcuyer. The cross-entropy method for optimization. In V. Govindaraju and C.R. Rao, editors, *Handbook of Statistics*, volume 31: Machine Learning, pages 19–34. Elsevier, Chennai, 2013.

9. Z. I. Botev, D. P. Kroese, and T. Taimre. Generalized cross-entropy methods for rare-event simulation and optimization. *Simulation: Transactions of the Society for Modeling and Simulation International*, 83(11):785–806, 2008.

10. J. C. C. Chan and D. P. Kroese. Improved cross-entropy method for estimation. *Statistics and Computing*, 22(5):1031–1040, 2012.

11. K. Chepuri and T. Homem de Mello. Solving the vehicle routing problem with stochastic demands using the cross entropy method. *Annals of Operations Research*, 134(1):153–181, 2005.

12. I. Cohen, B. Golany, and A. Shtub. Managing stochastic finite capacity multi-project systems through the cross-entropy method. *Annals of Operations Research*, 134(1):183–199, 2005.

13. A. Costa, J. Owen, and D. P. Kroese. Convergence properties of the cross-entropy method for discrete optimization. *Operations Research Letters*, 35(5):573–580, 2007.

14. T. M. Cover and J. A. Thomas. *Elements of Information Theory.* John Wiley & Sons, New York, 1991.

15. P. T. de Boer. *Analysis and Efficient Simulation of Queueing Models of Telecommunication Systems.* PhD thesis, University of Twente, 2000.

16. P. T. de Boer, D. P. Kroese, S. Mannor, and R. Y. Rubinstein. A tutorial on the cross-entropy method. *Annals of Operations Research*, 134(1):19–67, 2005.

17. P. T. de Boer, D. P. Kroese, and R. Y. Rubinstein. A fast cross-entropy method for estimating buffer overflows in queueing networks. *Management Science*, 50(7):883–895, 2004.

18. B. E. Helvik and O. Wittner. Using the cross-entropy method to guide/govern mobile agent's path finding in networks. In S. Pierre and R. Glitho, editors, *Mobile Agents for Telecommunication Applications: Third International Workshop, MATA 2001, Montreal*, pages 255–268, New York, 2001. Springer-Verlag.

19. K-P. Hui, N. Bean, M. Kraetzl, and D.P. Kroese. The cross-entropy method for network reliability estimation. *Annals of Operations Research*, 134:101–118, 2005.

20. E. T. Jaynes. *Probability Theory: The Logic of Science.* Cambridge University Press, Cambridge, 2003.

21. J. N. Kapur and H. K. Kesavan. The generalized maximum entropy principle. *IEEE Transactions on Systems, Man, and Cybernetics*, 19:1042–1052, 1989.

22. J. N. Kapur and H. K. Kesavan. *Entropy Optimization Principles with Applications.* Academic Press, New York, 1992.

23. J. Keith and D. P. Kroese. Sequence alignment by rare event simulation. In *Proceedings of the 2002 Winter Simulation Conference*, pages 320–327, San Diego, 2002.

24. D. P. Kroese and K. P. Hui. In: *Computational Intelligence in Reliability Engineering*, chapter 3: Applications of the Cross-Entropy Method in Reliability. Springer-Verlag, New York, 2006.

25. D. P. Kroese, S. Nariai, and K. P. Hui. Network reliability optimization via the cross-entropy method. *IEEE Transactions on Reliability*, 56(2):275–287, 2007.

26. D. P. Kroese, S. Porotsky, and R. Y. Rubinstein. The cross-entropy method for continuous multi-extremal optimization. *Methodology and Computing in Applied Probability*, 8:383–407, 2006.

27. D. P. Kroese and R. Y. Rubinstein. The transform likelihood ratio method for rare event simulation with heavy tails. *Queueing Systems*, 46:317–351, 2004.

28. D. P. Kroese, R. Y. Rubinstein, and T. Taimre. Application of the cross-entropy method to clustering and vector quantization. *Journal of Global Optimization*, 37:137–157, 2007.

29. D.P. Kroese, R. Y. Rubinstein, and P. W. Glynn. The cross-entropy method for estimation. In V. Govindaraju and C.R. Rao, editors, *Handbook of Statistics*, volume 31: Machine Learning, pages 35–59. Elsevier, Chennai, 2013.

30. S. Kullback. *Information Theory and Statistics.* John Wiley & Sons, New York, 1959.

31. Z. Liu, A. Doucet, and S. S. Singh. The cross-entropy method for blind multiuser detection. In *IEEE International Symposium on Information Theory*, Chicago, 2004. Piscataway.

32. S. Mannor, R. Y. Rubinstein, and Y. Gat. The cross-entropy method for fast policy search. In *The 20th International Conference on Machine Learning (ICML-2003)*, Washington, DC, 2003.

33. L. Margolin. *Cross-Entropy Method for Combinatorial Optimization.* Master's thesis, The Technion, Israel Institute of Technology, Haifa, July 2002.

34. L. Margolin. On the convergence of the cross-entropy method. *Annals of Operations Research*, 134(1):201–214, 2005.

35. I. Menache, S. Mannor, and N. Shimkin. Basis function adaption in temporal difference reinforcement learning. *Annals of Operations Research*, 134(1):215–238, 2005.

36. S. Nariai and D. P. Kroese. On the design of multi-type networks via the cross-entropy method. In *Proceedings of the Fifth International Workshop on the Design of Reliable Communication Networks (DRCN)*, pages 109–114, 2005.

37. S. Nariai, D. P. Kroese, and K. P. Hui. Designing an optimal network using the cross-entropy method. In *Intelligent Data Engineering and Automated Learning*, Lecture Notes in Computer Science, pages 228–233, New York, 2005. Springer-Verlag.

38. A. Ridder. Importance sampling simulations of Markovian reliability systems using cross-entropy. *Annals of Operations Research*, 134(1):119–136, 2005.

39. R. Y. Rubinstein. Optimization of computer simulation models with rare events. *European Journal of Operational Research*, 99:89–112, 1997.

40. R. Y. Rubinstein. The cross-entropy method for combinatorial and continuous optimization. *Methodology and Computing in Applied Probability*, 2:127–190, 1999.

41. R. Y. Rubinstein. Combinatorial optimization, cross-entropy, ants and rare events. In S. Uryasev and P. M. Pardalos, editors, *Stochastic Optimization: Algorithms and Applications*, pages 304–358, Dordrecht, 2001. Kluwer.

42. R. Y. Rubinstein. Combinatorial optimization via cross-entropy. In S. Gass and C. Harris, editors, *Encyclopedia of Operations Research and Management Sciences*, pages 102–106. Kluwer, 2001.

43. R. Y. Rubinstein. The cross-entropy method and rare-events for maximal cut and bipartition problems. *ACM Transactions on Modelling and Computer Simulation*, 12(1):27–53, 2002.

44. R. Y. Rubinstein. The stochastic minimum cross-entropy method for combinatorial optimization and rare-event estimation. *Methodology and Computing in Applied Probability*, 7:5–50, 2005.

45. R. Y. Rubinstein and D. P. Kroese. *The Cross-Entropy Method: A Unified Approach to Combinatorial Optimization, Monte Carlo Simulation and Machine Learning*. Springer-Verlag, New York, 2004.

46. R. Y. Rubinstein and B. Melamed. *Modern Simulation and Modeling*. John Wiley & Sons, New York, 1998.

47. C. E. Shannon. The mathematical theory of communications. *Bell Systems Technical Journal*, 27:623–656, 1948.

48. D. H. Wolpert. Information theory: the bridge connecting bounded rational game theory and statistical physics. In D. Braha and Y. Bar-Yam, editors, *Complex Engineering Systems*. Perseus Books, New York, 2004.

49. D. H. Wolpert and S. R. Bieniawski. Distributed control by Lagrangian steepest descent. In *IEEE Conference on Decision and Control*, volume 2, pages 1562–1567, 2004.

CHAPTER 9

SPLITTING METHOD

9.1 INTRODUCTION

The *splitting method* is a highly useful and versatile Monte Carlo method that uses a sequential sampling plan to decompose a "difficult" estimation problem into a sequence of "easy" problems. The method has been rediscovered and extended many times since its basic idea was proposed as far back as the 1950s, and it is closely related to particle Monte Carlo and sequential importance resampling techniques.

Splitting is particularly suitable for rare-event simulation. In contrast to importance sampling, the splitting method does not involve a change of probability law; instead, it replicates successful realizations of the simulated process in order to make the rare event more likely to occur. For this reason splitting can be used for many different purposes, including rare-event probability estimation, sampling from arbitrary high-dimensional distributions, Monte Carlo counting, and randomized optimization. The purpose of this chapter is to explain the fundamental ideas behind the splitting method, demonstrate a variety of splitting algorithms, and showcase a wide range of applications.

The rest of this chapter is organized as follows. Section 9.2 introduces the splitting idea as a Monte Carlo method for counting the number of self-avoiding walks. Section 9.3 builds on these ideas, and presents a general splitting algorithm involving a fixed splitting factor. Section 9.4 presents an important modification, called fixed effort splitting. In Section 9.5 we show that the splitting method not

Simulation and the Monte Carlo Method, Third Edition. By R. Y. Rubinstein and D. P. Kroese

only can deal with *dynamic* simulation models (involving random processes that evolve over time) but also with *static* (time-independent) models, thus leading to the *generalized splitting method*. An adaptive version of the (generalized) splitting method is given in Section 9.6. The splitting method is used in Section 9.7 for the efficient estimation of the reliability of complex networks. Section 9.8 explains how splitting can be used as a powerful Monte Carlo method for counting combinatorial objects, and Section 9.9 demonstrates the simplicity and effectiveness of the splitting method for counting objects such as the number of solutions to a satisfiability (SAT) problem, independent sets, binary contingency tables, and vertex colorings. In Section 9.10 we clarify how the splitting method can be viewed as a generalization of the acceptance–rejection method to (approximately) sample from arbitrary probability distributions. Similar to the CE method, the splitting method for rare-event simulation can be conveniently converted into a randomized optimization method. This is explained in Section 9.11.

9.2   COUNTING SELF-AVOIDING WALKS VIA SPLITTING

As a motivating example for the splitting method, consider the estimation of the number of self-avoiding walks of a fixed length on the two-dimensional lattice. We discussed this counting problem previously in Example 5.17, and used sequential importance sampling to obtain an unbiased estimator for the number of self-avoiding walks. Assuming that each walk starts at the origin $(0,0)$, a walk of length T can be represented by a vector $\mathbf{X} = (X_1, \ldots, X_T)$, where X_t is the t-th position visited. An alternative is to let X_t be the *direction* taken. (Note that for consistency with the rest of the notation in this chapter, we denote the length of the walk by T rather than n.) The partial walk $(X_1, \ldots, X_t)$ of length t is denoted by $\mathbf{X}_t$. For example, two possible outcomes of $\mathbf{X}_2$ are $((0,1),(1,1))$ and $((0,1),(0,0))$. The first is a self-avoiding walk of length 2, and the second is a walk of length 2 that intersects itself. Let $\mathscr{X}_t$ be the set of all possible self-avoiding walks of length t, $t = 1, \ldots, T$. We denote the set of all 4^T walks by $\mathscr{X}$. We are interested in estimating the cardinality of the set $\mathscr{X}_T$; that is, the number of self-avoiding walks of length T. Note that if $\mathbf{X} = (X_1, \ldots, X_T)$ is drawn uniformly from $\mathscr{X}$, then the number of self-avoiding walks can be written as

$$|\mathscr{X}_T| = \underbrace{\mathbb{P}(\mathbf{X} \in \mathscr{X}_T)}_{\ell} |\mathscr{X}|,$$

where $\ell = \mathbb{P}(\mathbf{X} \in \mathscr{X}_T)$ is the probability of uniformly drawing a self-avoiding walk from the set of all possible walks — which is very small for large T.

The idea behind splitting is to sequentially split sample paths of the random walk into multiple copies during the simulation, so as to make the occurrence of the rare event more frequent. The key observation is that the events $A_1 = \{\mathbf{X}_1 \in \mathscr{X}_1\}$, $A_2 = \{\mathbf{X}_2 \in \mathscr{X}_2\}, \ldots, A_T = \{\mathbf{X}_n \in \mathscr{X}_T\}$ form a decreasing sequence: $A_1 \supset A_2 \supset \cdots \supset A_T = \{\mathbf{X} \in \mathscr{X}_T\}$, since for a path of length t to be self-avoiding, the path of length $t - 1$ must also be self-avoiding. As a consequence, using the product rule (1.4), we can write $\ell = \mathbb{P}(\mathbf{X} \in \mathscr{X}_T)$ as

$$\ell = \underbrace{\mathbb{P}(A_1)}_{p_1} \underbrace{\mathbb{P}(A_2 \mid A_1)}_{p_2} \cdots \underbrace{\mathbb{P}(A_T \mid A_{T-1})}_{p_T} . \tag{9.1}$$

Even though direct estimation of ℓ might be meaningless when $\{\mathbf{X} \in \mathscr{X}_T\}$ is a rare event, estimating each p_t separately is viable, even in a rare-event setting, provided that the $\{p_t\}$ are not too small, such as $p_t = 10^{-2}$ or larger.

This can be achieved by dividing, at each stage t, the sample paths that have reached set $\mathscr{X}_t$ (thus the first t-steps are self-avoiding) into a fixed number, r (e.g., $r = 2$), identical copies and then continuing the simulation in exactly the same way that a single trajectory is generated. For example, starting with $N_1 = N$ trajectories of length 1, at stage $t = 1$ there are $r\,N$ copied/split trajectories. Of these, $N_2 \sim \text{Bin}(r\,N, 3/4)$ paths will reach stage $t = 2$, as there is a 1 in 4 chance that a path will return to $(0,0)$. These N_2 successful paths will be split into rN_2 copies, and so on. The number r is called the *splitting factor*. Denote the multi-set[1] of successful (self-avoiding) paths of length t by $\mathcal{X}_t$. Note that each element of $\mathcal{X}_t$ lies in $\mathscr{X}_t$. Further, let $\mathbf{e}_1 = (1,0)$ and $\mathbf{e}_2 = (0,1)$ be the unit vectors in the plane. The SAW splitting algorithm is summarized as follows:

Algorithm 9.2.1: SAW Splitting Algorithm

---

input : Initial sample size N, splitting factor r.
output: Estimator of $|\mathscr{X}_T|$: the number of self-avoiding walks of length T.

1   $\mathcal{X}_1 \leftarrow \emptyset$
2   **for** $i = 1$ **to** N **do**
3   Draw $\mathbf{X}_1 = X_1$ uniformly from $\{\mathbf{e}_1, \mathbf{e}_2, -\mathbf{e}_1, -\mathbf{e}_2\}$ and add $\mathbf{X}_1$ to $\mathcal{X}_1$.
4   **for** $t = 1$ **to** $T - 1$ **do**
5       $\mathcal{X}_{t+1} \leftarrow \emptyset$
6   **for** $\mathbf{X}_t \in \mathcal{X}_t$ **do**
7   **for** $i = 1$ **to** r **do**
8   Draw X_{t+1} uniformly from $\{\mathbf{e}_1, \mathbf{e}_2, -\mathbf{e}_1, -\mathbf{e}_2\}$. // next component
9               $\mathbf{X}_{t+1} \leftarrow (\mathbf{X}_t, X_{t+1})$
10   **if** $\mathbf{X}_{t+1} \in \mathscr{X}_{t+1}$ **then** add $\mathbf{X}_{t+1}$ to $\mathcal{X}_{t+1}$.

11 **return** $\widehat{\mathscr{X}_T} = 4^T |\mathcal{X}_T| / (N\,r^{T-1})$

We now show why Algorithm 9.2.1 returns a good estimator for $\mathscr{X}_T$. Having generated the sets $\{\mathcal{X}_t\}_{t=1}^T$, we can estimate the conditional probabilities $\{p_t\}$ in (9.1) as follows: Let N_t denote the number of elements in $\mathcal{X}_t$, $t = 1, \ldots, T$. At stage t there are N_t successful paths. These are split into rN_t copies, of which N_{t+1} reach the next level (are self-avoiding of length $t + 1$). Hence an estimator of p_{t+1} is $N_{t+1}/(r\,N_t)$, and a natural estimator of ℓ is

$$\widehat{\ell} = \frac{N_1}{N} \frac{N_2}{rN_1} \cdots \frac{N_T}{rN_{T-1}} = \frac{N_T}{N\,r^{T-1}}.$$

Multiplying $\widehat{\ell}$ by 4^T, we obtain an unbiased estimator of the number of self-avoiding paths $|\mathscr{X}_T|$. It turns out (see Remark 9.3.1) that this is an unbiased estimator of ℓ.

A standard way to derive confidence intervals is to run the algorithm independently for M iterations, which obtains independent copies $\widehat{\ell}_1, \ldots, \widehat{\ell}_M$. The sample mean of these provides the final estimate, and the relative error and confidence

---

[1] A multi-set can have duplicate elements

interval are determined in the usual way via (4.1) and (4.6). Note that if many independent replications are used, N may be taken to be small; for example, $N = 1$.

■ **EXAMPLE 9.1**

As a numerical illustration, consider the estimation of the number of self-avoiding walks of length $T = 19$. Using Algorithm 9.2.1 with splitting factor $r = 2$, an initial sample size of $N = 1$, and $M = 100$ independent replications, we obtain an estimate of $3.46 \cdot 10^8$ with an estimated relative error of 0.073. The 95% confidence interval is $(2.97, 3.95) \cdot 10^8$, which contains the true value 335116620.

Starting with one path at the first level, some of the 100 replications yield no full-length paths at the final level; others had more than 1000 such paths. Figure 9.1 shows the histogram of the number of paths of length T obtained at the final level of the splitting algorithm.

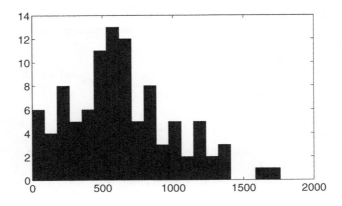

Figure 9.1: Histogram of N_T: the number of paths in the final level of the splitting algorithm.

9.3 SPLITTING WITH A FIXED SPLITTING FACTOR

In this section we generalize the splitting example of Section 9.2. Let $\mathbf{X} = \mathbf{X}_T = (X_1, \dots, X_T)$ be a random vector and let $\mathbf{X}_t = (X_1, \dots, X_t)$, $t = 1, 2, \dots, T$, be the corresponding partial vectors. Note that the stochastic process $\mathbf{X}_1, \mathbf{X}_2, \dots, \mathbf{X}_T$, may be viewed as a Markov chain. It is assumed that it is easy to sample from

1. the initial distribution (i.e., the distribution of $\mathbf{X}_1 = X_1$) and

2. the distribution of X_{t+1} conditional on $\mathbf{X}_t$.

Suppose that we wish to estimate the probability

$$\ell = \mathbb{P}(\mathbf{X}_T \in \mathscr{X}_T)$$

for some set $\mathscr{X}_T$. Let $\mathscr{X}_1, \ldots, \mathscr{X}_T$ be sets such that $\{\mathbf{X}_1 \in \mathscr{X}_1\}$, $\{\mathbf{X}_2 \in \mathscr{X}_2\}, \ldots, \{\mathbf{X}_T \in \mathscr{X}_T\}$ form a decreasing sequence of events. Let $\{r_t\}_{t=1}^{T-1}$ be a set of integers, fixed in advance, known as the *splitting factors*. We take N independent samples $X_1^{(1)}, \ldots, X_1^{(N)}$ according to the distribution of X_1, which we refer to as *particles*. The particles for which $X_1^{(i)} \in \mathscr{X}_1$ are copied r_1 times, and the particles for which $X_1^{(i)} \notin \mathscr{X}_1$ are discarded. For each retained copy $X_1^{(i)}$ we then construct $\mathbf{X}_2^{(i)}$ by simulating its second component according to the conditional distribution of X_2 given X_1. Particles for which $\mathbf{X}_2^{(i)} \in \mathscr{X}_2$ are copied r_2 times and particles for which $\mathbf{X}_2^{(i)} \notin \mathscr{X}_2$ are discarded. This process repeats for T steps. The collection (multi-set) of successful particles at step t is denoted by $\mathcal{X}_t$, with size N_t. The final estimate is $N^{-1} \prod_{t=1}^{T-1} r_t^{-1}$ multiplied by N_T. This leads to the following generalization of Algorithm 9.2.1:

Algorithm 9.3.1: Splitting with Fixed Splitting Factors

input : Initial sample size N, splitting factors $r_1, \ldots, r_{T-1}$.
output: Estimator of $\ell = \mathbb{P}(\mathbf{X}_T \in \mathscr{X}_T)$.

1 $\mathcal{X}_1 \leftarrow \emptyset$
2 **for** $i = 1$ **to** N **do**
3 $\quad$ Simulate X_1.   // simulate first component
4 $\quad$ **if** $X_1 \in \mathscr{X}_1$ **then** add X_1 to $\mathcal{X}_1$.   // retain values in $\mathscr{X}_1$
5 **for** $t = 1$ **to** $T - 1$ **do**
6 $\quad$ $\mathcal{X}_{t+1} \leftarrow \emptyset$
7 $\quad$ **for** $\mathbf{X}_t \in \mathcal{X}_t$ **do**
8 $\quad\quad$ **for** $i = 1$ **to** r_t **do**
9 $\quad\quad\quad$ Simulate $\mathbf{X}_{t+1}$ conditional on $\mathbf{X}_t$.
10 $\quad\quad\quad$ **if** $\mathbf{X}_{t+1} \in \mathscr{X}_{t+1}$ **then** add $\mathbf{X}_{t+1}$ to $\mathcal{X}_{t+1}$.

11 **return** $\widehat{\ell} = \dfrac{N_T}{N \prod_{t=1}^{T-1} r_t}$

■ **EXAMPLE 9.2 Estimating Hitting Probabilities**

Many rare-event estimation problems are of the following form. Let $\{Z_u, u \geqslant 0\}$ be a Markov process taking values in some state space $\mathscr{Z}$. The time index can be discrete or continuous. Consider two sets $E = \{z : S(z) \geqslant \gamma\}$ and $F = \{z : S(z) = 0\}$ for some positive function S on $\mathscr{Z}$, and $\gamma > 0$. Let τ be the first time that the Markov process enters either E or F, assuming, for simplicity, that it starts outside both of these sets. Suppose that E is much more difficult to enter than F and that we wish to estimate the rare-event probability that this happens:

$$\ell = \mathbb{P}(S(Z_\tau) \geqslant \gamma) .$$

This problem can be formulated in the splitting framework by using a sequence of levels $0 < \gamma_1 \leqslant \gamma_2 \leqslant \cdots \leqslant \gamma_T = \gamma$. Define nested sets $E_t = \{z : S(z) \geqslant \gamma_t\}$, $t = 1, \ldots, T$, and let τ_t be the first time that the Markov process hits either F or E_t. Let $\mathbf{X}_t$ be the path of the Markov process up to time τ_t, so that $\mathbf{X}_t = (Z_u, u \leqslant \tau_t)$, $t = 1, \ldots, T$. Some or all the $\{\tau_t\}$ can be infinite if some or all of the sets $\{E_t\}$ and F are not hit. Let $\mathscr{X}_t$ be the set of possible paths $\mathbf{x}_t = (z_u, u \leqslant \tau_t)$ for which $S(z_{\tau_t}) \geqslant \gamma_t$; that is, paths that enter set E_t before F. Then, ℓ can be written as $\mathbb{P}(\mathbf{X}_T \in \mathscr{X}_T)$ and can be estimated via Algorithm 9.3.1. Notice that, in order to continue a path $\mathbf{x}_t = (z_u, u \leqslant \tau_t)$, it suffices to only know (record) the final value z_{τ_t} at which it entered set E_t.

■ **EXAMPLE 9.3   Gambler's Ruin**

As an instance of the framework of Example 9.2, consider the random walk $Z = \{Z_u, u = 0, 1, \ldots\}$ on the integers, in Example 5.16. Let p and $q = 1 - p$ be the probabilities for upward and downward jumps, respectively. The goal is to estimate the rare-event probability ℓ that state K is reached before 0, starting from some intermediate state k. In the framework of Example 9.2, we have $\mathscr{Z} = \{0, \ldots, K\}$ and $S(z) = z$, and for the levels we can take $\gamma_1 = k + 1, \ldots, \gamma_{K-k} = K$, giving $T = K - k$ levels. The problem is known as the *gambler's ruin* problem, where Z_u represents the fortune of a gambler at time u, and ℓ is the probability that the gambler's fortune disappears before hitting the goal K. At any time when the gambler's fortune is $0 < z < K$, his/her next fortune is $z - 1$ or $z + 1$ with probability p or $q = 1 - p$, respectively.

At each stage t of Algorithm 9.3.1, there is only one way in which the process Z can enter the set $E_t = \{k + t, \ldots, K\}$ — namely via state $k + t$. Hence, in order to continue the paths of the Markov chain once it has reached level γ_t, it suffices to only know (record) the total number of paths that reached that level, N_t. Consequently, at each stage t, we run $r_t N_t$ independent Markov chains Z, starting from state $k + t$, and count how many of those (i.e., N_{t+1}) hit either level 0 or $k + t + 1$.

As a numerical illustration, we applied Algorithm 9.3.1 to the gambler's ruin problem with parameters, $p = 0.2$, $k = 8$, and $K = 20$. The exact probability is (e.g., see [32])

$$\ell = \frac{1 - (q/p)^k}{1 - (q/p)^K} \approx 5.9604 \cdot 10^{-8} \ .$$

Using a fixed splitting factor $r = 5$ for all levels, an initial sample size of $N = 100$, and $M = 100$ independent replications of the algorithm, we found the estimate $\widehat{\ell} = 5.8 \cdot 10^{-8}$ with an estimated relative error of 0.04, and a 95% confidence interval $(0.53, 0.62) \cdot 10^{-8}$.

Remark 9.3.1 (Splitting and Sequential Importance Resampling) The splitting method in Algorithm 9.3.1 can be seen as a sequential importance resampling (SIR) procedure with enrichment resampling. To see this, compare the algorithm with SIR Algorithm 5.9.2. Note that the notation is slightly different in Section 5.9. For example, d was used instead of T and $\mathbf{X}_{1:t}$ instead of $\mathbf{X}_t$. Also, here $H(\mathbf{X})$ is simply the indicator $I_{\{\mathbf{X}_T \in \mathscr{X}_T\}}$. The sequential construction of $\mathbf{X}_t$ is evident. The importance sampling density g is here simply the original density f. Let f_t be the density defined by $c_t f_t(\mathbf{x}_t) = I_{\{\mathbf{x}_t \in \mathscr{X}_t\}} g(\mathbf{x}_t)$. Note that $c_T = \ell$. If the densities $\{f_t\}$ are used as auxiliary densities, then the definition of the importance weight W_t becomes

$$
\begin{aligned}
W_t &= W_{t-1} \frac{c_t f_t(\mathbf{x}_t)}{c_{t-1} f_{t-1}(\mathbf{x}_{t-1}) g(x_t \mid \mathbf{x}_{t-1})} \\
&= W_{t-1} \frac{I_{\{\mathbf{x}_t \in \mathscr{X}_t\}} f(\mathbf{x}_t)}{I_{\{\mathbf{x}_{t-1} \in \mathscr{X}_{t-1}\}} f(\mathbf{x}_{t-1}) f(x_t \mid \mathbf{x}_{t-1})} = \frac{W_{t-1} I_{\{\mathbf{x}_t \in \mathscr{X}_t\}}}{I_{\{\mathbf{x}_{t-1} \in \mathscr{X}_{t-1}\}}}.
\end{aligned}
$$

That is, the importance weights of the particles are either unchanged or are set to zero. If enrichment resampling is applied with splitting factors $\{r_t\}$, then at step t the weight of every particle is either 0 or $\prod_{k=1}^{t-1} r_k^{-1}$. The probability ℓ can be estimated as the average of the importance weights. This gives exactly the estimator returned by Algorithm 9.3.1. There are N_T non-zero importance weights and all the non-zero weights are equal to $\prod_{t=1}^{T-1} r_t^{-1}$. As a consequence, unbiasedness results for splitting estimators follow directly from such results for SIR estimators.

As discussed in Section 5.9, the splitting factors and levels must be judiciously chosen to avoid either having a very large number of particles at the final level or none at all. This may not always be easy or feasible. The next section provides an alternative where the number of particles at each level is kept fixed.

9.4 SPLITTING WITH A FIXED EFFORT

Algorithm 9.3.1 produces a branching tree of "particles" where $N_t = |\mathscr{X}_t|$, the random number of successful particles at stage t, can be difficult to control. An alternative is to resample/split, at each stage t, the N_t successful particles into a fixed number, say N, particles. This ensures a fixed simulation effort at every stage. The difference between this fixed effort splitting and splitting with a fixed splitting factor is described next.

Let $\mathscr{X}_1, \ldots, \mathscr{X}_T$ be as given in Section 9.3. We begin, as previously, by simulating iid copies $X_1^{(1)}, \ldots, X_1^{(N)}$ of the first component X_1. Of these, a random collection $\mathscr{X}_1$ of size N_1 will be successful (those $X_1^{(i)}$ that belong to $\mathscr{X}_1$). From these we construct a new sample $Y_1^{(1)}, \ldots, Y_1^{(N)}$ of size N via some sampling mechanism, two common choices being bootstrap sampling and stratified sampling (both discussed in Section 5.9).

Bootstrap sampling is simply uniform sampling with replacement from $\mathscr{X}_1$, since all the weights are equal. A stratified way of distributing the total effort N over N_t different elite paths is to first split each path $\left\lfloor \frac{N}{N_t} \right\rfloor$ times and then assign the remaining effort randomly without replacement over the N_t paths. That is, we choose

$$
R^{(i)} = \left\lfloor \frac{N}{N_t} \right\rfloor + B^{(i)},
$$

where $(B^{(1)}, \ldots, B^{(N_t)})$ is uniformly distributed over the set of binary vectors with exactly $(N \bmod N_t)$ unities. This approach ensures that the elements in $\mathscr{X}_1$ are "evenly" resampled. For example, if $N_1 = 10$ and $N = 100$, each sample in $\mathscr{X}_1$ is copied exactly 10 times.

Having constructed the sample $\{Y_1^{(i)}\}$, we simulate for each $Y_1^{(i)}$ a random vector $\mathbf{X}_2^{(i)}$ according to the distribution of $(\mathbf{X}_2 \mid X_1 = Y_1^{(i)})$. The subset $\mathscr{X}_2$ of $\{\mathbf{X}_2^{(i)}\}$ for which $\mathbf{X}_2^{(i)} \in \mathscr{X}_2$ is used to construct a sample $\mathbf{Y}_2^{(1)}, \ldots, \mathbf{Y}_2^{(N)}$. This process repeats for T steps. The full algorithm is given as Algorithm 9.4.1. The final estimator is a product of the $\{N_t/N\}_{t=1}^T$. Note that for fixed effort splitting N cannot just be set to 1, as is common in Algorithm 9.3.1.

Algorithm 9.4.1: Splitting with Fixed Effort

---

   input : Sample size N.
   output: Estimator of $\ell = \mathbb{P}(\mathbf{X}_T \in \mathscr{X}_T)$.
1 $\mathcal{X}_1 \leftarrow \emptyset$
2 **for** $i = 1$ **to** N **do**
3 | Simulate X_1.   // simulate first component
4 | **if** $X_1 \in \mathscr{X}_1$ **then** add X_1 to $\mathcal{X}_1$.   // retain values in $\mathscr{X}_1$
5 $N_1 \leftarrow |\mathcal{X}_1|$
6 **for** $t = 1$ **to** $T - 1$ **do**
7 | $\mathbf{Y}_t^{(1)}, \ldots, \mathbf{Y}_t^{(N)} \leftarrow$ sample of size N from $\mathcal{X}_t$.   // resample
8 | $\mathcal{X}_{t+1} \leftarrow \emptyset$
9 | **for** $i = 1$ **to** N **do**
10 | | Simulate $\mathbf{X}_{t+1}$ conditional on $\mathbf{X}_t = \mathbf{Y}_t^{(i)}$.
11 | | **if** $\mathbf{X}_{t+1} \in \mathscr{X}_{t+1}$ **then** add $\mathbf{X}_{t+1}$ to $\mathcal{X}_{t+1}$. //retain values in $\mathscr{X}_{t+1}$
12 | |__
13 | $N_{t+1} \leftarrow |\mathcal{X}_{t+1}|$
14 **return** $\prod_{t=1}^T \frac{N_t}{N}$

9.5 GENERALIZED SPLITTING

Thus far we have assumed that $\mathbf{X}_T$ is a random vector of the form $\mathbf{X}_T = (X_1, X_2, \ldots, X_T)$, whose components can be simulated sequentially. But all that is really needed to apply a splitting approach is that the random object $\mathbf{X}$ be constructed via a sequence of T intermediate objects $\mathbf{X}_1, \mathbf{X}_2, \ldots, \mathbf{X}_T = \mathbf{X}$. These objects can have different or the same dimensions. Let f_t be the known or unknown pdf of $\mathbf{X}_t$, $t = 1, \ldots, T$.

Given subsets $\{\mathscr{X}_t\}$ such that $\{\mathbf{X}_1 \in \mathscr{X}_1\}, \ldots, \{\mathbf{X}_T \in \mathscr{X}_T\}$ form a decreasing sequence of events, a very generic framework for splitting is as follows:

Algorithm 9.5.1: A Generic Splitting Algorithm

---

   input : $\mathscr{X}_1, \ldots, \mathscr{X}_T$.
   output: Estimator of $\ell = \mathbb{P}(\mathbf{X}_T \in \mathscr{X}_T)$.
1 Create a multi-set $\mathcal{X}_1$ of samples from f_1.
2 **for** $t = 1$ **to** $T - 1$ **do**
3 | Create a multi-set $\mathcal{Y}_t$ of samples from f_t by splitting the elements of $\mathcal{X}_t$.
4 | $\mathcal{X}_{t+1} \leftarrow \mathcal{Y}_t \cap \mathscr{X}_{t+1}$
5 **return** Estimator $\widehat{\ell}$ of ℓ.

This framework, which is similar to that in Gilks and Berzuini [18], includes all splitting scenarios in Sections 9.2–9.4. For instance, to obtain the collection $\mathcal{Y}_t$ of paths in Example 9.2, each path in $\mathcal{X}_t = (Z_u, u \leqslant \tau_t)$ is split by *continuing* the

Markov process (Z_u) until time τ_{t+1}, using the standard generation algorithm for the Markov process. If the splitting factor is fixed, r_t, this gives r_t dependent paths where the first part of the paths (up to time τ_t) are identical.

The preceding splitting framework also includes the *generalized splitting (GS) method* of Botev and Kroese [7], which deals with static (i.e., not dependent on a time parameter) rare-event estimation problems. The setting for the GS algorithm follows.

Let S be a positive function of a random vector $\mathbf{X}$ with some known or unknown pdf f and let γ be a positive number. Suppose that we are interested in estimating the quantity

$$\ell = \mathbb{P}(S(\mathbf{X}) \geqslant \gamma) \, .$$

For large γ this is a rare-event probability, and hence this may be difficult to estimate. The splitting approach involves constructing a nested sequence of events

$$\{S(\mathbf{X}_1) \geqslant \gamma_1\} \supset \{S(\mathbf{X}_2) \geqslant \gamma_2\} \supset \cdots \supset \{S(\mathbf{X}_T) \geqslant \gamma_T\} = \{S(\mathbf{X}) \geqslant \gamma\},$$

where $\gamma_1 \leqslant \gamma_2 \leqslant \cdots \leqslant \gamma_T = \gamma$ and $\mathbf{X}_T$ is constructed sequentially via intermediate objects $\mathbf{X}_1, \ldots, \mathbf{X}_T$. Typically, these all take values in the same space. Let $\mathscr{X}_t = \{\mathbf{x} : S(\mathbf{x}) \geqslant \gamma_t\}$, and let f_t be the conditional pdf of $\mathbf{X}$ given $S(\mathbf{X}) \geqslant \gamma$; that is,

$$f_t(\mathbf{x}) = \frac{f(\mathbf{x}) I_{\{S(\mathbf{x}) \geqslant \gamma_t\}}}{\ell_t},$$

where ℓ_t is the normalization constant. Note that $\ell_T = \ell$.

The key insight is that if a sample $\mathscr{X}_t$ from density f_t is given, a sample $\mathscr{Y}_t$ from the same pdf can be obtained via any Markov chain whose invariant distribution is f_t. Moreover, if $\mathbf{X}$ is distributed according to f_t, then $(\mathbf{X} \,|\, \mathbf{X} \in \mathscr{X}_{t+1})$ is distributed according to f_{t+1}; this is simply the acceptance–rejection principle. This leads to the following algorithm:

Algorithm 9.5.2: Generalized Splitting with Fixed Splitting Factors

---

input : Sample size N, splitting factors $r_1, \ldots, r_{T-1}$, Markov transition density K_t whose invariant pdf is f_t, $t = 1, \ldots, T-1$.

output: Estimator of $\ell = \mathbb{P}(\mathbf{X} \in \mathscr{X}_T)$.

1 $\mathscr{X}_1 \leftarrow \emptyset$
2 **for** $i = 1$ **to** N **do**
3 $\quad$ Draw $\mathbf{X} \sim f$
4 $\quad$ **if** $\mathbf{X} \in \mathscr{X}_1$ **then** add $\mathbf{X}$ to $\mathscr{X}_1$.

5 **for** $t = 1$ **to** $T - 1$ **do**
6 $\quad$ $\mathscr{X}_{t+1} \leftarrow \emptyset$
7 $\quad$ **for** $\mathbf{X} \in \mathscr{X}_t$ **do**
8 $\quad\quad$ $\mathbf{Y} \leftarrow \mathbf{X}$
9 $\quad\quad$ **for** $i = 1$ **to** r_t **do**
10 $\quad\quad\quad$ Draw $\mathbf{Y}' \sim K_t(\mathbf{y} \,|\, \mathbf{Y})$.
11 $\quad\quad\quad$ **if** $\mathbf{Y}' \in \mathscr{X}_{t+1}$ **then** add $\mathbf{Y}'$ to $\mathscr{X}_{t+1}$.
12 $\quad\quad\quad$ $\mathbf{Y} \leftarrow \mathbf{Y}'$

13 **return** $\widehat{\ell} = |\mathscr{X}_T| N^{-1} \prod_{t=1}^{T-1} r_t^{-1}$

Algorithm 9.5.2 can be easily modified to a fixed effort splitting algorithm by replacing lines 7–12 with lines 7–15 below, and returning the estimator $\widehat{\ell} = \prod_{t=1}^{T} \frac{N_t}{N}$. Note that we use here stratified sampling.

Algorithm 9.5.3: Generalized Splitting with Fixed Effort

---

input : Sample size N, Markov transition density K_t whose invariant pdf is f_t, $t = 1, \ldots, T-1$.

output: Estimator of $\ell = \mathbb{P}(\mathbf{X} \in \mathscr{X}_T)$.

1   $\mathcal{X}_1 \leftarrow \emptyset$

2   **for** $i = 1$ **to** N **do**

3   Draw $\mathbf{X} \sim f$

4   **if** $\mathbf{X} \in \mathscr{X}_1$ **then** add $\mathbf{X}$ to $\mathcal{X}_1$.

5   **for** $t = 1$ **to** $T - 1$ **do**

6      $\mathcal{X}_{t+1} \leftarrow \emptyset$

7   Denote the elements of $\mathcal{X}_t$ by $\mathbf{X}_t^{(1)}, \ldots, \mathbf{X}_t^{(N_t)}$.

8   Draw $B^{(i)} \sim \mathsf{Bernoulli}(\frac{1}{2})$, $i = 1, \ldots, N_t$, such that $\sum_{i=1}^{N_t} B^{(i)} = N \bmod N_t$.

9   **for** $i = 1$ **to** N_t **do**

10         $R^{(i)} \leftarrow \left\lfloor \frac{N}{N_t} \right\rfloor + B^{(i)}$

11         $\mathbf{Y} \leftarrow \mathbf{X}_t^{(i)}$

12   **for** $j = 1$ **to** $R^{(i)}$ **do**

13   Draw $\mathbf{Y}' \sim K_t(\mathbf{y} \,|\, \mathbf{Y})$

14   **if** $\mathbf{Y}' \in \mathscr{X}_{t+1}$ **then** add $\mathbf{Y}'$ to $\mathcal{X}_{t+1}$.

15            $\mathbf{Y} \leftarrow \mathbf{Y}'$

16 **return** $\widehat{\ell} = \prod_{t=1}^{T} \frac{N_t}{N}$

There is considerable freedom in the choice of the transition density K_t. Ideally, $K_t(\mathbf{y} \,|\, \mathbf{x}) \approx f_t(\mathbf{y})$ for all $\mathbf{x}$. Note that a single transition according to K_t could consist of M transitions of a Markov chain with a transition density whose invariant distribution is also f_t. This amounts to carrying out an MCMC step with burn-in period M. For large M, if the chain is irreducible (and aperiodic), the sample would approximately be distributed according to f_t. However, this may be unnecessarily time-consuming. It is important to realize that, in order to apply the GS algorithm, the only requirement is that K_t have an invariant pdf f_t, not that it be close to f_t. Often, in practice, only a *single* ($M = 1$) MCMC move is performed, and the easiest way to ensure invariance is to sample from the conditional distribution of one or a few of the components of $\mathbf{X} \sim f_t$. In our applications, we typically use the *Gibbs move* described in Algorithm 9.5.4.

Algorithm 9.5.4: Gibbs Move for Generalized Splitting

---

input : A performance function S, a threshold level γ_t and a vector $\mathbf{x} = (x_1, \ldots, x_n)$ satisfying $S(\mathbf{x}) \geqslant \gamma_t$.

output: A random vector $\mathbf{Y} = (Y_1, \ldots, Y_n)$ with $S(\mathbf{Y}) \geqslant \gamma_t$.

1   Draw Y_1 from the conditional pdf $f_t(y_1 \,|\, x_2, \ldots, x_n)$.

2   **for** $i = 2$ **to** n **do**

3   Draw Y_i from the conditional pdf $f_t(y_i \,|\, Y_1, \ldots, Y_{i-1}, x_{i+1}, \ldots, x_n)$.

4   **return** $\mathbf{Y} = (Y_1, \ldots, Y_n)$

There are several variants of the Gibbs move possible. One modification is to update the components in a random order. Another is to update two or more of them simultaneously.

■ **EXAMPLE 9.4   Stochastic Shortest Path**

In Example 8.1 we used the CE method to estimate the probability, ℓ, that the shortest path between points A and B in the bridge network of Figure 8.1 exceeds 6. Here, the lengths $X_1, \ldots, X_5$ of the links are exponentially distributed (and independent) with means $(u_1, \ldots, u_5) = (1, 1, 0.3, 0.2, 0.1)$. We wish to apply the GS method to this problem, using the same levels as obtained in the CE example; that is, $(\gamma_1, \ldots, \gamma_5) = (1.1656, 2.1545, 3.1116, 4.6290, 6)$.

For the transition step in the GS algorithm, we apply the Gibbs move in Algorithm 9.5.4, which in this case consists of 5 steps. At the first step Y_1 is drawn from the conditional pdf

$$f_t(y_1 \mid x_2, \ldots, x_5) \propto f((y_1, x_2, \ldots, x_5)) I_{\{S((y_1, x_2, \ldots, x_5)) \geqslant \gamma\}}$$
$$\propto e^{-y_1/u_1} I_{\{S((y_1, x_2, \ldots, x_5)) \geqslant \gamma\}} \,,$$

where

$$S((y_1, x_2, \ldots, x_5)) = \min\{y_1 + x_4, y_1 + x_3 + x_5, x_2 + x_5, x_2 + x_3 + x_4\} \,.$$

It follows that Y_1 is drawn from an $\mathsf{Exp}(1/u_1)$ distribution truncated to the set $\{y_1 \geqslant 0 : y_1 + x_4 \geqslant \gamma \text{ and } y_1 + x_3 + x_5 \geqslant \gamma\} = (y_1^*, \infty)$, where $y_1^* = \max\{\gamma - x_4, \gamma - x_3 - x_5, 0\}$. Similarly, $Y_i, i = 2, \ldots, 5$ is drawn from an $\mathsf{Exp}(1/u_i)$ distribution truncated to the interval (y_i^*, ∞), with

$$y_2^* = \max\{\gamma - x_5, \gamma - x_3 - x_4, 0\} \,,$$
$$y_3^* = \max\{\gamma - Y_1 - x_5, \gamma - Y_2 - x_4, 0\} \,,$$
$$y_4^* = \max\{\gamma - Y_1, \gamma - Y_2 - Y_3, 0\} \,,$$
$$y_5^* = \max\{\gamma - Y_2, \gamma - Y_1 - Y_3, 0\} \,.$$

We ran 800 independent runs of Algorithm 9.5.3, each with a sample size of $N = 100$. A typical outcome was $\widehat{\ell} = 8.0 \cdot 10^{-6}$ with an estimated relative error of 0.03, giving a 95% confidence interval $(7.6, 8.5) \cdot 10^{-6}$. This is comparable to what was obtained via the CE method.

Figure 9.2 shows the histogram of the 800 estimates. We also carried out the Gibbs move in a random order (rather than 1,2,...,5). This did not lead to further improvements in accuracy.

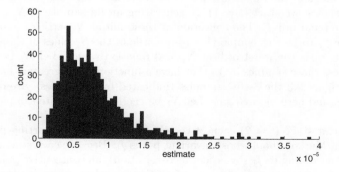

Figure 9.2: Histogram of 800 estimates returned by the splitting algorithm for the stochastic shortest path example.

9.6 ADAPTIVE SPLITTING

In the previous section we did not specify how the levels $\gamma_1, \ldots, \gamma_T$ should be chosen, or indeed how many levels are desired. A general rule of thumb is to choose level γ_t such that the conditional probability $p_t = \mathbb{P}(S(\mathbf{X}) \geqslant \gamma_{t+1} \mid S(\mathbf{X}) \geqslant \gamma_t)$ is not too small. Theoretical investigations [16, 19] indicate that the choices $T = \lceil \ln(\gamma)/2 \rceil$ and $p_t = e^{-2} \approx 0.135$ are optimal, under highly simplified assumptions, where $\lceil \cdot \rceil$ denotes rounding to the largest integer. Still this does not solve the problem of how the levels should be chosen.

A more practical approach is to determine the levels *adaptively*, similar to what is done in the CE method. The adaptive splitting algorithm presented in this section is quite similar to the CE algorithm. In particular, the adaptive splitting algorithm generates a sequence of pairs

$$(\gamma_1, \mathcal{X}_1), \; (\gamma_2, \mathcal{X}_2), \ldots, (\gamma_T, \mathcal{X}_T),$$

where $\mathcal{X}_t$ is a collection of dependent particles that are approximately distributed according to $f_t(\mathbf{x}) \propto f(\mathbf{x})I_{\{S(\mathbf{x}) \geqslant \gamma_t\}}$, whereas the CE algorithm generates a sequence of pairs

$$(\gamma_1, \mathbf{v}_1), \; (\gamma_1, \mathbf{v}_1), \ldots, (\gamma_T, \mathbf{v}_T),$$

where each $\mathbf{v}_t$ is a parameter of a parametric pdf $f(\mathbf{x}; \mathbf{v}_t)$ that is close (in cross-entropy sense) to f_t. The CE method has the desirable property that the elite samples at stage t (those samples that are drawn from $f(\mathbf{x}; \mathbf{v}_{t-1})$ and have a performance greater than or equal to γ_t) are *independent*, whereas in splitting the elite samples (the particles in $\mathcal{X}_t$) are not. The advantage of the splitting method is that it enables approximate sampling from f_T rather than direct sampling from a good approximation of f_t in some specified parametric family $f(\mathbf{x}; \mathbf{v})$; the latter of course also requires an additional importance sampling step.

Below we present the main steps of the adaptive splitting algorithm; see also [7] and [15]. The principal idea is, as mentioned before, to choose the thresholds such that the conditional probabilities $\{p_t\}$ of reaching the "next" level are not too small; for example, such that $p_t \approx \varrho, t = 1, \ldots, T-1$ for a given *rarity parameter* ϱ. Typically, ϱ is chosen between 0.1 and 0.3. Other inputs to the algorithm are the number of samples N at each level and the final level γ. The resulting splitting algorithm is also called *splitting with a fixed probability of success* [26].

The workings of the algorithm are illustrated via Figure 9.3. Let $\varrho = 0.15$. As the figure shows, we start ($t = 1$) by generating an iid sample of $N = 20$ points from the nominal pdf f. The collection of these initial N particles is denoted by $\mathcal{Y}_0$. Choose γ_1 to be the sample $(1 - \varrho)$-quantile of the S-values of the samples in $\mathcal{Y}_0$. That is, γ_1 is the worst of best $N\varrho = 3$ function values. Let $\mathcal{X}_1$ be the set of elite samples: those samples in $\mathcal{Y}_0$ that have a function value greater than or equal to γ_1. In Figure 9.3 the 3 elite samples (indicated by red circles), are denoted by $\mathbf{X}_1, \mathbf{X}_2, \mathbf{X}_3$, and here $\gamma_1 = S(\mathbf{X}_1)$. Let N_1 be the number of elite samples at stage $t = 1$.

At the next stage ($t = 2$) each of the elite particles is the starting point of an independent Markov chain whose invariant pdf is f_t. Because each starting point is distributed according to f_1 (by acceptance–rejection), all consecutive points of each Markov chain are so as well. The length of these Markov chains is determined by the total effort N. In particular, distributing this effort evenly among the Markov chains, the length of chain is at least $\lfloor N/N_1 \rfloor = 6$ and at most $\lfloor N/N_1 \rfloor + 1 = 7$. Which of the Markov chains receive the $(N \mod N_1) = 2$ extra steps can be decided by a random lottery. For example, in Figure 9.3 the Markov chains starting from $\mathbf{X}_1$ and $\mathbf{X}_2$ have length 7 and the Markov chain starting from $\mathbf{X}_3$ has length 6. Let $\mathcal{Y}_1$ be the set of all N particles thus generated, and let γ_2 be the worst of the best $N\varrho$ function values of the particles in $\mathcal{Y}_1$. Let $\mathcal{X}_2$ be the set of elite samples at stage $t = 2$ (indicated by blue dashed circles in Figure 9.3). From these, create multi-set $\mathcal{Y}_2$ via Markov sampling, then determine γ_3 and $\mathcal{X}_3$, and so on. The algorithm continues until some γ_t exceeds the target γ, at which point γ_t is reset to γ and the final level $T = t$ is returned, as well as the estimator $\widehat{\ell} = \prod_{t=1}^{T}(N_t/N)$ of $\ell = \mathbb{P}(S(\mathbf{X}) \geqslant \gamma)$. The full specification is given in Algorithm 9.6.1.

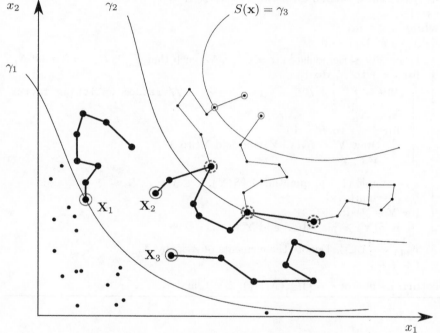

Figure 9.3: Illustration of adaptive generalized splitting with fixed effort.

Remark 9.6.1 When the levels are determined in an adaptive way, we will use the notation $\widehat{\gamma}_t$ instead of γ_t to emphasize that the $\{\widehat{\gamma}_t\}$ are random.

Despite the advantage of not having to determine the levels beforehand, the use of the adaptive splitting algorithm can have some disadvantages. Unlike the fixed-level splitting algorithms, the estimator $\widehat{\ell}$ is *biased*; see, for example, [6]. However, it is asymptotically unbiased as $N \to \infty$. To avoid complications, we could use the adaptive splitting method in a pilot run, to determine appropriate levels, and then use fixed splitting (repeated several times, using the same levels), to obtain an unbiased estimate and confidence interval. The alternative is to use a significantly larger sample size in the adaptive case than in the fixed-level case, but this could make the computation of confidence intervals more costly if many repetitions are required.

Algorithm 9.6.1: Adaptive Generalized Splitting with Fixed Effort

---

input : Sample size N, rarity parameter ϱ, and final level γ.
output: Estimator of $\ell = \mathbb{P}(S(\mathbf{X}) \geqslant \gamma)$, level number T, and levels $\widehat{\gamma}_1, \ldots, \widehat{\gamma}_T$.

1 $\mathcal{Y}_0 \leftarrow \emptyset$; $\mathcal{X}_1 \leftarrow \emptyset$
2 **for** $j = 1$ **to** N **do** draw $\mathbf{Y} \sim f$ and add $\mathbf{Y}$ to $\mathcal{Y}_0$.
3 $s \leftarrow$ sample $(1 - \varrho)$-quantile of $\{S(\mathbf{Y}), \mathbf{Y} \in \mathcal{Y}_0\}$.
4 $\widehat{\gamma}_1 \leftarrow \min\{\gamma, s\}$
5 **for** $\mathbf{Y} \in \mathcal{Y}_0$ **do**
6 $\quad \lfloor$ **if** $S(\mathbf{Y}) \geqslant \widehat{\gamma}_1$ **then** add $\mathbf{Y}$ to $\mathcal{X}_1$.

7 $N_1 \leftarrow |\mathcal{X}_1|$. Denote the elements of $\mathcal{X}_1$ by $\mathbf{X}_1^{(1)}, \ldots, \mathbf{X}_1^{(N_1)}$.
8 **if** $\widehat{\gamma}_1 = \gamma$ **then return** estimator $\widehat{\ell} = N_1/N$, $T = 1$, $\widehat{\gamma}_1 = \gamma$
9 $t \leftarrow 1$
10 **while** $\widehat{\gamma}_t < \gamma$ **do**
11 $\quad$ $\mathcal{Y}_t \leftarrow \emptyset$; $\mathcal{X}_{t+1} \leftarrow \emptyset$
12 $\quad$ Draw $B^{(i)} \sim \text{Bernoulli}(\frac{1}{2})$, $i = 1, \ldots, N_t$, such that $\sum_{i=1}^{N_t} B^{(i)} = N \bmod N_t$.
13 $\quad$ **for** $i = 1$ **to** N_t **do**
14 $\quad\quad$ $R^{(i)} \leftarrow \lfloor \frac{N}{N_t} \rfloor + B^{(i)}$ $\qquad$ // random splitting factor
15 $\quad\quad$ $\mathbf{Y} \leftarrow \mathbf{X}_t^{(i)}$
16 $\quad\quad$ **for** $j = 1$ **to** $R^{(i)}$ **do**
17 $\quad\quad\quad$ Draw $\mathbf{Y}' \sim K_t(\mathbf{y} \mid \mathbf{Y})$ and add $\mathbf{Y}'$ to $\mathcal{Y}_t$.
18 $\quad\quad\quad\lfloor$ $\mathbf{Y} \leftarrow \mathbf{Y}'$

19 $\quad$ $s \leftarrow$ sample $(1 - \varrho)$-quantile of $\{S(\mathbf{Y}), \mathbf{Y} \in \mathcal{Y}_t\}$
20 $\quad$ $\widehat{\gamma}_{t+1} \leftarrow \min\{\gamma, s\}$
21 $\quad$ **for** $\mathbf{Y} \in \mathcal{Y}_t$ **do**
22 $\quad\quad\lfloor$ **if** $S(\mathbf{Y}) \geqslant \widehat{\gamma}_{t+1}$ **then** add $\mathbf{Y}$ to $\mathcal{X}_{t+1}$.

23 $\quad$ $N_{t+1} \leftarrow |\mathcal{X}_{t+1}|$. Denote the elements of $\mathcal{X}_{t+1}$ by $\mathbf{X}_{t+1}^{(1)}, \ldots, \mathbf{X}_{t+1}^{(N_{t+1})}$.
24 $\quad$ $t \leftarrow t + 1$
25 **return** Estimator $\widehat{\ell} = \prod_{t=1}^{T}(N_t/N)$, $T = t$, and $\widehat{\gamma}_1, \ldots, \widehat{\gamma}_T = \gamma$.

■ **EXAMPLE 9.5   Stochastic Shortest Path (Continued)**

We continue Example 9.4 by applying the adaptive splitting algorithm to the estimation of ℓ. We first employ Algorithm 9.6.1 as a pilot algorithm to determine the levels. Using $\varrho = 0.1$ and $N = 1000$, we find the levels $(\widehat{\gamma}_1, \ldots, \widehat{\gamma}_5) = (1.2696, 2.4705, 3.6212, 4.8021, 6.0000)$. Repeating the fixed level Algorithm 9.5.3 800 times with these levels and $N = 100$ produces very similar results to those in Example 9.4: $\widehat{\ell} = 7.9 \cdot 10^{-6}$ with an estimated relative error of 0.03.

In contrast, repeating Algorithm 9.6.1 800 times with $N = 100$ gives an average estimate of $1.5 \cdot 10^{-5}$ with an estimated relative error of 0.025. Clearly, the estimator is biased in this case. However, using 80 repetitions with $N = 1000$ gives reliable estimates, with an overall estimate of $8.1 \cdot 10^{-6}$ and an estimated relative error of 0.025.

9.7   APPLICATION OF SPLITTING TO NETWORK RELIABILITY

In Example 4.2 and Section 5.4.1 we discussed the network (un)reliability estimation problem and showed how permutation Monte Carlo (PMC) can be used to greatly reduce the computational effort when the unreliability is very small. In this section we demonstrate that the generalized splitting method can provide a powerful (and simpler) alternative.

To recapitulate the network reliability estimation problem, consider a network (graph) consisting of n unreliable links (edges). The random states of the links are represented by a binary vector $\mathbf{X} = (X_1, \ldots, X_n)$ of independent Bernoulli random variables with $\mathbb{P}(X_i = 1) = p_i$, $i = 1, \ldots, n$. The system state is $H(\mathbf{X})$ for some *structure function* $H : \{0, 1\}^n \to \{0, 1\}$. The objective is to estimate the system unreliability $\bar{r} = \mathbb{P}(H(\mathbf{X}) = 0)$, which is difficult to estimate if the components are highly reliable (all $\{p_i\}$ close to 1).

Just as in the PMC method, consider the link states vector $\mathbf{X}_t, t \geqslant 0$, of a *dynamic* network process, where the network starts with all links down, and each link is repaired independently according to an $\mathsf{Exp}(-\ln(1 - p_i))$ distribution. Then $\mathbf{X}$ has the same distribution as $\mathbf{X}_1$, and hence can be viewed as a snapshot of the dynamic process at time $t = 1$.

Let $\mathbf{Y} = (Y_1, \ldots, Y_n)$ be the vector of repair times of the dynamic network, and let $S(\mathbf{Y})$ be the first time that the network becomes operational. Then,

$$\bar{r} = \mathbb{P}(S(\mathbf{Y}) > 1)$$

is a rare-event probability (for large $\{p_i\}$) and is amenable to the generalized splitting method. For example, given a sequence of levels $0 < \gamma_1 < \ldots < \gamma_T = 1$, we can apply Algorithm 9.5.3 with $\mathscr{X}_t = \{\mathbf{x} : S(\mathbf{x}) \geqslant \gamma_t\}$. Good choices for the levels can be obtained via a pilot run with the adaptive splitting Algorithm 9.6.1. It remains to specify the Markov move. For this, we can use the Gibbs move in Algorithm 9.5.4. In particular, each Y_i is drawn from a (truncated) exponential distribution. The specific form of the Gibbs move is given in Algorithm 9.7.1 below. Note that the components are updated in a random order.

Algorithm 9.7.1: Gibbs Move for Network Reliability

---

   input : A performance function S, a threshold level γ and a vector
          $\mathbf{y} = (y_1, \ldots, y_n)$ satisfying $S(\mathbf{y}) \geqslant \gamma$.
   output: A random vector $\mathbf{Y} = (Y_1, \ldots, Y_n)$ with $S(\mathbf{Y}) \geqslant \gamma$.
1   $\mathbf{Y} \leftarrow \mathbf{y}$
2   Draw a random uniform permutation $(\pi_1, \ldots, \pi_n)$ of $(1, \ldots, n)$.
3   **for** $i = 1$ **to** n **do**
4      $k \leftarrow \pi_i$
5   **if** $S(Y_1, \ldots, Y_{k-1}, 0, Y_{k+1}, \ldots, Y_n) \leqslant \gamma$ **then**
6         $Y_k \leftarrow \gamma + Z$, $Z \sim \mathsf{Exp}(-\ln(1 - p_k))$
7   **else**
8         $Y_k \leftarrow Z$, $Z \sim \mathsf{Exp}(-\ln(1 - p_k))$
9   **return** $\mathbf{Y}$

■ **EXAMPLE 9.6 Dodecahedron Network**

Figure 9.4 depicts a dodecahedron network, with 30 links (not labeled in the figure) and 20 nodes. Suppose that the network functions if all highlighted nodes (1, 4, 7, 10, 13, 16, 20) are connected via paths of functioning links.

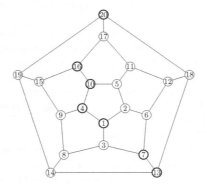

Figure 9.4: Dodecahedron network.

Consider the case where all link failure probabilities are $q = 1 - p = 0.01$. We ran Algorithm 9.6.1 with $N = 10,000$ and $\varrho = 0.5$ to obtain the 18 levels given in Table 9.1.

Table 9.1: Levels of the splitting algorithm for the dodecahedron reliability problem.

0.229623	0.421039	0.606999	0.801514	0.996535
0.284589	0.465779	0.655211	0.84999	1.000000
0.331599	0.511747	0.703615	0.898574	
0.376894	0.559533	0.752544	0.947693	

Using these levels and the same N and ϱ, 10 independent runs with Algorithm 9.5.3 were performed, giving the following estimates: $(6.89, 7.11, 7.07, 7.72, 7.16, 7.30, 7.38, 6.94, 7.70, 7.59) \cdot 10^{-6}$, and giving an average estimate of $7.29 \cdot 10^{-6}$ with an estimated relative error of 0.013.

9.8 APPLICATIONS TO COUNTING

In Section 9.2 we saw how the problem of counting self-avoiding walks can be viewed as an estimation problem, to which Monte Carlo techniques such as importance sampling and splitting can be applied. In this section we show how many other counting problems, such as the satisfiability counting problem and many graph-related counting problems, can be dealt with using the splitting method. It is interesting to note [31, 44] that in many cases the counting problem is hard to solve, while the associated decision or optimization problem is easier to solve. For example, finding the shortest path between two fixed vertices in a graph is easy, whereas finding the total number of paths between the two vertices is difficult.

The latter problem, for example, belongs to the $\#P$-*complete* complexity class — a concept related to the familiar class of NP-hard problems [30].

The key to applying the splitting method is to decompose a difficult counting problem — counting the number of elements in some set $\mathscr{X}^*$ — into a sequence of easier ones, using the following steps. First, find a sequence of decreasing sets

$$\mathscr{X} = \mathscr{X}_0 \supset \mathscr{X}_1 \supset \cdots \supset \mathscr{X}_T = \mathscr{X}^* \,, \tag{9.2}$$

where the elements in $\mathscr{X}$ are easy to count. Next, consider the problem of estimating

$$\ell = \mathbb{P}(\mathbf{X} \in \mathscr{X}^*) \,,$$

where $\mathbf{X}$ is uniformly distributed over $\mathscr{X}$. Note that $|\mathscr{X}^*| = \ell\,|\mathscr{X}|$ and hence an accurate estimate of ℓ will lead to an accurate estimate of $|\mathscr{X}^*|$. We can use any Monte Carlo algorithm to estimate ℓ. Similar to (9.1), we can write

$$\ell = \mathbb{P}(\mathbf{X} \in \mathscr{X}_1 \mid \mathbf{X} \in \mathscr{X}_0) \cdots \mathbb{P}(\mathbf{X} \in \mathscr{X}_T \mid \mathbf{X} \in \mathscr{X}_{T-1}) = \prod_{t=1}^{T} \frac{|\mathscr{X}_t|}{|\mathscr{X}_{t-1}|}. \tag{9.3}$$

The above decomposition of ℓ is motivated by that idea that estimating each $p_t = |\mathscr{X}_t|/|\mathscr{X}_{t-1}|$ may be much easier than estimating ℓ directly.

In general, to deliver a meaningful estimator of $|\mathscr{X}^*|$, we have to address the following two problems:

1. Construct the sequence $\mathscr{X}_0 \supset \mathscr{X}_1 \supset \cdots \supset \mathscr{X}_T = \mathscr{X}^*$ such that each $p_t = |\mathscr{X}_t|/|\mathscr{X}_{t-1}|$ is not a rare-event probability.

2. Obtain a low variance unbiased estimator $\widehat{p}_t$ of each p_t.

The first task can often be resolved by introducing a positive performance function $S(\mathbf{x})$ and levels $\gamma_1, \ldots, \gamma_T$ such that

$$\mathscr{X}_t = \{\mathbf{x} \in \mathscr{X} \,:\, S(\mathbf{x}) \geqslant \gamma_t\}, \quad t = 0, 1, \ldots, T.$$

The second task, in contrast, can be quite complicated, and involves sampling *uniformly* from each subset $\mathscr{X}_t$. Note that if we could sample uniformly from each subset, we could simply take the proportion of samples from $\mathscr{X}_{t-1}$ that fall in $\mathscr{X}_t$ as the estimator for p_t. For such an estimator to be efficient (have low variance), the subset $\mathscr{X}_t$ must be relatively "dense" in $\mathscr{X}_{t-1}$. In other words, p_t should not be too small. The generalized splitting method in Section 9.5 is tailor-made for this situation. Namely in this case the target pdf $f(\mathbf{x})$ is the uniform pdf on $\mathscr{X}_0$ and each pdf f_t corresponds to the uniform pdf on $\mathscr{X}_t$, $t = 1, \ldots, T$. A typical generalized splitting procedure could involve the adaptive version in Algorithm 9.6.1, the fixed-level version in Algorithm 9.5.3, or a combination of the two. For example, the former can be used in a pilot run to determine the levels for the latter, whereas the latter provides an unbiased estimator.

For convenience, we repeat the main steps of the adaptive version for counting, allowing for level-dependent rarity parameters $\varrho_t, t = 1, 2, \ldots$ and adding one extra "screening" step. Screening simply deletes duplicate particles in the elite multi-sets. Such a modification enriches the elite sets in subsequent iterations. Of course, duplications are only an issue when the sample space is discrete (as in counting problems).

Algorithm 9.8.1 (Adaptive Splitting Algorithm for Counting)

- Input: performance function S, final level γ, rarity parameters ϱ_t, $t = 1, 2, \ldots$, sample size N.

- Output: estimator $\widehat{\ell}$ of $\ell = \mathbb{P}(S(\mathbf{X}) \geqslant \gamma) = \mathbb{P}(\mathbf{X} \in \mathscr{X}^*)$, estimator $\widehat{|\mathscr{X}^*|}$ of $|\mathscr{X}^*| = |\{\mathbf{x} \in \mathscr{X} : S(\mathbf{x}) \geqslant \gamma)\}|$, level number T and levels $\widehat{\gamma}_1, \ldots, \widehat{\gamma}_T$.

1. **Acceptance–Rejection.** Set a counter $t = 1$. Generate a sample $\mathcal{Y}_0 = \{\mathbf{Y}^{(1)}, \ldots, \mathbf{Y}^{(N)}\}$ uniformly on $\mathscr{X} = \mathscr{X}_0$. Compute the threshold $\widehat{\gamma}_1$ as the sample $(1 - \varrho_1)$-quantile of the values of $S(\mathbf{Y}^{(1)}), \ldots, S(\mathbf{Y}^{(N)})$. Let $\mathcal{X}_1 = \{\mathbf{X}_1^{(1)}, \ldots, \mathbf{X}_1^{(N_1)}\}$ be the collection (multi-set) of elements $\mathbf{X}_1 \in \mathcal{Y}_0$ for which $S(\mathbf{X}_1) \geqslant \widehat{\gamma}_1$ (the elite points). Set $\widehat{p}_1 = N_1/N$ as an unbiased estimator of $p_1 = |\mathcal{X}_1|/|\mathscr{X}_0|$. Note that $\mathbf{X}_1, \ldots, \mathbf{X}_{N_1} \sim \mathsf{U}(\mathscr{X}_1)$, the uniform distribution on the set $\mathscr{X}_1 = \{\mathbf{x} : S(\mathbf{x}) \geqslant \widehat{\gamma}_1\}$. If $\widehat{\gamma}_1 > \gamma$, return $\widehat{\ell} = N_1/N, T = 1$, and $\widehat{\gamma}_1 = \gamma$ and stop.

2. **Screening.** Reset $\mathcal{X}_t$ and N_t by screening out (removing) any duplicate particles.

3. **Splitting.** To each particle in the elite multi-set $\mathcal{X}_t$ apply a Markov chain sampler with $\mathsf{U}(\mathscr{X}_t)$ as its invariant (stationary) distribution, and length $\lfloor N/N_t \rfloor$. Apply the Markov chain sampler to $N - \lfloor N/N_t \rfloor = N \mod N_1$ randomly chosen endpoints for one more period. Denote the new entire sample (of size N) by $\mathcal{Y}_t$.

4. **Selecting elites.** Compute level $\widehat{\gamma}_{t+1}$ as the sample $(1 - \varrho_t)$ quantile of the values of $S(\mathbf{Y}), \mathbf{Y} \in \mathcal{Y}_t$. If $\widehat{\gamma}_{t+1} > \gamma$, reset it to γ. Determine the elite sample, i.e., the largest subset $\mathcal{X}_{t+1} = \{\mathbf{X}_{t+1}^{(1)}, \ldots, \mathbf{X}_{t+1}^{(N_{t+1})}\}$ of $\mathcal{Y}_t$ consisting of N_{t+1} points for which $S(\mathbf{X}_{t+1}) \geqslant \widehat{\gamma}_{t+1}$.

5. **Estimating p_{t+1}.** Take $\widehat{p}_{t+1} = N_{t+1}/N$ as an estimator of $p_{t+1} = |\mathcal{X}_{t+1}|/|\mathcal{X}_t|$.

6. **Stopping rule and estimation.** Increase the counter $t = t+1$. If $\widehat{\gamma}_{t+1} < \gamma$, repeat from Step 2; otherwise return $T = t$, $\widehat{\gamma}_1, \ldots, \widehat{\gamma}_T$ and the estimators $\widehat{\ell} = \prod_{t=1}^{T}(N_t/N)$ and $\widehat{|\mathscr{X}^*|} = \widehat{\ell}|\mathscr{X}|$.

The main ingredient that remains to be specified is the Markov move in Step 3. A convenient choice is to use a Gibbs move, one variant of which is described in Algorithm 9.5.4. That is, given a vector $\mathbf{x} = (x_1, \ldots, x_n)$ and an intermediate threshold γ_t, the transition to $\mathbf{Y} = (Y_1, \ldots, Y_n)$ is carried out by drawing one or more components Y_i from the uniform distribution of $\mathbf{Y}$ conditional on the remaining components *and* conditional on $S(\mathbf{Y}) \geqslant \gamma_t$.

Remark 9.8.1 (Adaptive Choice of ϱ_t) Choosing in advance a fixed rarity parameter $\varrho_t = \varrho$ for all iterations t may not always be desirable, especially in discrete problems. The reason is that in Step 3 of Algorithm 9.8.1 many points $\mathbf{y} \in \mathcal{Y}_t$ could have identical function values. As a consequence, in Step 4 the number of elite samples, N_{t+1}, could be much larger than ϱN. For example, if at iteration t the ordered performance values are $1, 1, 2, 2, 3, 3, 3, 3, 3$, and $\varrho = 0.1$, and $N = 10$, then $\widehat{\gamma}_{t+1} = 3$

and $N_{t+1} = 4$, whereas $N\varrho = 1$. In an extreme case this could even lead to $N_t = N$ and the algorithm would lock. To eliminate such undesirable behavior, Step 4 in Algorithm 9.8.1 can be modified as follows: Choose a fixed ϱ; say $\varrho = 0.1$. Next, let a_1 and a_2 be such that $a_1 \leqslant \varrho \leqslant a_2$; say $a_1 = 0.1$ and $a_2 = 0.3$. Initially, let $\widehat{\gamma}_{t+1}$, $\mathcal{X}_{t+1}$ and N_{t+1} be exactly as in Step 4 of Algorithm 9.8.1. If $N_{t+1} \leqslant a_2 N$ (not too many elite samples), then continue with Step 5. Otherwise, increase $\widehat{\gamma}_{t+1}$ by some fixed amount, say 1, and reset $\mathcal{X}_{t+1}$ and N_{t+1} *unless* the new number of elite samples is too small, $N_{t+1} < a_1 N$, in which case the original choices are used.

Remark 9.8.2 (Other Counting Strategies) Algorithm 9.8.1 can be combined with various other counting strategies. An obvious one is to count directly how many unique particles are encountered in the final level, giving the *direct estimator*

$$\widehat{|\mathcal{X}^*|}_{\text{dir}} = |\texttt{unique}(\mathcal{X}_T)| \,, \tag{9.4}$$

where $\texttt{unique}$ screens out the duplicate elements of $\mathcal{X}_T$. To increase further the accuracy of $\widehat{|\mathcal{X}^*|}_{\text{dir}}$ we can take a larger sample after reaching the final level γ, or pool the results of several independent runs of Algorithm 9.8.1. We found numerically that the direct estimator is very useful and accurate, as compared to the estimator

$$\widehat{|\mathcal{X}^*|} = |\mathcal{X}| \prod_{t=1}^{T} \frac{N_t}{N},$$

when $|\mathcal{X}^*|$ is not too large and less than the sample size N.

Another possibility is to combine Algorithm 9.8.1 with the classic *capture-recapture* (CAP-RECAP) method [36], by running the algorithm twice and counting how many unique final-level particles in the second run also appear in the first run. For example, if the first run has n_1 unique particles and the second has n_2 unique particles, R of whom are also in the first set, then (because $n_2/|\mathcal{X}^*| \approx R/n_1$), a (biased) estimator of $|\mathcal{X}^*|$ is $\frac{n_1 n_2}{R}$.

9.9 CASE STUDIES FOR COUNTING WITH SPLITTING

In this section we illustrate the splitting method for counting via a selection of difficult counting problems. For each case we (1) describe the problem and indicate how it fits into the framework of Section 9.8, (2) specify how the Markov transition step is carried out, and (3) provide numerical examples.

9.9.1 Satisfiability (SAT) Problem

The Boolean satisfiability (SAT) problem plays a central role in combinatorial optimization and computational complexity. Any NP-complete problem, such as the max-cut problem, the graph-coloring problem, and the TSP, can be translated *in polynomial time* into a SAT problem. The SAT problem plays a central role in solving large-scale computational problems, such as planning and scheduling, integrated circuit design, computer architecture design, computer graphics, image processing, and finding the folding state of a protein.

There are different formulations for the SAT problem, but the most common one comprises the following two components [22]:

- A vector of n Boolean variables $\mathbf{x} = (x_1, \ldots, x_n)^\top$, called a *truth assignment*, representing statements that can either be TRUE $(=1)$ or FALSE $(=0)$. The negation (the logical NOT) of a variable x_i is denoted by $\overline{x}_i$. A variable or its negation is called a *literal*.

- A set of m distinct *clauses* $\{C_1, C_2, \ldots, C_m\}$ of the form $C_i = l_{i_1} \vee l_{i_2} \vee \cdots \vee l_{i_k}$, where the l's are literals and $\vee$ denotes the logical OR operator. For example, $0 \vee 1 = 1$.

The simplest SAT problem can now be formulated as: find a truth assignment $\mathbf{x}$ such that *all* clauses are true. Denoting the logical AND operator by $\wedge$, we can represent the above SAT problem via a single formula as

$$ F = \bigwedge_{i=1}^{m} \bigvee_{j=1}^{|C_i|} l_{i_j} \ . $$

The SAT formula is said to be in *conjunctive normal form* (CNF).

As an illustration of the SAT problem and the corresponding SAT counting problem, consider the following toy example of coloring the nodes of the graph in Figure 9.5. Is it possible to color the nodes either black or white in such a way that no two adjacent nodes have the same color? If so, how many such colorings are there?

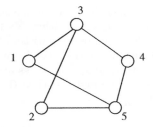

Figure 9.5: Can the graph be colored with two colors so that no two adjacent nodes have the same color?

We can translate this graph-coloring problem into a SAT problem in the following way: Let x_j be the Boolean variable representing the statement "the j-th node is colored black". Obviously, x_j is either TRUE or FALSE, and we wish to assign truth to either x_j or $\overline{x}_j$, for each $j = 1, \ldots, 5$. The restriction that adjacent nodes cannot have the same color can be translated into a number of clauses that must all hold. For example, "node 1 and node 3 cannot both be black" can be translated as clause $C_1 = \overline{x}_1 \vee \overline{x}_3$. Similarly, the statement "node 1 and node 3 cannot both be white" is translated as $C_2 = x_1 \vee x_3$. The same holds for all other pairs of adjacent nodes. The clauses can now be conveniently summarized as in Table 9.2. Here, in the left-hand table, for each clause C_i a 1 in column j means that the clause contains x_j; a -1 means that the clause contains the negation $\overline{x}_j$; and a 0 means that the clause does not contain either of them. Let us call the corresponding $(m \times n)$ matrix $A = (a_{ij})$ the *clause matrix*. For example, $a_{75} = -1$ and $a_{42} = 0$. A common representation of a SAT in CNF form is to list for each clause only the indexes of all Boolean variables present in that clause. In addition, each index that corresponds to a negation of a variable is preceded by a minus sign. This is given in the right-hand panel of Table 9.2.

Table 9.2: SAT table and an alternative representation of the clause matrix.

	1	2	3	4	5				
$\mathbf{x}^\top$	0	1	0	1	0				
C_1	-1	0	-1	0	0	1	C_1	-1	-3
C_2	1	0	1	0	0	0	C_2	1	3
C_3	-1	0	0	0	-1	1	C_3	-1	-5
C_4	1	0	0	0	1	0	C_4	1	5
C_5	0	-1	-1	0	0	1	C_5	-2	-3
C_6	0	1	1	0	0	1	C_6	2	3
C_7	0	-1	0	0	-1	1	C_7	-2	-5
C_8	0	1	0	0	1	1	C_8	2	5
C_9	0	0	-1	-1	0	1	C_9	-3	-4
C_{10}	0	0	1	1	0	1	C_{10}	3	4
C_{11}	0	0	0	-1	-1	1	C_{11}	-4	-5
C_{12}	0	0	0	1	1	1	C_{12}	4	5

Now let $\mathbf{x} = (x_1, \ldots, x_n)^\top$ be a truth assignment. The question is whether there exists an $\mathbf{x}$ such that all clauses $\{C_k\}$ are satisfied. Define the clause value $C_i(\mathbf{x}) = 1$ if clause C_i is TRUE with truth assignment $\mathbf{x}$ and $C_i(\mathbf{x}) = 0$ if it is FALSE. For example, for truth assignment $(0, 1, 0, 1, 0)$ the corresponding clause values are given in the rightmost column of the left-hand panel in Table 9.2. We see that the second and fourth clauses are violated. However, the assignment $(1, 1, 0, 1, 0)$ does indeed render all clauses true, and this therefore gives a way in which the nodes can be colored: $1 = $ black, $2 = $ black, $3 = $ white, $4 = $ black, $5 = $ white. It is easy to see that $(0, 0, 1, 0, 1)$ is the only other assignment that renders all the clauses true.

In general, let $n_i = \sum_{j=1}^{n} I_{\{a_{ij}=-1\}}$ be the total number of (-1)- entries in the i-th row of the clause matrix A. If $x_j = 0$ for all $(+1)$-entries a_{ij} and $x_j = 1$ for all (-1)-entries, then $C_i(\mathbf{x}) = 0$ and $\sum_{i=1}^{n} a_{ij}x_j = -n_i$; otherwise, this sum is at least $b_i = 1 - n_i$ and $C_i(\mathbf{x}) = 1$. Consequently, the SAT problem can be formulated as a search problem in a linearly constrained set:

$$\text{Find} \quad \mathbf{x} \in \{0, 1\}^n \quad \text{such that} \quad A\mathbf{x} \geqslant \mathbf{b},$$

where $\mathbf{b} = (b_1, \ldots, b_n)^\top$.

The problem of deciding whether there *exists* a valid assignment, and indeed providing such a vector, is called the *SAT-assignment* problem. A SAT-assignment problem in which each clause contains exactly K literals is called a *K-SAT* problem. Finding the total number of such valid assignments is called the *SAT-counting* problem. The latter is more difficult and is the focus our our case study.

In the framework of Section 9.8, let $\mathscr{X} = \{0, 1\}^n$ be the set of all truth vectors, and define for each $\mathbf{x} \in \mathscr{X}$,

$$S(\mathbf{x}) = \sum_{i=1}^{m} I_{\{\sum_{j=1}^{n} a_{ij}x_j \geqslant b_i\}},$$

which is the total number of clauses that are satisfied. Our objective is to estimate the number of elements in the set $\mathscr{X}^* = \{\mathbf{x} \in \mathscr{X} : S(\mathbf{x}) \geqslant m\}$.

We will use the following Markov move for the splitting algorithm. Note that this algorithm can be applied to any counting problem involving binary vectors.

Algorithm 9.9.1: Gibbs Move for Binary Vector Counting

---

input : A performance function S, a threshold level γ and a binary vector
$\mathbf{x} = (x_1, \ldots, x_n) \in \{0, 1\}^n$ satisfying $S(\mathbf{x}) \geqslant \gamma$.
output: A random binary vector $\mathbf{Y} = (Y_1, \ldots, Y_n)$ with $S(\mathbf{Y}) \geqslant \gamma$.
1 $\mathbf{Y}^0 \leftarrow (0, x_2, \ldots, x_n); \ \mathbf{Y}^1 \leftarrow (1, x_2, \ldots, x_n)$
2 **if** $S(\mathbf{Y}^0) \geqslant \gamma$ and $S(\mathbf{Y}^1) \geqslant \gamma$ **then**
3 $\quad$ Choose $Y_1 \in \{0, 1\}$ with probability $\frac{1}{2}$.
4 **else**
5 $\quad \lfloor \ Y_1 \leftarrow x_1$
6 **for** $i = 2$ **to** n **do**
7 $\quad$ $\mathbf{Y}^0 \leftarrow (Y_1, \ldots, Y_{i-1}, 0, x_{i+1} \ldots, x_n); \ \mathbf{Y}^1 \leftarrow (Y_1, \ldots, Y_{i-1}, 1, x_{i+1} \ldots, x_n)$
8 $\quad$ **if** $S(\mathbf{Y}^0) \geqslant \gamma$ and $S(\mathbf{Y}^1) \geqslant \gamma$ **then**
9 $\quad \quad$ Choose $Y_i \in \{0, 1\}$ with probability $\frac{1}{2}$.
10 $\quad$ **else**
11 $\quad \quad \lfloor \ Y_i \leftarrow x_i$
12 **return** $\mathbf{Y} = (Y_1, \ldots, Y_n)$

9.9.1.1 Numerical Experiments Although K-SAT counting problems for $K \geqslant 2$ are NP-hard, numerical studies nevertheless indicate that most K-SAT problems are easy to solve for certain values of n and m. To study this phenomenon, Mézard and Montanari [29] define a family of *random K-SAT* problems, where each clause is drawn uniformly from the set of $\binom{n}{K} 2^K$ clauses, independently of the other clauses. It has been observed empirically that a crucial parameter characterizing this problem is the *clause density* $\beta = m/n$, and that, depending on K, the most difficult problems occur around a critical clause density β_K. For $K = 3$, for example, $\beta_3 \approx 4.3$.

We present data from experiments with two different instances of random 3-SAT models, both of which can be found at the SATLIB website `http://www.cs.ubc.ca/~hoos/SATLIB/benchm.html`. In the tables we use the following notation.

- N_t and N_t^{scr} denote the number of elites before and after screening, respectively;

- $\overline{S}_t$ and $\underline{S}_t$ denote the upper and the lower elite levels reached, respectively (the $\underline{S}_t$ levels are the same as the $\widehat{\gamma}_t$ levels in the description of the algorithm);

- $\widehat{p}_t = N_t/N$ is the estimator of the t-th conditional probability;

- (intermediate) product estimator after the t-th iteration $\widehat{|\mathscr{X}_t^*|} = |\mathscr{X}| \prod_{i=1}^{t} \widehat{p}_i$;

- (intermediate) direct estimator after the t-th iteration $\widehat{|\mathscr{X}_t^*|}_{\mathrm{dir}}$, which is obtained by counting directly the number of distinct points satisfying all clauses.

First Model: 3-SAT (20 × 91) model. This instance of the 3-SAT problem (file: `uf20-01.cnf` on the SATLIB website) consists of $n = 20$ variables and $m = 91$ clauses. The number of solutions is $|\mathscr{X}^*| = 8$.

A typical dynamics of a run of the adaptive splitting Algorithm 9.8.1 with $N = 1000$ and $\varrho = 0.1$ is given in Table 9.3.

Table 9.3: Dynamics of Algorithm 9.8.1 for 3-SAT (20×91) model.

t	$\widehat{\lvert \mathscr{X}_t^* \rvert}$	$\widehat{\lvert \mathscr{X}_t^* \rvert}_{\text{dir}}$	N_t	N_t^{scr}	$\overline{S}_t$	$\underline{S}_t$	$\widehat{p}_t$
1	1.69E+05	0	161	161	89	84	0.161
2	2.00E+04	0	118	117	89	87	0.118
3	5.50E+03	1	276	268	91	88	0.276
4	8.85E+02	1	161	144	91	89	0.161
5	9.03E+01	5	102	61	91	90	0.102
6	7.40E+00	8	82	8	91	91	0.0820

We ran the algorithm 10 times with $N = 1000$ and $\varrho = 0.1$. The average estimate was 8.40 with an estimated relative error (calculated for the 10 independent runs) of 0.07. The direct estimator $\widehat{\lvert \mathscr{X}^* \rvert}_{\text{dir}}$ gave always the exact result $\lvert \mathscr{X} \rvert^* = 8$.

Second model: 3-SAT (75×325) model. This instance of the 3-SAT problem (file: `uf75-01.cnf` on the SATLIB website) consists of $n = 75$ variables and $m = 325$ clauses. The number of solutions is $\lvert \mathscr{X}^* \rvert = 2258$.

Table 9.4 presents the dynamics of the adaptive Algorithm 9.8.1 for this case, using $N = 10,000$ and $\varrho = 0.1$.

Table 9.4: Dynamics of Algorithm 9.8.1 for the random 3-SAT (75×325) model.

t	$\widehat{\lvert \mathscr{X}_t^* \rvert}$	$\widehat{\lvert \mathscr{X}_t^* \rvert}_{\text{dir}}$	N_t	N_t^{scr}	$\overline{S}_t$	$\underline{S}_t$	$\widehat{p}_t$
1	3.82E+21	0	1011	1011	305	292	0.101
2	5.00E+20	0	1309	1309	306	297	0.131
3	5.00E+19	0	1000	1000	309	301	0.100
4	6.08E+18	0	1216	1216	310	304	0.122
5	1.12E+18	0	1843	1843	314	306	0.184
6	1.63E+17	0	1455	1455	312	308	0.146
7	1.94E+16	0	1188	1188	315	310	0.119
8	5.97E+15	0	3083	3083	317	311	0.308
9	1.68E+15	0	2808	2808	317	312	0.281
10	4.45E+14	0	2654	2654	318	313	0.265
11	1.10E+14	0	2473	2473	319	314	0.247
12	2.57E+13	0	2333	2333	319	315	0.233
13	5.09E+12	0	1983	1983	319	316	0.198
14	9.19E+11	0	1806	1806	320	317	0.181
15	1.63E+11	0	1778	1778	321	318	0.178
16	2.49E+10	0	1523	1523	322	319	0.152
17	3.17E+09	0	1273	1273	322	320	0.127
18	3.73E+08	0	1162	1162	323	321	0.116
19	3.56E+07	0	967	967	324	322	0.097
20	2.73E+06	1	766	766	325	323	0.077
21	1.37E+05	0	501	500	324	324	0.050
22	2.35E+03	168	172	168	325	325	0.017

Running the splitting algorithm 10 times independently (using sample size $N = 10,000$ and $\varrho = 0.1$), we obtained an average estimate 2343.1 with an estimated relative error of 0.03. All 2258 solutions were found by pooling the unique particles of these 10 runs.

9.9.2 Independent Sets

Consider a graph $G = (V, E)$ with m edges and n vertices. Our goal is to count the number of independent vertex sets of the graph. A vertex set is called *independent* if no two vertices are connected by an edge, that is, if no two vertices are adjacent; see Figure 9.6 for an illustration of this concept.

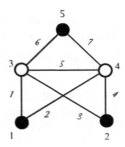

Figure 9.6: Black vertices form an independent set, since they are not adjacent to each other.

Label the vertices and edges as $1, \ldots, n$ and $1, \ldots, m$, respectively, and let $A = (a_{ij})$ be the $m \times n$ matrix with $a_{ij} = 1$ if edge i is connected to vertex j; and $a_{ij} = 0$ otherwise. The i-th row of A thus has exactly two nonzero elements that identify the vertices of the i-th edge. For the graph in Figure 9.6 we have

$$A = \begin{pmatrix} 1 & 0 & 1 & 0 & 0 \\ 1 & 0 & 0 & 1 & 0 \\ 0 & 1 & 1 & 0 & 0 \\ 0 & 1 & 0 & 1 & 0 \\ 0 & 0 & 1 & 1 & 0 \\ 0 & 0 & 1 & 0 & 1 \\ 0 & 0 & 0 & 1 & 1 \end{pmatrix}.$$

The transpose of A is known as the *incidence matrix* of the graph. Any set $\mathscr{V}$ of vertices can be represented by a binary column vector $\mathbf{x} = (x_1, \ldots, x_n)^\top \in \{0, 1\}^n$, where $x_j = 1$ means that vertex j belongs to $\mathscr{V}$. For this set to be an independent set, it cannot contain both end vertices of any edge. In terms of matrix A and vector $\mathbf{x}$, this can be written succinctly as $A\mathbf{x} \leqslant \mathbf{1}$, where $\mathbf{1}$ is the $n \times 1$ vector of 1s.

For a vertex set representation $\mathbf{x}$, let $S(\mathbf{x})$ be the number of edges whose end vertices do not *both* belong to the vertex set represented by $\mathbf{x}$; that is,

$$S(\mathbf{x}) = \sum_{i=1}^{m} I_{\{\sum_{j=1}^{n} a_{ij} x_j \leqslant 1\}}.$$

We wish to estimate the number of independent sets $|\mathscr{X}^*| = |\{\mathbf{x} \in \{0,1\}^n : S(\mathbf{x}) \geqslant m\}|$. As this is a counting problem involving fixed-length binary vectors, we can use Algorithm 9.9.1 for the Markov move in splitting Algorithm 9.8.1.

9.9.2.1 Numerical Experiment We wish to estimate the number of independent sets in a hypercube graph of order n. The vertices of such a graph can be identified with a binary vector of length n — there are thus 2^n vertices. Pairs of vertices are connected by an edge if and only if the corresponding binary vectors differ by one component. Each vertex therefore has n neighbors. Figure 9.7 depicts the hypercube graphs of order $n = 3$ and $n = 4$. It is known that there are 2, 3, 7, 35, 743, 254475, 19768832143, independent sets for hypergraphs of order $n = 0, 1, 2 \ldots, 6$, respectively. For example, for $n = 2$ (the square graph), the independent sets are $\{(0,0),(1,1)\}, \{(1,0),(0,1)\}, \{(0,0)\}, \{(0,1)\}, \{(1,0)\}, \{(1,1)\}$, and the empty set $\emptyset$. Exact results are not known for $n \geqslant 7$.

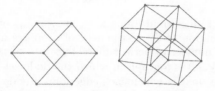

Figure 9.7: Three- and four-dimensional hypercube graphs.

Table 9.5 gives the dynamics of the splitting algorithm for the 6-dimensional hypercube graph. To ensure unbiasedness of the estimator, we used *fixed levels*, which were determined by a pilot run of Algorithm 9.8.1 with parameters $N = 19200$ (100 times the number of edges) and $\varrho = 0.2$. The same sample size was used in the fixed-level run.

Table 9.5: Dynamics of Algorithm 9.5.3 for counting the number of independent sets of the 6-dimensional hypercube graph.

t	$\widehat{|\mathscr{X}_t^*|}$	$\widehat{|\mathscr{X}_t^*|}_{\text{dir}}$	N_t	N_t^{scr}	$\overline{S}_t$	$\underline{S}_t$	$\widehat{p}_t$
1	3.75E+18	0	3908	3907	184	155	0.204
2	8.61E+17	0	4403	4402	183	164	0.229
3	1.92E+17	0	4283	4282	187	170	0.223
4	5.15E+16	0	5151	5150	187	174	0.268
5	1.50E+16	0	5579	5578	189	177	0.291
6	3.38E+15	0	4339	4338	189	180	0.226
7	1.05E+15	2	5943	5942	192	182	0.310
8	2.72E+14	0	4995	4994	191	184	0.260
9	1.26E+14	1	8869	8868	192	185	0.462
10	5.43E+13	4	8286	8285	192	186	0.432
11	2.13E+13	4	7516	7515	192	187	0.391
12	7.60E+12	18	6866	6865	192	188	0.358
13	2.37E+12	38	5977	5976	192	189	0.311
14	6.15E+11	153	4987	4986	192	190	0.260
15	1.23E+11	643	3826	3825	192	191	0.199
16	1.23E+11	3182	19200	19183	192	191	1.000
17	1.99E+10	3116	3116	3115	192	192	0.162

Repeating the fixed-level algorithm 15 times (with the same levels) gave an average estimate of $2.00 \cdot 10^{10}$ with an estimated relative error of 0.014.

For the 7-dimensional case we found an estimate of the number of independent sets of $7.87 \cdot 10^{19}$ with an estimated relative error of 0.001, using 1577 replications of the fixed-level splitting Algorithm 9.5.3, with $N = 10,000$ samples. The levels were determined by a pilot run of the adaptive splitting algorithm, with $N = 10,000$ and $\varrho = 0.2$.

9.9.3   Permanent and Counting Perfect Matchings

The *permanent* of a general $n \times n$ matrix $A = (a_{ij})$ is defined as

$$\mathrm{per}(A) = |\mathscr{X}^*| = \sum_{\mathbf{x} \in \mathscr{X}} \prod_{i=1}^{n} a_{ix_i} , \qquad (9.5)$$

where $\mathscr{X}$ is the set of all permutations $\mathbf{x} = (x_1, \ldots, x_n)$ of $(1, \ldots, n)$. It is well known that the calculation of the permanent of a *binary* matrix is equivalent to the calculation of the number of perfect matchings in a certain bipartite graph. A *bipartite graph* $G = (V, E)$ is a graph in which the node set V is the union of two disjoint sets V_1 and V_2, and in which each edge joins a node in V_1 to a node in V_2. A *matching* of size m is a collection of m edges in which each node occurs at most once. A *perfect matching* is a matching of size n.

To see the relation between the permanent of a binary matrix $A = (a_{ij})$ and the number of perfect matchings in a graph, consider the bipartite graph G, where V_1 and V_2 are disjoint copies of $\{1, \ldots, n\}$ and $(i, j) \in E$ if and only if $a_{ij} = 1$, for all i and j. As an example, let A be the 3×3 matrix

$$A = \begin{pmatrix} 1 & 1 & 1 \\ 1 & 1 & 0 \\ 0 & 1 & 1 \end{pmatrix} . \qquad (9.6)$$

The corresponding bipartite graph is given in Figure 9.8. The graph has three perfect matchings, one of which is displayed in the figure with bold lines. Each such matching corresponds to a permutation $\mathbf{x}$ for which the product $\prod_{i=1}^{n} a_{ix_i}$ is equal to 1.

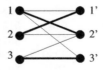

Figure 9.8: Bipartite graph. The bold edges form a perfect matching.

For a general $n \times n$ matrix A, we can apply the splitting algorithm to find the permanent of A by way of the performance function

$$S(\mathbf{x}) = \sum_{i=1}^{n} a_{ix_i} .$$

Let $\mathscr{X}_j$ denote the set of permutations with a performance (number of matchings) greater than or equal to t, $t = 1, \ldots, n$. We are interested in calculating $|\mathscr{X}_n| = \mathrm{per}(A)$. Taking into account that $\mathscr{X}_0$ is the set of all permutations of length n, we have the product form (9.3). We apply the following Gibbs move:

Algorithm 9.9.2: Gibbs Move for Permutation Sampling

input : A permutation $\mathbf{x} = (x_1, \ldots, x_n)$ with $S(\mathbf{x}) \geqslant \gamma$.
output: A random permutation $\mathbf{Y}$ with the same property.

1 $\mathbf{Y} \leftarrow \mathbf{x}$
2 **for** $i = 1$ **to** $n - 1$ **do**
3   **for** $j = i + 1$ **to** n **do**
4   Let $\mathbf{Y}'$ be the equal to $\mathbf{Y}$ with the I-th and J-th elements swapped.
5   **if** $S(\mathbf{Y}') \geqslant \gamma$ **then**
6   Draw $U \sim \mathsf{U}(0, 1)$.
7   **if** $U < 1/2$ **then** $\mathbf{Y} \leftarrow \mathbf{Y}'$

8 **return Y**

9.9.3.1 Numerical Experiment We wish to compute the permanent of the following 30×30 sparse matrix A:

$$
A = \begin{pmatrix}
1 & 0 & 0 & 0 & 0 & 0 & 0 & 0 & 0 & 0 & 0 & 0 & 0 & 0 & 0 & 0 & 0 & 0 & 0 & 0 & 0 & 0 & 0 & 0 & 0 & 0 & 0 & 0 & 0 & 0 \\
0 & 1 & 0 & 0 & 1 & 0 & 0 & 0 & 0 & 0 & 0 & 0 & 0 & 0 & 0 & 0 & 0 & 0 & 1 & 0 & 0 & 0 & 0 & 0 & 0 & 0 & 0 & 0 & 0 & 0 \\
0 & 0 & 1 & 0 & 0 & 0 & 0 & 0 & 0 & 0 & 0 & 0 & 0 & 0 & 0 & 0 & 0 & 1 & 1 & 0 & 1 & 0 & 1 & 0 & 0 & 0 & 0 & 0 & 0 & 0 \\
0 & 0 & 0 & 1 & 0 & 1 & 0 & 0 & 0 & 0 & 0 & 0 & 1 & 0 & 0 & 0 & 1 & 0 & 0 & 0 & 0 & 0 & 0 & 0 & 0 & 0 & 1 & 0 & 0 & 0 \\
0 & 0 & 0 & 1 & 1 & 0 & 0 & 0 & 1 & 0 & 0 & 0 & 0 & 0 & 0 & 0 & 0 & 0 & 0 & 0 & 0 & 0 & 0 & 0 & 0 & 0 & 0 & 1 & 0 & 0 \\
0 & 0 & 1 & 0 & 0 & 1 & 0 & 1 & 0 & 0 & 0 & 0 & 0 & 0 & 0 & 0 & 0 & 1 & 0 & 0 & 0 & 0 & 0 & 0 & 0 & 0 & 0 & 0 & 0 & 0 \\
0 & 0 & 0 & 0 & 0 & 0 & 1 & 0 & 0 & 0 & 0 & 0 & 0 & 0 & 0 & 0 & 0 & 0 & 0 & 0 & 0 & 0 & 0 & 0 & 0 & 0 & 0 & 0 & 0 & 0 \\
0 & 0 & 0 & 0 & 0 & 0 & 1 & 0 & 0 & 0 & 0 & 1 & 0 & 0 & 0 & 0 & 0 & 1 & 0 & 0 & 0 & 0 & 0 & 0 & 0 & 0 & 0 & 0 & 0 & 0 \\
1 & 1 & 0 & 0 & 0 & 0 & 0 & 0 & 1 & 0 & 0 & 0 & 0 & 0 & 0 & 0 & 0 & 0 & 0 & 0 & 0 & 0 & 0 & 0 & 0 & 0 & 0 & 0 & 0 & 0 \\
1 & 0 & 0 & 0 & 0 & 0 & 0 & 0 & 0 & 0 & 0 & 0 & 0 & 0 & 0 & 0 & 1 & 0 & 0 & 0 & 0 & 0 & 0 & 0 & 0 & 0 & 0 & 0 & 0 & 0 \\
0 & 0 & 0 & 0 & 0 & 1 & 0 & 1 & 0 & 1 & 1 & 0 & 0 & 0 & 0 & 0 & 0 & 1 & 0 & 0 & 0 & 0 & 0 & 0 & 0 & 0 & 0 & 0 & 0 & 0 \\
1 & 0 & 0 & 0 & 0 & 0 & 1 & 0 & 0 & 0 & 1 & 0 & 0 & 1 & 0 & 0 & 0 & 0 & 1 & 0 & 0 & 0 & 0 & 0 & 0 & 1 & 0 & 0 & 0 & 0 \\
1 & 0 & 0 & 0 & 0 & 0 & 0 & 0 & 0 & 1 & 0 & 1 & 0 & 0 & 0 & 0 & 0 & 0 & 0 & 0 & 0 & 0 & 0 & 0 & 0 & 0 & 0 & 0 & 0 & 0 \\
0 & 0 & 0 & 0 & 0 & 0 & 0 & 0 & 0 & 0 & 0 & 1 & 1 & 0 & 0 & 0 & 0 & 0 & 0 & 0 & 0 & 0 & 0 & 0 & 0 & 0 & 0 & 0 & 0 & 0 \\
0 & 0 & 1 & 0 & 0 & 0 & 0 & 0 & 0 & 0 & 0 & 0 & 0 & 1 & 0 & 0 & 0 & 0 & 0 & 0 & 0 & 0 & 0 & 0 & 0 & 0 & 0 & 0 & 0 & 0 \\
0 & 0 & 0 & 0 & 0 & 0 & 0 & 1 & 0 & 0 & 0 & 0 & 0 & 0 & 1 & 0 & 0 & 0 & 0 & 0 & 0 & 0 & 0 & 0 & 0 & 0 & 0 & 0 & 0 & 0 \\
0 & 0 & 0 & 0 & 0 & 1 & 0 & 0 & 0 & 0 & 0 & 0 & 0 & 0 & 1 & 1 & 0 & 0 & 0 & 0 & 0 & 0 & 0 & 0 & 0 & 0 & 0 & 0 & 0 & 0 \\
0 & 0 & 0 & 0 & 0 & 1 & 0 & 0 & 0 & 0 & 1 & 0 & 0 & 0 & 0 & 1 & 0 & 0 & 0 & 0 & 0 & 0 & 0 & 0 & 0 & 0 & 0 & 0 & 0 & 0 \\
0 & 0 & 0 & 0 & 0 & 0 & 0 & 0 & 0 & 0 & 0 & 0 & 0 & 0 & 0 & 1 & 0 & 0 & 0 & 1 & 0 & 0 & 0 & 0 & 0 & 0 & 0 & 0 & 0 & 0 \\
1 & 0 & 0 & 0 & 0 & 0 & 0 & 1 & 0 & 0 & 1 & 0 & 0 & 0 & 0 & 0 & 0 & 0 & 1 & 0 & 0 & 0 & 0 & 0 & 0 & 1 & 0 & 0 & 0 & 0 \\
0 & 0 & 0 & 0 & 0 & 0 & 0 & 0 & 0 & 0 & 0 & 0 & 0 & 0 & 0 & 0 & 0 & 0 & 0 & 1 & 0 & 0 & 1 & 0 & 0 & 0 & 0 & 0 & 0 & 0 \\
0 & 0 & 0 & 0 & 0 & 0 & 0 & 0 & 0 & 0 & 0 & 0 & 0 & 0 & 0 & 0 & 0 & 0 & 0 & 1 & 0 & 0 & 0 & 0 & 0 & 0 & 0 & 0 & 0 & 0 \\
0 & 0 & 0 & 0 & 0 & 0 & 0 & 0 & 0 & 0 & 0 & 0 & 0 & 0 & 0 & 0 & 0 & 0 & 1 & 0 & 0 & 0 & 0 & 0 & 0 & 0 & 0 & 0 & 0 & 0 \\
1 & 0 & 0 & 0 & 0 & 0 & 0 & 0 & 0 & 0 & 0 & 0 & 0 & 0 & 0 & 0 & 0 & 0 & 0 & 1 & 1 & 0 & 0 & 0 & 0 & 0 & 0 & 0 & 0 & 0 \\
0 & 0 & 0 & 0 & 0 & 0 & 0 & 0 & 0 & 0 & 0 & 0 & 0 & 0 & 0 & 0 & 0 & 0 & 1 & 0 & 1 & 0 & 0 & 0 & 0 & 0 & 0 & 0 & 0 & 0 \\
0 & 0 & 0 & 0 & 0 & 1 & 0 & 0 & 0 & 0 & 0 & 1 & 1 & 0 & 0 & 0 & 0 & 0 & 0 & 0 & 0 & 0 & 0 & 0 & 1 & 0 & 0 & 0 & 0 & 0 \\
0 & 1 & 0 & 0 & 0 & 0 & 0 & 0 & 1 & 0 & 0 & 0 & 0 & 0 & 1 & 0 & 0 & 0 & 0 & 0 & 0 & 0 & 0 & 0 & 0 & 1 & 0 & 0 & 0 & 0 \\
0 & 0 & 0 & 1 & 0 & 0 & 0 & 0 & 0 & 0 & 0 & 0 & 0 & 0 & 0 & 0 & 0 & 0 & 0 & 0 & 0 & 0 & 0 & 0 & 0 & 0 & 1 & 0 & 0 & 0 \\
0 & 0 & 0 & 0 & 0 & 0 & 0 & 0 & 0 & 1 & 0 & 0 & 0 & 0 & 0 & 0 & 0 & 0 & 0 & 0 & 0 & 0 & 0 & 0 & 0 & 0 & 0 & 1 & 0 & 0 \\
0 & 0 & 0 & 0 & 0 & 0 & 0 & 0 & 0 & 0 & 0 & 0 & 0 & 0 & 0 & 0 & 0 & 0 & 1 & 0 & 0 & 0 & 0 & 0 & 0 & 0 & 0 & 0 & 0 & 1
\end{pmatrix}.
$$

Table 9.6 presents the performance of 10 runs of splitting Algorithm 9.8.1 for this problem. We chose $N = 100,000$ and $\varrho = 0.1$. In this case the direct estimator always returns the true value, 266.

Table 9.6: Performance of splitting Algorithm 9.8.1 for the permanent of the 30×30 matrix A.

Run	Iterations	$\widehat{|\mathscr{X}^*|}$	$\widehat{|\mathscr{X}^*|}_{\text{dir}}$	CPU
1	21	261.14	266	115.68
2	21	254.45	266	115.98
3	21	268.04	266	115.65
4	21	272.20	266	117.68
5	21	261.50	266	118.38
6	21	255.03	266	117.10
7	21	261.36	266	116.58
8	21	266.82	266	115.82
9	21	264.76	266	115.84
10	21	254.13	266	116.13
Average	21	261.94	266	116.48

9.9.4 Binary Contingency Tables

Given are two vectors of positive integers $\mathbf{r} = (r_1, \ldots, r_m)$ and $\mathbf{c} = (c_1, \ldots, c_n)$ such that $r_i \leqslant n$ for all i, $c_j \leqslant n$ for all j, and $\sum_{i=1}^m r_i = \sum_{j=1}^n c_j$. A binary contingency table with row sums $\mathbf{r}$ and column sums $\mathbf{c}$ is an $m \times n$ matrix $\mathbf{x} = (x_{ij})$ of zero-one entries satisfying $\sum_{j=1}^n x_{ij} = r_i$ for every row i and $\sum_{i=1}^m x_{ij} = c_j$ for every column j. The problem is to count all such contingency tables. Wang and Zhang [43] give explicit expressions for the number of $(0,1)$-tables with fixed row and column totals (containing the class of binary contingency tables), which can be used to exactly compute these numbers for small matrices (say $m, n < 15$) and small row/column totals. We wish to apply instead splitting Algorithm 9.8.1.

We define the configuration space $\mathscr{X}$ as the space of binary $m \times n$ matrices where all column sums are satisfied:

$$\mathscr{X} = \left\{ \mathbf{x} \in \{0, 1\}^{m+n} : \sum_{i=1}^m x_{ij} = c_j, \; j = 1, \ldots, n \right\}.$$

Clearly, sampling uniformly on $\mathscr{X}$ is straightforward. For the performance function $S : \mathscr{X} \to \mathbb{Z}_-$, we take

$$S(\mathbf{x}) = -\sum_{i=1}^m \left| \sum_{j=1}^n x_{ij} - r_i \right|.$$

That is, we take the negative difference of the row sums $\sum_{j=1}^n x_{ij}$ with the target r_i if the column sums are right.

The Markov (Gibbs) move for the splitting algorithm is specified in Algorithm 9.7. In particular, for each element $x_{ij} = 1$ we check if the performance function does not become worse if we set x_{ij} to 0 and x_{kj} to 1 for any 0-entry x_{kj} (in the same column). We then uniformly select among these entries (including i) and make the swap. Note that in this way we keep the column sums correct.

Algorithm 9.9.3: Gibbs Move for Random Contingency Table Sampling

input : A matrix $\mathbf{x} = (x_{ij})$ with performance $S(\mathbf{x}) \geqslant \gamma$.

output: A random matrix $\mathbf{Y} = (Y_{ij})$ with the same property.

1   $\mathbf{Y} \leftarrow \mathbf{x}$

2   **for** $j = 1$ **to** n **do**

3     $\mathcal{I} \leftarrow \{i : Y_{ij} = 1\}$

4   **for** $i \in \mathcal{I}$ **do**

5       $\mathcal{K} \leftarrow \{k : Y_{kj} = 0\} \cup \{i\}$.

6       $\mathcal{M} \leftarrow \emptyset$

7   **for** $k \in \mathcal{K}$ **do**

8         $\mathbf{Y}' \leftarrow \mathbf{Y}$; $Y'_{ij} \leftarrow 0$; $Y'_{kj} \leftarrow 1$

9   **if** $S(\mathbf{Y}') \geqslant \gamma$ **then** $\mathcal{M} \to \mathcal{M} \cup \{k\}$

10   Choose M uniformly from $\mathcal{M}$.

11      $Y_{ij} \leftarrow 0$; $Y_{Mj} \leftarrow 1$

12 **return** $\mathbf{Y}$

9.9.4.1 Numerical Experiments   In Chen et al. [10] a delicate sequential importance sampling procedure is used to estimate the number of binary contingency tables. To demonstrate the efficacy and simplicity of the splitting algorithm, we apply the splitting algorithm to two of their numerical examples.

Model 1. Consider the case $m = 12, n = 12$ with row and column sums

$$\mathbf{r} = (2,2,2,2,2,2,2,2,2,2,2,2), \ \mathbf{c} = (2,2,2,2,2,2,2,2,2,2,2,2).$$

The true count value is known to be $21,959,547,410,077,200$ (reported in [43]). Table 9.7 presents a typical dynamics of the adaptive splitting Algorithm 9.8.1 for Model 1 using $N = 50,000$ and $\varrho = 0.5$.

Table 9.7: Typical dynamics of the splitting Algorithm 9.8.1 for Model 1 using $N = 50,000$ and $\varrho = 0.5$.

t	$\widehat{|\mathscr{X}^*|}$	N_t	N_t^{scr}	$\overline{S}_t$	S_t	$\widehat{p}_t$
1	4.56E+21	13361	13361	-2	-24	0.6681
2	2.68E+21	11747	11747	-2	-12	0.5874
3	1.10E+21	8234	8234	-2	-10	0.4117
4	2.76E+20	5003	5003	-2	-8	0.2502
5	3.45E+19	2497	2497	0	-6	0.1249
6	1.92E+18	1112	1112	0	-4	0.0556
7	2.08E+16	217	217	0	-2	0.0109

The average of the 10 such estimates was $2.17 \cdot 10^{16}$ with an estimated relative error 0.0521. The average CPU time was around 4 seconds.

Model 2. Consider the problem of counting tables with Darwin's finch data as marginal sums given in Chen et al. [10]. The data are $m = 12, n = 17$, with row and columns sums

$$\mathbf{r} = (14, 13, 14, 10, 12, 2, 10, 1, 10, 11, 6, 2), \quad \mathbf{c} = (3, 3, 10, 9, 9, 7, 8, 9, 7, 8, 2, 9, 3, 6, 8, 2, 2).$$

The true count value [10] is $67, 149, 106, 137, 567, 600$. The average estimate based on 10 independent experiments using sample size $N = 200,000$ and rarity parameter $\varrho = 0.5$, was $6.71 \cdot 10^{16}$ with an estimated relative error of 0.08.

9.9.5 Vertex Coloring

Given a graph $G = (V, E)$ consisting of $n = |V|$ nodes and $m = |E|$ edges, a *q-coloring* is a coloring of the vertices with q colors is such that no two adjacent vertices have the same edge. We showed in Section 9.9.1 how a 2-coloring problem can be translated into a SAT problem. In this section we apply the splitting algorithm to counting the number of q-colorings of a graph.

Let $\mathscr{X} = \{1, \ldots, q\}^n = \{(x_1, \ldots, x_n) : x_i \in \{1, \ldots, q\}, i = 1, \ldots, n\}$ be the set of possible vertex colorings. For a vertex coloring $\mathbf{x} \in \mathscr{X}$, let $S(\mathbf{x})$ be minus the number of adjacent vertices that share the same color. The objective is to count the number of elements in the set $\mathscr{X}^* = \{\mathbf{x} \in \mathscr{X} : S(\mathbf{x}) \geqslant 0\}$.

We use the following Gibbs move in Algorithm 9.8.1:

Algorithm 9.9.4: Gibbs Move for Vertex Coloring

---
input : A graph $G = (V, E)$, the number of available colors q, a threshold value γ, and a vertex coloring $\mathbf{x} = (x_1, \ldots, x_n)$ with performance $S(\mathbf{x}) \geqslant \gamma$.
output: A random vertex coloring $\mathbf{Y}$ with $S(\mathbf{Y}) \geqslant \gamma$.
1 **for** $i = 1$ **to** n **do**
2 $\quad$ $\mathcal{M} \leftarrow \emptyset$
3 $\quad$ **for** $c = 1$ **to** q **do**
4 $\quad\quad$ **if** $S((Y_1, \ldots, Y_{i-1}, c, x_{i+1} \ldots, x_n)) \geqslant \gamma$ **then** $\mathcal{M} \rightarrow \mathcal{M} \cup \{c\}$
5 $\quad$ Choose Y_i uniformly at random from $\mathcal{M}$.
6 **return** $\mathbf{Y} = (Y_1, \ldots, Y_n)$

9.9.5.1 Numerical Experiment Consider the 3D order-4 grid graph in Figure 9.9. This graph has $n = 64$ vertices and $m = 3 \times 4^2 \times 3 = 144$ edges.

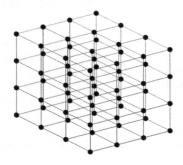

Figure 9.9: 3D grid graph of order 4.

How many 5-colorings are there? We used a pilot run (Algorithm 9.8.1) with sample size $N = 1440$ and $\varrho = 0.2$. The pilot run resulted in 27 levels, which were subsequently used in multiple independent repetitions of the fixed-level Algorithm 9.5.3. In particular, we kept repeating until an estimated relative error of 0.03 was obtained. This took 92 repetitions in 40 seconds. The final estimate was $4.19 \cdot 10^{31}$.

9.10 SPLITTING AS A SAMPLING METHOD

In this section we investigate how the splitting algorithm can be used to sample from complicated distributions. We focus on the fixed-level generalized splitting algorithms in Section 9.5.

We can reason inductively as follows: By acceptance–rejection, each element in $\mathcal{X}_1$ is distributed according to f_1; that is, according to the conditional pdf of $\mathbf{X} \sim f$ given $\mathbf{X} \in \mathcal{X}_1$. Consequently, all elements $\mathbf{Y}'$ generated during the Markov sampling phase at $t = 1$ also have the same density, because this Markov chain has invariant density f_1. It follows, again by acceptance–rejection, that all elements in $\mathcal{X}_2$ (those $\mathbf{Y}'$ that fall in $\mathcal{X}_2$) have density f_2. Markov sampling at phase $t = 2$ creates other particles from f_2, and those that fall in $\mathcal{X}_3$ have density f_3, and so on. As a result, the elements in $\mathcal{X}_T$ are all distributed according to f_T — the conditional pdf of $\mathbf{X} \sim f$ given $\mathbf{X} \in \mathcal{X}_T$. In particular, for applications to counting (Section 9.8), f is typically the uniform pdf on a set $\mathcal{X}$ and f_T the uniform distribution on a much smaller set $\mathcal{X}_T$. In this way the splitting method can be viewed as a multi-level generalization of the acceptance–rejection method.

There is, however, a subtle but important point to make about sampling. The reasoning above does not hold for randomly selected points from the final set, as the distribution of the points in $\mathcal{X}_T$ points depends on the size of the final set N_T: particles corresponding to a small N_T may be quite differently distributed to particles corresponding to a large N_T. So, if we select at random one of the returned points, its distribution differs in general from the conditional distribution given the rare event. Nevertheless, by repeating the algorithm n times, the empirical distribution of the set of all points returned over all n replicates converges to the conditional distribution given the rare event. The following example illustrates this illusive behavior of the splitting algorithm.

■ EXAMPLE 9.7   A Bivariate Discrete Uniform Example

We want to apply a generalized splitting algorithm to estimate $\ell = \mathbb{P}(S(\mathbf{X}) \geqslant 8)$, where $\mathbf{X} = (X_1, X_2)$ has a discrete uniform distribution on the square grid $\{1, \ldots, 8\}^2$ and $S(\mathbf{x}) = \max\{x_1, x_2\}$. In addition we want to compare the conditional distribution of the samples returned by a GS algorithm to the true distribution of $\mathbf{X}$ given $S(\mathbf{X}) \geqslant 8$, being the uniform distribution on the set $\{(1, 8), \ldots, (8, 8), (8, 1), \ldots, (8.7)\}$, which has 15 elements.

For our algorithm we take Algorithm 9.5.2 with fixed splitting factors, using two levels: the final level $\gamma_2 = 8$ and the intermediate level $\gamma_1 = 7$; see Figure 9.10. Note that in Algorithm 9.5.2, $\mathcal{X}_t = \{\mathbf{x} : S(\mathbf{x}) \geqslant \gamma_t\}, t = 1, 2$.

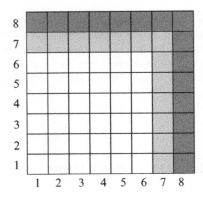

Figure 9.10: Apply the splitting algorithm to sample uniformly from the level set $\{\mathbf{x} : S(\mathbf{x}) \geqslant 8\}$ (dark shading) using the intermediate level set $\{\mathbf{x} : S(\mathbf{x}) \geqslant 7\}$ (light shading).

A point $\mathbf{x} = (x_1, x_2)$ with $S(\mathbf{x}) \geqslant 7$ is split into $r = 2$ points, $\mathbf{Y}^{(1)} = (Y_1^{(1)}, x_2)$ and $\mathbf{Y}^{(2)} = (Y_1^{(2)}, x_2)$, where $Y_1^{(1)}$ and $Y_1^{(2)}$ are drawn independently from the uniform distribution of X_1 conditional on $S((X_1, x_2)) \geqslant 7$. Note that the second coordinate is kept the same in a deliberate attempt to bias the distribution of the (one or two) final states. However, this is a valid Markov move for the splitting algorithm.

The splitting algorithm thus simplifies to the following. We first generate $\mathbf{X} = (X_1, X_2)$ according to the uniform distribution on $\{1, \ldots, 8\}^2$. If $\max\{X_1, X_2\} \geqslant 7$, we resample X_1 twice, conditional on $\max(X_1, X_2) \geqslant 7$, to obtain two states $\mathbf{Y}^{(1)}$ and $\mathbf{Y}^{(2)}$. Among these two states, we retain those for which $S(\mathbf{Y}) \geqslant 8$, if any. Let $M = N_2$ be the number of particles in the final level. So the algorithm returns either $M = 0$ (no points), $M = 1$ and one point $\mathbf{X}^*$, or $M = 2$ and two points $\mathbf{X}_1^*$ and $\mathbf{X}_2^*$.

Simple calculations give the following possible outcomes of this scheme. With probability $1/8$, we have $X_2 = 8$, in which case the splitting algorithm outputs $M = 2$ states $\mathbf{X}_1^*$ and $\mathbf{X}_2^*$ that are independent and uniformly distributed over $\{(1, 8), \ldots, (8, 8)\}$. With probability $1/8$, we have $X_2 = 7$, in which case the algorithm outputs two copies of state $(8, 7)$ with probability $1/64$ and one copy of $(8, 7)$ with probability $14/64$. For $j \in \{1, \ldots, 6\}$ with probability $1/32$, we have $X_2 = j$ and $X_1 \geqslant 7$. Conditional on this happening, the splitting algorithm outputs two copies of $(8, j)$ with probability $1/4$, and one copy with probability $1/2$.

Let $\mathbf{Z}$ be a randomly selected state returned by the splitting algorithm (if any). $\mathbf{Z}$ takes values in the set $\mathscr{X}_2$, augmented by an extra state $\emptyset$. Each state $(i, 8)$ has probability $1/64 = 8/512$ of being selected, state $(8, 7)$ has probability $(1/8)(15/64) = 15/512$, state $(8, j)$ has probability $(1/8)(3/4)(1/4) = 3/128 = 12/512$ for $1 \leqslant j \leqslant 6$, and with the remaining probability $361/512$ nothing is returned. Table 9.8 summarizes the situation. We see, on one hand, that conditional on $M > 0$, the states do not have the same probability.

Table 9.8: Sampling probabilities for the splitting algorithm.

x	$\mathbb{P}(\mathbf{Z} = \mathbf{x})$	$\mathbb{P}(\mathbf{Z} = \mathbf{x} \mid M > 0)$	$\mathbb{P}(\mathbf{X} = \mathbf{x} \mid S(\mathbf{X}) \geqslant 8)$
$(1,8),\ldots,(8,8)$	$8/512$	$8/151 \approx 0.053$	$1/15 \approx 0.0667$
$(8,1),\ldots,(8,6)$	$12/512$	$12/151 \approx 0.079$	$1/15$
$(8,7)$	$15/512$	$15/151 \approx 0.1$	$1/15$
$\emptyset$	$361/512$	0	0

On the other hand, $\mathbb{P}(M = 2) = 89/512$ and $\mathbb{P}(M = 1) = 62/512$, so that

$$\mathbb{E}[M]/2 = 15/64.$$

It follows that the probability $\ell = \mathbb{P}(S(\mathbf{X}) \geqslant 8)$ can be estimated in an unbiased way via the splitting estimator $\widehat{\ell} = M/2$.

Suppose now that we run the preceding splitting algorithm $n = 100$ times, collect the set (multiset) of all states returned over these n replications, and pick a state at random from it. The second panel of Table 9.8 gives the probabilities of the 15 different states (estimated by simulation) in that case. Now they are all much closer to $1/15 \approx 0.066666667$. These probabilities actually converge exponentially fast to $1/15$ when $n \to \infty$.

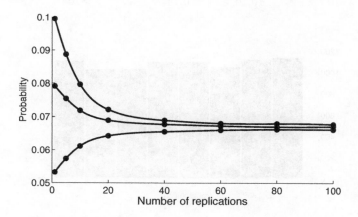

Figure 9.11: Let $\mathbf{Z}$ be a particle selected at random from all particles produced by n independent replications of the splitting algorithm. The graphs show how $\mathbb{P}(\mathbf{Z} = (1,8))$ (bottom), $\mathbb{P}(\mathbf{Z} = (8,1))$ (middle), $\mathbb{P}(\mathbf{Z} = (8,7))$ (top) all converge to $1/15$ as n increases.

9.10.0.2 Numerical Experiment We test the uniformity of the sample returned by the splitting algorithm for the 3-SAT (20×91) model in Section 9.9.1.1. By uniformity we mean that the sample passes the standard χ^2 test. In particular, let $C_1, \ldots, C_8$ be the counts for the 8 solutions, out of $C_1 + \cdots + C_8 = m$ samples.

Under the hypothesis of uniformity, the statistic

$$\sum_{i=1}^{8} \frac{(C_i - m/8)^2}{m/8},$$

has approximately a χ_7^2 distribution for large m.

We repeated Algorithm 9.8.1 (with $N = 1000$ and $\varrho = 0.1$) a number of times, pooling the particles in the final multi-set. Table 9.9 shows the results of four χ^2 tests. The first column gives the number of independent replications of the splitting algorithm, and the last column the total (pooled) number of particles in the final multi-set. The middle columns give the value of the test statistic and the corresponding p-value: the probability that under the null-hypothesis (uniformity) the test statistic takes a value greater than or equal to the one observed. Since all p-values are large, the null-hypothesis is not rejected. A histogram of the counts for the last case (10 replications) is given in Figure 9.12.

Table 9.9: Results of the χ^2 test for uniformity.

Replications	Test statistic	p-Value	Total number of samples
1	3.62	0.82	853
2	3.70	0.84	1729
5	5.50	0.60	4347
10	10.10	0.14	8704

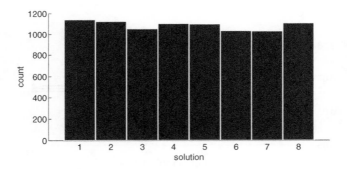

Figure 9.12: Histogram of the number of times, out of 8704, that each of the 8 solutions in the 3-SAT (20×91) problem was found.

9.11 SPLITTING FOR OPTIMIZATION

The adaptive splitting algorithm for counting the number of elements in sets of the form $\{\mathbf{x} : S(\mathbf{x}) \geqslant \gamma\}$ can be modified, without much effort, to an optimization algorithm for the performance function S. The resulting algorithm is suitable for solving both discrete and continuous optimization problems, and thus can be considered as an alternative to the standard cross-entropy and MinxEnt methods.

The main difference with the counting version is that the splitting method for optimization does not require knowledge of the final level γ or estimation of the level crossing probabilities $\{p_t\}$.

Algorithm 9.11.1 (Splitting Algorithm for Optimization)

- Input: performance function S, rarity parameter ϱ, sample size N.

- Output: estimators $\widehat{\gamma}^*$ and $\widehat{\mathbf{x}}^*$ of the maximal performance γ^* and the corresponding optimal solution $\mathbf{x}^*$.

1. **Acceptance–Rejection.** Set a counter $t = 1$. Generate a sample $\mathcal{Y}_0 = \{\mathbf{Y}^{(1)}, \ldots, \mathbf{Y}^{(N)}\}$ uniformly on $\mathscr{X} = \mathscr{X}_0$. Compute the threshold $\widehat{\gamma}_1$ as the sample $(1 - \varrho_1)$ quantile of the values of $S(\mathbf{Y}^{(1)}), \ldots, S(\mathbf{Y}^{(N)})$. Let $\mathcal{X}_1 = \{\mathbf{X}_1^{(1)}, \ldots, \mathbf{X}_1^{(N_1)}\}$ be the collection (multi-set) of elements $\mathbf{X}_1 \in \mathcal{Y}_0$ for which $S(\mathbf{X}_1) \geqslant \widehat{\gamma}_1$ (the elite points).

2. **Screening.** Reset $\mathcal{X}_t$ and N_t by screening out (removing) any duplicate particles.

3. **Splitting.** To each particle in the elite multi-set $\mathcal{X}_t$ apply a Markov chain sampler with $\mathsf{U}(\mathscr{X}_t)$ as its invariant (stationary) distribution, and length $\lfloor N/N_t \rfloor$. Apply the Markov chain sampler to $N - \lfloor N/N_t \rfloor = N \mod N_t$ randomly chosen endpoints for one more period. Denote the new entire sample (of size N) by $\mathcal{Y}_t$.

4. **Selecting elites.** Compute level $\widehat{\gamma}_{t+1}$ as the sample $(1 - \varrho_t)$ quantile of the values of $S(\mathbf{Y}), \mathbf{Y} \in \mathcal{Y}_t$. Determine the elite sample, that is, the largest subset $\mathcal{X}_{t+1} = \{\mathbf{X}_{t+1}^{(1)}, \ldots, \mathbf{X}_{t+1}^{(N_{t+1})}\}$ of $\mathcal{Y}_t$ consisting of N_{t+1} points for which $S(\mathbf{X}_{t+1}) \geqslant \widehat{\gamma}_{t+1}$.

5. **Stopping rule.** If for some $t \geqslant d$, say $d = 5$, the overall best performance $\widehat{\gamma}^*$ has not changed in the last d iterations, then **stop** and deliver $\widehat{\gamma}^*$ as the estimator for the optimal function value, and also return the corresponding solution $\widehat{\mathbf{x}}^*$; otherwise, increase the counter $t = t + 1$ and repeat from Step 2.

The main ingredient to be specified is the Markov move in Step 3. As in the rare-event and counting cases, Gibbs moves are convenient; that is, one or more components are chosen from the uniform distribution on a level set $\{\mathbf{x} : S(\mathbf{x}) \geqslant \gamma\}$ while the other components are kept fixed.

▪ EXAMPLE 9.8 Knapsack Problem

An intrepid explorer has to decide which of n possible items to pack into his/her knapsack. Associated with each item are m attributes: volume, weight, and so on. Let a_{ij} represent the i-th attribute of item j, and let $A = (a_{ij})$ be the corresponding $m \times n$ matrix. Associated with each attribute i is a maximal capacity c_i; for example, c_1 could be the maximum volume of the knapsack and c_2 the maximum weight the explorer can carry. Similarly, associated with each item j is a reward r_j if this item is taken into the knapsack. Let $\mathbf{c} = (c_i)$ and $\mathbf{r} = (r_j)$ denote the $m \times 1$ vector of capacities and

$n \times 1$ vector of rewards, respectively. Finally, let $\mathbf{x} = (x_j)$ be the $n \times 1$ decision vector of the explorer: $x_j = 1$ if item j is packed and $x_j = 0$ otherwise, $j = 1, \ldots, n$.

The binary knapsack problem can be formulated as: find which items can be packed so as to optimize the total reward, while all attribute constraints are satisfied. In mathematical terms this becomes

$$\max_{\mathbf{x} \in \{0,1\}^n} \mathbf{r}^{\mathsf{T}} \mathbf{x}$$

subject to

$$A\mathbf{x} \leqslant \mathbf{c}.$$

To solve this constrained minimization problem, we can maximize the function

$$S(\mathbf{x}) = \sum_{j=1}^{n} r_j\, x_j + \left(-\sum_{j=1}^{n} r_j \right) \sum_{i=1}^{m} I_{\{\sum_{j=1}^{n} a_{ij} x_j > c_i\}},$$

where the second term in $S(\mathbf{x})$ is a penalty term. As a numerical example, consider the Sento1.dat knapsack problem given in http://people.brunel. ac.uk/~mastjjb/jeb/orlib/files/mknap2.txt. The problem has 30 constraints and 60 variables. The 30×60 attribute matrix A has to be read from the data file (180 rows of 10 numbers) in row format. In particular, the first row of A starts with 47, 774, ... and ends with ..., 898, 37.

We ran the splitting Algorithm 9.11.1 with $\varrho = 0.1$ and $N = 1000$. In 10 out of 10 trials the optimal value 7772 was found. A typical dynamics is given in Table 9.10. The notation in the table is the same as in Section 9.9.1.1.

Table 9.10: Typical dynamics of the splitting Algorithm 9.11.1 for the knapsack problem, using $\varrho = 0.1$ and $N = 1000$.

t	N_t	N_t^{scr}	$\overline{S}_t$	$\underline{S}_t$	p_t
1	100	100	-62496	-202462	0.100
2	100	100	-15389	-109961	0.100
3	100	100	3535	-43463	0.100
4	100	100	4082	-7648	0.100
5	100	100	5984	2984	0.100
6	100	100	6017	4407	0.100
7	100	100	6730	5245	0.100
8	100	100	6664	5867	0.100
9	100	100	7086	6366	0.100
10	100	100	7277	6798	0.100
11	101	101	7475	7114	0.101
12	100	100	7495	7337	0.100
13	100	100	7592	7469	0.100
14	103	101	7695	7563	0.103
15	104	97	7739	7639	0.104
16	108	59	7772	7697	0.108
17	107	8	7772	7739	0.107
18	157	1	7772	7772	0.157

9.11.1 Continuous Optimization

For continuous problems, sampling a component of $\mathbf{X}$ uniformly from a level set $\{\mathbf{x} : S(\mathbf{x}) \geqslant \gamma\}$ may not be easy or fast. An alternative is to replace at some stage t the uniform sampling step with a simpler sampling mechanism; for example, sampling from a normal distribution with standard deviation that depends on the positions of the particles in $\mathcal{X}_t$. Such an approach is taken in [14], giving a very fast and accurate splitting algorithm for continuous optimization. The detailed description of this algorithm for minimizing a continuous function follows.

Algorithm 9.11.2: Splitting for Continuous Optimization

input : Objective function S, sample size N, rarity parameter ϱ, scale factor w, and maximum number of attempts MaxTry.

output: Final iteration number t and sequence $(\mathbf{X}_{\text{best},1}, b_1), \ldots, (\mathbf{X}_{\text{best},t}, b_t)$ of best solutions and function values at each iteration.

1 Generate $\mathcal{Y}_0 = \{\mathbf{Y}_1, \ldots, \mathbf{Y}_N\}$ uniformly. Set $t \leftarrow 0$ and $N^{\text{e}} \leftarrow \lceil N\varrho \rceil$.

2 **while** stopping condition is not satisfied **do**

3   Determine the N^{e} smallest values, $S_{(1)} \leqslant \cdots \leqslant S_{(N^{\text{e}})}$, of $\{S(\mathbf{X}), \mathbf{X} \in \mathcal{Y}_t\}$, and store the corresponding vectors, $\mathbf{X}_{(1)}, \ldots, \mathbf{X}_{(N^{\text{e}})}$, in $\mathcal{X}_{t+1}$. Set $b_{t+1} \leftarrow S_1$ and $\mathbf{X}_{\text{best},t+1} \leftarrow X_{(1)}$.

4   Draw $B_i \sim \mathsf{Bernoulli}(\frac{1}{2})$, $i = 1, \ldots, N^{\text{e}}$, with $\sum_{i=1}^{N^{\text{e}}} B_i = N \bmod N^{\text{e}}$.

5   **for** $i = 1$ **to** N^{e} **do**

6          $R_i \leftarrow \lfloor \frac{N}{N^{\text{e}}} \rfloor + B_i$ // random splitting factor

7          $\mathbf{Y} \leftarrow \mathbf{X}_{(i)}$; $\mathbf{Y}' \leftarrow \mathbf{Y}$

8   **for** $j = 1$ **to** R_i **do**

9   Draw $I \in \{1, \ldots, N^{\text{e}}\} \setminus \{i\}$ uniformly and let $\boldsymbol{\sigma}_i \leftarrow w|\mathbf{X}^{(i)} - \mathbf{X}^{(I)}|$.

10   Generate a uniform permutation $\boldsymbol{\pi} = (\pi_1, \ldots, \pi_n)$ of $(1, \ldots, n)$.

11   **for** $k = 1$ **to** n **do**

12   **for** Try $= 1$ **to** MaxTry **do**

13                     $\mathbf{Y}'(\pi_k) \leftarrow \mathbf{Y}(\pi_k) + \boldsymbol{\sigma}_i(r_k)Z$, $Z \sim \mathsf{N}(0,1)$

14   **if** $S(\mathbf{Y}') > S(\mathbf{Y})$ **then** $\mathbf{Y} \leftarrow \mathbf{Y}'$ and **break**.

15   Add $\mathbf{Y}$ to $\mathcal{Y}_{t+1}$

16      $t \leftarrow t + 1$

17 **return** $\{(\mathbf{X}_{\text{best},k}, b_k), k = 1, \ldots, t\}$

In Line 9, $\boldsymbol{\sigma}_i$ is a n-dimensional vector of standard deviations obtained by taking the component-wise absolute difference between the vectors $\mathbf{X}^{(i)}$ and $\mathbf{X}^{(I)}$, multiplied by a constant w. The input variable MaxTry governs how much computational time is dedicated to updating a component. In most cases the choices $w = 0.5$ and MaxTry $= 5$ work well. It was found that a relatively high ϱ value works well, such as $\varrho = 0.4, 0.8$, or even $\varrho = 1$. The latter case means that at each stage t *all* samples from $\mathcal{Y}_{t-1}$ carry over to the elite set $\mathcal{X}_t$.

A possible stopping condition is to stop if the overall best found function value does not change over d (e.g., $d = 5$) consecutive iterations.

■ **EXAMPLE 9.9 Minimizing the Rastrigin Function**

The n-dimensional *Rastrigin* function is defined by

$$S(\mathbf{x}) = \sum_{i=1}^{n} \left[x_i^2 - 10\cos(2\pi x_i) + 10 \right], \quad \mathbf{x} \in \mathbb{R}^n .$$

It has served as a benchmark function for many optimizers (e.g., see [14]). In most test cases the function is minimized over the n-dimensional cube $[-5.12, 5.12]^n$.

For the case $n = 50$, Figure 9.13 shows the maximum and minimum function values obtained in each iteration of Algorithm 9.11.2, using the parameters $N = 50$, $\varrho = 0.8$, and $w = 0.5$. We see an exponential rate of convergence to the true minimum, 0.

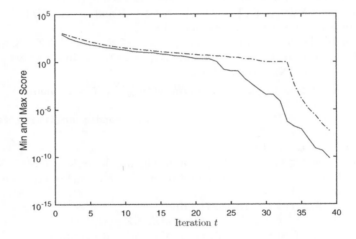

Figure 9.13: Performance of the continuous splitting algorithm for the 50-dimensional Rastrigin function.

PROBLEMS

9.1 Implement Algorithm 9.2.1 and verify the results in Example 9.1.

9.2 For the self-avoiding walk counting problem in Example 9.1 compare the efficiency of splitting Algorithm 9.2.1 with that of Algorithm 5.8.3 on Page 164, which uses sequential importance sampling. Improve both algorithms by combining them into a sequential importance resampling method.

9.3 Apply Algorithm 9.3.1 to the gambler's ruin problem in Example 9.3 and verify the numerical results given in the example.

9.4 Implement and run a splitting algorithm on a synthetic max-cut problem described in Problem 8.10 on Page 299, for a network with $n = 400$ nodes, with $m = 200$. Compare with the CE algorithm.

9.5 Let X_1 and X_2 be independent exponentially distributed random variables with mean 1. Implement the generalized splitting Algorithm 9.5.3 to estimate the probability $\ell = \mathbb{P}(X_1 + X_2 \geqslant 30)$. Use $T = 10$, with levels $\gamma_i = 3i$ for $i = 1, \ldots, 10$, and sample size (fixed effort) $N = 100$. For the Markov transition use the Gibbs move of Algorithm 9.5.4. Generate 800 independent runs of the splitting algorithm, and compare the overall estimate with the true probability. Also estimate the relative error of the overall estimator. Finally, plot all points obtained in the 800 replications whose sum is greater than or equal to 30.

9.6 Repeat Problem 9.5, but now determine the levels adaptively via a pilot run with Algorithm 9.6.1, using $\varrho = 0.1$ and $N = 10^3$. Is there any improvement in the relative error compared with using the levels $1, 2, \ldots, 5$? Also run Algorithm 9.6.1 100 times with an effort of $N = 10^4$ and compare with the previous results.

9.7 Suppose in the bridge network of Example 4.2 on Page 110 the link states $X_1, \ldots, X_5$ all have failure probability $q = 1 - p = 0.001$. Implement Algorithm 9.5.3 with $N = 10^2$ using the Gibbs move in Algorithm 9.7.1. Run the algorithm 1000 times to obtain a 95% confidence interval for the system unreliability. Specify which intermediate levels were used.

9.8 Let $\{A_i\}$ be an arbitrary collection of subsets of some finite set $\mathscr{X}$. Prove the following *inclusion–exclusion* principle, which can be useful for decomposing a complicated counting problem into possibly simpler ones.

$$\left| \cup_i A_i \right| = \sum |A_i| - \sum_{i<j} |A_i \cap A_j| + \sum_{i<j<k} |A_i \cap A_j \cap A_k| - \cdots .$$

9.9 A famous problem in combinatorics is the *distinct representatives* problem, which is formulated as follows: Given a set $\mathscr{A}$ and subsets $\mathscr{A}_1, \ldots, \mathscr{A}_n$ of $\mathscr{A}$, is there a vector $\mathbf{x} = (x_1, \ldots, x_n)$ such that $x_i \in \mathscr{A}_i$ for each $i = 1, \ldots, n$ and the $\{x_i\}$ are all distinct (that is, $x_i \neq x_j$ if $i \neq j$)?

a) Suppose, for example, that $\mathscr{A} = \{1, 2, 3, 4, 5\}$, $\mathscr{A}_1 = \{1, 2, 5\}$, $\mathscr{A}_2 = \{1, 4\}$, $\mathscr{A}_3 = \{3, 5\}$, $\mathscr{A}_4 = \{3, 4\}$, and $\mathscr{A}_5 = \{1\}$. Count the total number of distinct representatives.

b) Argue why the total number of distinct representatives in the problem above is equal to the *permanent* of the following matrix A:

$$A = \begin{pmatrix} 1 & 1 & 0 & 0 & 1 \\ 1 & 0 & 0 & 1 & 0 \\ 0 & 0 & 1 & 0 & 1 \\ 0 & 0 & 1 & 1 & 0 \\ 1 & 0 & 0 & 0 & 0 \end{pmatrix}.$$

9.10 Let $X_1, \ldots, X_n$ be independent Bernoulli random variables, each with success probability $1/2$. Devise a splitting algorithm to efficiently estimate $\ell = \mathbb{P}(X_1 + \cdots + X_n \geqslant \gamma)$ for large γ (e.g., $\gamma = 0.9n$). For the Markov move, use Algorithm 9.9.1. Show the dynamics of the algorithm similar to the tables in Section 9.9.

9.11   Let $\mathscr{X}$ be the set of permutations $\mathbf{x} = (x_1, \ldots, x_n)$ of the numbers $1, \ldots, n$, and let

$$S(\mathbf{x}) = \sum_{j=1}^{n} j\, x_j \,. \tag{9.7}$$

Let $\mathscr{X}^* = \{\mathbf{x} : S(\mathbf{x}) \geqslant \gamma\}$, where γ is chosen such that $|\mathscr{X}^*|$ is very small relative to $|\mathscr{X}| = n!$. Implement splitting Algorithm 9.8.1 to estimate $|\mathscr{X}^*|$, for $n = 8$ and $\gamma = 201$, using $N = 10^4$ and rarity parameter $\varrho = 0.1$ for all levels. Repeat the algorithm 1000 times and report the overall estimate $\widehat{|\mathscr{X}^*|}$, its estimated relative error, and also the direct estimator (9.4). Use the following Markov move at each iteration t. Given a permutation $\mathbf{x} = (x_1, \ldots, x_n)$, select I and J uniformly from $\{1, \ldots, n\}$, swap x_I and x_J to obtain a new permutation $\mathbf{x}'$, and check whether $S(\mathbf{x}') \geqslant \gamma_t$. If so, accept the move to $\mathbf{x}'$; otherwise, repeat (with the same $\mathbf{x}$). Note that $I = J$ results in $\mathbf{x}' = \mathbf{x}$.

9.12   The volume V_n of the n-dimensional simplex $\mathscr{Y}_n = \{(y_1, \ldots, y_n) : y_i \geqslant 0, i = 1, \ldots, n, \sum_{i=1}^{n} y_i \leqslant 1\}$ is equal to $1/n!$. Devise a splitting algorithm to estimate V_n for $n = 5$, $n = 10$, and $n = 20$. Compare the method with the crude Monte Carlo (acceptance–rejection) approach.

9.13   Write a splitting algorithm for the TSP, using the 2-opt neighborhood structure. For various test cases, obtain a table similar to Table 8.11 on Page 291.

9.14   How many *Hamiltonian cycles* exist in a given graph? That is, how many different tours (cycles) does the graph contain that visit each node (vertex) exactly once, apart for the beginning/end node? Note that the problem of finding a particular Hamiltonian cycle is a special case of the TSP in which the distances between adjacent nodes are 1, and other distances are ∞, and the objective is to find a tour of length n.

A possible way to solve the Hamiltonian cycles counting problem via splitting is as follows:

1. Run Algorithm 9.8.1 to estimate the permanent of the adjacency matrix $A = (a_{ij})$ of the graph, as in Section 9.9.3. Let $\mathcal{X}_T$ be the screened elite samples (permutations of $(1, \ldots, n)$) at the final level, and let $\widehat{P}$ be the estimator of the permanent. Note that, by construction, for each such permutation $(\pi_1, \ldots, \pi_n)$ it holds that $\sum_{i=1}^{n} a_{ix_i} = n$.

2. Proceed with one more iteration of Algorithm 9.8.1 to produce a set $\mathcal{Y}_T$ of permutations with $\sum_{i=1}^{n} a_{ix_i} = n$. Each such permutation corresponds to an incidence matrix $B = (b_{ij})$ defined by $b_{ij} = a_{ij}$ for $j = \pi_i$, $i = 1, \ldots, n$ and $b_{ij} = 0$ otherwise. This matrix B can be viewed as the one-step transition matrix of a Markov chain in which from each state only one transition is possible. If, starting from 1, the Markov chain returns to 1 in exactly n steps, then the corresponding path is a Hamiltonian cycle; otherwise, it is not.

   Let ζ be the proportion of Hamiltonian cycles in $\mathcal{Y}_T$.

3. Deliver $\widehat{P}\zeta$ as an estimator of the number Hamiltonian cycles in the graph.

Implement this method for the graph in Figure 9.14, using $N = 100,000$ and $\varrho = 0.1$. Also provide the direct estimator of the number of Hamiltonian cycles.

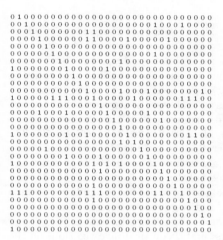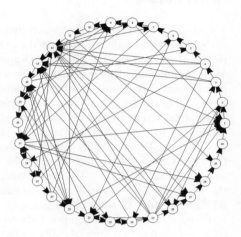

Figure 9.14: How many Hamiltonian cycles does the directed graph on the right have? The adjacency matrix is given on the left.

9.15 The *degree* of a vertex in a graph $G = (V, E)$ is the number of edges starting from that vertex. How many graphs with vertices labeled $1, \ldots, n$ are such that vertex i has a prescribed degree d_i, $i = 1, \ldots, n$? The problem can be viewed as of choosing those edges in the complete graph K_n such that the resulting graph G matches the given sequence $\mathbf{d} = (d_1, \ldots, d_n)$. In K_n all vertices are connected to each other, so there are $m = n(n-1)/2$ edges, labeled $1, \ldots, m$. Let $\mathbf{x} = (x_1, \ldots, x_m)^\top$ be such that $x_j = 1$ when edge j is chosen, and $x_j = 0$ otherwise, $j = 1, \ldots, m$. In order for such an assignment $\mathbf{x} \in \{0, 1\}^m$ to match the given degree sequence $(d_1, \ldots, d_n)$, it is necessary that $\sum_{j=1}^m x_j = \frac{1}{2} \sum_{i=1}^n d_i$, since this is the total number of edges. In other words, the configuration space is

$$\mathscr{X} = \left\{ \mathbf{x} \in \{0, 1\}^m : \sum_{j=1}^m x_j = \tfrac{1}{2} \sum_{i=1}^n d_i \right\}.$$

For the random graph problem we define the score function $S : \mathscr{X} \to \mathbb{Z}_-$ by

$$S(\mathbf{x}) = -\sum_{i=1}^n |\deg(i(\mathbf{x})) - d_i|,$$

where $\deg(i(\mathbf{x}))$ is the degree of vertex i under the configuration $\mathbf{x}$. Each configuration that satisfies the degree sequence will have a performance function equal to 0.

   a) Describe a simple way to sample uniformly from $\mathscr{X}$.
   b) Provide a Gibbs move for the splitting algorithm
   c) Implement the splitting algorithm, and apply it to the degree sequence $\mathbf{d} = (3, 2, 1, 1, 1, 1, 1)$.

9.16 Apply Algorithm 9.11.2 to the minimization of the Rosenbrock function, defined in (8.64) on Page 301.

Further Reading

The splitting method was first published in [24] as a sequential procedure in which paths are split in order to increase the occurrence of some rare event. A similar *enrichment* method was used in [42] to improve the estimation of polymer dimensions. The fundamental ideas of sequential sampling, importance sampling, and splitting (enrichment, branching, cloning, resampling) have been used in many forms ever since. All of these methods may be viewed as special cases of sequential importance resampling, although such a broad generalization does no justice to the great variety of Monte Carlo methods that combine these ideas [8, 9, 12, 13, 21, 27, 28, 33, 37]. The class of SIR algorithms can itself be generalized to include Markov chain Monte Carlo (MCMC) steps. This larger class of algorithms is probably best described as particle MCMC [1]. It includes generalized splitting [6, 7], cloning, [34, 35] and the resample-move algorithm [18]. Other algorithms similar to splitting include RESTART [40, 41], subset simulation [2, 3], and stopping-time resampling [11].

Discussions on the difference between various splitting algorithms (e.g., fixed effort versus fixed number of splitting factors) and theoretical investigations on the optimal choices of the splitting parameters may be found in [16, 17, 19, 25].

As demonstrated in Section 9.8, the splitting method is well suited to solve difficult counting problems, using the key decomposition (9.3). For good references on #P-complete problems with emphasis on SAT problems, see [29, 30]. The counting class #P was defined by Valiant [39]. Randomized algorithms for approximating the solutions of some well-known counting #P-complete problems and their relation to MCMC are treated in [20, 23, 30, 31, 38, 44]. Bayati and Saberi [4] propose an efficient importance sampling algorithm for generating graphs uniformly at random. Chen et al. [10] discuss the efficient estimation, via sequential importance sampling, of the number of 0-1 tables with fixed row and column sums. Blanchet [5] provides the first importance sampling estimator with bounded relative error for this problem.

REFERENCES

1. C. Andrieu, A. Doucet, and R. Holenstein. Particle Markov chain Monte Carlo methods. *Journal of the Royal Statistical Society: Series B (Statistical Methodology)*, 72(3):269–342, 2010.

2. S.-K. Au and J. L. Beck. Estimation of small failure probabilities in high dimensions by subset simulation. *Probabilistic Engineering Mechanics*, 16(4):263–277, 2001.

3. S.-K. Au, J. Ching, and J. L. Beck. Application of subset simulation methods to reliability benchmark problems. *Structural Safety*, 29(3):183 – 193, 2007.

4. M. Bayati and A. Saberi. Fast generation of random graphs via sequential importance sampling. Manuscript, Stanford University, 2006.

5. J. Blanchet. Importance sampling and efficient counting for binary contingency tables. *Annals of Applied Probability*, 19:949–982, 2009.

6. Z. I. Botev. *The Generalized Splitting Method for Combinatorial Counting and Static Rare-Event Probability Estimation.* PhD thesis, The University of Queensland, 2009.

7. Z. I. Botev and D. P. Kroese. Efficient Monte Carlo simulation via the generalized splitting method. *Statistics and Computing*, 22:1–16, 2012.

8. H.-P. Chan and T.-L. Lai. A sequential Monte Carlo approach to computing tail probabilities in stochastic models. *Annals of Applied Probabability*, 21(6):2315–2342, 12 2011.

9. H.-P. Chan and T.-L. Lai. A general theory of particle filters in hidden Markov models and some applications. *Annals of Statistics*, 41(6):2877–2904, 12 2013.

10. Y. Chen, P. Diaconis, S. P. Holmes, and J. Liu. Sequential Monte Carlo methods for statistical analysis of tables. *Journal of the American Statistical Association*, 100:109–120, 2005.

11. Y. Chen, J. Xie, and J. S. Liu. Stopping-time resampling for sequential Monte Carlo methods. *Journal of the Royal Statistical Society: Series B (Statistical Methodology)*, 67(2):199–217, 2005.

12. P. Del Moral, A. Doucet, and A. Jasra. Sequential Monte Carlo samplers. *Journal of the Royal Statistical Society: Series B (Statistical Methodology)*, 68(3):411–436, 2006.

13. A. Doucet, N. de Freitas, and N. Gordon, editors. *Sequential Monte Carlo methods in practice*. Statistics for Engineering and Information Science. Springer New York, 2001.

14. Q. Duan and D. P. Kroese. Splitting for optimization. *Computers and Operations Research*, 73:119–131, 2016.

15. P. Dupuis, B. Kaynar, R. Y. Rubinstein, A. Ridder, and R. Vaisman. Counting with combined splitting and capture–recapture methods. *Stochastic Models*, 28:478–502, 2012.

16. M. J. J. Garvels. *The Splitting Method in Rare Event Simulation*. PhD thesis, Universiteit Twente, 2000.

17. M. J. J. Garvels and D. P. Kroese. A comparison of restart implementations. In *Simulation Conference Proceedings, 1998. Winter*, volume 1, pages 601–608 vol.1, 1998.

18. W. R. Gilks and C. Berzuini. Following a moving target: Monte Carlo inference for dynamic Bayesian models. *Journal of the Royal Statistical Society: Series B (Statistical Methodology)*, 63(1):127–146, 2001.

19. P. Glasserman, P. Heidelberger, P. Shahabuddin, and T. Zajic. Multilevel splitting for estimating rare event probabilities. *Operations Research*, 47(4):585–600, 1999.

20. C. P. Gomes and B. Selman. Satisfied with physics. *Science*, pages 784–785, 2002.

21. N. J. Gordon, D. J. Salmond, and A. F. M. Smith. Novel approach to nonlinear/non-Gaussian Bayesian state estimation. *Radar and Signal Processing, IEE Proceedings F*, 140(2):107–113, 1993.

22. J. Gu, P. W. Purdom, J. Franco, and B. W. Wah. Algorithms for the satisfiability (SAT) problem: A survey. In *Satisfiability Problem: Theory and Applications*, volume 35 of DIMACS Series in Discrete Mathematics. American Mathematical Society, 1996.

23. M. Jerrum. *Counting, Sampling and Integrating: Algorithms and Complexity*. Birkhauser Verlag, Basel, 2003.

24. H. Kahn and T. E. Harris. Estimation of particle transmission by random sampling. *National Bureau of Standards Applied Mathematics Series*, 12:27–30, 1951.

25. P. L'Ecuyer, V. Demers, and B. Tuffin. Rare events, splitting, and quasi-Monte Carlo. *ACM Trans. Model. Comput. Simul.*, 17(2), 2007.

26. P. L'Ecuyer, V. Demers, and B. Tuffin. Splitting for rare-event simulation. *ACM Transactions on Modeling and Computer Simulation (TOMACS)*, 17(2):1–44, 2007.

27. P. L'Ecuyer, F. Le Gland, P. Lezaud, and B. Tuffin. Splitting techniques. In G. Rubino and B. Tuffin, editors, *Rare Event Simulation*. John Wiley & Sons, New York, 2009.

28. J. S. Liu. *Monte Carlo strategies in Scientific Computing*. Springer, New York, 2001.

29. M. Mézard and A. Montanari. *Constraint Satisfaction Networks in Physics and Computation*. Oxford University Press, Oxford, 2006.

30. M. Mitzenmacher and E. Upfal. *Probability and Computing: Randomized Algorithms and Probabilistic Analysis*. Cambridge University Press, Cambridge, 2005.

31. R. Motwani and R. Raghavan. *Randomized Algorithms*. Cambridge University Press, Cambridge, 1997.

32. S. M. Ross. *A First Course in Probability*. Prentice Hall, Englewood Cliffs, NJ, 7th edition, 2005.

33. D. B. Rubin. The calculation of posterior distributions by data augmentation: Comment: A noniterative sampling/importance resampling alternative to the data augmentation algorithm for creating a few imputations when fractions of missing information are modest: The SIR algorithm. *Journal of the American Statistical Association*, 82(398):pp. 543–546, 1987.

34. R. Y. Rubinstein. Randomized algorithms with splitting: Why the classic randomized algorithms do not work and how to make them work. *Methodology and Computing in Applied Probability*, 12(1):1–50, 2010.

35. R. Y. Rubinstein. Stochastic enumeration method for counting NP-hard problems. *Methodology and Computing in Applied Probability*, 15(2):249–291, 2013.

36. G.A.F. Seber. *Estimation of Animal Abundance and Related Parameters*. Blackburn Press, Caldwell, N.J., second edition, 2002.

37. Ø. Skare, E. Bølviken, and L. Holden. Improved sampling-importance resampling and reduced bias importance sampling. *Scandinavian Journal of Statistics*, 30(4):719–737, 2003.

38. R. Tempo, G. Calafiore, and F. Dabbene. *Randomized Algorithms for Analysis and Control of Uncertain Systems*. Springer-Verlag, Berlin, 2004.

39. L.G. Valiant. The complexity of enumeration and reliability problems. *SIAM Journal on Computing*, 8:410–421, 1979.

40. M. Villén-Altamirano and J. Villén-Altamirano. RESTART: A method for accelerating rare events simulations. In *Proceedings of the 13th International Teletraffic Congress*, pages 71–76. North-Holland, 1991.

41. M. Villén-Altamirano and J. Villén-Altamirano. On the efficiency of RESTART for multidimensional systems. *ACM Transactions on Modeling and Computer Simulation*, 16(3):251–279, 2006.

42. F. T. Wall and J. J. Erpenbeck. New method for the statistical computation of polymer dimensions. *Journal of Chemical Physics*, 30(3):634–637, 1959.

43. B.-Y. Wang and F. Zhang. On the precise number of (0,1)-matrices in $\mathcal{U}(R, S)$. *Discrete Mathematics*, 187(13):211–220, 1998.

44. D. J. A. Welsh. *Complexity: Knots, Colouring and Counting*. Cambridge University Press, Cambridge, 1993.

CHAPTER 10

STOCHASTIC ENUMERATION METHOD

10.1 INTRODUCTION

Many counting problems can be formulated in terms of estimating the cost of a tree. In this chapter we introduce a new generic Monte Carlo technique, called *stochastic enumeration* (SE), to solve such counting problems effectively. The SE method is a sequential importance sampling technique in which random paths are generated through the tree, in a parallel fashion.

The rest of this chapter is organized as follows: Section 10.2 provides background on tree search and tree counting, and reviews the basic depth-first and breadth-first search algorithms. Knuth's algorithm for estimating the cost of a tree is given in Section 10.3. Section 10.4 describes the SE method and discusses how it can be combined with fast decision-making algorithms, called oracles. In Section 10.5 it is shown how SE can be applied to solve various counting problems, including counting the number of paths in graphs, the number of valid assignments for a satisfiability problem, and the number of perfect matchings in a bipartite graph. For each case several numerical experiments are given, showing the effectiveness of the SE method. Finally, Section 10.6 provides an application of SE to network reliability.

10.2 TREE SEARCH AND TREE COUNTING

Before introducing the stochastic enumeration algorithm, it is prudent to review a few fundamental concepts related to searching and counting trees.

Two of the most useful algorithms for searching trees (or more generally, graphs or networks) are the *depth-first search* (DFS) and *breadth-first search* (BFS) methods. Starting from a root of the tree, the algorithm enumerates all the vertices of the tree in a certain order. In the BFS case all vertices at the first level of the tree are visited first, then all vertices at the second level, and so on. The DFS algorithm, in contrast, tends to descend to low levels of the tree before visiting the upper level vertices. To keep track of which vertices still need to be visited, the DFS algorithm places vertices in a *stack*. This is a data structure where elements are added to the top and are taken out from the top, in a last-in–first-out manner. Similarly, the BFS algorithm places vertices in a *queue*. Here the elements are added to the back of the queue and taken out from the front. Figure 10.1 illustrates these two data structures.

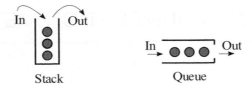

Figure 10.1: Items are added to (pushed onto) the top of the stack and are taken (popped off) from the top in a last-in–first-out manner. In contrast, items enter a queue from the back and leave from the front, in a first-in–first-out manner.

Algorithms 10.2.1 and 10.2.2 describe the DFS and BFS methods, respectively. For simplicity, we represent both the stack and the queue objects as ordered lists (vectors), where new elements are always added to the back of the list. Elements are either taken out from the front of the list (for a queue) or from the back (for a stack).

Algorithm 10.2.1: Depth-First Search

input : Tree $\mathcal{T}_u$ with root u.
output: List V of all vertices of the tree.

1   $V \leftarrow ()$ // initialize the list of visited vertices
2   $S \leftarrow (u)$ // initialize the stack
3   **while** $|S| > 0$ **do**
4      $v \leftarrow S(\text{end})$ // take the last element of stack
5      $S \leftarrow S(1 : \text{end} - 1)$ // reset the stack by removing last element
6   **if** $v \notin V$ **then**
7          $V \leftarrow (V, v)$ // add v to the list of visited vertices
8   **for** $w \in \mathcal{N}(v)$ **do**
9              $S \leftarrow (S, w)$ // add all neighbors of v to the stack

10 **return** V

Algorithm 10.2.2: Breadth-First Search

input : Tree $\mathcal{T}_u$ with root u.

output: List V of all vertices of the tree.

1 $V \leftarrow ()$

2 $Q \leftarrow (u)$

3 **while** $|Q| > 0$ **do**

4      $v \leftarrow Q(1)$ `// take the first element in the queue`

5      $Q \leftarrow Q(2:\text{end})$ `// reset the queue by removing first element`

6   **if** $v \notin V$ **then**

7         $V \leftarrow (V, v)$

8   **for** $w \in \mathcal{N}(v)$ **do**

9            $Q \leftarrow (Q, w)$ `// add all neighbors of v to the queue`

10 **return** V

■ **EXAMPLE 10.1**

As an illustration of the workings of the DFS and BFS algorithms, consider the trees in Figure 10.2. Starting from the root A, the DFS algorithm visits the remaining vertices in the order C, H, G, F, B, E, I, D, assuming that the neighbors of a vertex are placed onto the stack from left to right in the figure. For example, the neighbors of A are B and C, and if the B is placed onto the stack before C, then C is visited first. Similarly, the BFS algorithm visits the vertices in the order A, B, C, D, E, F, G, H, I.

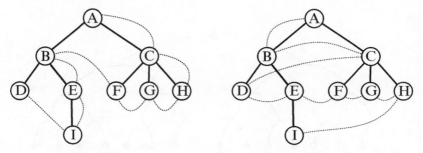

Figure 10.2: Dashed lines show the order in which the vertices are visited for the DFS (left) and BFS (right) algorithms.

We now turn our attention to tree counting. Consider a rooted tree $\mathcal{T} = (\mathcal{V}, \mathcal{E})$ with vertex set $\mathcal{V}$ and edge set $\mathcal{E}$ (so that $|\mathcal{E}| = |\mathcal{V}| - 1$). With each vertex v is associated a nonnegative cost $c(v)$. A quantity of interest is the total cost of the tree,

$$\text{Cost}(\mathcal{T}) = \sum_{v \in \mathcal{V}} c(v) \,. \tag{10.1}$$

If total enumeration of the tree is feasible, then the cost can be determined directly via DFS and BFS.

Many counting problems can be formulated as tree-counting problems. In particular, let $\mathbf{x}$ be an object in some discrete set $\mathcal{X}$ that can be constructed sequentially via intermediate objects $\mathbf{x}_0, \mathbf{x}_1, \ldots, \mathbf{x}_n$. Think of $\mathbf{x}$ as being a vector $(x_1, \ldots, x_n)$ and $\mathbf{x}_t$ as the partial vector $(x_1, \ldots, x_t)$, $t = 1, \ldots, n$. Each $\mathbf{x}$ is thus associated with a path $\mathbf{x}_0, \mathbf{x}_1, \ldots, \mathbf{x}_n$ in the tree of all such paths, with the "empty" vector $\mathbf{x}_0 = (\ )$ being the root of the tree. Determining $|\mathscr{X}|$, that is, the number of elements in $\mathscr{X}$, corresponds to counting the cost of the tree where all leaves of the tree have cost 1 and the internal vertices have cost 0. Frequently the situation arises that $|\mathscr{X}|$ is known but the number of elements in a much smaller subset $\mathscr{X}^*$ of $\mathscr{X}$ is not known. This amounts to counting the total costs of the tree in which only the leaves in $\mathscr{X}^*$ have non-zero cost equal to 1.

■ **EXAMPLE 10.2 SAT Counting**

Consider the 2-SAT CNF formula (see Section 9.9.1)

$$(l_1 \vee l_2) \wedge (l_2 \vee l_3) \wedge (l_2 \vee l_4) \, ,$$

and let $\mathscr{X}^*$ be the set of truth assignments $\mathbf{x} = (x_1, \ldots, x_4) \in \mathscr{X} = \{0, 1\}^4$ for which the formula is satisfied. Direct inspection shows that $\mathscr{X}^* = \{0100, 0101, 0110, 0111, 1011, 1100, 1101, 1110, 1111\}$ (omitting brackets and commas for simplicity). The corresponding binary tree is given in Figure 10.3. Any vertex at the t-level of the tree corresponds to a partial assignment $(x_1, \ldots, x_t)$. Its left-child corresponds to $(x_1, \ldots, x_t, 0)$ and its right-child to $(x_1, \ldots, x_t, 1)$. The black leaves, at tree level $t = 4$, are the valid truth assignments.

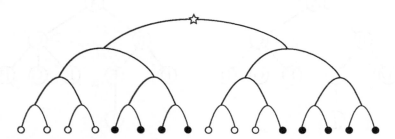

Figure 10.3: Binary counting tree for the 2-SAT counting problem.

■ **EXAMPLE 10.3 Counting Paths in Graphs**

Figure 10.4 shows the *Petersen graph* of order $(7, 4)$. How many (non-intersecting) paths does the graph have between the two highlighted vertices at the top and bottom-left? A systematic way to determine this is to sequentially construct a tree of paths of various lengths, all starting from the top vertex and ending either at the bottom-left vertex or at a vertex whose neighbors have already been visited. Note that the lengths of these paths vary from 3 to 13.

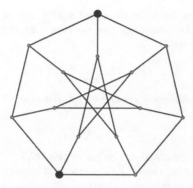

Figure 10.4: The (7,4)-Petersen graph. How many paths are there between the two highlighted vertices?

A similar type of problems is counting self-avoiding walks (see Example 5.17 on Page 162 and Section 9.2 on Page 308). Here the underlying graph is the grid graph on $\mathbb{Z}^2$ and the objective is to count the number of non-intersecting walks of a fixed length n, starting from the origin $(0,0)$. If a walk is represented as the vector $\mathbf{x} = (x_1, \ldots, x_n)$, where x_i can take four possible values (up, down, left, right), then the corresponding counting tree is a *quadtree* (each parent has exactly four children). The tree leaves correspond to the 4^n walks of length n, and only the non-intersecting ones have non-zero costs 1.

10.3 KNUTH'S ALGORITHM FOR ESTIMATING THE COST OF A TREE

Knuth [10] proposed a randomized algorithm to *estimate* the cost of a tree, when full enumeration of the tree is not feasible. Consider the cost of a tree $\mathcal{T}$ as given in (10.1). For each vertex v we denote the set of children or *successors* of v by $\mathscr{S}(v)$.

Knuth's algorithm constructs a random path $X_0, X_1, \ldots$, through the tree, starting from the root vertex and ending at a leaf, selecting at each step t at random (and uniformly) a successor vertex from the set $\mathscr{S}(X_t)$. In addition, two variables are updated at each step: (1) the product D of the degrees of all vertices encountered and (2) the estimated cost C of the tree. Algorithm 10.3.1 shows the details.

Algorithm 10.3.1: Knuth's Algorithm for Estimating the Cost of a Tree

input : Tree $\mathcal{T}$ with root u.
output: Estimator C of the total cost of $\mathcal{T}$.

1 $C \leftarrow c(u), D \leftarrow 1, X \leftarrow u$
2 **while** $|\mathscr{S}(X)| > 0$ **do**
3     $D \leftarrow |\mathscr{S}(X)|D$
4     $X \leftarrow$ uniformly selected successor in $\mathscr{S}(X)$
5     $C \leftarrow C + c(X)D$
6 **return** C

In practice, Algorithm 10.3.1 is repeated via N independent runs, giving estimators $C_1, \ldots, C_N$. In the usual way (see Section 4.2), the combined estimator for these runs is the sample mean

$$\overline{C} = \frac{C_1 + \cdots + C_N}{N}, \tag{10.2}$$

and its accuracy can be assessed via the estimated relative error

$$\widehat{\mathrm{RE}} = \frac{S_C}{\sqrt{N} \times \overline{C}}, \tag{10.3}$$

where S_C is the sample standard deviation of the $\{C_i\}$.

Knuth's algorithm can be viewed as a sequential importance sampling algorithm with a one-step–look-ahead sampling strategy, similar to the algorithm for counting self-avoiding walks in Example 5.17 on Page 162. To see this, extend the tree with 0-cost vertices such that each vertex at depth $t < n$ has at least one successor. Hence, each path returned by the algorithm now has length n, but the total cost of the extended tree is still the same as the original one. Let $\mathscr{V}_t$ be the set of all vertices at the t-th level of the extended tree (hence, $\mathscr{V}_0$ has only the root vertex, and $\mathscr{V}_n$ has all the leaf vertices). For each $t = 0, 1, \ldots, n$, Knuth's algorithm defines a density g_t on the set $\mathscr{X}_t$ of all paths $\mathbf{x}_t = (x_0, x_1, \ldots, x_t)$ from the root to level t. Specifically,

$$g_t(\mathbf{x}_t) = \frac{1}{d(x_0)} \frac{1}{d(x_1)} \cdots \frac{1}{d(x_{t-1})} \overset{\text{def}}{=} \frac{1}{D(\mathbf{x}_t)},$$

where $d(x) = |\mathscr{S}(x)|$ denotes the vertex degree of x. The total cost of the vertices at level t of the tree can thus be written as

$$\sum_{v \in \mathscr{V}_t} c(v) = \sum_{\mathbf{x}_t} \frac{c(x_t)}{g_t(\mathbf{x}_t)} g_t(\mathbf{x}_t) = \mathbb{E}_{g_t}\left[\frac{c(X_t)}{g_t(\mathbf{X}_t)}\right] = \mathbb{E}_{g_t}\left[c(X_t) D(\mathbf{X}_t)\right],$$

so that $c(X_t) D(\mathbf{X}_t)$ is an unbiased estimator of this cost, and the sum $\sum_{t=0}^{n} c(X_t) D(\mathbf{X}_t)$ is an unbiased estimator of the total cost of the tree.

Knuth's algorithm is very effective as long as the costs are evenly distributed over the tree. Consider, for example, a binary tree of depth (maximum level) n, where all the leaves have cost 1 and all the internal vertices have cost 0. In this case a *single* run ($N = 1$) of Algorithm 10.3.1 provides the total cost of the tree, 2^n, regardless of which path $x_0, x_1, \ldots, x_n$ is generated. However, Knuth's algorithm can be very poor when the tree has little symmetry. An extreme example is the "hairbrush" tree in Figure 10.5 of depth n. Suppose that the cost of all vertices is zero except for the right-most vertex at depth n, which has a cost of unity.

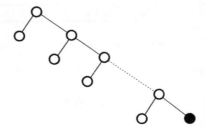

Figure 10.5: Hairbrush tree.

Knuth's algorithm reaches the vertex of interest (colored black in the figure) with probability $1/2^n$ and with $D = C = 2^n$. In all other cases, the algorithm terminates with $C = 0$. It follows that the expectation and variance of Knuth's estimator are, respectively,

$$\mathbb{E}\left[C\right] = \frac{1}{2^n} \cdot 2^n = 1$$

and

$$\mathrm{Var}\left(C\right) = \frac{1}{2^n} \cdot \left(2^n\right)^2 - 1 = 2^n - 1 \ .$$

Hence, although the estimator is unbiased, its variance grows exponentially with n.

10.4 STOCHASTIC ENUMERATION

The *stochastic enumeration* (SE) algorithm generalizes Knuth's algorithm by simulating multiple paths through the tree simultaneously. These paths are also allowed to interact, in the sense that paths avoid each other and therefore explore a larger part of the tree.

Let $B \geqslant 1$ be a fixed parameter, representing the computational budget. We will refer to a set of vertices at the same level of the tree as a *hypervertex*. If $\mathcal{V}$ is a hypervertex, define the collection of hyperchildren of $\mathcal{V}$ by

$$\mathscr{H}(\mathcal{V}) = \left\{ \begin{array}{ll} \{\mathscr{S}(\mathcal{V})\} & \text{if } |\mathscr{S}(\mathcal{V})| \leqslant B \ , \\ \{\mathcal{W} \subseteq \mathscr{S}(\mathcal{V}) : |\mathcal{W}| = B\} & \text{if } |\mathscr{S}(\mathcal{V})| > B \ . \end{array} \right.$$

Note that if the total number of successors of all vertices in $\mathcal{V}$ is less than or equal to the budget B, then $\mathcal{V}$ has only *one* hyperchild, namely $\mathscr{S}(\mathcal{V})$; otherwise, it has $\binom{|\mathscr{S}(\mathcal{V})|}{B}$ hyperchildren, all containing exactly B vertices. We define the cost $c(\mathcal{V})$ of a hypervertex to be the sum of the costs of the individual vertices, and the successors $\mathscr{S}(\mathcal{V})$ to be the union of the successors of the individual vertices in $\mathcal{V}$.

The stochastic enumeration algorithm is shown in Algorithm 10.4.1 and is very similar in form to Algorithm 10.3.1. The crucial difference is that it constructs a random sequence $\mathcal{X}_0, \mathcal{X}_1, \ldots,$ of hypervertices (collections of vertices at the same levels $1, 2, \ldots$) rather than vertices. Note that for $B = 1$ the SE algorithm reduces to Knuth's algorithm.

Algorithm 10.4.1: Stochastic Enumeration Algorithm for Estimating the Cost of a Tree

---

input : Tree $\mathscr{T}$ with root u and a budget $B \geqslant 1$.
output: Estimator of the total cost of $\mathscr{T}$.

1   $D \leftarrow 1, \mathcal{X} = \{u\}, C \leftarrow c(u)$
2   **while** $|\mathscr{S}(\mathcal{X})| > 0$ **do**
3      $D \leftarrow \frac{|\mathscr{S}(\mathcal{X})|}{|\mathcal{X}|} D$
4      $\mathcal{X} \leftarrow$ uniformly selected successor (hyperchild) in $\mathscr{H}(\mathcal{X})$
5      $C \leftarrow C + \frac{c(\mathcal{X})}{|\mathcal{X}|} D$
6   **return** C

Using similar reasoning as for Knuth's estimator, the SE estimator can be shown to be unbiased; see also [16]. In practice, SE is repeated a number of times to assess the accuracy of the estimate via the corresponding relative error and confidence interval.

Because both Knuth's algorithm and SE use importance sampling, the resulting estimator can misjudge the cost of a tree if the distribution of vertex costs is very skewed, even though the estimator is unbiased, as poignantly demonstrated by the hairbrush tree example in Figure 10.5. By increasing the budget, SE is better able to deal with such trees. Indeed, for the hairbrush tree a budget of size $B = 2$ suffices to produce a zero-variance estimator! However, for some trees Knuth's method may already provide a reliable estimate of the tree cost, so that there is no need to increase B, as the computational effort grows linearly with B. Our advice is to gradually increase the budget, starting with $B = 1$, until the estimates become stable (no longer significantly change).

Example 10.4 illustrates the workings of Algorithm 10.4.1. Note that while Knuth's algorithm operates on vertices at each step, SE operates on collections of vertices, that is, on hypervertices.

■ **EXAMPLE 10.4**

Consider the tree $\mathscr{T}$ displayed on the left-hand side of Figure 10.6. Suppose that we wish to count the total number of vertices; so $c(v) = 1$ for all $v \in \mathscr{V}$.

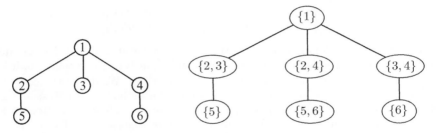

Figure 10.6: Applying SE to the tree on the left with a budget $B = 2$ gives one path through the hypertree on the right.

We step through Algorithm 10.4.1 using a budget $B = 2$. For clarity we index the objects D, $\mathcal{X}$, and C according to the iteration number, starting at iteration 0. The SE algorithm starts at the root and finds that there are 3 successors. Hence, we set $D_0 \leftarrow 1$, $\mathcal{X}_0 \leftarrow \{1\}$, and $C_0 \leftarrow 1$. Next, we determine $\mathscr{S}(\mathcal{X}_0)$ — the set of all successors of $\mathcal{X}_0$, which is $\{2, 3, 4\}$. Since the budget is 2, the hyperchildren of $\mathcal{X}_0$ are all subsets of size 2 of $\{2, 3, 4\}$, giving $\mathscr{H}(\mathcal{X}_0) = \{\{2, 3\}, \{2, 4\}, \{3, 4\}\}$.

D_1 is obtained by multiplying D_0 with $|\mathscr{S}(\mathcal{X}_0)|/|\mathcal{X}_0|$, which may be viewed as the "average degree" of hypervertex $\mathcal{X}_0$. In this case $\mathcal{X}_0$ only contains one vertex, which has degree 3, so $D_1 \leftarrow \frac{3}{1} \cdot 1 = 3$. One of the hyperchildren is chosen at random, say $\mathcal{X}_1 \leftarrow \{2, 3\}$. C_1 is set to $C_0 + \frac{c(\mathcal{X}_1)}{|\mathcal{X}_1|} D_1 = 1 + \frac{2}{2} \cdot 3 = 4$.

At the second iteration D_2 is obtained by multiplying D_1 with the average degree of $\{2, 3\}$, which is $1/2$, since in the original tree the degree of vertex 2 is 1 and the degree of vertex 3 is 0. So $D_2 \leftarrow \frac{1}{2} \cdot 3 = \frac{3}{2}$. As $\mathcal{X}_1$ only has one hyperchild, $\mathcal{X}_2 \leftarrow \{5\}$, and C_2 becomes to $4 + \frac{1}{1} \cdot \frac{3}{2} = 11/2$. The algorithm then terminates, as there are no more hyperchildren.

Each run of Algorithm 10.4.1 will choose one of 3 possible paths through the hypertree. Below we consider all the paths, the obtained estimator and the corresponding probabilities:

1. $\{1\} \rightarrow \{2,3\} \rightarrow \{5\}$, $C = 1 + 3 + 3/2 = 11/2$, with probability $1/3$.

2. $\{1\} \rightarrow \{2,4\} \rightarrow \{5,6\}$, $C = 1 + 3 + 3 = 7$, with probability $1/3$.

3. $\{1\} \rightarrow \{3,4\} \rightarrow \{6\}$, $C = 1 + 3 + 3/2 = 11/2$, with probability $1/3$.

It follows that

$$\mathbb{E}\,[C] = \frac{1}{3}\left(\frac{11}{2} + 7 + \frac{11}{2}\right) = 6$$

and

$$\mathrm{Var}(C) = \frac{1}{3}\left[\left(\frac{11}{2}\right)^2 + 7^2 + \left(\frac{11}{2}\right)^2\right] - 6^2 = \frac{1}{2}.$$

The reader may verify (see Problem 10.2) that for a budget $B = 1$ the variance of C is 2. Thus, although running two paths in parallel doubles the computation time (compared with running a single path), the variance is reduced by a factor of 4. For a budget $B = 3$ the SE algorithm will even give the zero-variance estimator, $C = 6$, as the hypertree consists, in this case, of a single path $\{1\} \rightarrow \{2,3,4\} \rightarrow \{5,6\}$.

10.4.1 Combining SE with Oracles

When running SE one would like to generate only paths in the tree that are useful for estimating the cost of the tree. Suppose that v is a tree vertex that is visited during the execution of the SE algorithm. If it is known that any a path continued via a successor w of v will have cost 0, then it makes sense to delete w from the list of successors of v. The YES–NO decision of keeping w or not can often be calculated quickly via a so-called *oracle*. Example 10.5 illustrates the idea of using SE with an oracle.

■ **EXAMPLE 10.5**

We can also use SE or Knuth's algorithm to count the number of non-intersecting paths, as we do from vertex 1 to 7 in the directed graph of Figure 10.7. Starting from root 1, we construct a tree where at the first level there are 3 potential successors. However, we can quickly establish that paths going through vertex 4 can never reach vertex 7. Hence vertex 4 should be eliminated from the successor list at iteration 1. Similarly 4 should be eliminated at various other possible stages of the algorithm, reducing the original search tree considerably; see Figure 10.8.

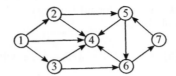

Figure 10.7: A directed graph.

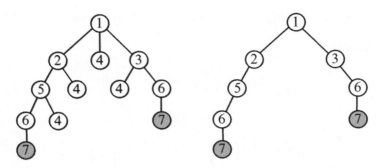

Figure 10.8: Original search tree and the reduced tree obtained by using oracles.

Interestingly, there are many problems in computational combinatorics for which the decision making is easy (polynomial) but the counting is hard. As mentioned in Section 9.8, finding out how many paths there are between two vertices in a graph is #P-complete, while the corresponding YES–NO decision problem (is there such a path?) is easy and can be solved via fast algorithms such as Dijkstra's algorithm [1] or simply via BFS or DFS. Other examples include finding a solution to a 2-SAT problem (easy) versus finding the number of such solutions (hard), and finding perfect matchings in a bipartite graph (easy) versus finding the number of such perfect matching (hard).

Our strategy is thus to incorporate fast polynomial decision-making oracles into SE to solve #P-complete problems.

10.5 APPLICATION OF SE TO COUNTING

In this section we demonstrate three applications of SE involving difficult (#P-complete) counting problems. For each case we provide a worked example, indicate which oracle can be used, and carry out various numerical experiments that show the effectiveness of the method.

10.5.1 Counting the Number of Paths in a Network

Here we use SE to count the number of paths in a network with a fixed source and sink. We start with a toy example.

■ EXAMPLE 10.6   Extended Bridge

Our objective is to apply SE to count how many paths are there from s to t in the graph of Figure 10.9. A quick inspection shows there are 7 such paths:

$$(e_1, e_3, e_5, e_7), \ (e_1, e_3, e_6), \ (e_1, e_4, e_5, e_6), (e_1, e_4, e_7) \\ (e_2, e_3, e_4, e_7), \ (e_2, e_5, e_7), \ (e_2, e_6) \,. \tag{10.4}$$

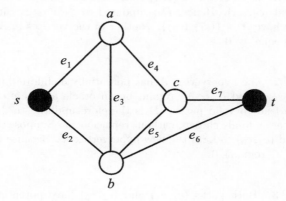

Figure 10.9: Extended bridge.

The associated counting tree is given in Figure 10.10. The SE algorithm only constructs a part of the tree, by running in parallel B paths until a leaf vertex is reached. Each vertex at level t corresponds to a (partial) path $(x_1, x_2, \ldots, x_t)$ through the graph, starting from s. The bottom-left leaf vertex, for example, corresponds to the path (e_1, e_3, e_5, e_7). The leaf vertices have cost 1; all other vertices have cost 0. The root vertex (path of length 0) has been left out in the notation of a path, but could be denoted by the empty vector ( ).

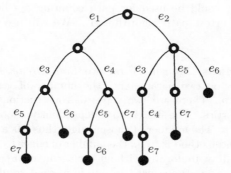

Figure 10.10: Cunting tree corresponding to the extended bridge.

Stepping through a typical run of Algorithm 10.4.1, with a budget $B = 2$, could give the following result:

1. **Initialization**. Initialize D to 1, the cost C to 0, and let $\mathcal{X}$ contain the root of the tree, ( ).

2. **Iteration 1**. The root vertex has potentially 2 children, (e_1) and (e_2) of length 1. We use an oracle to verify that there indeed exists paths of the form $(e_1, \ldots)$ and $(e_2, \ldots)$. Hence, D is updated to $\frac{2}{1} \cdot 1 = 2$. Since the budget $B = 2$, we have $\mathcal{X} = \{(e_1), (e_2)\}$. Neither of the vertices in $\mathcal{X}$ are leaves, so C remains equal to 0.

3. **Iteration 2**. The hypervertex $\mathcal{X}$ has potentially 5 children, $(e_1, e_3), (e_1, e_4)$, $(e_2, e_3), (e_2, e_5)$, and (e_2, e_6). Again, by an oracle all these children can be verified to lead to valid paths. D is therefore updated to $\frac{5}{2} \cdot 2 = 5$. Two children are randomly chosen, without replacement. Suppose that we selected (e_1, e_4) and (e_2, e_5) to form the new hypervertex $\mathcal{X}$. Because the vertices are not leaves, C remains 0.

4. **Iteration 3**. Both paths (e_1, e_4) and (e_2, e_5) have potentially 2 children. However, of these 4 children, only 3 lead to valid paths. In particular, (e_2, e_5, e_4) is not valid, and is therefore discarded by the oracle. Hence, D is updated to $\frac{3}{2} \cdot 5 = 15/2$. The new hypervertex $\mathcal{X}$ is formed by selecting 2 of the 3 children at random, without replacement; say (e_1, e_4, e_5) and (e_2, e_5, e_7) are chosen. Since the last vertex is a leaf, C becomes $0 + \frac{1}{2} \cdot \frac{15}{2} = \frac{15}{4}$.

5. **Iteration 4**. The oracle determines that there is now only one valid child of $\mathcal{X}$, namely (e_1, e_4, e_5, e_6). Hence, D is updated to $\frac{1}{2} \cdot \frac{15}{2} = \frac{15}{4}$, $\mathcal{X} = \{(e_1, e_4, e_5, e_6)\}$, and since the child vertex is a leaf, C becomes $\frac{15}{4} + \frac{1}{1} \cdot \frac{15}{4} = 7.5$. Because there are no further children, the algorithm terminates.

Typical oracles that could be used for path counting are breadth-first search (BFS) or Dijkstra's shortest path algorithm [1]. We will use the former in our numerical experiments.

10.5.1.1 Numerical Results   We run SE on two *Erdös–Rényi random graphs* [5]. Such a graph depends on two parameters: the number of vertices, n, and the *density* of the graph, p. There are two types of constructions giving graphs of slightly different properties. The first is to independently choose each of the $\binom{n}{2}$ edges with probability p. The number of chosen edges thus has a $\mathrm{Bin}(n(n-1)/2, \, p)$ distribution. The second method is to fix the number of edges to $m = np$ (assuming this is an integer) and draw uniformly without replacement m edges from the total of $\binom{n}{2}$ possible edges. The graphs used in our numerical results were generated using the latter method.

Model 1. The first graph, with $n = 24$ vertices, is taken from [12]. The adjacency matrix of the graph is

$$A = \begin{pmatrix}
0&0&0&1&0&0&0&0&0&0&0&1&0&0&1&1&1&0&1&1&0&0&0&0\\
0&0&0&0&0&0&0&0&0&0&1&0&1&1&1&0&0&0&0&0&0&1&0&0\\
0&0&0&0&1&0&1&0&0&0&0&1&0&0&0&0&0&0&0&0&0&0&0&0\\
1&0&0&0&0&0&1&0&0&0&0&1&0&0&0&0&1&0&1&0&0&0&0&0\\
0&0&1&0&0&0&0&0&0&0&0&0&1&1&0&1&0&0&0&0&0&0&0&0\\
0&0&0&0&0&0&1&0&0&0&1&0&1&0&1&0&0&0&0&0&0&0&0&0\\
0&0&1&1&0&1&0&0&1&0&0&1&0&1&0&0&0&0&0&0&0&0&0&0\\
0&0&0&0&0&0&0&1&1&0&0&0&1&0&0&1&0&0&0&0&0&0&0&0\\
0&0&0&0&0&1&1&0&0&0&0&0&0&1&1&0&0&0&0&1&0&0&0&0\\
0&1&0&0&0&1&0&0&0&0&0&0&0&0&0&1&1&0&0&1&1&0&0&0\\
1&0&0&1&0&0&1&0&0&0&0&1&0&0&0&0&0&0&0&0&0&0&1&0\\
0&1&1&0&0&1&1&0&0&0&1&0&0&0&0&0&0&0&1&0&1&0&1&0\\
0&1&0&0&0&0&0&1&0&0&0&0&0&0&0&0&0&0&0&0&0&0&0&0\\
1&1&0&0&1&0&0&0&1&0&0&0&0&0&0&0&0&0&0&0&0&0&0&0\\
1&0&0&0&1&0&0&0&1&0&1&0&0&0&0&0&0&0&1&0&0&1&1&1\\
1&0&0&1&1&0&1&1&1&1&0&1&0&0&0&0&0&0&0&1&1&0&1&0\\
0&0&0&0&0&0&0&0&0&0&0&0&0&0&0&1&1&0&0&0&0&1&1&0\\
0&0&0&1&1&0&0&0&0&0&0&1&1&0&1&0&0&0&0&0&0&0&0&0\\
1&0&0&0&0&0&0&0&0&1&0&0&0&1&1&0&0&0&0&0&1&0&0&0\\
0&0&0&0&0&0&0&1&0&1&0&1&0&0&0&0&0&0&0&0&0&0&0&0\\
0&1&0&0&0&0&0&0&1&0&0&0&0&0&0&0&0&0&0&0&0&0&0&0\\
0&0&0&0&0&0&0&0&1&1&0&0&1&1&1&0&1&0&0&0&0&0&0&1\\
0&0&0&0&0&0&0&0&0&0&0&0&0&0&1&0&0&0&0&0&0&1&0\\
\end{pmatrix}.$$

The objective is to estimate the total number of paths between vertex 1 and 24. In this case the exact solution is 1892724.

Ten independent runs of the SE Algorithm 10.4.1, using a budget of $B = 50$ and $N = 400$ repetitions per run, gave the following results, times 10^6:

$$1.81, 1.93, 1.98, 1.82, 1.90, 1.94, 1.86, 1.87, 1.90, 1.92, 1.89,$$

which gives an estimated relative error per run (i.e., the sample standard deviation divided by the sample mean) of 0.0268. The overall result of the 10 runs gives an estimate of $1.8927 \cdot 10^6$ with an estimated relative error $0.0268/\sqrt{10} \approx 0.0085$. The CPU time was about 4 seconds per run.

Model 2. For our second model, we randomly generated an Erdös–Rényi graph with $n = 200$ vertices and $m = 200$ edges. Ten independent runs of the SE Algorithm 10.4.1, using a budget of $B = 100$ and $N = 500$ repetitions per run, gave the following results, times 10^7:

$$7.08, 7.39, 7.28, 7.38, 7.12, 7.05, 7.22, 7.48, 7.47, 7.40, 7.29,$$

which gives an estimated relative error per run of 0.021.

For $B = 1$ and $N = 50000$, we obtained instead the estimates, times 10^7:

$$7.40, 7.37, 7.38, 7.30, 7.20, 7.40, 7.15, 7.31, 7.27, 7.34, 7.31,$$

which gives an estimated relative error per run of 0.011. This gives confidence that Knuth's algorithm works well on this instance.

10.5.2   Counting SATs

To show how SE works for counting SATs, we illustrate it first for the 2-SAT toy problem in Example 10.2, concerning the CNF formula $(l_1 \vee l_2) \wedge (l_2 \vee l_3) \wedge (l_2 \vee l_4)$.

■ **EXAMPLE 10.7   Example 10.2 (Continued)**

The SE algorithm generates B paths through the binary counting tree of Figure 10.3, while also employing an oracle to see if a partial vector $\mathbf{x}_t$ can lead to a valid assignment. In effect, by using an oracle, the SE generates paths in the reduced tree given in Figure 10.11. Left branching again corresponds to adding a 0, and right branching to adding a 1.

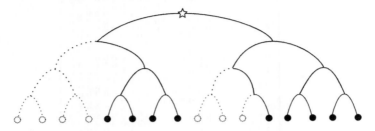

Figure 10.11: Reduced tree for the 2-SAT counting problem

Using a budget $B = 2$, a typical run of the SE algorithm (with an oracle) could be as follows. Note that we leave out the brackets and commas in the notation for the binary vectors, so, for example, 010 is shorthand for (0,1,0).

1. **Initialization.** Initialize D to 1, the cost C to 0, and let $\mathcal{X}$ contain the root of the tree.

2. **Iteration 1.** The root vertex has potentially 2 children: 0 and 1. We apply the oracle twice to verify that there indeed exist assignments of the form $0\cdots$ and $1\cdots$. Hence D is updated to $\frac{2}{1} \cdot 1 = 2$. Since the budget $B = 2$, we have $\mathcal{X} = \{0, 1\}$. Neither of the vertices in $\mathcal{X}$ are leaves, so C remains equal to 0.

3. **Iteration 2.** The hypervertex $\mathcal{X}$ has potentially 4 children: $00, 01, 10, 11$. But the oracle rules out 00, so 3 children remain. D is therefore updated to $\frac{3}{2} \cdot 2 = 3$, and two children are randomly chosen, without replacement; say 01 and 10. Because neither are leaves, C remains 0.

4. **Iteration 3.** We again have 4 potential children, but 100 is ruled out, leaving 010, 011, and 101. Hence, D is updated to $\frac{3}{2} \cdot 3 = 9/2$. Suppose that we select $\mathcal{X} = \{010, 101\}$. Again C remains 0.

5. **Iteration 4.** After applying the oracle, the children of $\mathcal{X}$ are found to be $0100, 0101$, and 1011. All of these are leaf vertices with cost 1. D is updated to $\frac{3}{2} \cdot \frac{9}{2} = \frac{27}{4}$ and C becomes $0 + \frac{2}{2} \cdot \frac{27}{4} = 6.75$. Since there are no further children, the algorithm terminates.

Although decision making for K-SATs is NP-hard for $K \geqslant 3$, there are several very fast heuristics for this purpose. The most popular one is the famous DPLL solver [2, 3]. The basic backtracking algorithm runs by choosing a literal, assigning a truth value to it, simplifying the formula, and then recursively checking if the simplified formula is satisfiable. If this is the case, the original formula is satisfiable; otherwise, the same recursive check is done assuming the opposite truth value. This is known as the splitting rule, as it splits the problem into two simpler sub-problems. The simplification step essentially removes all clauses that become true under the assignment from the formula, and all literals that become false from the remaining clauses.

For the 2-SAT decision problem DPLL has polynomial complexity in the number of clauses. In our experiments we used the freely available `minisat` solver from `http://minisat.se/`, which is based on DPLL.

10.5.2.1 Numerical Results Here, we present numerical results for several SAT models.

We first revisit the 3-SAT (75×325) problem from Section 9.9.1.1, which has 2258 solutions. Running 10 independent replications of SE, with $B = 20$ and $N = 100$, produced the following estimates:

$$2359.780, 2389.660, 2082.430, 2157.850, 2338.100,$$
$$2238.940, 2128.920, 2313.390, 2285.910, 2175.790.$$

This gives an estimated relative error per run of 0.047, and the combined estimate is 2247 with an estimated relative error of 0.015.

Comparison with other methods. We compared the SE method with splitting and also with `SampleSearch` [6, 7]. To make a fair comparison, each method was run for the same amount of time. This also determines the number of replications. For example, if the time limit is 300 seconds, and a run of SE takes 3 seconds, then 100 independent runs can be performed.

We consider three 3-SAT instances:

1. 75×325. This is the instance given above.

2. 75×270. This is the 75×270 instance with the last 55 clauses removed.

3. 300×1080. This is the 75×270 instance "replicated" 4 times. Specifically, its clause matrix is given by the block-diagonal matrix

$$B = \begin{pmatrix} A & O & O & O \\ O & A & O & O \\ O & O & A & O \\ O & O & O & A \end{pmatrix},$$

where A is the clause matrix corresponding to the 75×270 instance.

Regarding the choice of parameters, for SE we used $B = 20$ for the 75×325 instance, $B = 100$ for the 75×270 instance and $B = 300$ for the 300×1080 instance. For splitting, we used $N = 10,000$ and $\varrho = 0.1$ for all instances. `SampleSearch` does not require specific input parameters.

The results of the comparison are given in Table 10.1. For each of the methods, we list the estimate and the estimated relative error.

Table 10.1: Comparison of the efficiencies `SampleSearch` (SaS), stochastic enumeration (SE), and splitting (Split).

Instance	Time	SaS	SaS RE	SE	SE RE	Split	Split RE
75×325	137 sec	2212	2.04E-02	2248.8	9.31E-03	2264.3	6.55E-02
75×270	122 sec	1.32E+06	2.00E-02	1.34E+06	1.49E-02	1.37E+06	3.68E-02
300×1080	1600 sec	1.69E+23	9.49E-01	3.32E+24	3.17E-02	3.27E+24	2.39E-01

For the 300×1080 problem, for example, SE has a relative error that is 30 times smaller than `SampleSearch`, and 7 times smaller than splitting. In terms of computational times, this means a speedup by factors of 900 and 49, respectively.

10.5.3   Counting the Number of Perfect Matchings in a Bipartite Graph

Here, we deal with the application of SE to the estimation of the number of perfect matchings in a bipartite graph. Recall from Section 9.9.3 that in such a graph the vertices can be partitioned into two independent sets such that each edge has one vertex in each of the sets. The adjacency matrix can therefore be decomposed as

$$B = \begin{pmatrix} O_{p \times p} & A \\ A^\mathsf{T} & O_{q \times q} \end{pmatrix}$$

for some $p \times q$ matrix B (called the *biadjacency* matrix) and zero-matrices $O_{p \times p}$ and $O_{q \times q}$. A matching is a subset of the edges with the property that no two edges share the same vertex. For the case where $p = q$, a perfect matching is a matching of all the vertices. The number of perfect matchings in a bipartite graph is identical to the permanent of the biadjacency matrix A; see Section 9.9.3.

A natural implementation of SE involves ordering the vertices from 1 to p and constructing partial vectors of the form $(x_1, x_2, \ldots, x_t)$, where $x_i \neq x_j, i \neq j$. This vector identifies the first t edges (i, x_i), $i = 1, \ldots, t$ of the perfect matching.

■ **EXAMPLE 10.8   Bipartite Graph**

Consider the bipartite graph in Figure 10.12, whose biadjacency is given on the left of the graph.

$$A = \begin{pmatrix} 1 & 1 & 1 & 1 \\ 1 & 1 & 0 & 1 \\ 1 & 0 & 1 & 0 \\ 1 & 0 & 0 & 1 \end{pmatrix},$$

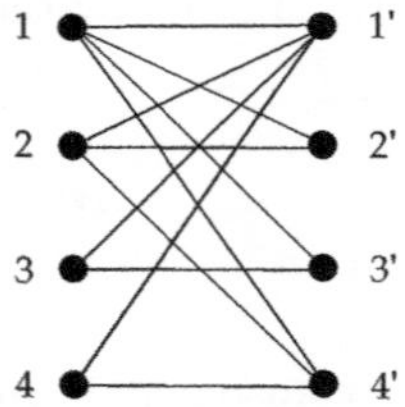

Figure 10.12: Bipartite graph with its biadjacency matrix.

The associated counting tree is given in Figure 10.13.

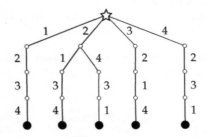

Figure 10.13: Counting tree associated with the bipartite graph in Figure 10.12.

The tree is highly balanced, and hence Knuth's algorithm (SE with a budget of $B = 1$) will perform well. A typical run is as follows:

1. **Initialization**. Initialize D to 1, the cost C to 0, and let X be the root of the tree.

2. **Iteration 1**. The root vertex has potentially 4 children: $(1), (2), (3)$, and (4). We apply the oracle four times to verify that there indeed exist permutations (matchings) for all of these cases. Hence D is updated to $\frac{4}{1} \cdot 1 = 4$. Let X be one of the children, selected at random, say $X = (2)$. C remains equal to 0.

3. **Iteration 2**. X has potentially 3 children (permutation vectors of length 2): $(2, 1), (2, 2)$ and $(2, 4)$. However, the permutation vector $(2, 2)$ is ruled out, because it corresponds to the edges $(1, 2)$ and $(2, 2)$, which violates the requirement that no two edges can have the same end vertices. D is therefore updated to $2 \cdot 4 = 8$, and X becomes a randomly chosen child, say $(2, 1)$. C remains 0.

4. **Iteration 3**. $X = (2, 1)$ only has one child $(2, 1, 3)$, which is not a leaf. So D and C remain unaltered, and X becomes $(2, 1, 3)$.

5. **Iteration 4**. Again, there is only one child: $(2, 1, 3, 4)$, which is a leaf. D remains equal to 8, and C becomes $0 + 8 = 8$. The algorithm then terminates.

We see that the algorithm returns $C = 8$ with probability $1/4$ and $C = 4$ with probability $3/4$, which has an expectation of 5 — the total number of perfect matchings.

For our oracle we use the well-known *Hungarian method* [11] to check (in polynomial time) if a partial matching can be extended to a perfect matching.

10.5.3.1 Numerical Results We present here numerical results for two models, one small and one large.

Model 1 (Small Model): Consider the biadjacency matrix A given in Section 9.9.3.1 on Page 333. The number of perfect matchings for the corresponding bipartite graph (i.e., the permanent of A) is $|\mathscr{X}^*| = 266$ obtained using full enumeration.

To assess the performance of SE, we repeated Algorithm 10.4.1 ten times with parameter values ($B = 50$ and $N = 10$), giving estimates

$$264.21, 269.23, 270.16, 268.33, 272.10, 259.81, 271.62, 269.47, 264.86, 273.77 .$$

The overall mean is 268.4, with an estimated relative error of 0.005. The estimated relative error per run is $\sqrt{10}$ larger; that is, 0.0158. Each run took about 2 seconds.

Model 2 (Large Model): Here we generate a random 100×100 biadjacency matrix $A = (a_{ij})$ by setting $a_{ij} = 1$ for 200 uniformly chosen index pairs (i, j). The other 9800 entries are set to 0. We perform two SE experiments: one with $B = 100$ and $N = 100$; the other with $B = 1$ and $R = 10000$. For the first case the results were (times 10^7)

$$4.06, 3.95, 3.63, 3.72, 3.93, 3.93, 3.48, 3.70, 3.80, 3.90,$$

with an overall mean of $3.81 \cdot 10^7$ and an estimated relative error per run of 0.063 (overall 0.0147). The average CPU time per run was 340 seconds. The second case (Knuth's estimator) gave the estimates (times 10^7):

$$3.48, 2.98, 4.46, 3.48, 3.43, 3.97, 5.02, 4.27, 3.72, 3.61,$$

with an average CPU time per run of 400 seconds. The overall mean is $3.84 \cdot 10^7$ and the estimated relative error per run is 0.1555 (overall 0.0492).

Clearly, running 100 paths in parallel is beneficial here, as the relative error is less than half of that of Knuth's estimator. Note that after compensating for the increased budget (by taking N 100 times larger) Knuth's estimator actually takes longer to run.

10.6   APPLICATION OF SE TO NETWORK RELIABILITY

Consider a network reliability problem as discussed in Sections 5.4.1 and 9.7 where all edge failure probabilities are equal to some small number q. Under this simplified setting the *Spectra* method gives an efficient approach to calculate the network unreliability $\bar{r}(q)$. We briefly describe the Spectra approach.

Let the network edges (links) be labeled $1, \ldots, n$. Edges can be working (*up*) or failed (*down*). Similarly, the network is either *UP* or *DOWN*. Suppose, that initially all links are down and thus the network is in the *DOWN* state, and let $\boldsymbol{\pi} = (i_1, \ldots, i_n)$ be a permutation of edges. Given the permutation $\boldsymbol{\pi}$, start "repairing" edges (change the edges state from *down* to *up*) moving through the permutation from left to right and check the *UP/DOWN* state of the network after each step. Find the index k of the first edge i_k, $k = 1, \ldots, n$, for which the network switches from *DOWN* to *UP*. This index k is called the *construction anchor* of $\boldsymbol{\pi}$ and is denoted by $a^{\mathrm{C}}(\boldsymbol{\pi})$. The total number of edge permutations that have construction anchor k is denoted by $A^{\mathrm{C}}(k)$. The probability vector

$$\frac{1}{n!} \left(A^{\mathrm{C}}(1), \ldots, A^{\mathrm{C}}(n) \right)$$

is called the *construction spectra*, or simply *C-spectra* of the network. A similar notion is that of *destruction spectra* or *D-spectra*, which is obtained by starting with a network in which all links are up and in which edges are destroyed in the order of the permutation. In particular, given an edge permutation $\boldsymbol{\pi} = (i_1, \ldots, i_n)$, let a^{D} be the first edge i_j for which the system becomes *DOWN* and let $A^{\mathrm{D}}(k)$ be the total number of such *destruction anchors* of level k. Then the D-Spectra is defined as the probability vector

$$\frac{1}{n!} \left(A^{\mathrm{D}}(1), \ldots, A^{\mathrm{D}}(n) \right) .$$

The D-spectra can be obtained from the C-spectra (and vice versa) by reversing the order of the vector: $A^{\mathrm{D}}(k) = A^{\mathrm{C}}(n-k+1)$, $k = 1, \ldots, n$. Namely, by reversing each edge permutation $(i_1, \ldots, i_n)$ to $(i_n, \ldots, i_1)$, we see that the number of permutations for which the network switches from *UP* to *DOWN* after k steps is equal to the number of permutations for which the network switches from *DOWN* to *UP* after $n - k + 1$ steps.

Once the D- or C-spectra is available, one can directly calculate the network unreliability $\bar{r}(q)$. Let us define a *failure set* to be a set of edges such that their failure causes the network to be *DOWN*, and denote by $\mathcal{N}_k$ the number of failure sets of size k. It is readily seen that

$$\mathcal{N}_k = \binom{n}{k} \underbrace{\frac{1}{n!} \sum_{j=1}^{k} A^{\mathrm{D}}(j)}_{F_k} . \tag{10.5}$$

This statement has a simple combinatorial explanation: F_k is the fraction of all failure sets of size k among all subsets of size k taken from the set of n components. The vector $(F_1, \dots, F_N)$ is called the *cumulative D-spectra*.

The probability that in a failure set of size k all of its edges are *down* while all other edges are *up* is $q^k(1-q)^{n-k}$. Because the network is *DOWN* if and only if there is such a failure set, it follows that

$$\bar{r}(q) = \sum_{k=1}^{n} \mathcal{N}_k\, q^k (1-q)^{n-k}. \tag{10.6}$$

Thus to calculate the network reliability for any q it suffices to calculate the D-spectra. The latter can be treated as a tree-counting problem, as illustrated in the following example.

■ **EXAMPLE 10.9**

Consider the simple reliability network with 4 links in Figure 10.14. The system is *UP* if vertices s and t are connected by functioning links.

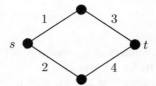

Figure 10.14: Simple network.

To determine the C-spectra, we can construct a permutation tree as in Figure 10.15. Each path in the tree corresponds to a permutation $(i_1, \dots, i_k)$ up to the corresponding construction anchor k. Let the cost of the leaf that ends this path be equal to $1/(n(n-1)\cdots(n-k)) = (n-k)!/n!$. Then the k-th component of the C-spectra, $A^{\mathrm{C}}(k)/n!$, is equal to the total cost of the leaves in level k of the tree. For example, the 4 leaves at level $k=2$ all have cost $1/12$, and the 16 leaves at level $k=3$ all have cost $1/24$. It is clear that none of the construction anchors are 1 or 4. It follows that the C-spectra is given by $(0, 1/3, 2/3, 0)$, the D-spectra by $(0, 2/3, 1/3, 0)$, and the cumulative D-spectra by $(0, 2/3, 1, 1)$.

Using (10.6), the unreliability is given by

$$\bar{r}(q) = 6 \cdot \frac{2}{3} \cdot q^2(1-q)^2 + 4 \cdot 1 \cdot q^3(1-q)^2 + 1 \cdot 1 \cdot q^4 = 4q^2 - 4q^3 + q^4 = (1 - (1-q)^2)^2 .$$

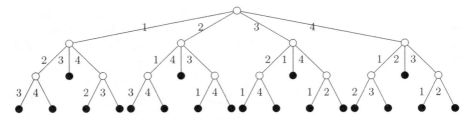

Figure 10.15: Permutation tree of the simple network.

For a general network, estimating the unreliability thus involves the following steps:

1. Construct the permutation tree of the network, where each path ends in a construction anchor point.

2. For each level k set the cost of each leaf to $(n-k)!/n!$, and add all these costs to find the k-th C-spectra.

3. From the C-spectra derive the D-spectra and the cumulative D-spectra.

4. Compute the unreliability via (10.6).

For large n constructing the whole permutation tree is infeasible. Instead, SE can be used to *estimate* the spectra, by estimating the cost of the tree at each level $k = 1, \ldots, n$. In particular, Algorithm 10.4.1 can be directly applied, with the obvious modification that for each level (iteration) the estimated cost at that level, $c(\mathcal{X})/|\mathcal{X}|$ has to be returned.

10.6.1 Numerical Results

In this section we present a numerical comparison between the PMC Algorithm 5.4.2 and the SE Algorithm 10.4.1 for two models.

Recall that the crucial parameter of the SE algorithm is its budget B. Clearly, increasing B means an increase in computation effort (linear in B). In both examples we choose $B = 10$ and repeat the algorithm for $N = 1000$ times. We find numerically that this parameter selection is sufficient for reliable estimation of the spectra. Based on numerous experiments with different models, our general advice for choosing the budget is as follows: Start from small budget, say $B \leqslant 5$ and increase it gradually. As soon as the rare-event probability of interest stops to increase, fix this budget and perform the main experiment. This follows from the fact that the SE tends to underestimate the rare-event probabilities.

Model 1 (Dodecahedron Network): Consider the dodecahedron network in Figure 9.4. Suppose that the network is functioning if vertices 10 and 19 are connected. For this example the estimation of the components of the D-spectra is not a rare-event estimation problem. For such models PMC and SE deliver accurate and comparable estimations for small failure probabilities.

We run the SE algorithm with $N = 1000$ and $B = 10$ and the spectra algorithm with $N = 350000$. The running time for both algorithms is about 4.7 seconds. In this example the minimal D-spectra component has the value $\approx 5 \cdot 10^{-4}$, so both algorithms deliver excellent performances in a reasonable amount of time. Table 10.2 provides typical values obtained during the algorithm's execution, and Table 10.3 summarizes the corresponding probability that the network is *DOWN*.

Table 10.2: PMC and SE spectra estimations for the dodecahedron graph.

Cum. D-spectra	SE	PMC	Cum. D-spectra	SE	PMC
$F(1)$	0	0	$F(16)$	$7.90 \cdot 10^{-1}$	$7.91 \cdot 10^{-1}$
$F(2)$	0	0	$F(17)$	$8.48 \cdot 10^{-1}$	$8.49 \cdot 10^{-1}$
$F(3)$	$5.01 \cdot 10^{-4}$	$4.90 \cdot 10^{-4}$	$F(18)$	$8.91 \cdot 10^{-1}$	$8.93 \cdot 10^{-1}$
$F(4)$	$2.22 \cdot 10^{-3}$	$2.18 \cdot 10^{-3}$	$F(19)$	$9.24 \cdot 10^{-1}$	$9.26 \cdot 10^{-1}$
$F(5)$	$6.30 \cdot 10^{-3}$	$6.33 \cdot 10^{-3}$	$F(20)$	$9.49 \cdot 10^{-1}$	$9.51 \cdot 10^{-1}$
$F(6)$	$1.41 \cdot 10^{-2}$	$1.43 \cdot 10^{-2}$	$F(21)$	$9.67 \cdot 10^{-1}$	$9.68 \cdot 10^{-1}$
$F(7)$	$2.90 \cdot 10^{-2}$	$2.86 \cdot 10^{-2}$	$F(22)$	$9.80 \cdot 10^{-1}$	$9.80 \cdot 10^{-1}$
$F(8)$	$5.53 \cdot 10^{-2}$	$5.45 \cdot 10^{-2}$	$F(23)$	$9.89 \cdot 10^{-1}$	$9.88 \cdot 10^{-1}$
$F(9)$	$9.80 \cdot 10^{-2}$	$9.68 \cdot 10^{-2}$	$F(24)$	$9.94 \cdot 10^{-1}$	$9.94 \cdot 10^{-1}$
$F(10)$	$1.65 \cdot 10^{-1}$	$1.65 \cdot 10^{-1}$	$F(25)$	$9.97 \cdot 10^{-1}$	$9.97 \cdot 10^{-1}$
$F(11)$	$2.62 \cdot 10^{-1}$	$2.60 \cdot 10^{-1}$	$F(26)$	$9.99 \cdot 10^{-1}$	$9.99 \cdot 10^{-1}$
$F(12)$	$3.81 \cdot 10^{-1}$	$3.79 \cdot 10^{-1}$	$F(27)$	1	1
$F(13)$	$5.03 \cdot 10^{-1}$	$5.03 \cdot 10^{-1}$	$F(28)$	1	1
$F(14)$	$6.19 \cdot 10^{-1}$	$6.18 \cdot 10^{-1}$	$F(29)$	1	1
$F(15)$	$7.15 \cdot 10^{-1}$	$7.15 \cdot 10^{-1}$	$F(30)$	1	1

Table 10.3: PMC and SE reliability estimations for the dodecahedron graph.

q	SE		PMC	
	$\widehat{\overline{r}(q)}$	$\widehat{\mathrm{Var}\,(\overline{r}(q))}$	$\widehat{\overline{r}(q)}$	$\widehat{\mathrm{Var}\,(\overline{r}(q))}$
10^{-5}	$2.01 \cdot 10^{-15}$	$2.20 \cdot 10^{-32}$	$1.96 \cdot 10^{-15}$	$1.54 \cdot 10^{-32}$
10^{-4}	$2.00 \cdot 10^{-12}$	$2.20 \cdot 10^{-26}$	$1.96 \cdot 10^{-12}$	$1.53 \cdot 10^{-26}$
10^{-3}	$2.01 \cdot 10^{-9}$	$2.18 \cdot 10^{-20}$	$1.97 \cdot 10^{-9}$	$1.48 \cdot 10^{-20}$
10^{-2}	$2.07 \cdot 10^{-6}$	$2.02 \cdot 10^{-14}$	$2.03 \cdot 10^{-6}$	$1.03 \cdot 10^{-14}$
10^{-1}	$2.86 \cdot 10^{-3}$	$1.49 \cdot 10^{-8}$	$2.84 \cdot 10^{-3}$	$1.90 \cdot 10^{-9}$

Model 2 (Hypercube Graph): For our second model we consider the hypercube graph of order 5, denoted H_5. This is a regular graph with $2^5 = 32$ vertices and $5 \cdot 2^4 = 80$ edges; see also Figure 9.7 on Page 331. Each vertex corresponds to a binary vector $(x_1, \ldots, x_5)$, and two vertices are connected by an edge whenever their Hamming distance (minimum number of substitutions required to change one

Table 10.5: Estimated spectra components.

Algorithm	$F(4)$	$F(5)$	$F(6)$	$F(7)$	$F(8)$	$F(9)$
SE	$8.93 \cdot 10^{-8}$	$4.96 \cdot 10^{-7}$	$1.67 \cdot 10^{-6}$	$4.42 \cdot 10^{-6}$	$9.60 \cdot 10^{-6}$	$2.03 \cdot 10^{-5}$
PMC	0	0	0	0	$1.00 \cdot 10^{-5}$	$2.67 \cdot 10^{-5}$

vector into the other) is 1. Suppose that the network functions when vertices $(0,0,0,0,0)$ and $(1,1,0,0,0)$ are connected.

We run the SE Algorithm 10.4.1 with $N = 1000$ and $B = 10$ and the PMC Algorithm 5.4.2 with $N = 300000$. The running time for both SE and PMC is 28 seconds. Table 10.4 provides the estimated network unreliability with both PMC and SE algorithms.

Table 10.4: PMC and SE based reliability estimations for H_5 graph.

q	SE		PMC	
	$\widehat{\overline{r}(q)}$	$\widehat{\mathrm{Var}\,(\overline{r}(q))}$	$\widehat{\overline{r}(q)}$	$\widehat{\mathrm{Var}\,(\overline{r}(q))}$
10^{-5}	$1.94 \cdot 10^{-25}$	$7.26 \cdot 10^{-52}$	$2.32 \cdot 10^{-39}$	$1.15 \cdot 10^{-48}$
10^{-4}	$1.94 \cdot 10^{-20}$	$7.24 \cdot 10^{-42}$	$2.31 \cdot 10^{-30}$	$1.14 \cdot 10^{-38}$
10^{-3}	$1.93 \cdot 10^{-15}$	$7.10 \cdot 10^{-32}$	$2.20 \cdot 10^{-21}$	$1.01 \cdot 10^{-28}$
10^{-2}	$2.05 \cdot 10^{-10}$	$5.94 \cdot 10^{-22}$	$1.38 \cdot 10^{-12}$	$3.45 \cdot 10^{-19}$
10^{-1}	$1.94 \cdot 10^{-5}$	$1.28 \cdot 10^{-12}$	$2.07 \cdot 10^{-5}$	$1.25 \cdot 10^{-11}$

Note the difference in the *DOWN* probabilities for small values of q. The PMC Algorithm 5.4.2 cannot estimate rare-event probabilities; hence the delivered estimate is underestimating the true probability. Table 10.5 summarizes the first components of the obtained spectra with PMC and SE algorithms respectively. We do not report $F(0), \ldots, F(3)$, because they were evaluated as zero by full enumeration. With the same procedure, we obtain the exact value of $F(4)$, which is equal to $8.3195 \cdot 10^{-8}$, close (about 7%) to the reported value of $8.93 \cdot 10^{-8}$.

Next, we run the PMC Algorithm 5.4.2 with sample sizes $10^6, 10^7$, and 10^8, respectively. It is not surprising that $N = 10^6$ and $N = 10^7$ are insufficient to provide a meaningful estimator for the rare-event probability of $8.3195 \cdot 10^{-8}$. Table 10.6 summarizes the results obtained for $N = 10^8$. The running time of PMC is about 9000 seconds and the estimated variance for the rare-event probabilities is larger than the one obtained by SE.

Table 10.6: PMC reliability estimation for $N = 10^8$.

q	$\widehat{\overline{r}(q)}$	$\widehat{\text{Var}(\overline{r}(q))}$
10^{-5}	$1.80 \cdot 10^{-25}$	$5.80 \cdot 10^{-51}$
10^{-4}	$1.80 \cdot 10^{-20}$	$5.74 \cdot 10^{-41}$
10^{-3}	$1.81 \cdot 10^{-15}$	$5.18 \cdot 10^{-31}$
10^{-2}	$1.88 \cdot 10^{-13}$	$1.85 \cdot 10^{-21}$
10^{-1}	$2.00 \cdot 10^{-5}$	$2.66 \cdot 10^{-14}$

PROBLEMS

10.1   As a persuasive application of the DFS algorithm, consider the 8-queens problem of Example 6.13.   Construct a search tree for all the ways the queens can be arranged such that no queen can capture another. Each vertex $(x_1, \ldots, x_t)$ of the tree represents a valid arrangement of the queens in rows $1, \ldots, t$. At the first level of the tree there are 8 possible positions for the first queen, but to avoid symmetric solutions, you may assume that the first queen can only occupy positions in the first 4 columns, so that the children of the root ( ) vertex are $(1), (2), (3), (4)$. At the second level, the children of, for example, vertex (1) are $(2, 3), \ldots, (2, 8)$. And so on.   Implement a DFS algorithm to find all solutions $(x_1, \ldots, x_8)$ of the 8-queens problem. Note that two queens at grid positions (a, b) and (c, d), where $a, b, c, d \in \{1, \ldots, 8\}$, share the same diagonal if $(a - c)/(b - d)$ is equal to 1 or -1.

10.2   In Example 10.4 show that Knuth's algorithm returns an estimator C with $\mathbb{E}[C] = 6$ and $\text{Var}(C) = 2$.

10.3   It is difficult to guess in advance what B to choose, and also if it is better to perform one run of Algorithm 10.4.1 with budget B, or B runs with budget 1. For both trees in Figure 10.16, find out which scenario is better: two runs with $B = 1$ or one run with $B = 2$.

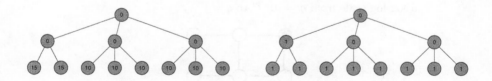

Figure 10.16: Example trees and weights.

10.4   Implement Algorithm 10.4.1 to estimate the number of paths in the $(7, 4)$-Petersen graph of Figure 10.4.

10.5   Consider the 2-SAT CNF formula $C_1 \wedge C_2 \wedge \ldots \wedge C_{n-1}$, where $C_i = l_i \vee \overline{l}_{i+1}$, $i = 1, \ldots, n - 1$.

   a) Show that the total number of solutions is $|\mathscr{X}^*| = n + 1$ and find these solutions.

b) For each $\mathbf{x} \in \mathscr{X}^*$ give the probability that Knuth's algorithm terminates at $\mathbf{x}$.

c) Implement SE for the cases $n = 21, n = 51, n = 101$, and $n = 151$, using the parameter choices $(B = 1, N = 500), (B = 5, N = 100), (B = 50, N = 10)$, and $(B = 100, N = 1)$. For each case report the estimated relative error and CPU time (for $N = 1$ leave the relative error blank).

10.6   Consider the 2-SAT formula $C_1 \wedge C_2 \wedge \ldots \wedge C_n$ where $C_i = l_i \vee l_{i+1}$, $i = 1, \ldots, n$. Implement an SE algorithm to estimate the number of valid assignments for any size m and any budget B. Give the number of valid assignments for $n = 1, \ldots, 10$. What is the pattern?

10.7   Implement an SE algorithm to find the permanent of the 50×50 binary matrix A defined by the following algorithm, which is based on a simple linear congruential generator. Compare the results for two sets of parameter choices: $B = 50$ and $N = 100$ versus $B = 1$ and $N = 5000$.

```
1  a ← 7⁵,  x ← 12345,  m ← 2³¹ − 1
2  for i = 1 to 50 do
3  │   for j = 1 to 50 do
4  │   │    x ← a x mod m
5  │   │    if x/m < 0.1 then
6  │   │    │    A(i, j) ← 1
7  │   │    else
8  │   │    └    A(i, j) ← 0
```

10.8   The 3×3 grid graph in Figure 10.17 is functioning if the black vertices are connected by functioning links. Each of the 12 links has a failure probability of q and the links fail independently of each other.

a) Compute the construction and destruction spectra via Algorithm 10.4.1, with the modification that the estimated cost at each level of the permutation tree is returned, rather than the total cost.

b) Using the spectra and formula (10.6), draw a graph of the unreliability on a log-log scale from $q = 10^{-30}$ to $q = 1$.

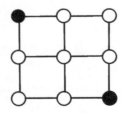

Figure 10.17: Grid graph.

Further Reading

Closely related accounts of stochastic enumeration appeared first in [13], [14], and [15]. It was realized in [16] that the approach is a generalization of Knuth's sequential importance sampling scheme for counting the cost of trees [10], which greatly simplifies the exposition of the algorithm and enhances its applicability.

The use of fast decision-making algorithms (oracles) for solving NP-hard problems is very common in Monte-Carlo methods; see, for example [8], which presents a *fully polynomial randomized approximation scheme* (FPRAS) for network unreliability based on the well-known DNF polynomial counting algorithm (oracle) of [9]. Another example is given in [4], where Kruskal's spanning trees algorithm is used for network reliability estimation. The use of SE for network reliability estimation was introduced in [17]. Wei and Selman proposed the `ApproxCount` algorithm in [18]. It is a *local search* method that uses Markov chain Monte Carlo (MCMC) sampling to approximate the true counting quantity. For SAT counting Gogate and Dechter [6, 7] devised an alternative counting technique called `SampleMinisat`, based on sampling from the search space of a Boolean formula through `SampleSearch`, using importance sampling.

REFERENCES

1. T. H. Cormen, C. E. Leiserson, and R. L. Rivest. *Introduction to Algorithms*. MIT Press, Cambridge, MA, 2009.

2. M. Davis, G. Logemann, and D. Loveland. A machine program for theorem proving. *Communications of the ACM*, 5:394–397, 1962.

3. M. Davis and H. Putnam. A computing procedure for quantification theory. *Journal of the ACM*, 7:201–215, 1960.

4. T. Elperin, I. B. Gertsbakh, and M. Lomonosov. Estimation of network reliability using graph evolution models. *IEEE Transactions on Reliability*, 40(5):572–581, 1991.

5. P. Erdös and A. Rényi. On the evolution of random graphs. In *Publication of the Mathematical Institute of the Hungarian Academy of Sciences*, pages 17–61, 1960.

6. V. Gogate and R. Dechter. Approximate counting by sampling the backtrack-free search space. In *Proceedings of the 22nd National Conference on Artificial Intelligence*, volume 1 of *AAAI'07*, pages 198–203. AAAI Press, 2007.

7. V. Gogate and R. Dechter. Samplesearch: Importance sampling in presence of determinism. *Artificial Intelligence*, 175(2):694–729, 2011.

8. D. R. Karger. A randomized fully polynomial time approximation scheme for the all terminal network reliability problem. In *Proceedings of the 27th Annual ACM Symposium on Theory of Computing*, STOC '95, pages 11–17, New York, 1995.

9. R. M. Karp, M. Luby, and N. Madras. Monte-Carlo approximation algorithms for enumeration problems. *Journal of Algorithms*, 10(3):429–448, 1989.

10. D. E. Knuth. Estimating the efficiency of backtrack programs. *Mathematics of Computation*, 29, 1975.

11. H. W. Kuhn. The Hungarian method for the assignment problem. *Naval Research Logistics Quarterly*, 2:83–97, 1955.

12. B. Roberts and D.P. Kroese. Estimating the number of s-t paths in a graph. *Journal of Graph Algorithms and Applications*, 11(1):195–214, 2007.

13. R. Y. Rubinstein. Stochastic enumeration method for counting NP-hard problems. *Methodology and Computing in Applied Probability*, 15(2):249–291, 2011.

14. R. Y. Rubinstein, A. Ridder, and R. Vaisman. *Fast Sequential Monte Carlo Methods.* John Wiley & Sons, New York, 2013.

15. R. Vaisman. *Stochastic Enumeration Method for Counting, Rare-Events and Optimization.* PhD thesis, Technion, Israel Institute of Technology, Haifa, September 2013.

16. R. Vaisman and D. P. Kroese. Stochastic enumeration method for counting trees. *Methodology and Computing in Applied Probability,* pages 1–43, 2015. doi: 10.1007/s11009-015-9457-4.

17. R. Vaisman, D. P. Kroese, and I. B. Gertsbakh. Improved sampling plans for combinatorial invariants of coherent systems. *IEEE Transactions on Reliability,* 2015. doi: 10.1109/TR.2015.2446471.

18. W. Wei and B. Selman. A new approach to model counting. In *Proceedings of the 8th International Conference on Theory and Applications of Satisfiability Testing,* SAT'05, pages 324–339, Berlin, 2005. Springer-Verlag.

APPENDIX

A.1 CHOLESKY SQUARE ROOT METHOD

Let Σ be a covariance matrix. We wish to find a matrix B such that $\Sigma = BB^\top$. The *Cholesky square root method* computes a lower triangular matrix B via a set of recursive equations as follows: From (1.23) we have

$$Z_1 = b_{11}X_1 + \mu_1 . \tag{A.1}$$

Therefore, $\mathrm{Var}(Z_1) = \sigma_{11} = b_{11}^2$ and $b_{11} = \sigma_{11}^{1/2}$. Proceeding with the second component of (1.23), we obtain

$$Z_2 = b_{21}X_1 + b_{22}X_2 + \mu_2 \tag{A.2}$$

and thus

$$\sigma_{22} = \mathrm{Var}(Z_2) = \mathrm{Var}(b_{21}X_1 + b_{22}X_2) = b_{21}^2 + b_{22}^2 . \tag{A.3}$$

Further, from (A.1) and (A.2),

$$\sigma_{12} = \mathbb{E}[(Z_1 - \mu_1)(Z_2 - \mu_2)] = \mathbb{E}[b_{11}X_1(b_{21}X_1 + b_{22}X_2)] = b_{11}b_{21} . \tag{A.4}$$

Hence, from (A.3) and (A.4) and the symmetry of Σ,

$$b_{21} = \frac{\sigma_{12}}{b_{11}} = \frac{\sigma_{12}}{\sigma_{11}^{1/2}} \tag{A.5}$$

$$b_{22} = \left(\sigma_{22} - \frac{\sigma_{21}^2}{\sigma_{11}}\right)^{1/2} . \tag{A.6}$$

Simulation and the Monte Carlo Method, Third Edition. By R. Y. Rubinstein and D. P. Kroese   **377**
Copyright © 2017 John Wiley & Sons, Inc. Published 2017 by John Wiley & Sons, Inc.

Generally, the b_{ij} can be found from the recursive formula

$$b_{ij} = \frac{\sigma_{ij} - \sum_{k=1}^{j-1} b_{ik}b_{jk}}{\left(\sigma_{jj} - \sum_{k=1}^{j-1} b_{jk}^2\right)^{1/2}} , \tag{A.7}$$

where, by convention,

$$\sum_{k=1}^{0} b_{ik}b_{jk} = 0, \quad 1 \leqslant j \leqslant i \leqslant n .$$

A.2   EXACT SAMPLING FROM A CONDITIONAL BERNOULLI DISTRIBUTION

Suppose the vector $\mathbf{X} = (X_1, \ldots, X_n)$ has independent components, with $X_i \sim$ Ber(p_i), $i = 1, \ldots, n$. It is not difficult to see (see Problem A.1) that the conditional distribution of $\mathbf{X}$ given $\sum_i X_i = k$ is given by

$$\mathbb{P}\left(X_1 = x_1, \ldots, X_n = x_n \,\middle|\, \sum_{i=1}^{n} X_i = k\right) = \frac{\prod_{i=1}^{n} w_i^{x_i}}{c} , \tag{A.8}$$

where c is a normalization constant and $w_i = p_i/(1 - p_i)$, $i = 1, \ldots, n$. Generating random variables from this distribution can be done via the so-called *drafting* procedure, described, for example, in [3]. The Matlab code below provides a procedure for calculating the normalization constant c and drawing from the conditional joint pdf above.

■ EXAMPLE A.1

Suppose $\mathbf{p} = (1/2, 1/3, 1/4, 1/5)$ and $k = 2$. Then $\mathbf{w} = (w_1, \ldots, w_4) = (1, 1/2, 1/3, 1/4)$. The first element of Rgens(k,w), with k = 2 and w = $\mathbf{w}$ is $35/24 \approx 1.45833$. This is the normalization constant c. Thus, for example,

$$\mathbb{P}\left(X_1 = 0, X_2 = 1, X_3 = 0, X_4 = 1 \,\middle|\, \sum_{i=1}^{4} X_i = 2\right) = \frac{w_2\, w_4}{35/24} = \frac{3}{35} \approx 0.08571 .$$

To generate random vectors according to this conditional Bernoulli distribution call condbern(p,k), where k is the number of unities (here 2) and p is the probability vector $\mathbf{p}$. This function returns the positions of the unities, such as (1, 2) or (2,4).

```
function sample = condbern(k,p)
% k = no of units in each sample, P = probability vector
W=zeros(1,length(p));
sample=zeros(1,k);
ind1=find(p==1);
sample(1:length(ind1))=ind1;
k=k-length(ind1);
```

```
ind=find(p<1 & p>0);
W(ind)=p(ind)./(1-p(ind));
for i=1:k
    Pr=zeros(1,length(ind));
    Rvals=Rgens(k-i+1,W(ind));
    for j=1:length(ind)
        Pr(j)=W(ind(j))*Rvals(j+1)/((k-i+1)*Rvals(1));
    end
    Pr=cumsum(Pr);
    entry=ind(min(find(Pr>rand)));
    ind=ind(find(ind~=entry));
    sample(length(ind1)+i)=entry;
end
sample=sort(sample);
return


function Rvals = Rgens(k,W)
N=length(W);
T=zeros(k,N+1);
R=zeros(k+1,N+1);
for i=1:k
    for j=1:N, T(i,1)=T(i,1)+W(j)^i; end
    for j=1:N, T(i,j+1)=T(i,1)-W(j)^i; end
end
R(1,:)=ones(1,N+1);
for j=1:k
    for l=1:N+1
        for i=1:j
            R(j+1,l)=R(j+1,l)+(-1)^(i+1)*T(i,l)*R(j-i+1,l);
        end
    end
    R(j+1,:)=R(j+1,:)/j;
end
Rvals=[R(k+1,1),R(k,2:N+1)];
return
```

A.3 EXPONENTIAL FAMILIES

Exponential families play an important role in statistics; see, for example, [2]. Let $\mathbf{X}$ be a random variable or vector (in this section, vectors will always be interpreted as *column* vectors) with pdf $f(\mathbf{x}; \boldsymbol{\theta})$ (with respect to some measure), where $\boldsymbol{\theta} = (\theta_1, \ldots, \theta_m)^\top$ is an m-dimensional parameter vector. $\mathbf{X}$ is said to belong to an m-parameter *exponential family* if there exist real-valued functions $t_i(\mathbf{x})$ and $h(\mathbf{x}) > 0$ and a (normalizing) function $c(\boldsymbol{\theta}) > 0$ such that

$$f(\mathbf{x}; \boldsymbol{\theta}) = c(\boldsymbol{\theta})\, e^{\boldsymbol{\theta} \cdot \mathbf{t}(\mathbf{x})}\, h(\mathbf{x})\,, \tag{A.9}$$

where $\mathbf{t}(\mathbf{x}) = (t_1(\mathbf{x}), \ldots, t_m(\mathbf{x}))^\top$ and $\boldsymbol{\theta} \cdot \mathbf{t}(\mathbf{x})$ is the inner product $\sum_{i=1}^m \theta_i t_i(\mathbf{x})$. The representation of an exponential family is in general not unique.

Remark A.3.1 (Natural Exponential Family) The standard definition of an exponential family involves a family of densities $\{g(\mathbf{x}; \mathbf{v})\}$ of the form

$$g(\mathbf{x}; \mathbf{v}) = d(\mathbf{v}) \, e^{\boldsymbol{\theta}(\mathbf{v}) \cdot \mathbf{t}(\mathbf{x})} \, h(\mathbf{x}) \,, \tag{A.10}$$

where $\boldsymbol{\theta}(\mathbf{v}) = (\theta_1(\mathbf{v}), \ldots, \theta_m(\mathbf{v}))^\top$, and the $\{\theta_i\}$ are real-valued functions of the parameter $\mathbf{v}$. By *reparameterization* — by using the θ_i as parameters — we can represent (A.10) in so-called *canonical form* (A.9). In effect, $\boldsymbol{\theta}$ is the natural parameter of the exponential family. For this reason, a family of the form (A.9) is called a *natural exponential family*.

Table A.1 displays the functions $c(\boldsymbol{\theta})$, $t_k(x)$, and $h(x)$ for several commonly used distributions (a dash means that the corresponding value is not used).

Table A.1: The functions $c(\boldsymbol{\theta})$, $t_k(x)$ and $h(x)$ for commonly used distributions.

Distr.	$t_1(x)$, $t_2(x)$	$c(\boldsymbol{\theta})$	θ_1, θ_2	$h(x)$
Gamma(α, λ)	x, $\ln x$	$\dfrac{(-\theta_1)^{\theta_2+1}}{\Gamma(\theta_2+1)}$	$-\lambda$, $\alpha - 1$	1
N(μ, σ^2)	x, x^2	$\dfrac{e^{\theta_1^2/(4\theta_2)}}{\sqrt{-\pi/\theta_2}}$	$\dfrac{\mu}{\sigma^2}$, $-\dfrac{1}{2\sigma^2}$	1
Weib(α, λ)	x^α, $\ln x$	$-\theta_1(\theta_2+1)$	$-\lambda^\alpha$, $\alpha - 1$	1
Bin(n, p)	x, $-$	$(1 + e^{\theta_1})^{-n}$	$\ln\left(\dfrac{p}{1-p}\right)$, $-$	$\dbinom{n}{x}$
Poi(λ)	x, $-$	$e^{-e^{\theta_1}}$	$\ln \lambda$, $-$	$\dfrac{1}{x!}$
G(p)	$x - 1$, $-$	$1 - e^{\theta_1}$	$\ln(1-p)$, $-$	1

As an important instance of a natural exponential family, consider the univariate, single-parameter $(m = 1)$ case with $t(x) = x$. Thus we have a family of densities $\{f(x; \theta), \theta \in \Theta \subset \mathbb{R}\}$ given by

$$f(x; \theta) = c(\theta) \, e^{\theta x} \, h(x) \,. \tag{A.11}$$

If $h(x)$ is a pdf, then $c^{-1}(\theta)$ is the corresponding *moment generating function*:

$$c^{-1}(\theta) = \int e^{\theta x} \, h(x) \, \mathrm{d}x \,.$$

It is sometimes convenient to introduce instead the logarithm of the moment generating function:

$$\zeta(\theta) = \ln \int e^{\theta x} h(x) \, \mathrm{d}x \,,$$

which is called the *cumulant function*. We can now write (A.11) in the following convenient form:

$$f(x; \theta) = e^{\theta x - \zeta(\theta)} \, h(x) \,. \tag{A.12}$$

■ **EXAMPLE A.2**

If we take h as the density of the $N(0, \sigma^2)$-distribution, $\theta = \lambda/\sigma^2$ and $\zeta(\theta) = \sigma^2 \theta^2/2$, then the family $\{f(\cdot; \theta), \theta \in \mathbb{R}\}$ is the family of $N(\lambda, \sigma^2)$ densities, where σ^2 is fixed and $\lambda \in \mathbb{R}$.

Similarly, if we take h as the density of the $\mathsf{Gamma}(a, 1)$-distribution, and let $\theta = 1 - \lambda$ and $\zeta(\theta) = -a\ln(1 - \theta) = -a\ln\lambda$, we obtain the class of $\mathsf{Gamma}(a, \lambda)$ distributions, with a fixed and $\lambda > 0$. Note that in this case $\Theta = (-\infty, 1)$.

Starting from any pdf f_0, we can easily generate a natural exponential family of the form (A.12) in the following way: Let Θ be the largest interval for which the cumulant function ζ of f_0 exists. This includes $\theta = 0$, since f_0 is a pdf. Now define

$$f(x; \theta) = e^{\theta x - \zeta(\theta)} f_0(x) . \tag{A.13}$$

Then $\{f(\cdot; \theta), \theta \in \Theta\}$ is a natural exponential family. We say that the family is obtained from f_0 by an *exponential twist* or *exponential change of measure* (ECM) with a *twisting* or *tilting* parameter θ.

Remark A.3.2 (Reparameterization) It may be useful to reparameterize a natural exponential family of the form (A.12) into the form (A.10). Let $X \sim f(\cdot; \theta)$. It is not difficult to see that

$$\mathbb{E}_\theta[X] = \zeta'(\theta) \quad \text{and} \quad \mathrm{Var}_\theta(X) = \zeta''(\theta) . \tag{A.14}$$

$\zeta'(\theta)$ is increasing in θ, since its derivative, $\zeta''(\theta) = \mathrm{Var}_\theta(X)$, is always greater than 0. Thus, we can reparameterize the family using the mean $v = \mathbb{E}_\theta[X]$. In particular, to the above natural exponential family there corresponds a family $\{g(\cdot; v)\}$ such that for each pair (θ, v) satisfying $\zeta'(\theta) = v$ we have $g(x; v) = f(x; \theta)$.

■ **EXAMPLE A.3**

Consider the second case in Example A.2. Note that we constructed in fact a natural exponential family $\{f(\cdot; \theta), \theta \in (-\infty, 1)\}$ by exponentially twisting the $\mathsf{Gamma}(\alpha, 1)$ distribution, with density $f_0(x) = x^{\alpha-1}e^{-x}/\Gamma(a)$. We have $\zeta'(\theta) = \alpha/(1 - \theta) = v$. This leads to the reparameterized density

$$g(x; v) = \exp\left(\theta x + \alpha\ln(1 - \theta)\right) f_0(x) = \frac{\exp\left(-\frac{\alpha}{v} x\right) \left(\frac{\alpha}{v}\right)^\alpha x^{\alpha-1}}{\Gamma(\alpha)} ,$$

corresponding to the $\mathsf{Gamma}(\alpha, \alpha v^{-1})$ distribution, $v > 0$.

CE Updating Formulas for Exponential Families

We now obtain an *analytic* formula for a general one-parameter exponential family. Let $X \sim f(x; u)$ for some nominal reference parameter u. For simplicity, assume that $\mathbb{E}_u[H(X)] > 0$ and that X is nonconstant. Let $f(x; u)$ be a member of a one-parameter exponential family $\{f(x; v)\}$. Suppose the parameterization $\eta = \psi(v)$ puts the family in canonical form. That is,

$$f(x; v) = g(x; \eta) = e^{\eta x - \zeta(\eta)} h(x) .$$

Moreover, let us assume that v corresponds to the expectation of X. This can always be established by reparameterization; see Remark A.3.2. Note that, in particular, $v = \zeta'(\eta)$. Let $\theta = \psi(u)$ correspond to the nominal reference parameter. Since $\max_v \mathbb{E}_u[H(X) \ln f(X; v)] = \max_\eta \mathbb{E}_\theta[H(X) \ln g(X; \eta)]$, we can obtain the optimal solution v^* to (5.60) by finding, as in (5.61), the solution η^* to

$$\mathbb{E}_\theta \left[H(X) \frac{d}{d\eta} \ln g(X; \eta) \right] = 0$$

and putting $v^* = \psi^{-1}(\eta^*)$. Since $(\ln g(X; \eta))' = x - \zeta'(\eta)$, and $\zeta'(\eta) = v$, we see that v^* is given by the solution of $\mathbb{E}_u[H(X)(-v + X)] = 0$. Hence v^* is given by

$$v^* = \frac{\mathbb{E}_u[H(X) X]}{\mathbb{E}_u[H(X)]} = \frac{\mathbb{E}_w[H(X) W(X; u, w) X]}{\mathbb{E}_w[H(X) W(X; u, w)]} \tag{A.15}$$

for any reference parameter w. It is not difficult to check that v^* is indeed a unique global maximum of $D(v) = \mathbb{E}_u[H(X) \ln f(X; v)]$. The corresponding estimator $\widehat{v}$ of v^* in (A.15) is

$$\widehat{v} = \frac{\sum_{i=1}^N H(X_i) W(X_i; u, w) X_i}{\sum_{i=1}^N H(X_i) W(X_i; u, w)}, \tag{A.16}$$

where $X_1, \ldots, X_N$ is a random sample from the density $f(\cdot; w)$.

A similar explicit formula can be found for the case where $\mathbf{X} = (X_1, \ldots, X_n)$ is a vector of *independent* random variables such that each component X_j belongs to a one-parameter exponential family parameterized by the mean; that is, the density of each X_j is given by

$$f_j(x; u_j) = e^{x\theta(u_j) - \zeta(\theta(u_j))} h_j(x),$$

where $\mathbf{u} = (u_1, \ldots, u_n)$ is the nominal reference parameter. It is easy to see that problem (5.63) under the independence assumption becomes "separable," that is, it reduces to n subproblems of the form above. Thus, we find that the optimal reference parameter vector $\mathbf{v}^* = (v_1^*, \ldots, v_n^*)$ is given as

$$v_j^* = \frac{\mathbb{E}_\mathbf{u}[H(\mathbf{X}) X_j]}{\mathbb{E}_\mathbf{u}[H(\mathbf{X})]} = \frac{\mathbb{E}_\mathbf{w}[H(\mathbf{X}) W(\mathbf{X}; \mathbf{u}, \mathbf{w}) X_j]}{\mathbb{E}_\mathbf{w}[(\mathbf{X}) W(\mathbf{X}; \mathbf{u}, \mathbf{w})]}. \tag{A.17}$$

Moreover, we can estimate the j-th component of $\mathbf{v}^*$ as

$$\widehat{v}_j = \frac{\sum_{i=1}^N H(\mathbf{X}_i) W(\mathbf{X}_i; \mathbf{u}, \mathbf{w}) X_{ij}}{\sum_{i=1}^N H(\mathbf{X}_i) W(\mathbf{X}_i; \mathbf{u}, \mathbf{w})}, \tag{A.18}$$

where $\mathbf{X}_1, \ldots, \mathbf{X}_N$ is a random sample from the density $f(\cdot; \mathbf{w})$ and X_{ij} is the j-th component of $\mathbf{X}_i$.

A.4   SENSITIVITY ANALYSIS

The crucial issue in choosing a good importance sampling density $f(\mathbf{x}; \mathbf{v})$ to estimate $\nabla^k \ell(\mathbf{u})$ via (7.16) is to ensure that the corresponding estimators have low variance. We consider this issue for the cases $k = 0$ and $k = 1$. For $k = 0$, this means minimizing the variance of $\widehat{\ell}(\mathbf{u}; \mathbf{v})$ with respect to $\mathbf{v}$, which is equivalent to solving the minimization program

$$\min_{\mathbf{v}} \mathcal{L}^0(\mathbf{v}; \mathbf{u}) = \min_{\mathbf{v}} \mathbb{E}_{\mathbf{v}}[H^2(\mathbf{X}) W^2(\mathbf{X}; \mathbf{u}, \mathbf{v})] \; . \tag{A.19}$$

For the case $k = 1$, note that $\nabla \widehat{\ell}(\mathbf{u}; \mathbf{v})$ is a vector rather than a scalar. To obtain a good reference vector $\mathbf{v}$, we now minimize the *trace of the associated covariance matrix*, which is equivalent to minimizing

$$\min_{\mathbf{v}} \mathcal{L}^1(\mathbf{v}; \mathbf{u}) = \min_{\mathbf{v}} \mathbb{E}_{\mathbf{v}} \left[H^2(\mathbf{X}) W^2(\mathbf{X}; \mathbf{u}, \mathbf{v}) \operatorname{tr}\left(\mathcal{S}(\mathbf{u}; \mathbf{x})\mathcal{S}(\mathbf{u}; \mathbf{x})^\top\right) \right], \tag{A.20}$$

where tr denotes the trace. For exponential families both optimization programs are *convex*, as demonstrated in the next proposition. To conform with our earlier notation for exponential families in Section A.3, we use $\boldsymbol{\theta}$ and $\boldsymbol{\eta}$ instead of $\mathbf{u}$ and $\mathbf{v}$, respectively.

A.4.1 Convexity Results

Proposition A.4.1 *Let $\mathbf{X}$ be a random vector from an m-parameter exponential family of the form (A.9). Then $\mathcal{L}^k(\boldsymbol{\eta}; \boldsymbol{\theta})$, $k = 0, 1$, defined in (A.19) and (A.20), are convex functions of $\boldsymbol{\eta}$.*

Proof: Consider first the case $k = 0$. We have (see (7.23))

$$\mathcal{L}^0(\boldsymbol{\eta}; \boldsymbol{\theta}) = c(\boldsymbol{\theta})^2 \int \frac{H^2(\mathbf{x})}{c(\boldsymbol{\eta})} \, \mathrm{e}^{(2\boldsymbol{\theta}-\boldsymbol{\eta}) \cdot \mathbf{t}(\mathbf{x})} h(\mathbf{x}) \mathrm{d}\mathbf{x} \, , \tag{A.21}$$

where

$$c(\boldsymbol{\eta})^{-1} = \int \mathrm{e}^{\boldsymbol{\eta} \cdot \mathbf{t}(\mathbf{z})} h(\mathbf{z}) \, \mathrm{d}\mathbf{z} \, .$$

Substituting the above into (A.21) yields

$$\mathcal{L}^0(\boldsymbol{\eta}; \boldsymbol{\theta}) = c(\boldsymbol{\theta})^2 \int \int H^2(\mathbf{x}) \, \mathrm{e}^{2\boldsymbol{\theta} \cdot \mathbf{t}(\mathbf{x}) + \boldsymbol{\eta} \cdot (\mathbf{t}(\mathbf{z}) - \mathbf{t}(\mathbf{x}))} h(\mathbf{x}) \, h(\mathbf{z}) \, \mathrm{d}\mathbf{x} \, \mathrm{d}\mathbf{z} \, . \tag{A.22}$$

Now, for any linear function, $a(\boldsymbol{\eta})$ of $\boldsymbol{\eta}$, the function $\mathrm{e}^{a(\boldsymbol{\eta})}$ is convex. Since $H^2(\mathbf{x})$ is nonnegative, it follows that for any fixed $\boldsymbol{\theta}$, $\mathbf{x}$, and $\mathbf{z}$, the function under the integral sign in (A.22) is convex in $\boldsymbol{\eta}$. This implies the convexity of $\mathcal{L}^0(\boldsymbol{\eta}; \boldsymbol{\theta})$.

The case $k = 1$ follows in exactly the same way, noting that the trace $\operatorname{tr}\left(\mathcal{S}(\boldsymbol{\theta}; \mathbf{x})\mathcal{S}(\boldsymbol{\theta}; \mathbf{x})^\top\right)$ is a nonnegative function for $\mathbf{x}$ for any $\boldsymbol{\theta}$. □

Remark A.4.1 Proposition A.4.1 can be extended to the case where

$$\ell(\mathbf{u}) = \varphi(\ell_1(\mathbf{u}), \dots, \ell_k(\mathbf{u}))$$

and

$$\ell_i(\mathbf{u}) = \mathbb{E}_{\mathbf{u}}[H_i(\mathbf{X})] = \mathbb{E}_{\mathbf{v}}[H_i(\mathbf{X})W(\mathbf{X}; \mathbf{u}; \mathbf{v})] = \mathbb{E}_{\mathbf{v}}[H_i W], \quad i = 1, \dots, k \, .$$

Here the $\{H_i(\mathbf{X})\}$ are sample functions associated with the same random vector $\mathbf{X}$ and $\varphi(\cdot)$ is a real-valued differentiable function. We prove its validity for the case $k = 2$. In this case the estimators of $\ell(\mathbf{u})$ can be written as

$$\widehat{\ell}(\mathbf{u}; \mathbf{v}) = \varphi(\widehat{\ell}_1(\mathbf{u}; \mathbf{v}), \widehat{\ell}_2(\mathbf{u}; \mathbf{v})) \,,$$

where $\widehat{\ell}_1(\mathbf{u}; \mathbf{v})$ and $\widehat{\ell}_2(\mathbf{u}; \mathbf{v})$ are the usual importance sampling estimators of $\ell_1(\mathbf{u})$ and $\ell_2(\mathbf{u})$, respectively. By virtue of the delta method (see Problem 7.11), $N^{1/2}(\widehat{\ell}(\mathbf{u}; \mathbf{v}) - \ell(\mathbf{u}))$ is asymptotically normal, with mean 0 and variance

$$
\begin{aligned}
\sigma^2(\mathbf{v}; \mathbf{u}) &= a^2 \operatorname{Var}_{\mathbf{v}}(H_1 W) + b^2 \operatorname{Var}_{\mathbf{v}}(H_2 W) + 2\, a\, b \operatorname{Cov}_{\mathbf{v}}(H_1 W, H_2 W) \\
&= \mathbb{E}_{\mathbf{v}}\left[(a H_1 + b H_2)^2 W^2\right] + R(\mathbf{u}) \,. \quad\quad\quad\quad \text{(A.23)}
\end{aligned}
$$

Here $R(\mathbf{u})$ consists of the remaining terms that are independent of $\mathbf{v}$, $a = \partial\varphi(x_1, x_2)/\partial x_1$ and $b = \partial\varphi(x_1, x_2)/\partial x_2$ at $(x_1, x_2) = (\ell_1(\mathbf{u}), \ell_2(\mathbf{u}))$. For example, for $\varphi(x_1, x_2) = x_1/x_2$, one gets $a = 1/\ell_2(\mathbf{u})$ and $b = -\ell_1(\mathbf{u})/\ell_2(\mathbf{u})^2$.

The convexity of $\sigma^2(\mathbf{v}; \mathbf{u})$ in $\mathbf{v}$ now follows similarly to the proof of Proposition A.4.1.

A.4.2 Monotonicity Results

Consider optimizing the functions $\mathcal{L}^k(\mathbf{v}; \mathbf{u})$, $k = 0, 1$ in (A.19) and (A.20) with respect to $\mathbf{v}$. Let $\mathbf{v}^*(k)$ be the optimal solutions for $k = 0, 1$. The following proposition states that the optimal reference parameter always leads to a "fatter" tail for $f(\mathbf{x}; \mathbf{v}^*)$ than that of the original pdf $f(\mathbf{x}; \mathbf{u})$. This important phenomenon is the driving force for all of our beautiful results in this book, as well as for preventing the degeneracy of our importance sampling estimates. For simplicity, the result is given for the gamma distribution only. Similar results can be established with respect to some other parameters of the exponential family and for the CE approach.

Proposition A.4.2 *Let $X \sim \mathsf{Gamma}(\alpha, u)$. Suppose that $H^2(x)$ is a monotonically increasing function on the interval $[0, \infty)$. Then*

$$v^*(k) < u, \ k = 0, 1 \,. \quad\quad\quad\quad \text{(A.24)}$$

Proof: The proof will be given for $k = 0$ only. The proof for $k = 1$, using the trace operator, is similar. To simplify the notation, we write $\mathcal{L}(v)$ for $\mathcal{L}^0(v; u)$.

Since $\mathcal{L}(v)$ is convex, it suffices to prove that its derivative with respect to v is positive at $v = u$. To this end, represent $\mathcal{L}(v)$ as

$$\mathcal{L}(v) = c \int_0^\infty v^{-\alpha} H^2(x)\, x^{\alpha-1} \mathrm{e}^{-(2u-v)x}\, \mathrm{d}x \,,$$

where the constant $c = u^{2\alpha} \Gamma(\alpha)^{-1}$ is independent of v. Differentiating $\mathcal{L}(v)$ above with respect to v at $v = u$, gives

$$\mathcal{L}'(v)|_{v=u} = \mathcal{L}'(u) = c \int_0^\infty \left(x - \alpha u^{-1}\right) u^{-\alpha} H^2(x)\, x^{\alpha-1}\, \mathrm{e}^{-ux}\, \mathrm{d}x \,.$$

Integrating by parts yields

$$
\begin{aligned}
\mathcal{L}'(u) &= \lim_{R \to \infty} -c\,u^{-\alpha-1} R^{\alpha} e^{-uR} H^2(R) + c\,u^{-\alpha-1} \int_0^{\infty} x^{\alpha} e^{-ux} \, dH^2(x) \\
&= c\,u^{-\alpha-1} \int_0^{\infty} x^{\alpha} e^{-ux} \, dH^2(x) \,, \tag{A.25}
\end{aligned}
$$

provided $H^2(R)R^{\alpha} \exp(-uR)$ tends to 0 as $R \to \infty$. Finally, since $H^2(x)$ is monotonically increasing in x, we conclude that the integral (A.25) is positive, and consequently, $\mathcal{L}'(u) > 0$. This fact, and the convexity of $\mathcal{L}(v)$, imply that $v^*(0) < u$.

$\square$

Proposition A.4.2 can be extended to the multidimensional gamma distribution, as well as to some other exponential family distributions. For details see [13].

A.5 A SIMPLE CE ALGORITHM FOR OPTIMIZING THE PEAKS FUNCTION

The following Matlab code provides a simple implementation of a CE algorithm to solve the peaks function; see Example 8.12 on page 292.

```
n = 2;                           % dimension
mu = [-3,-3]; sigma = 3*ones(1,n); N = 100; eps = 1E-5; rho = 0.1;

while max(sigma) > eps
   X = randn(N,n)*diag(sigma)+ mu(ones(N,1),:);
   SX= S(X);                     %Compute the performance
   sortSX = sortrows([X, SX],n+1);
   Elite = sortSX((1-rho)*N:N,1:n); % elite samples
   mu = mean(Elite,1);           % take sample mean row-wise
   sigma = std(Elite,1);         % take sample st.dev. row-wise
   [S(mu),mu,max(sigma)]         % output the result
end

function out = S(X)
out =   3*(1-X(:,1)).^2.*exp(-(X(:,1).^2) - (X(:,2)+1).^2) ...
   - 10*(X(:,1)/5 - X(:,1).^3 - X(:,2).^5).*exp(-X(:,1).^2-X(:,2).^2) ...
   - 1/3*exp(-(X(:,1)+1).^2 - X(:,2).^2);
end
```

A.6 DISCRETE-TIME KALMAN FILTER

Consider the hidden Markov model

$$
\begin{aligned}
X_t &= A\,X_{t-1} + \varepsilon_{1t} \\
Y_t &= B\,X_t + \varepsilon_{2t}, \qquad t = 1, 2, \ldots, \tag{A.26}
\end{aligned}
$$

where A and B are matrices (B does not have to be a square matrix). We adopt the notation of Section 5.10. The initial state X_0 is assumed to be $\mathsf{N}(\mu_0, \Sigma_0)$ distributed.

The objective is to obtain the filtering pdf $f(x_t \,|\, \mathbf{y}_{1:t})$ and the *predictive* pdf $f(x_t \,|\, \mathbf{y}_{1:t-1})$. Observe that the joint pdf of $\mathbf{X}_{1:t}$ and $\mathbf{Y}_{1:t}$ must be Gaussian, since these random vectors are linear transformations of independent standard Gaussian random variables. It follows that $f(x_t \,|\, \mathbf{y}_{1:t}) \sim \mathsf{N}(\mu_t, \Sigma_t)$ for some mean vector μ_t and covariance matrix Σ_t. Similarly, $f(x_t \,|\, \mathbf{y}_{1:t-1}) \sim \mathsf{N}(\widetilde{\mu}_t, \widetilde{\Sigma}_t)$ for some mean vector $\widetilde{\mu}_t$ and covariance matrix $\widetilde{\Sigma}_t$. We wish to compute μ_t, $\widetilde{\mu}_t$, Σ_t, and $\widetilde{\Sigma}_t$ recursively. The argument goes as follows: By assumption, $(X_{t-1} \,|\, \mathbf{y}_{1:t-1}) \sim \mathsf{N}(\mu_{t-1}, \Sigma_{t-1})$. Combining this with the fact that $X_t = A\,X_{t-1} + \varepsilon_{1t}$ yields

$$(X_t \,|\, \mathbf{y}_{1:t-1}) \sim \mathsf{N}(A\,\mu_{t-1},\ A\Sigma_{t-1}A^\top + C_1)\ .$$

In other words,

$$
\begin{aligned}
\widetilde{\mu}_t &= A\,\mu_{t-1}, \\
\widetilde{\Sigma}_t &= A\Sigma_{t-1}A^\top + C_1\ .
\end{aligned}
\tag{A.27}
$$

Next, we determine the joint pdf of X_t and Y_t given $\mathbf{Y}_{1:t-1} = \mathbf{y}_{1:t-1}$. Decomposing $\widetilde{\Sigma}_t$ and C_2 as $\widetilde{\Sigma}_t = RR^\top$ and $C_2 = QQ^\top$, respectively (e.g., via the Cholesky square root method), we can write (see (1.23))

$$
\begin{pmatrix} X_t \\ Y_t \end{pmatrix}\ \Big|\ \mathbf{y}_{1:t-1} = \begin{pmatrix} \widetilde{\mu}_t \\ B\widetilde{\mu}_t \end{pmatrix} + \begin{pmatrix} R & 0 \\ BR & Q \end{pmatrix} \begin{pmatrix} U \\ V \end{pmatrix},
$$

where, conditional on $Y_{t-1} = \mathbf{y}_{1:t-1}$, U and V are independent standard normal random vectors. The corresponding covariance matrix is

$$
\begin{pmatrix} R & 0 \\ BR & Q \end{pmatrix} \begin{pmatrix} R^\top & R^\top B^\top \\ 0 & Q^\top \end{pmatrix} = \begin{pmatrix} RR^\top & RR^\top B^\top \\ BRR^\top & BRR^\top B^\top + QQ^\top \end{pmatrix},
$$

so that we have

$$
\begin{pmatrix} X_t \\ Y_t \end{pmatrix}\ \Big|\ \mathbf{y}_{1:t-1} \sim \mathsf{N}\left(\begin{pmatrix} \widetilde{\mu}_t \\ B\widetilde{\mu}_t \end{pmatrix},\ \begin{pmatrix} \widetilde{\Sigma}_t & \widetilde{\Sigma}_t B^\top \\ B\widetilde{\Sigma}_t & B\widetilde{\Sigma}_t B^\top + C_2 \end{pmatrix} \right)
\tag{A.28}
$$

(note that $\widetilde{\Sigma}_t$ is symmetric).

The result (A.28) enables us to find the conditional pdf $f(x_t \,|\, \mathbf{y}_t)$ with the aid of the following general result (see Problem A.2 below for a proof): If

$$
\begin{pmatrix} X \\ Y \end{pmatrix} \sim \mathsf{N}\left(\begin{pmatrix} m_1 \\ m_2 \end{pmatrix},\ \begin{pmatrix} S_{11} & S_{12} \\ S_{21} & S_{22} \end{pmatrix} \right),
$$

then

$$(X \,|\, Y = y) \sim \mathsf{N}\left(m_1 + S_{12}S_{22}^{-1}(y - m_2),\ S_{11} - S_{12}S_{22}^{-1}S_{12}^\top \right)\ .
\tag{A.29}$$

Because $f(x_t \,|\, \mathbf{y}_{1:t}) = f(x_t \,|\, \mathbf{y}_{1:t-1}, y_t)$, an immediate consequence of (A.28) and (A.29) is

$$
\begin{aligned}
\mu_t &= \widetilde{\mu}_t + \widetilde{\Sigma}_t B^\top (B\widetilde{\Sigma}_t B^\top + C_2)^{-1}(y_t - B\widetilde{\mu}_t)\ , \\
\Sigma_t &= \widetilde{\Sigma}_t - \widetilde{\Sigma}_t B^\top (B\widetilde{\Sigma}_t B^\top + C_2)^{-1} B\widetilde{\Sigma}_t\ .
\end{aligned}
\tag{A.30}
$$

Updating formulas (A.27) and (A.30) form the (discrete-time) *Kalman filter*. Starting with some known μ_0 and Σ_0, we determine first $\widetilde{\mu}_1$ and $\widetilde{\Sigma}_1$, then μ_1 and Σ_1, and so on. Notice that $\widetilde{\Sigma}_t$ and Σ_t do not depend on the observations $y_1, y_2, \ldots$ and can therefore be determined *off-line*. The Kalman filter discussed above can be extended in many ways, for example, by including control variables and time-varying parameter matrices. The nonlinear filtering case is often dealt with by linearizing the state and observation equations via a Taylor expansion. This leads to an approximative method called the *extended Kalman filter*.

A.7 BERNOULLI DISRUPTION PROBLEM

As an example of a finite-state hidden Markov model, we consider the following *Bernoulli disruption problem*. In Example 6.8 a similar type of "change point" problem is discussed in relation to the Gibbs sampler. However, the crucial difference is that in the present case the detection of the change point can be done *sequentially*.

Let $Y_1, Y_2, \ldots$ be Bernoulli random variables, and let T be a geometrically distributed random variable with parameter r. Conditional on T, the $\{Y_i\}$ are mutually independent, and $Y_1, Y_2, \ldots, Y_{T-1}$ all have a success probability a, whereas $Y_T, Y_{T+1}, \ldots$ all have a success probability b. Thus, T is the change or disruption point. Suppose that T cannot be observed, but only $\{Y_t\}$. We wish to decide if the disruption has occurred based on the outcome $\mathbf{y}_{1:t} = (y_1, \ldots, y_t)$ of $\mathbf{Y}_{1:t} = (Y_1, \ldots, Y_t)$. An example of the observations is depicted in Figure A.1, where the dark lines indicate the times of successes $(Y_i = 1)$.

Figure A.1: Observations for the disruption problem.

The situation can be described via the HMM illustrated in Figure A.2. Namely, let $\{X_t, t = 0, 1, 2, \ldots\}$ be a Markov chain with state space $\{0, 1\}$, transition matrix

$$P = \begin{pmatrix} 1 - r & r \\ 0 & 1 \end{pmatrix},$$

and initial state $X_0 = 0$. Then the objective is to find $\mathbb{P}(T \leqslant t \mid \mathbf{Y}_{1:t} = \mathbf{y}_{1:t}) = \mathbb{P}(X_t = 1 \mid \mathbf{Y}_t = \mathbf{y}_{1:t})$.

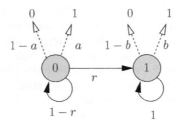

Figure A.2: HMM diagram for the disruption problem.

This conditional probability can be computed efficiently by introducing

$$\alpha_t(j) = \mathbb{P}(X_t = j, \mathbf{Y}_{1:t} = \mathbf{y}_{1:t}) \,.$$

By conditioning on X_{t-1}, we have

$$
\begin{aligned}
\alpha_t(j) &= \sum_i \mathbb{P}(X_t = j, X_{t-1} = i, \mathbf{Y}_{1:t} = \mathbf{y}_{1:t}) \\
&= \sum_i \mathbb{P}(X_t = j, Y_t = y_t \mid X_{t-1} = i, \mathbf{Y}_{1:t-1} = \mathbf{y}_{1:t-1}) \, \alpha_{t-1}(i) \\
&= \sum_i \mathbb{P}(X_t = j, Y_t = y_t \mid X_{t-1} = i) \, \alpha_{t-1}(i). \\
&= \sum_i \mathbb{P}(Y_t = y_t \mid X_t = j) \, \mathbb{P}(X_t = j \mid X_{t-1} = i) \, \alpha_{t-1}(i) \,.
\end{aligned}
$$

In particular, we find the recurrence relation

$$\alpha_t(0) = a_{0\,y_t} \, (1-r) \, \alpha_{t-1}(0) \quad \text{and} \quad \alpha_t(1) = a_{1\,y_t} \{ r \, \alpha_{t-1}(0) + \alpha_{t-1}(1) \} \,,$$

with $a_{ij} = \mathbb{P}(Y = j \mid X = i)$, $i, j \in \{0, 1\}$ (thus $a_{00} = 1 - a$, $a_{01} = a$, $a_{10} = 1 - b$, $a_{11} = b$), and initial values

$$\alpha_1(0) = a^{y_1} (1-a)^{1-y_1} (1-r) \quad \text{and} \quad \alpha_1(1) = b^{y_1} (1-b)^{1-y_1} r \,.$$

In Figure A.3 a plot is given of the probability $\mathbb{P}(X_t = 1 \mid \mathbf{Y}_{1:t} = \mathbf{y}_{1:t}) = \alpha_t(1)/(\alpha_t(1) + \alpha_t(2))$, as a function of t, for a test case with $a = 0.4$, $b = 0.6$, and $r = 0.01$. In this particular case $T = 49$. We see a dramatic change in the graph after the disruption takes effect.

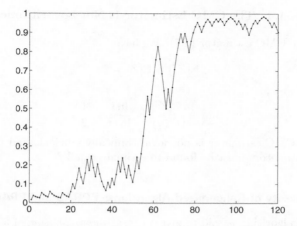

Figure A.3: Probability $\mathbb{P}(X_t = 1 \mid \mathbf{Y}_{1:t} = \mathbf{y}_{1:t})$ as a function of t.

A.8 COMPLEXITY

In this section we consider various *complexity* concepts that describe the efficiency of algorithms using some form of randomness during their execution.

A.8.1 Complexity of Rare-Event Algorithms

The theoretical framework in which one typically examines rare-event probability estimation is based on *complexity theory*, as introduced in [1, 10]. In particular, the estimators are classified either as *polynomial-time* or as *exponential-time*. It is shown in [1, 12] that for an arbitrary estimator, $\widehat{\ell}$ of ℓ, to be polynomial-time as a function of some γ, it suffices that its squared coefficient of variation, κ^2, or its relative error, κ, is bounded in γ by some polynomial function, $p(\gamma)$. For such polynomial-time estimators, the required sample size to achieve a fixed relative error does not grow too fast as the event becomes rarer.

Consider the estimator (4.7) and assume that ℓ becomes very small as $\gamma \to \infty$. Note that

$$\mathbb{E}[Z^2] \geqslant (\mathbb{E}[Z])^2 = \ell^2 .$$

Hence, the best one can hope for with such an estimator is that its second moment of Z^2 decreases proportionally to ℓ^2 as $\gamma \to \infty$. We say that the rare-event estimator (4.7) has *bounded relative error* if for all γ

$$\mathbb{E}[Z^2] \leqslant c\,\ell^2 \tag{A.31}$$

for some fixed $c \geqslant 1$. Because bounded relative error is not always easy to achieve, the following weaker criterion is often used. We say that the estimator (4.7) is *logarithmically efficient* (sometimes called *asymptotically optimal*) if

$$\lim_{\gamma \to \infty} \frac{\ln \mathbb{E}[Z^2]}{\ln \ell^2} = 1 . \tag{A.32}$$

■ **EXAMPLE A.4 The CMC Estimator Is Not Logarithmically Efficient**

Consider the CMC estimator (4.7). We have

$$\mathbb{E}[Z^2] = \mathbb{E}[Z] = \ell \, ,$$

so that

$$\lim_{\gamma \to \infty} \frac{\ln \mathbb{E}[Z^2]}{\ln \ell^2(\gamma)} = \frac{\ln \ell}{\ln \ell^2} = \frac{1}{2} \, .$$

Hence, the CMC estimator is not logarithmically efficient, and therefore alternative estimators must be found to estimate small ℓ.

A.8.2 Complexity of Randomized Algorithms: FPRAS and FPAUS

A randomized algorithm is said to give an (ε, δ)-*approximation* of a parameter z if its output Z satisfies

$$\mathbb{P}(|Z - z| \leqslant \varepsilon z) \geq 1 - \delta \, ; \tag{A.33}$$

that is, the "relative error" $|Z - z|/z$ of the approximation Z lies with high probability ($> 1 - \delta$) below some small number ε.

One of the main tools in proving (A.33) for various randomized algorithms is the so-called *Chernoff bound*, which states that for any random variable Y and any number a

$$\mathbb{P}(Y \leqslant a) \leqslant \min_{\theta > 0} e^{\theta a} \, \mathbb{E}[e^{-\theta Y}] \, . \tag{A.34}$$

Namely, for any fixed a and $\theta > 0$, define the functions $H_1(z) = I_{\{z \leqslant a\}}$ and $H_2(z) = e^{\theta(a-z)}$. Then, clearly, $H_1(z) \leqslant H_2(z)$ for all z. As a consequence, for any θ,

$$\mathbb{P}(Y \leqslant a) = \mathbb{E}[H_1(Y)] \leqslant \mathbb{E}[H_2(Y)] = e^{\theta a} \, \mathbb{E}[e^{-\theta Y}] \, .$$

The bound (A.34) now follows by taking the smallest such θ. An important application is the following.

Theorem A.8.1 *Let $X_1, \ldots, X_n$ be iid* Ber(p) *random variables. Then their sample mean provides an (ε, δ)-approximation for p, that is,*

$$\mathbb{P}\left(\left| \frac{1}{n} \sum_{i=1}^{n} X_i - p \right| \leqslant \varepsilon p \right) \geq 1 - \delta \, , \tag{A.35}$$

provided that $n \geq 3 \ln(2/\delta)/(p\varepsilon^2)$.

Proof: Let $Y = X_1 + \cdots + X_n$, and $\ell_L = \mathbb{P}(Y \leqslant (1 - \varepsilon)np)$. Because $\mathbb{E}[e^{-\theta Y}] = \mathbb{E}[e^{-\theta X_1}]^n = (1 - p + pe^{-\theta})^n$, the Chernoff bound gives

$$\ell_L \leqslant e^{\theta np(1-\varepsilon)} (1 - p + pe^{-\theta})^n \, , \tag{A.36}$$

for any $\theta > 0$. By direct differentiation we find that the optimal θ^* (giving the smallest upper bound) is

$$\theta^* = \ln \left(\frac{\varepsilon p - p + 1}{(\varepsilon - 1)(p - 1)} \right) \, .$$

It is not difficult to verify (see Problem 1.3) that by substituting $\theta = \theta^*$ in the right-hand side of (A.36) and taking the logarithms on both sides, $\ln(\ell_L)$ can be upper-bounded by $np\,h(p,\varepsilon)$, where $h(p,\varepsilon)$ is given by

$$h(\varepsilon,p) = -\frac{1}{p}\ln\left(1+\frac{\varepsilon\,p}{1-p}\right) + (1-\varepsilon)\theta^* \, . \tag{A.37}$$

For fixed $0 < \varepsilon < 1$, the function $h(p,\varepsilon)$ is monotonically decreasing in p, $0 < p < 1$. Namely,

$$\frac{\partial h(\varepsilon,p)}{\partial p} = \frac{1}{p^2}\left[-\frac{\varepsilon p}{1-p} + \ln\left(1+\frac{\varepsilon\,p}{1-p}\right)\right] < 0 \, ,$$

since $-y + \ln(1+y) < 0$ for any $y > 0$. It follows that

$$h(\varepsilon,p) \leqslant h(\varepsilon,0) = -\varepsilon - (1-\varepsilon)\ln(1-\varepsilon) = -\sum_{k=2}^{\infty}\frac{\varepsilon^k}{(k-1)k} \leqslant \frac{-\varepsilon^2}{2} \, .$$

And therefore,

$$\ell_L \leqslant \exp\left(-\frac{np\,\varepsilon^2}{2}\right) \, .$$

Similarly, Chernoff's bound provides the following upper bound for $\ell_U = \mathbb{P}(Y \geqslant (1+\varepsilon)np) = \mathbb{P}(-Y \leqslant -(1+\varepsilon)np)$:

$$\ell_U \leqslant \exp\left(-\frac{np\,\varepsilon^2}{2+\frac{2}{3}\varepsilon}\right) \tag{A.38}$$

for all $0 < \varepsilon < 1$; see Problem 1.4. In particular, $\ell_L + \ell_U \leqslant 2\exp(-np\varepsilon^2/3)$. Combining these results gives

$$\mathbb{P}(|Y - np| \leqslant np\varepsilon) = 1 - \ell_L - \ell_U \geqslant 1 - 2\mathrm{e}^{-np\,\varepsilon^2/3},$$

so that by choosing $n \geqslant 3\ln(2/\delta)/(p\varepsilon^2)$, the probability above is guaranteed to be greater than or equal to $1-\delta$. $\qquad\qquad\qquad\square$

Definition A.8.1 (FPRAS) A randomized algorithm is said to provide a *fully polynomial randomized approximation scheme (FPRAS)* if, for any input vector **x** and any parameters $\varepsilon > 0$ and $0 < \delta < 1$, the algorithm outputs an (ε,δ)-approximation to the desired quantity $z(\mathbf{x})$ in time that is polynomial in $\varepsilon^{-1}, \ln\delta^{-1}$ and the size n of the input vector **x**.

Thus, the sample mean in Theorem A.8.1 provides an FPRAS for estimating p. Note that the input vector **x** consists of the Bernoulli variables $X_1, \ldots, X_n$.

To illustrate the usefulness of the FPRAS concept, consider the counting problem of SATs in *disjunctive normal form* (DNF), where the objective is to count how many truth assignments satisfy a SAT formula of the form

$$F = \bigvee_{i=1}^{m}\bigwedge_{j=1}^{n_i} l_{i_j} \, .$$

Let $\mathscr{X}$ be the set of all assignments, and let $\mathscr{X}_i$ be the subset of all assignments satisfying clause C_i, $i = 1, \ldots, m$. Denote by $\mathscr{X}^*$ the set of assignments satisfying *at least one* of the clauses $C_1, \ldots, C_m$, that is, $\mathscr{X}^* = \cup_{i=1}^m \mathscr{X}_i$. The DNF counting problem is to compute $|\mathscr{X}^*|$. It is readily seen that if a clause C_i has n_i literals, then the number of truth assignments is $|\mathscr{X}_i| = 2^{n-n_i}$. Clearly, $0 \leqslant |\mathscr{X}^*| \leqslant |\mathscr{X}| = 2^n$ and, because an assignment can satisfy more than one clause, also $|\mathscr{X}^*| \leqslant \sum_{i=1}^m |\mathscr{X}_i|$.

Next we show how to construct a randomized algorithm for this #P-complete problem. The first step is to *augment* the state space $\mathscr{X}$ with an index set $\{1, \ldots, m\}$. Specifically, define

$$\mathscr{A} = \{(i, \mathbf{x}) : \mathbf{x} \in \mathscr{X}_i, \ i = 1, \ldots, m\} . \tag{A.39}$$

This set is illustrated in Figure A.4. In this case we have $m = 7$, $|\mathscr{X}_1| = 3$, $|\mathscr{X}_2| = 2$, and so on.

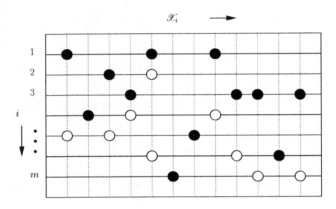

Figure A.4: Sets $\mathscr{A}$ (formed by all points) and $\mathscr{A}^*$ (formed by the black points).

For a fixed i we can identify the subset $\mathscr{A}_i = \{(i, \mathbf{x}) : \mathbf{x} \in \mathscr{X}_i\}$ of $\mathscr{A}$ with the set $\mathscr{X}_i$. In particular, the two sets have the same number of elements. Next, we construct a subset $\mathscr{A}^*$ of $\mathscr{A}$ with size exactly equal to $|\mathscr{X}^*|$. This is done by associating with each assignment in $\mathscr{X}^*$ exactly one pair $(i, \mathbf{x})$ in $\mathscr{A}$. In particular, we can use the pair with the smallest clause index number; that is, we can define $\mathscr{A}^*$ as

$$\mathscr{A}^* = \{(i, \mathbf{x}) : \mathbf{x} \in \mathscr{X}_i, \ \mathbf{x} \notin \mathscr{X}_k \ \text{for} \ k < i, \ i = 1, \ldots, m\} .$$

In Figure A.4, $\mathscr{A}^*$ is represented by the black points. Note that each element of $\mathscr{X}^*$ is represented once in $\mathscr{A}$; that is, each "column" has exactly one black point.

Since $|\mathscr{A}| = \sum_{i=1}^m |\mathscr{X}_i| = \sum_{i=1}^m 2^{n-n_i}$ is available, we can estimate $|\mathscr{X}^*| = |\mathscr{A}^*| = |\mathscr{A}| \ell$ by estimating $\ell = |\mathscr{A}^*|/|\mathscr{A}|$. Note that this is a simple application of (9.3). The ratio ℓ can be estimated by generating pairs *uniformly* in $\mathscr{A}$ and counting how often they occur in $\mathscr{A}^*$. It turns out that, for the union of sets and, in particular, for the DNF problem, generating pairs uniformly in $\mathscr{A}$ is straightforward and can be done in two stages using the composition method. Namely, first choose an index i, $i = 1, \ldots, m$ with probability

$$p_i = \frac{|\mathscr{A}_i|}{|\mathscr{A}|} = \frac{|\mathscr{X}_i|}{\sum_{i=1}^m |\mathscr{X}_i|} = \frac{2^{n-n_i}}{\sum_{i=1}^m 2^{n-n_i}} ;$$

next, choose an assignment **x** uniformly from $\mathscr{X}_i$. This can be done by choosing a value 1 or 0 with equal probability and independently for each literal that is *not in clause i*. The resulting probability of choosing the pair $(i, \mathbf{x})$ can be found via conditioning as

$$\mathbb{P}(I = i, \mathbf{X} = \mathbf{x}) = \mathbb{P}(I = i)\,\mathbb{P}(\mathbf{X} = \mathbf{x}\,|\,I = i) = \frac{|\mathscr{A}_i|}{|\mathscr{A}|}\,\frac{1}{|\mathscr{A}_i|} = \frac{1}{|\mathscr{A}|}\,,$$

which corresponds to the uniform distribution on $\mathscr{A}$. The DNF counting algorithm can be written as follows [11]:

Algorithm A.8.1: DNF Counting Algorithm

---

input : DNF formula with m clauses and n literals.
output: Estimator of the number of satisfying assignments, $|\mathscr{X}^*|$.
1 $Z \leftarrow 0$
2 **for** $k = 1$ **to** N **do**
3 $\quad$ With probability $p_i \propto |\mathscr{X}_i|$, choose uniformly and randomly an assignment $\mathbf{X} \in \mathscr{X}_i$.
4 $\quad$ **if** **X** *is not in any* $\mathscr{X}_i$ *for* $i > k$ **then** $Z \leftarrow Z + 1$
5 $\widehat{|\mathscr{X}^*|} \leftarrow \frac{Z}{N}\sum_{i=1}^{m}|\mathscr{X}_i|$
6 **return** $\widehat{|\mathscr{X}^*|}$

Below we present a theorem [11] stating that Algorithm A.8.1 provides an FPRAS for counting the number of satisfying assignments in a DNF formula. Its proof is based on the fact that $\mathscr{A}^*$ is relatively *dense* in $\mathscr{A}$. Specifically, it uses the fact that for the union of sets $\ell = |\mathscr{A}^*|/|\mathscr{A}| \geq 1/m$, which follows directly from the fact that each assignment can satisfy at most m clauses.

Theorem A.8.2 (DNF Counting Theorem) *The DNF counting Algorithm A.8.1 is an FPRAS, provided that* $N \geq 3m \ln(2/\delta)/\varepsilon^2$.

Proof: In Line 3 of Algorithm A.8.1 an element is chosen uniformly from $\mathscr{A}$. The probability that this element belongs to $\mathscr{A}^*$ is at least $1/m$. Choose

$$N = \frac{3m}{\varepsilon^2}\ln\frac{2}{\delta}\,, \tag{A.40}$$

where $\varepsilon > 0$ and $\delta > 0$. Then N is polynomial in m, ε^{-1} and $\ln\frac{1}{\delta}$, and the processing time of each sample is polynomial in m. By Theorem A.8.1, we find that with the number of samples N as in (A.40), the quantity Z/N (see Algorithm A.8.1) provides an (ε, δ)-approximation to ℓ and thus $\widehat{|\mathscr{X}^*|}$ provides an (ε, δ)-approximation to $|\mathscr{X}^*|$.

$\square$

Since exact uniform sampling is not always feasible, MCMC or splitting techniques can be used to sample *approximately* from a uniform distribution.

Let Z be the random output of a sampling algorithm on a finite sample space $\mathscr{X}$. We say that the sampling algorithm generates an ε-*uniform sample* from $\mathscr{X}$ if, for any $\mathscr{Y} \subset \mathscr{X}$,

$$\left|\mathbb{P}(Z \in \mathscr{Y}) - \frac{|\mathscr{Y}|}{|\mathscr{X}|}\right| \leqslant \varepsilon\,. \tag{A.41}$$

Definition A.8.2 (FPAUS) A sampling algorithm is called a *fully polynomial almost uniform sampler (FPAUS)* if, given an input vector $\mathbf{x}$ and a parameter $\varepsilon > 0$, the algorithm generates an ε-uniform sample from $\mathscr{X}(\mathbf{x})$ and runs in time that is polynomial in $\ln \varepsilon^{-1}$ and the size of the input vector $\mathbf{x}$.

■ **EXAMPLE A.5   FPAUS for Independent Sets: Example 9.9.2 (Continued)**

An FPAUS for independent sets takes as input a graph $G = (V, E)$ and a parameter $\varepsilon > 0$. The sample space $\mathscr{X}$ consists of all independent sets in G, with the output being an ε-uniform sample from $\mathscr{X}$. The time required to produce such an ε-uniform sample should be polynomial in the size of the graph and $\ln \varepsilon^{-1}$. The final goal is to prove that given an FPAUS, one can construct a corresponding FPRAS. Such a proof is based on the product formula (9.3) and is given in Theorem 10.5 of [11].

For the knapsack problem, it can be shown that there is an FPRAS *provided* that there exists an FPAUS; see also Exercise 10.6 of [11]. However, the existence of such a method is still an open problem [7].

A.8.3   SATs in CNF

A CNF SAT counting problem can be translated into a DNF one by using De Morgan's law,

$$\left(\bigcap \mathscr{X}_i\right)^c = \bigcup \mathscr{X}_i^c \quad \text{and} \quad \left(\bigcup \mathscr{X}_i\right)^c = \bigcap \mathscr{X}_i^c. \tag{A.42}$$

Thus, if the $\{\mathscr{X}_i\}$ are subsets of some set $\mathscr{X}$, then

$$\left|\bigcap \mathscr{X}_i\right| = |\mathscr{X}| - \left|\bigcup \mathscr{X}_i^c\right|. \tag{A.43}$$

In particular, consider a CNF SAT counting problem and let $\mathscr{X}_i$ be the set of all assignments that satisfy the i-th clause, C_i, $i = 1, \ldots, m$. Recall that C_i is of the form $l_{i1} \vee l_{i_2} \vee \cdots \vee l_{i_{n_i}}$. The set of assignments satisfying *all* clauses is $\mathscr{X}^* = \cap \mathscr{X}_i$. In view of (A.43), to count $\mathscr{X}^*$ one could instead count the number of elements in $\cup \mathscr{X}_i^c$. Now $\mathscr{X}_i^c$ is the set of all assignments that satisfy the clause $\bar{l}_{i_1} \wedge \bar{l}_{i2} \wedge \cdots \wedge \bar{l}_{i_{n_i}}$. Thus, the problem is translated into a DNF SAT counting problem.

As an example, consider the CNF SAT counting problem with clause matrix

$$A = \begin{pmatrix} -1 & 1 & 0 \\ 0 & -1 & 1 \\ -1 & 0 & -1 \end{pmatrix}.$$

In this case $\mathscr{X}^*$ comprises three assignments, namely, $(1, 0, 0)$, $(1, 1, 0)$, and $(1, 1, 1)$. Consider next the DNF SAT counting problem with clause matrix $-A$. Then the set of assignments that satisfy at least one clause for this problem is $\{(0, 0, 0), (0, 0, 1), (0, 1, 0), (0, 1, 1), (1, 0, 1)\}$, which is exactly the complement of $\mathscr{X}^*$.

The results above seem to imply that there exists an FPRAS for SAT counting in CNF form. Unfortunately, this is not the case, because the translation step does not scale polynomially with the size of the problem.

A.8.4 Complexity of Stochastic Programming Problems

Consider the following optimization problem:

$$\ell^* = \min_{\mathbf{u} \in \mathscr{U}} \ell(\mathbf{u}) = \min_{\mathbf{u} \in \mathscr{U}} \mathbb{E}_f[H(\mathbf{X}; \mathbf{u})], \qquad (A.44)$$

where it is assumed that $\mathbf{X}$ is a random vector with known pdf f having support $\mathscr{X} \subset \mathbb{R}^n$, and $H(\mathbf{X}; \mathbf{u})$ is the sample function depending on $\mathbf{X}$ and the decision vector $\mathbf{u} \in \mathbb{R}^m$.

As an example, consider a two-stage stochastic programming problem with recourse, which is an optimization problem that is divided into two stages. At the first stage, one has to make a decision on the basis of some available information. At the second stage, after a realization of the uncertain data becomes known, an optimal second-stage decision is made. Such a stochastic programming problem can be written in the form (A.44), with $H(\mathbf{X}; \mathbf{u})$ being the optimal value of the second-stage problem.

We now discuss the issue of how difficult it is to solve a stochastic program of type (A.44). We should expect that this problem is at least as difficult as minimizing $\ell(\mathbf{u})$, $\mathbf{u} \in \mathscr{U}$ in the case where $\ell(\mathbf{u})$ is given *explicitly*, say by a closed-form analytic expression or, more generally, by an "oracle" capable of computing the values and the derivatives of $\ell(\mathbf{u})$ at every given point. As far as problems of minimization of $\ell(\mathbf{u})$, $\mathbf{u} \in \mathscr{U}$, with an explicitly given objective are concerned, the solvable case is known: this is the convex programming case. That is, $\mathscr{U}$ is a closed convex set and $\ell : \mathscr{U} \to \mathbb{R}$ is a convex function. It is known that generic convex programming problems satisfying mild computability and boundedness assumptions can be solved in polynomial time. In contrast to this, typical nonconvex problems turn out to be NP-hard.

We should also stress that a claim that "such and such problem is difficult" relates to a *generic* problem and does *not* imply that the problem has no solvable particular cases. When speaking about conditions under which the stochastic program (A.44) is efficiently solvable, it makes sense to assume that $\mathscr{U}$ is a closed convex set and $\ell(\cdot)$ is convex on $\mathscr{U}$. We gain from a technical viewpoint (and do not lose much from a practical viewpoint) by assuming $\mathscr{U}$ to be bounded. These assumptions, plus mild technical conditions, would be sufficient to make (A.44) easy (manageable) if $\ell(\mathbf{u})$ were given explicitly. However, in stochastic programming, it makes no sense to assume that we can compute efficiently the expectation in (A.44), thus arriving at an explicit representation of $\ell(\mathbf{u})$. If this were the case, there would be no necessity to treat (A.44) as a stochastic program.

We argue now that stochastic programming problems of the form (A.44) can be solved reasonably efficiently by using Monte Carlo sampling techniques, provided that the probability distribution of the random data is not "too bad" and certain general conditions are met. In this respect, we should explain what we mean by "solving" stochastic programming problems. Let us consider, for example, two-stage linear stochastic programming problems with recourse. Such problems can be written in the form (A.44) with

$$\mathscr{U} = \{\mathbf{u} : A\mathbf{u} = \mathbf{b}, \ \mathbf{u} \geqslant \mathbf{0}\} \text{ and } H(\mathbf{X}; \mathbf{u}) = \langle \mathbf{c}, \mathbf{u} \rangle + Q(\mathbf{X}; \mathbf{u}),$$

where $\langle \mathbf{c}, \mathbf{u} \rangle$ is the cost of the first-stage decision and $Q(\mathbf{X}; \mathbf{u})$ is the optimal value of the second-stage problem:

$$\min_{\mathbf{y} \geqslant 0} \langle \mathbf{q}, \mathbf{y} \rangle \text{ subject to } \mathbf{Tu} + \mathbf{Wy} \geqslant \mathbf{h} . \tag{A.45}$$

Here, $\langle \cdot, \cdot \rangle$ denotes the inner product. $\mathbf{X}$ is a vector whose elements are composed from elements of vectors $\mathbf{q}$ and $\mathbf{h}$ and matrices $\mathbf{T}$ and $\mathbf{W}$, which are assumed to be random.

If we assume that the random data vector $\mathbf{X} = (\mathbf{q}, \mathbf{W}, \mathbf{T}, \mathbf{h})$ takes K different values (called *scenarios*) $\{\mathbf{X}_k, k = 1, \ldots, K\}$, with respective probabilities $\{p_k, k = 1, \ldots, K\}$, then the obtained two-stage problem can be written as one large linear programming problem:

$$\min_{\mathbf{u}, \mathbf{y}_1, \ldots, \mathbf{y}_K} \quad \langle \mathbf{c}, \mathbf{u} \rangle + \sum_{k=1}^{K} p_k \langle \mathbf{q}_k, \mathbf{y}_k \rangle$$
$$\text{s.t} \quad \mathbf{Au} = \mathbf{b}, \ \mathbf{T}_k \mathbf{u} + \mathbf{W}_k \mathbf{y}_k \geqslant \mathbf{h}_k, \ k = 1, \ldots, K, \tag{A.46}$$
$$\mathbf{u} \geqslant \mathbf{0}, \ \mathbf{y}_k \geqslant \mathbf{0}, \ k = 1, \ldots, K.$$

If the number of scenarios K is not too large, then the above linear programming problem (A.46) can be solved accurately in a reasonable period of time. However, even a crude discretization of the probability distribution of $\mathbf{X}$ typically results in an exponential growth of the number of scenarios with the increase of the dimension of $\mathbf{X}$. Suppose, for example, that the components of the random vector $\mathbf{X}$ are mutually independently distributed, each having a small number r of possible realizations. Then the size of the corresponding input data grows linearly in n (and r), while the number of scenarios $K = r^n$ grows exponentially.

We would like to stress that from a practical point of view, it does not make sense to try to solve a stochastic programming problem with high precision. A numerical error resulting from an inaccurate estimation of the involved probability distributions, modeling errors, and so on, can be far bigger than the optimization error. We argue now that two-stage stochastic problems can be solved efficiently with reasonable accuracy, provided that the following conditions are met:

(1) The feasible set $\mathscr{U}$ is fixed (deterministic).

(2) For all $\mathbf{u} \in \mathscr{U}$ and $\mathbf{X} \in \mathscr{X}$, the objective function $H(\mathbf{X}; \mathbf{u})$ is real-valued.

(3) The considered stochastic programming problem can be solved efficiently (by a deterministic algorithm) if the number of scenarios is not too large.

When applied to two-stage stochastic programming, the conditions (1) and (2) mean that the recourse is relatively complete and the second-stage problem is bounded from below. Note that it is said that the recourse is *relatively complete*, if for every $\mathbf{u} \in \mathscr{U}$ and every possible realization of random data, the second-stage problem is feasible. The above condition (3) certainly holds in the case of two-stage *linear* stochastic programming with recourse.

In order to proceed, let us consider the following Monte Carlo sampling approach. Suppose that we can generate an iid random sample $\mathbf{X}_1, \ldots, \mathbf{X}_N$ from $f(\mathbf{x})$, and we can estimate the expected value function $\ell(\mathbf{u})$ by the sample average

$$\widehat{\ell}(\mathbf{u}) = \frac{1}{N} \sum_{j=1}^{N} H(\mathbf{X}_j; \mathbf{u}) . \tag{A.47}$$

Note that $\widehat{\ell}$ depends on the sample size N and on the generated sample, and in that sense is random. Consequently, we approximate the true problem (A.44) by the following approximated one:

$$\min_{\mathbf{u} \in \mathscr{U}} \widehat{\ell}(\mathbf{u}) \ . \tag{A.48}$$

We refer to (A.48) as the *stochastic counterpart* or *sample average approximation* problem. The optimal value $\widehat{\ell}^*$ and the set $\widehat{\mathscr{U}}^*$ of optimal solutions of the stochastic counterpart problem (A.48) provide estimates of their true counterparts, ℓ^* and $\mathscr{U}^*$, of problem (A.44). It should be noted that once the sample is generated, $\widehat{\ell}(\mathbf{u})$ becomes a deterministic function and problem (A.48) becomes a stochastic programming problem with N scenarios $\mathbf{X}_1, \ldots, \mathbf{X}_N$ taken with equal probabilities $1/N$. It also should be mentioned that the stochastic counterpart method is *not* an algorithm. One still has to solve the obtained problem (A.48) by employing an appropriate (deterministic) algorithm.

By the law of large numbers (see Theorem 1.11.1) $\widehat{\ell}(\mathbf{u})$ converges (point-wise in $\mathscr{U}$) with probability 1 to $\ell(\mathbf{u})$ as N tends to infinity. Therefore, it is reasonable to expect for $\widehat{\ell}^*$ and $\widehat{\mathscr{U}}^*$ to converge to their counterparts of the true problem (A.44) with probability 1 as N tends to infinity. And indeed, such convergence can be proved under mild regularity conditions. However, for a fixed $\mathbf{u} \in \mathscr{U}$, convergence of $\widehat{\ell}(\mathbf{u})$ to $\ell(\mathbf{u})$ is notoriously slow. By the central limit theorem (see Theorem 1.11.2), it is of order $\mathcal{O}(N^{-1/2})$. The rate of convergence can be improved, sometimes significantly, by variance reduction methods. However, using Monte Carlo techniques, one cannot evaluate the expected value $\ell(\mathbf{u})$ very accurately.

The following analysis is based on the exponential bounds of the *large deviations* theory. Denote by $\mathscr{U}^\varepsilon$ and $\widehat{\mathscr{U}}^\varepsilon$ the sets of ε-optimal solutions of the true and stochastic counterpart problems, respectively, that is, $\bar{\mathbf{u}} \in \mathscr{U}^\varepsilon$ iff $\bar{\mathbf{u}} \in \mathscr{U}$ and $\ell(\bar{\mathbf{u}}) \leqslant \inf_{\mathbf{u} \in \mathscr{U}} \ell(\mathbf{u}) + \varepsilon$. Note that for $\varepsilon = 0$ the set $\mathscr{U}^0$ coincides with the set of the optimal solutions of the true problem. Choose accuracy constants $\varepsilon > 0$ and $0 \leqslant \delta < \varepsilon$ and the confidence (significance) level $\alpha \in (0, 1)$. Suppose, for the moment, that the set $\mathscr{U}$ is finite, although its cardinality $|\mathscr{U}|$ can be very large. Then, using Cramér's large deviations theorem, it can be shown [9] that there exists a constant $\eta(\varepsilon, \delta)$ such that

$$N \geqslant \frac{1}{\eta(\varepsilon, \delta)} \ln \left(\frac{|\mathscr{U}|}{\alpha} \right) \tag{A.49}$$

guarantees that the probability of the event $\{\widehat{\mathscr{U}}^\delta \subset \mathscr{U}^\varepsilon\}$ is at least $1 - \alpha$. That is, for any N bigger than the right-hand side of (A.49), we are guaranteed that any δ-optimal solution of the corresponding stochastic counterpart problem provides an ε-optimal solution of the true problem with probability at least $1 - \alpha$. In other words, solving the stochastic counterpart problem with accuracy δ guarantees solving the true problem with accuracy ε with probability at least $1 - \alpha$.

The number $\eta(\varepsilon, \delta)$ in the estimate (A.49) is defined as follows: Consider a mapping $\pi : \mathscr{U} \setminus \mathscr{U}^\varepsilon \to \mathscr{U}$ such that $\ell(\pi(\mathbf{u})) \leqslant \ell(\mathbf{u}) - \varepsilon$ for all $\mathbf{u} \in \mathscr{U} \setminus \mathscr{U}^\varepsilon$. Such mappings do exist, although not uniquely. For example, any mapping $\pi : \mathscr{U} \setminus \mathscr{U}^\varepsilon \to \mathscr{U}^0$ satisfies this condition. The choice of such a mapping gives a certain flexibility to the corresponding estimate of the sample size. For $\mathbf{u} \in \mathscr{U}$, consider the random variable

$$Y_{\mathbf{u}} = H(\mathbf{X}; \pi(\mathbf{u})) - H(\mathbf{X}; \mathbf{u}) \ ,$$

its moment generating function $M_{\mathbf{u}}(t) = \mathbb{E}\left[e^{tY_{\mathbf{u}}}\right]$, and the large deviations *rate function*

$$I_{\mathbf{u}}(z) = \sup_{t \in \mathbb{R}} \left\{ tz - \ln M_{\mathbf{u}}(t) \right\} .$$

Note that $I_{\mathbf{u}}(\cdot)$ is the conjugate of the function $\ln M_{\mathbf{u}}(\cdot)$ in the sense of convex analysis. Note also that, by construction of mapping $\pi(\mathbf{u})$, the inequality

$$\mu_{\mathbf{u}} = \mathbb{E}\left[Y_{\mathbf{u}}\right] = \ell(\pi(\mathbf{u})) - \ell(\mathbf{u}) \leqslant -\varepsilon \tag{A.50}$$

holds for all $\mathbf{u} \in \mathscr{U} \setminus \mathscr{U}^{\varepsilon}$. Finally, we define

$$\eta(\varepsilon, \delta) = \min_{\mathbf{u} \in \mathscr{U} \setminus \mathscr{U}^{\varepsilon}} I_{\mathbf{u}}(-\delta) . \tag{A.51}$$

Because of (A.50) and since $\delta < \varepsilon$, the number $I_{\mathbf{u}}(-\delta)$ is positive, provided that the probability distribution of $Y_{\mathbf{u}}$ is not too bad. Specifically, if we assume that the moment generating function $M_{\mathbf{u}}(t)$, of $Y_{\mathbf{u}}$, is finite valued for all t in a neighborhood of 0, then the random variable $Y_{\mathbf{u}}$ has finite moments and $I_{\mathbf{u}}(\mu_{\mathbf{u}}) = I'(\mu_{\mathbf{u}}) = 0$, and $I''(\mu_{\mathbf{u}}) = 1/\sigma_{\mathbf{u}}^2$ where $\sigma_{\mathbf{u}}^2 = \mathrm{Var}\left[Y_{\mathbf{u}}\right]$. Consequently, $I_{\mathbf{u}}(-\delta)$ can be approximated by using the second-order Taylor expansion, as follows:

$$I_{\mathbf{u}}(-\delta) \approx \frac{(-\delta - \mu_{\mathbf{u}})^2}{2\sigma_{\mathbf{u}}^2} \geqslant \frac{(\varepsilon - \delta)^2}{2\sigma_{\mathbf{u}}^2} .$$

This suggests that one can expect the constant $\eta(\varepsilon, \delta)$ to be of order $(\varepsilon - \delta)^2$. And indeed, this can be ensured by various conditions. Consider the following ones:

(A1) *There exists a constant $\sigma > 0$ such that for any $\mathbf{u} \in \mathscr{U} \setminus \mathscr{U}^{\varepsilon}$, the moment generating function $M_{\mathbf{u}}^*(t)$ of the random variable $Y_{\mathbf{u}} - \mathbb{E}\left[Y_{\mathbf{u}}\right]$ satisfies*

$$M_{\mathbf{u}}^*(t) \leqslant \exp\left(\sigma^2 t^2 / 2\right), \quad \forall t \in \mathbb{R} . \tag{A.52}$$

Note that the random variable $Y_{\mathbf{u}} - \mathbb{E}\left[Y_{\mathbf{u}}\right]$ has zero mean. Moreover, if it has a normal distribution, with variance $\sigma_{\mathbf{u}}^2$, then its moment generating function is equal to the right-hand side of (A.52). Condition (A.52) means that the tail probabilities $\mathbb{P}\left(|H(\mathbf{X}; \pi(\mathbf{u})) - H(\mathbf{X}; \mathbf{u})| > t\right)$ are bounded from above by $\mathcal{O}(1) \exp\left(-t^2 / (2\sigma_{\mathbf{u}}^2)\right)$. Note that by $\mathcal{O}(1)$ we denote generic absolute constants. This condition certainly holds if the distribution of the considered random variable has a bounded support. Condition (A.52) implies that $M_{\mathbf{u}}(t) \leqslant \exp(\mu_{\mathbf{u}} t + \sigma^2 t^2 / 2)$. It follows that

$$I_{\mathbf{u}}(z) \geqslant \sup_{t \in \mathbb{R}} \left\{ tz - \mu_{\mathbf{u}} t - \sigma^2 t^2 / 2 \right\} = \frac{(z - \mu_{\mathbf{u}})^2}{2\sigma^2} , \tag{A.53}$$

and hence, for any $\varepsilon > 0$ and $\delta \in [0, \varepsilon)$,

$$\eta(\varepsilon, \delta) \geqslant \frac{(-\delta - \mu_{\mathbf{u}})^2}{2\sigma^2} \geqslant \frac{(\varepsilon - \delta)^2}{2\sigma^2} . \tag{A.54}$$

It follows that, under assumption (A1), the estimate (A.49) can be written as

$$N \geqslant \frac{2\sigma^2}{(\varepsilon - \delta)^2} \ln\left(\frac{|\mathscr{U}|}{\alpha}\right) . \tag{A.55}$$

Remark A.8.1 Condition (A.52) can be replaced by a more general one,

$$M_{\mathbf{u}}^*(t) \leqslant \exp(\psi(t)), \ \ \forall\, t \in \mathbb{R}\,, \tag{A.56}$$

where $\psi(t)$ is a convex even function with $\psi(0) = 0$. Then $\ln M_{\mathbf{u}}(t) \leqslant \mu_{\mathbf{u}} t + \psi(t)$, and hence $I_{\mathbf{u}}(z) \geqslant \psi^*(z - \mu_{\mathbf{u}})$, where ψ^* is the conjugate of the function ψ. It follows that

$$\eta(\varepsilon, \delta) \geqslant \psi^*(-\delta - \mu_{\mathbf{u}}) \geqslant \psi^*(\varepsilon - \delta)\,. \tag{A.57}$$

For example, instead of assuming that the bound (A.52) holds for all $t \in \mathbb{R}$, we can assume that it holds for all t in a finite interval $[-a, a]$, where $a > 0$ is a given constant. That is, we can take $\psi(t) = \sigma^2 t/2$ if $|t| \leqslant a$ and $\psi(t) = +\infty$ otherwise. In that case, $\psi^*(z) = z^2/(2\sigma^2)$ for $|z| \leqslant a\sigma^2$ and $\psi^*(z) = a|z| - a^2\sigma^2/2$ for $|z| > a\sigma^2$.

A key feature of the estimate (A.55) is that the required sample size N depends *logarithmically* both on the size of the feasible set $\mathscr{U}$ and on the significance level α. The constant σ, postulated in assumption (A1), measures, in some sense, the variability of the considered problem. For example, for $\delta = \varepsilon/2$, the right-hand side of the estimate (A.55) is proportional to $(\sigma/\varepsilon)^2$. For Monte Carlo methods, such dependence on σ and ε seems to be unavoidable. In order to see this, consider a simple case where the feasible set $\mathscr{U}$ consists of just two elements: $\mathscr{U} = \{u_1, u_2\}$, with $\ell(u_2) - \ell(u_1) > \varepsilon > 0$. By solving the corresponding stochastic counterpart problem, we can ensure that u_1 is the ε-optimal solution if $\widehat{\ell}(u_2) - \widehat{\ell}(u_1) > 0$. If the random variable $H(X; u_2) - H(X; u_1)$ has a normal distribution with mean $\mu = \ell(u_2) - \ell(u_1)$ and variance σ^2, then $\widehat{\ell}(u_2) - \widehat{\ell}(u_1) \sim \mathsf{N}(\mu, \sigma^2/N)$ and the probability of the event $\{\widehat{\ell}(u_2) - \widehat{\ell}(u_1) > 0\}$ (i.e., of the correct decision) is $\Phi(\mu\sqrt{N}/\sigma)$, where Φ is the cdf of $\mathsf{N}(0, 1)$. We have that $\Phi(\varepsilon\sqrt{N}/\sigma) < \Phi(\mu\sqrt{N}/\sigma)$, and in order to make the probability of the incorrect decision less than α, we have to take the sample size $N > z_{1-\alpha}^2 \sigma^2/\varepsilon^2$, where $z_{1-\alpha}$ is the $(1 - \alpha)$-quantile of the standard normal distribution. Even if $H(X; u_2) - H(X; u_1)$ is not normally distributed, the sample size of order σ^2/ε^2 could be justified asymptotically, say by applying the central limit theorem.

Let us also consider a simplified variant of the estimate (A.55). Suppose the following:

(A2) *There is a positive constant C such that the random variable $\mathbf{Y_u}$ is bounded in absolute value by a constant C for all $\mathbf{u} \in \mathscr{U} \setminus \mathscr{U}^\varepsilon$.*

Under assumption (A2), we have that for any $\varepsilon > 0$ and $\delta \in [0, \varepsilon]$,

$$I_{\mathbf{u}}(-\delta) \geqslant \mathcal{O}(1)\frac{(\varepsilon - \delta)^2}{C^2}, \ \text{ for all } \mathbf{u} \in \mathscr{U} \setminus \mathscr{U}^\varepsilon\,, \tag{A.58}$$

and hence $\eta(\varepsilon, \delta) \geqslant \mathcal{O}(1)(\varepsilon - \delta)^2/C^2$. Consequently, the bound (A.49) for the sample size that is required to solve the true problem with accuracy $\varepsilon > 0$ and probability at least $1 - \alpha$, by solving the stochastic counterpart problem with accuracy $\delta = \varepsilon/2$, takes the form

$$N \geqslant \mathcal{O}(1)\left(\frac{C}{\varepsilon}\right)^2 \ln\left(\frac{|\mathscr{U}|}{\alpha}\right)\,. \tag{A.59}$$

Now let $\mathscr{U}$ be a bounded, not necessarily a finite, subset of $\mathbb{R}^m$ of diameter

$$D = \sup_{\mathbf{u}', \mathbf{u} \in \mathscr{U}} \|\mathbf{u}' - \mathbf{u}\|\,.$$

Then, for $\tau > 0$, we can construct a set $\mathscr{U}_\tau \subset \mathscr{U}$ such that for any $\mathbf{u} \in \mathscr{U}$ there is $\mathbf{u}' \in \mathscr{U}_\tau$ satisfying $\|\mathbf{u} - \mathbf{u}'\| \leqslant \tau$, and $|\mathscr{U}_\tau| = (\mathcal{O}(1)D/\tau)^m$.

Suppose next that the following condition holds:

(A3) *There exists a constant $\sigma > 0$ such that for any $\mathbf{u}', \mathbf{u} \in \mathscr{U}$ the moment generating function $M_{\mathbf{u}',\mathbf{u}}(t)$, of random variable $H(\mathbf{X}; \mathbf{u}') - H(\mathbf{X}; \mathbf{u}) - \mathbb{E}[H(\mathbf{X}; \mathbf{u}') - H(\mathbf{X}; \mathbf{u})]$, satisfies*

$$M_{\mathbf{u}',\mathbf{u}}(t) \leqslant \exp\left(\sigma^2 t^2 / 2\right), \quad \forall t \in \mathbb{R}. \tag{A.60}$$

Assumption (A3) is slightly stronger than assumption (A1); that is, assumption (A3) follows from (A1) by taking $\mathbf{u}' = \pi(\mathbf{u})$. Then, by (A.55), for $\varepsilon' > \delta$, we can estimate the corresponding sample size required to solve the reduced optimization problem, obtained by replacing $\mathscr{U}$ with $\mathscr{U}_\tau$, as

$$N \geqslant \frac{2\sigma^2}{(\varepsilon' - \delta)^2} \left[n \left(\ln D - \ln \tau \right) + \ln \left(\mathcal{O}(1)/\alpha \right) \right]. \tag{A.61}$$

Suppose further that there exists a function $\kappa : \mathscr{X} \to \mathbb{R}_+$ and $\varrho > 0$ such that

$$|H(\mathbf{X}; \mathbf{u}') - H(\mathbf{X}; \mathbf{u})| \leqslant \kappa(\mathbf{X}) \|\mathbf{u}' - \mathbf{u}\|^\varrho \tag{A.62}$$

holds for all $\mathbf{u}', \mathbf{u} \in \mathscr{U}$ and all $\mathbf{X} \in \mathscr{X}$. It follows, by (A.62), that

$$|\widehat{\ell}(\mathbf{u}') - \widehat{\ell}(\mathbf{u})| \leqslant N^{-1} \sum_{j=1}^{N} |H(\mathbf{X}_j; \mathbf{u}') - H(\mathbf{X}_j; \mathbf{u})| \leqslant \widehat{\kappa} \|\mathbf{u}' - \mathbf{u}\|^\varrho, \tag{A.63}$$

where $\widehat{\kappa} = N^{-1} \sum_{j=1}^{N} \kappa(\mathbf{X}_j)$. Let us further assume the following:

(A4) *The moment generating function $M_\kappa(t) = \mathbb{E}\left[e^{t\kappa(\mathbf{X})}\right]$ of $\kappa(\mathbf{X})$ is finite valued for all t in a neighborhood of 0.*

It follows then that the expectation $L = \mathbb{E}[\kappa(\mathbf{X})]$ is finite, and moreover, by Cramér's large deviations theorem, that for any $L' > L$ there exists a positive constant $\beta = \beta(L')$ such that

$$\mathbb{P}\left(\widehat{\kappa} > L'\right) \leqslant \mathrm{e}^{-N\beta}. \tag{A.64}$$

Let $\widehat{\mathbf{u}}$ be a δ-optimal solution of the stochastic counterpart problem and let $\tilde{\mathbf{u}} \in \mathscr{U}_\tau$ be a point such that $\|\widehat{\mathbf{u}} - \tilde{\mathbf{u}}\| \leqslant \tau$. Let us take $N \geqslant \beta^{-1} \ln(2/\alpha)$, so that by (A.64) we have

$$\mathbb{P}\left(\widehat{\kappa} > L'\right) \leqslant \alpha/2. \tag{A.65}$$

Then with probability of at least $1 - \alpha/2$, the point $\tilde{\mathbf{u}}$ is a $(\delta + L'\tau^\varrho)$-optimal solution of the reduced stochastic counterpart problem. Setting

$$\tau = [(\varepsilon - \delta)/(2L')]^{1/\varrho},$$

we find that with probability at least $1 - \alpha/2$, the point $\tilde{\mathbf{u}}$ is an ε'-optimal solution of the reduced stochastic counterpart problem with $\varepsilon' = (\varepsilon + \delta)/2$. Moreover, by taking a sample size satisfying (A.61), we find that $\tilde{\mathbf{u}}$ is an ε'-optimal solution of the reduced expected-value problem with probability at least $1 - \alpha/2$. It follows that $\widehat{\mathbf{u}}$ is an ε''-optimal solution of the stochastic counterpart problem (A.44) with probability at least $1 - \alpha$ and $\varepsilon'' = \varepsilon' + L\tau^\varrho \leqslant \varepsilon$. We obtain the estimate

$$N \geqslant \frac{4\sigma^2}{(\varepsilon - \delta)^2} \left[n \left(\ln D + \varrho^{-1} \ln \frac{2L'}{\varepsilon - \delta} \right) + \ln \left(\frac{\mathcal{O}(1)}{\alpha} \right) \right] \vee [\beta^{-1} \ln (2/\alpha)] \quad (A.66)$$

for the sample size, where $\vee$ denotes the maximum.

The result above is quite general and does not involve the convexity assumption. The estimate (A.66) of the sample size contains various constants and is too conservative for practical applications. However, it can be used as an estimate of the complexity of two-stage stochastic programming problems. In typical applications (e.g., in the convex case) the constant $\varrho = 1$, in which case condition (A.62) means that $H(\mathbf{X}; \cdot)$ is Lipschitz continuous on $\mathscr{U}$ with constant $\kappa(\mathbf{X})$. Note that there are also some applications where ϱ could be less than 1. We obtain the following basic result:

Theorem A.8.3 *Suppose that assumptions* (A3) *and* (A4) *hold and $\mathscr{U}$ has a finite diameter D. Then for $\varepsilon > 0$, $0 \leqslant \delta < \varepsilon$ and sample size N satisfying* (A.66)*, we are guaranteed that any δ-optimal solution of the stochastic counterpart problem is an ε-optimal solution of the true problem with probability at least $1 - \alpha$.*

In particular, if we assume that $\varrho = 1$ and $\kappa(\mathbf{X}) = L$ for all $\mathbf{X} \in \mathscr{X}$, that is, $H(\mathbf{X}; \cdot)$ is Lipschitz continuous on $\mathscr{U}$ with constant L independent of $\mathbf{X} \in \mathscr{X}$, then we can take $\sigma = \mathcal{O}(1)DL$ and remove the term $\beta^{-1} \ln(2/\alpha)$ on the right-hand side of (A.66). Further, by taking $\delta = \varepsilon/2$ we find in that case the following estimate of the sample size (compare with estimate (A.59)):

$$N \geqslant \mathcal{O}(1) \left(\frac{DL}{\varepsilon} \right)^2 \left[n \ln \left(\frac{DL}{\varepsilon} \right) + \ln \left(\frac{\mathcal{O}(1)}{\alpha} \right) \right] . \quad (A.67)$$

We can write the following simplified version of Theorem A.8.3.

Theorem A.8.4 *Suppose that $\mathscr{U}$ has a finite diameter D and condition* (A.62) *holds with $\varrho = 1$ and $\kappa(\mathbf{X}) = L$ for all $\mathbf{X} \in \mathscr{X}$. Then with sample size N satisfying* (A.67)*, we are guaranteed that every $(\varepsilon/2)$-optimal solution of the stochastic counterpart problem is an ε-optimal solution of the true problem with probability at least $1 - \alpha$.*

The foregoing estimates of the required sample size suggest complexity of order σ^2/ε^2 with respect to the desirable accuracy. This is in sharp contrast to deterministic (convex) optimization, where complexity usually is bounded in terms of $\ln(\varepsilon^{-1})$. In view of the discussion above, it should not be surprising that (even linear) two-stage stochastic programs usually cannot be solved with high accuracy. Nevertheless, the estimates (A.66) and (A.67) depend *linearly* on the dimension n of the first-stage decision vector. They also depend linearly on $\ln(\alpha^{-1})$. This means that by increasing confidence, say, from 99% to 99.99%, we need to increase the sample size by a factor of $\ln 100 \approx 4.6$ at most. This also suggests that, by using Monte Carlo sampling techniques, one can solve a two-stage stochastic program with reasonable accuracy, say with relative accuracy of 1% or 2%, in a reasonable time, provided that (1) its variability is not too large, (2) it has relatively complete recourse, and (3) the corresponding stochastic counterpart problem can be solved efficiently. And indeed, this was verified in numerical experiments with two-stage problems having a linear second-stage recourse. Of course, the estimate (A.66) of the sample size is far too conservative for the actual calculations. For practical applications, there are techniques that allow us to estimate the error of the feasible solution $\bar{\mathbf{u}}$ for a given sample size N; see, for example, [14].

The preceding estimates of the sample size are quite general. For convex problems, these bounds can be tightened in some cases. That is, suppose that the problem is convex, that is, the set $\mathscr{U}$ is convex and functions $H(\mathbf{X}; \cdot)$ are convex for all $\mathbf{X} \in \mathscr{X}$. Suppose further that $\kappa(\mathbf{X}) \equiv L$, the set $\mathscr{U}^0$, of optimal solutions of the true problem, is nonempty and bounded, and for some $r \geqslant 1$, $c > 0$, and $a > 0$, the following growth condition holds:

$$\ell(\mathbf{u}) \geqslant \ell^* + c\,[\mathrm{dist}(\mathbf{u}, \mathscr{U}^0)]^r, \quad \forall \mathbf{u} \in \mathscr{U}^a , \qquad (A.68)$$

where $a > 0$ and $\mathscr{U}^a = \{\mathbf{u} \in \mathscr{U} : \ell(\mathbf{u}) \leqslant \ell^* + a\}$ is the set of a-optimal solutions of the true problem. Then, for any $\varepsilon \in (0, a)$ and $\delta \in [0, \varepsilon/2)$, we have the following estimate of the required sample size:

$$N \geqslant \left(\frac{\mathcal{O}(1)L}{c^{1/r}\varepsilon^{(r-1)/r}} \right)^2 \left[n \ln \left(\frac{\mathcal{O}(1)LD_a^*}{\varepsilon} \right) + \ln \left(\frac{1}{\alpha} \right) \right] , \qquad (A.69)$$

where D_a^* is the diameter of $\mathscr{U}^a$. Note that if $\mathscr{U}^0 = \{\mathbf{u}^*\}$ is a singleton, then it follows from (A.68) that $D_a^* \leqslant 2(a/c)^{1/r}$.

In particular, if $r = 1$ and $\mathscr{U}^0 = \{\mathbf{u}^*\}$ is a singleton, that is, the solution $\mathbf{u}^*$ is *sharp*, then D_a^* can be bounded by $4c^{-1}\varepsilon$, and hence we obtain the estimate

$$N \geqslant \mathcal{O}(1)c^{-2}L^2 \left[n \ln \left(\mathcal{O}(1)c^{-1}L \right) + \ln \left(\alpha^{-1} \right) \right] , \qquad (A.70)$$

which does not depend on ε. That is, in that case, convergence to the exact optimal solution $\mathbf{u}^*$ happens with probability 1 in finite time.

For $r = 2$, condition (A.68) is called the *second-order* or *quadratic* growth condition. Under the quadratic growth condition, the first term on the right-hand side of (A.69) becomes of order $c^{-1}L^2\varepsilon^{-1}$.

PROBLEMS

A.1   Prove (A.8).

A.2   Let X and Y be Gaussian random vectors, with joint distribution given by

$$\begin{pmatrix} X \\ Y \end{pmatrix} \sim \mathsf{N} \left(\underbrace{\begin{pmatrix} \mu_1 \\ \mu_2 \end{pmatrix}}_{\mu}, \underbrace{\begin{pmatrix} \Sigma_{11} & \Sigma_{12} \\ \Sigma_{21} & \Sigma_{22} \end{pmatrix}}_{\Sigma} \right) .$$

a)   Defining $S = \Sigma_{12}\Sigma_{22}^{-1}$, show that

$$\begin{pmatrix} I & -S \\ 0 & I \end{pmatrix} \Sigma \begin{pmatrix} I & 0 \\ -S^{\mathsf{T}} & I \end{pmatrix} = \begin{pmatrix} \Sigma_{11} - S\Sigma_{21} & 0 \\ 0 & \Sigma_{22} \end{pmatrix} .$$

b) Using the above result, show that for any vectors u and v

$$(u^\top \ v^\top)\Sigma^{-1}\begin{pmatrix} u \\ v \end{pmatrix} = (u^\top - v^\top S^\top)\widetilde{\Sigma}^{-1}(u - Sv) + v^\top \Sigma_{22}^{-1}v \ ,$$

where $\widetilde{\Sigma} = (\Sigma_{11} - S\Sigma_{21})$.

c) The joint pdf of X and Y is given by

$$f(x,y) = c_1 \exp\left[-\frac{1}{2}(x^\top - \mu_1^\top \quad y^\top - \mu_2^\top)\Sigma^{-1}\begin{pmatrix} x - \mu_1 \\ y - \mu_2 \end{pmatrix}\right]$$

for some constant c_1. Using b), show that the conditional pdf $f(x \,|\, y)$ is of the form

$$f(x \,|\, y) = c_2(y) \exp\left[-\frac{1}{2}(x^\top - \widetilde{\mu}^\top)\,\widetilde{\Sigma}^{-1}\,(x - \widetilde{\mu})\right],$$

with $\widetilde{\mu} = \mu_1 + S(y - \mu_2)$, and where $c_2(y)$ is some function of y (need not be specified). This proves that

$$(X \,|\, Y = y) \sim \mathsf{N}\left(\mu_1 + \Sigma_{12}\Sigma_{22}^{-1}(y - \mu_2), \ \Sigma_{11} - \Sigma_{12}\Sigma_{22}^{-1}\Sigma_{12}^\top\right) \ .$$

1.3   Prove the upper bound (A.37).

1.4   Prove the upper bound (A.38).

Further Reading

More details on exponential families and their role in statistics may be found in [2]. An accessible account of hidden Markov models is [5].

The estimate (A.55) of the sample size, for finite feasible set $\mathscr{U}$, was obtained in [9]. For a general discussion of such estimates and extensions to the general case, see [14]. For a discussion of the complexity of *multistage* stochastic programming problems, see, for example, [16]. Finite time convergence in cases of sharp optimal solutions is discussed in [15].

The FPRAS for counting SATs in DNF is due to Karp and Luby [8], who also give the definition of FPRAS. The first FPRAS for counting the volume of a convex body was given by Dyer et al. [6]. See also [4] for a general introduction to random and nonrandom algorithms.

REFERENCES

1. S. Asmussen and R. Y. Rubinstein. Complexity properties of steady-state rare-events simulation in queueing models. In J. H. Dshalalow, editor, *Advances in Queueing: Theory, Methods and Open Problems*, pages 429–462, New York, 1995. CRC Press.

2. G. Casella and R. L. Berger. *Statistical Inference*. Duxbury Press, 2nd edition, 2001.

3. S. X. Chen and J. S. Liu. Statistical applications of the Poisson-binomial and conditional Bernoulli distributions. *Statistica Sinica*, 7:875–892, 1997.

4. T. H. Cormen, C. E. Leiserson, R. L. Rivest, and C. Stein. *Introduction to Algorithms.* MIT Press, Cambridge, MA, 2nd edition, 2001.

5. R. O. Duda, P. E. Hart, and D. G. Stork. *Pattern Classification.* John Wiley & Sons, New York, 2001.

6. M. Dyer, A. Frieze, and R. Kannan. A random polynomial-time algorithm for approximation the volume of convex bodies. *Journal of the ACM*, 38:1–17, 1991.

7. M. Jerrum and A. Sinclair. *Approximation Algorithms for NP-hard Problems*, chapter The Markov chain Monte Carlo method: An approach to approximate counting and integration. PWS, Boston, 1996.

8. R. M. Karp and M. Luby. Monte Carlo algorithms for enumeration and reliability problems. In *Proceedings of the 24-th IEEE Annual Symposium on Foundations of Computer Science*, pages 56–64, Tucson, 1983.

9. A. J. Kleywegt, A. Shapiro, and T. Homem de Mello. The sample average approximation method for stochastic discrete optimization. *SIAM Journal on Optimization*, 12:479–502, 2001.

10. V. Kriman and R. Y. Rubinstein. Polynomial time algorithms for estimation of rare events in queueing models. In J. Dshalalow, editor, *Frontiers in Queueing: Models and Applications in Science and Engineering*, pages 421–448, New York, 1995. CRC Press.

11. M. Mitzenmacher and E. Upfal. *Probability and Computing: Randomized Algorithms and Probabilistic Analysis.* Cambridge University Press, Cambridge, 2005.

12. R. Y. Rubinstein and B. Melamed. *Modern Simulation and Modeling.* John Wiley & Sons, New York, 1998.

13. R. Y. Rubinstein and A. Shapiro. *Discrete Event Systems: Sensitivity Analysis and Stochastic Optimization via the Score Function Method.* John Wiley & Sons, New York, 1993.

14. A. Shapiro. Monte Carlo sampling methods. In A. Ruszczyński and A. Shapiro, editors, *Handbook in Operations Research and Management Science*, volume 10. Elsevier, Amsterdam, 2003.

15. A. Shapiro and T. Homem de Mello. On the rate of convergence of optimal solutions of Monte Carlo approximations of stochastic programs. *SIAM Journal on Optimization*, 11(1):70–86, 2001.

16. A. Shapiro and A. Nemirovski. On complexity of stochastic programming problems. In V. Jeyakumar and A. M. Rubinov, editors, *Continuous Optimization: Current Trends and Application.* Springer-Verlag, New York, 2005.

ABBREVIATIONS AND ACRONYMS

BFS	breadth-first search
cdf	cumulative distribution function
CE	cross-entropy
CMC	crude Monte Carlo
CNF	conjunctive normal form
DEDS	discrete-event dynamical system
DESS	discrete-event statical system
DES	discrete-event simulation
DFS	depth-first search
DNF	disjunctive normal form
ECM	exponential change of measure
FPAUS	fully polynomial almost uniform sampler
FPRAS	fully polynomial randomized approximation scheme
GS	generalized splitting
HMM	hidden Markov model
iid	independent and identically distributed
ITLR	inverse-transform likelihood ratio
KKT	Karush–Kuhn–Tucker

Simulation and the Monte Carlo Method, Third Edition. By R. Y. Rubinstein and D. P. Kroese
Copyright © 2017 John Wiley & Sons, Inc. Published 2017 by John Wiley & Sons, Inc.

MRG   Multiple-recursive generator

max-cut   maximal cut

MCMC   Markov chain Monte Carlo

MinxEnt   minimum cross-entropy

pdf   probability density function (both discrete and continuous)

PERT   program evaluation and review technique

PMC   permutation Monte Carlo

RSAT   random SAT

SAT   satisfiability problem

SAW   self-avoiding walk

SDE   stochastic differential equation

SE   stochastic enumeration

SF   score function

SIS   sequential importance sampling

SIR   sequential importance resampling

SLR   standard likelihood ratio

TLR   transform likelihood ratio

TSP   traveling salesman problem

VM   variance minimization

LIST OF SYMBOLS

$\gg$	much greater than
$\propto$	proportional to
$\sim$	is distributed according to
$\approx$	approximately
$\top$	transpose
∇	∇f is the gradient of f
∇^2	$\nabla^2 f$ is the Hessian of f
$\mathbb{E}$	expectation
$\mathbb{N}$	set of natural numbers $\{0, 1, \ldots\}$
$\mathbb{P}$	probability measure
$\mathbb{R}$	the real line = one-dimensional Euclidean space
$\mathbb{R}^n$	n-dimensional Euclidean space
$\mathcal{D}$	Kullback–Leibler CE
$\mathcal{H}$	Shannon entropy
$\mathcal{M}$	mutual information
$\mathcal{S}$	score function

Simulation and the Monte Carlo Method, Third Edition. By R. Y. Rubinstein and D. P. Kroese
Copyright © 2017 John Wiley & Sons, Inc. Published 2017 by John Wiley & Sons, Inc.

Ber	Bernoulli distribution
Beta	beta distribution
Bin	binomial distribution
Exp	exponential distribution
G	geometric distribution
Gamma	gamma distribution
N	normal or Gaussian distribution
Pareto	Pareto distribution
Poi	Poisson distribution
U	uniform distribution
Weib	Weibull distribution
α	smoothing parameter or acceptance probability
γ	level parameter
ζ	cumulant function (log of moment generating function)
ϱ	rarity parameter
$D(\mathbf{v})$	objective function for CE minimization
f	probability density (discrete or continuous)
g	importance sampling density
I_A	indicator function of event A
$\ln$	(natural) logarithm
N	sample size
$\mathcal{O}$	big-O order symbol
S	performance function
$S_{(i)}$	i-th order statistic
$\mathbf{u}$	nominal reference parameter (vector)
$\mathbf{v}$	reference parameter (vector)
$\widehat{\mathbf{v}}$	estimated reference parameter
$\mathbf{v}^*$	CE optimal reference parameter
$_*\mathbf{v}$	VM optimal reference parameter
$V(\mathbf{v})$	objective function for VM minimization
W	likelihood ratio
$\mathbf{x}, \mathbf{y}$	vectors
$\mathbf{X}, \mathbf{Y}$	random vectors
$\mathscr{X}, \mathscr{Y}$	sets

INDEX

Printed and bound by CPI Group (UK) Ltd, Croydon, CR0 4YY

16/04/2025

14658520-0003